Ultrafast Dynamics of Chemical Systems

Understanding Chemical Reactivity

Volume 7

The titles published in this series are listed at the end of this volume.

Ultrafast Dynamics of Chemical Systems

Edited by

JOHN D. SIMON

Department of Chemistry, University of California,
San Diego, La Jolla, California, U.S.A.

Kluwer Academic Publishers

Dordrecht / Boston / London

Library of Congress Cataloging-in-Publication Data

Ultrafast dynamics of chemical systems / edited by John D. Simon.
p. cm. -- (Understanding chemical reactivity ; v. 7)
Includes bibliographical references and index.
ISBN 0-7923-2489-7 (acid-free)
1. Chemical kinetics. 2. Picosecond pulses. I. Simon, John D. (John Douglas), 1957- . II. Series.
QD502.U48 1994
541.3'94--dc20 93-45408

ISBN 0-7923-2489-7

Published by Kluwer Academic Publishers,
P.O. Box 17, 3300 AA Dordrecht, The Netherlands.

Kluwer Academic Publishers incorporates
the publishing programmes of
D. Reidel, Martinus Nijhoff, Dr W. Junk and MTP Press.

Sold and distributed in the U.S.A. and Canada
by Kluwer Academic Publishers,
101 Philip Drive, Norwell, MA 02061, U.S.A.

In all other countries, sold and distributed
by Kluwer Academic Publishers Group,
P.O. Box 322, 3300 AH Dordrecht, The Netherlands.

Printed on acid-free paper

Printed in the Netherlands

Table of Contents

Preface

The last decade has witnessed significant advances in the ability to generate short light pulses throughout the optical spectrum. These developments have had a tremendous impact on the field of chemical dynamics. Fundamental questions concerning chemical reactions, once thought to be unaddressable, are now easily studied in real-time experiments. Ultrafast spectroscopies are currently being used to study a variety of fundamental chemical phenomena. This book focuses on some of the experimental and associated theoretical studies of reactions in clusters, liquid and solid media.

Many of the advances in our understanding of the fundamental details of chemical reactivity result from the interplay of experiment and theory. This theme is present in many of the chapters, indicating the pervasiveness of a combined approach for eludicating molecular models of chemical reactions. With parallel developments in computer simulation, complex chemical systems are being studied at a molecular level. The discussions presented in this book recount many areas at the forefront of "ultrafast chemistry". They serve the purpose of both bringing the expert up to date with the work being done in many laboratories as well as introducing those not directly involved in this field to the diverse set of problems that can be studied.

I hope that this book conveys the excitement that both I and the other authors in this volume feel about the field of ultrafast chemistry.

John D. Simon
1993

J.D. Simon (ed.), Ultrafast Dynamics of Chemical Systems, vii.

1. Introduction to Ultrafast Laser Spectroscopic Techniques Used in the Investigation of Condensed Phase Chemical Reactivity

PEIJUN CONG and JOHN D. SIMON
Department of Chemistry and Institute for Nonlinear Science, University of California at San Diego, La Jolla, CA 92093–0341, U.S.A.

1. Introduction

The majority of chemical reactions occur in the condensed phase. This has spurred a tremendous theoretical and experimental effort towards understanding chemical reactivity in the condensed phase on the molecular level. While the half-life for a unimolecular reaction can span from picoseconds to years, the actual time that a molecule spends transversing the potential surface from reactant to product is very short. In understanding rate processes, the most crucial stage in a chemical reaction is generally 'the transitions state,' representing the structure of the reactant (and solvent) at the point where the reaction crosses from the reactant surface to that of the product. There have been several elegant studies of transitions states in molecular reactions; however, it is still the case that most of the knowledge about the structure of transition state species comes from theoretical calculation. Only with the development of picosecond and femtosecond laser technology has direct, real-time studies of the ultrafast dynamic processes in condensed phase become possible [1]. These light sources are used in a variety of clever ways to probe chemical dynamics.

The purpose of this chapter is to introduce various techniques that are often used in the field of ultrafast laser spectroscopy to those readers who are not familiar with the terminology or are new to this field. Special emphasis is placed on those methods that are relevant to the discussions in latter chapters of this book. Even though every effort is made to reflect the state of the art of the various forms of ultrafast laser spectroscopy, this chapter is not intended to be a comprehensive review of this field. Instead, we try to convey the essence of the most current technology, without getting into the exhaustive details, while leading interested readers to the representative original literature and reviews on the relevant techniques. As this book focuses on condensed phase chemistry, we further restrict the discussion to ultrafast

J.D. Simon (ed.), Ultrafast Dynamics of Chemical Systems, 1–36.

techniques used in the study of reactions in solution. Ultrafast technology has also been elegantly applied to gas phase reaction dynamics and various molecular processes on surfaces. Excellent reviews exist for these applications and the interested reader can find discussions of these techniques in [2, 3] respectively. The remaining part of this chapter is organized as follows. First, a short introduction on current methods for generating ultrashort laser pulses is given. This is followed by descriptions of various techniques of ultrafast spectroscopy. To organize the discussion, the laser spectroscopies are loosely classified into the following categories: *absorption*, *emission*, *Raman*, *polarization*, and *coherent spectroscopies*. Each section describes the essential features of the experiment. Examples of chemical studies that used these techniques are given.

2. Generation of Ultrashort Light Pulses

The spectral range covered by ultrafast laser spectroscopy is mainly limited by the ability to generate short pulses at the desired wavelength. Most of the techniques currently used to produce picosecond and femtosecond pulses were developed over the last three decades. A comprehensive review on the technical aspects of ultrashort pulse generation is beyond the scope of this chapter and interested readers should consult recent reviews [4] and proceedings of topical meetings on the subject [5]. We will limit our discussion to a few major classes of lasers that are most widely used.

Synchronously-pumped dye lasers are convenient source of tunable picosecond (10^{-12} s) and femtosecond (10^{-15} s) laser pulses. These lasers are generally pumped by a frequency-doubled mode-locked Nd:YAG or Nd:YLF laser. These dye lasers can be tuned from 550 nm to the near infrared. Output pulse durations are typically a few picoseconds without intracavity Group Velocity Dispersion (GVD) compensation [4]. With GVD compensation, sub-100 femtosecond pulses have been generated. The pulse energies generated by synchronously pumped dye lasers are on the order of 1 nJ. Higher energies are required for many chemical applications. To achieve this, the output from these lasers is normally amplified in a multi-stage power amplifier. The pump power for such optical amplifiers is generally derived from either a low repetition rate Q-switched and frequency-doubled Nd:YAG laser or the output of a Nd:YAG or Nd:YLF regenerative amplifier. Other optical amplifiers using copper vapor lasers and excimer laser have also been reported. Amplified pulse energies on the order of milli-joules are easily achieved and the repetition rate can be up to several kilohertz. In most designs, there is a trade off between pulse energy and repetition rate. The average output energy of an amplified, synchronously-pumped laser system is tens of milliwatts.

One major limitation of these laser systems is the dependability in generating sub-100 femtosecond pulses.

For the past several years, the colliding-pulse mode-locked (CPM) dye laser has been the main technique to obtain stable sub-100 femtosecond pulses [6]. Most CPM dye lasers are pumped by a continuous-wave (CW) Ar^+ laser and the mode-locking is induced by the counter propagating light fields colliding inside the saturable absorber jet. Only the pulses that are strong enough to bleach the saturable absorber can survive in the ring cavity, while all other modes of lasing are suppressed. Pulses as short as 50 fs can be routinely generated by the CPM technique. There are two main amplification schemes for the CPM laser, one is based on Cu vapor lasers which is capable of generating $\sim$50 μJ pulses at a few kilohertz repetition rate without much pulse broadening [7]. Another method is based on low repetition rate, high power, Q-switched Nd:YAG lasers, which can amplify the sub-nanojoule CPM pulses to 1 millijoule per pulse at repetition rates typically less than 40 Hz [8]. Pulse broadening is not very significant in either optical scheme. The major drawback of a CPM laser is its limited tunability, which can be complemented by white-light continuum generation (see following discussions on this technique).

The discovery of self mode-locking action in continuous-wave Ti:Sapphire lasers by Spence *et al.* in early 1991 advanced the science of ultrafast laser research [9]. The impact of Ti:Sapphire lasers is comparable to the introduction of CPM lasers in the early 1980's. Like the CPM laser, Ti:Sapphire lasers are also normally pumped by a CW Ar^+ laser. Mode-locking mechanism in these lasers is induced by Kerr lensing in the Ti:Sapphire crystal. With proper optical alignment and a small external perturbation, the Ti:Sapphire laser switches from CW to mode-locked operation. Provided the laser is housed in an environment without excessive external perturbations, mode-locking can be maintained for an extended period of time without any active element. Compared with the CPM lasers, these self mode-locked Ti:Sapphire lasers offer a number of distinctive advantages: tunability from 750 nm to the near infrared, high average power (which is at least one order of magnitude higher than the CPM laser), superior stability, and low operating cost due to their all-solid-state nature. Because of the wide gain profile of Ti:Sapphire materials, they can support very short optical pulses. The shortest pulses obtained to date from these lasers is on the order of 10 fs [10]. The possibility of generating sub-10 fs pulses from these devices is very promising.

Since the discovery of the Kerr-lens-induced mode-locked Ti:Sapphire laser, numerous methods have been developed to amplify the femtosecond pulses from a few nanojoules to the millijoule range. In some designs, energies up to a joule have been reported. At first the general approach was similar to how sync-pumped dye lasers and CPM lasers are amplified, i.e., a traveling-wave multi-stage power amplifier. More recently, Ti:Sapphire regenerative amplifiers, which can utilize the broad gain profile and long storage time of Ti:Sapphire, have been developed. Most of the regenerative amplifier designs involve pulse-stretching, chirped-pulse amplification, and pulse recompression [11]. Pulselengths of 150 fs with millijoule energies at multi-kilohertz repetition rate have been generated in a number of different laboratories. The average outputs from these devices are about an order of magnitude greater than what was achieved with the amplified synchronously-pumped technology. The development of optical amplifiers for Ti:Sapphire lasers is still in an early stage and even more powerful and versatile methods can be expected in the coming years.

To achieve the desired wavelength for an experimental study, the easiest approach would be to tune the laser to the needed frequency. However, direct tuning of an ultrafast laser is often impossible. In this case, non-linear optical techniques are used to obtain the necessary tunability [12]. Frequency doubling and sum frequency generation are frequently used to up-convert a tunable visible laser into a tunable UV or near-UV source. On the other hand, difference frequency generation and optical parametric amplification are often employed to down-convert laser radiation from the visible region into the infrared. When non-linear mixing techniques are still inadequate to achieve the desired frequency, white-light continuum generation has been utilized to extend the tunability of picosecond and especially femtosecond lasers [4, 5]. In this technique, an intense laser beam is focused tightly into a liquid medium, commonly, water, or ethylene glycol. A white light continuum emerges from the medium as a result of a collection of non-linear processes including self-phase modulation and various high-order Raman processes. Depending on the peak power and the overall energy of the input ultrashort light pulse, the white light continuum output can cover a spectral range of a few tens of nanometers for a picosecond laser to a few hundred nanometers for a femtosecond pulse. Spectral band-widths that range from the near-UV to the near-infrared are common for femtosecond continuum. A particular wavelength can easily be selected from the continuum by a narrow band interference filter. Selected pieces of continuum light can also be amplified using the various optical amplifiers described above. The tunability of ultrafast lasers has been greatly extended by this white light continuum technique.

3. Spectroscopic Techniques and Selected Applications

A. Absorption Spectroscopy

Transient absorption spectroscopy is probably the simplest form of pump-probe time-resolved spectroscopy, yet it is a powerful technique used by many researchers, principally due to its versatility and simplicity. It has been applied to study a number of important photochemical and photophysical processes in the condensed phase, e.g., photodissociation and recombination [13], optical bleaching and recovery [14], photodetachment [15], vibrational relaxation [16], solvation, and charge transfer (e.g., electron transfer and proton transfer) [17]. Specific examples will be discussed later.

1. Basic Experimental Principles

Figure 1 shows a typical arrangement for a two-pulse pump-probe transient absorption spectrometer. In this Michelson interferometric setup, the pump pulse at frequency ω_1 excites the sample. After a variable delay time, τ, the second pulse at frequency ω_2 is introduced to the medium to probe the changes of absorbance at ω_2. The variable time delay between the pump pulse and the probe pulse is normally achieved by varying the difference in optical paths through a high-precision mechanical translation stage controlled by a microcomputer.

After excitation by a short (pump) laser pulse, the transmitted intensity, I_{trans}, of the probe laser pulse, I_{probe}, in the small signal limit, can be expressed as follows.

$$I_{\text{trans}}(\tau) = \int I_{\text{probe}}(t - \tau)(1 - \Delta A(\tau))\, \mathrm{d}t. \tag{1}$$

In the above expression, $\Delta A(\tau)$ is the time-dependent transient absorption change induced by the pump pulse integrated over the pump-probe interaction volume. Equation (1) is only valid when the pump and probe beams are well separated temporally.

The detected transient absorption change may be classified into three categories: absorption, bleach, and gain. If the intensity of the probe pulse decreases in the presence of the pump pulse, then there is an increase in *absorption* by the sample. If the intensity of the probe beam increases in the presence of the pump pulse, but remains less than the intensity of the probe beam before it enters the sample, then th pump has caused a *bleach* in the density of absorbing molecules. Finally, if the intensity of the probe beam increases in the presence of the pump pulse beyond its incident intensity, the transient signal is referred to as *stimulated gain*. In this later case, This latter signal results from the molecules in the excited state being 'stimulated' back to a lower level by the probe pulse.

The typical pump-pulse-induced relative absorbance change amounts to far less than a few percent of the total absorbance of the sample. Unfortunately, this is generally less than the pulse-to-pulse fluctuations exhibited by the lasers used in such studies. As a result, a normalization and signal averaging

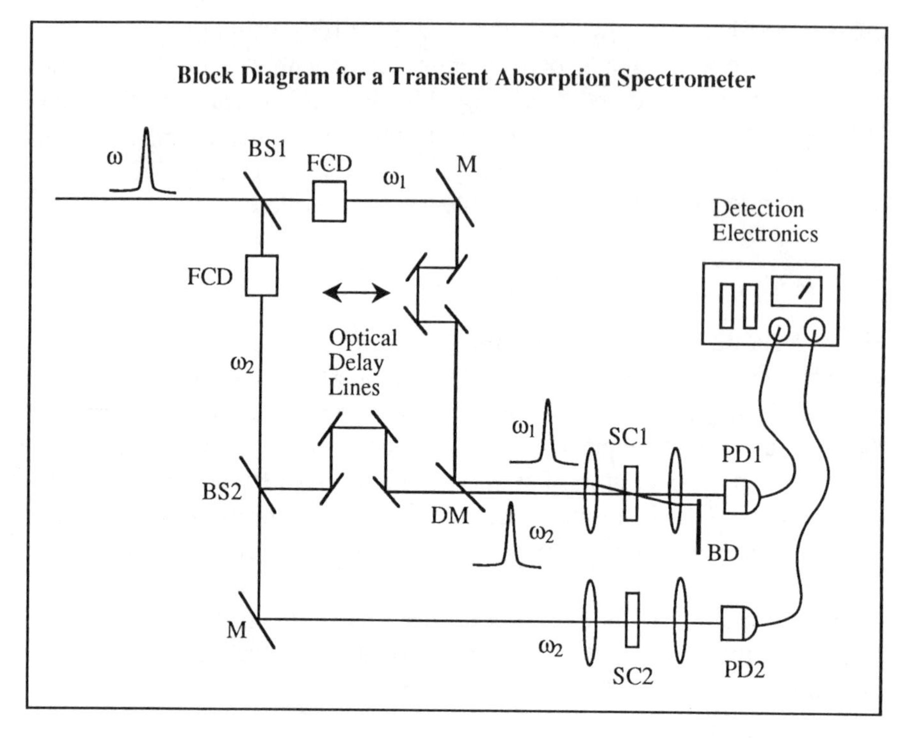

Fig. 1. Block diagram for a transient absorption spectrometer. BD, beam dump; BS, beam splitter; DM, dichroic mirror; FCD, frequency conversion device; M, mirror; PD, photodiode; SC, sample cell. A femtoseocnd (or picosecond) laser beam at frequency ω is split into two parts by BS1. These two beams are then converted into desired pump and probe pulses at frequency ω_1 and ω_2, respectively, through FCDs (optional). The FCDs can be a frequency-doubling nonlinear crystal, or a continuum generation and wavelength selection setup (see Section II for more details). The laser beam at ω_2 is further divided into two separate parts of equal intensity. One (probe) beam is recombined with the pump beam (ω_1) at DM and focussed into SC1, while the other (reference) beam traverses an identical sample cell (SC2) in absence of a pump beam. The relative timing of the pump and probe pulses incident on the sample is determined by the difference in optical paths of the two laser beams, which is controlled by a precision variable optical delay line. The translation stage is typically equiped with a computerized stepper motor (not shown). PD1 and PD2 detect the integrated intensities of the probe and reference beams, and the difference in the signals detected by these two photodiodes vs. pump-probe delay time measures the transient absorption induced by the pump pulse.

scheme is of vital importance to increase the signal-to-noise ratio in transient absorption experiments. In the setup shown in Figure 1, a reference beam is employed to reduce the shot-to-shot noise. In this case, the probe laser is spilt into two beams. One beam overlaps the region of the sample excited by the pump laser beam, while the other beam traverses through the sample cell in a region that is not perturbed by the pump pulse. The ratio of the intensities of the probe beam to the reference beam as a function of the delay time reflects the evolution of the transient absorbance of the sample. In general, multiple laser shots are averaged at a particular delay time in order to achieve an acceptable signal-to-noise ratio. This technique is ideal for low-repetition rate laser systems (i.e., less than 50 Hz) coupled with a gated integrating and averaging system.

A slight variation of this technique is needed for laser systems with medium to high repetition rates, i.e., greater than 500 Hz. In this case, the pump beam is modulated by a mechanical chopper to introduce an AC component in the transmitted intensity of the probe beam. This AC component is then detected by a lock-in amplifier as a function of the delay time. For a laser system with a repetition rate beyond a few hundred kilohertz, the mechanical chopper is often replaced by either an acousto-optic or electro-optic modulation device. It should be pointed out that the selection of the modulation frequency is not random. A judicious choice will place the modulation frequency at a position which is at or near the minimum of the intrinsic noise spectrum of the laser.

In kilohertz laser systems, techniques have been developed to take full advantage of both of the detection and averaging schemes. In a combination of these two techniques, a boxcar integrator is employed to integrate the individual laser pulses and the AC component in this output is then detected by a lock-in amplifier. Xie *et al.* [18] was able to achieve a detection limit of 10^{-6} using this technique.

Many detectors are used in transient absorption spectrometer. By far the most popular are silicon-based photodiodes. These are chosen for their large linear response range and cost-effectiveness. For cases where the probe beam is relatively weak, the diodes are replaced by photomultiplier tubes. Both of these detection schemes are limited to the UV to visible region. In addition to detecting the changes induced in the light beam, techniques which center on measuring the acoustic wave generated from the heat released by the sample cell have been reported [19]. This approach also provides a direct measure of the sample absorbance. In this case the detector is generally a piezoelectric microphone that is either attached to or submerged in the sample cell.

For pump-probe absorption experiments in which an infrared probe beam is used, pyroelectric detectors such as HgCdTe are used. Compared with the photomultipliers used in the ultraviolet and visible region, these infrared detectors have a rather low-quantum efficiency and high thermal noise level. Most infrared detectors consequently need to be operated at liquid nitrogen temperature. In order to perform transient infrared spectroscopy and avoid the deficiencies of infrared detectors, Hochstrasser and his co-workers developed a non-linear optically-gated upconversion technique [20, 21]. In their experimental arrangement, the pump light is derived from a pulsed pico- or femto-second light source, but the probe light is a continuous beam from an infrared diode laser. Time resolution is achieved by mixing the transmitted infrared light with an ultrashort light pulse in a non-linear optical crystal.

The resulting sum frequency, which is in the visible or near UV region, is then detected by a conventional photomultiplier. This technique has been successfully implemented in a variety of studies including the study of photodissociation of hemoglobin [22, 23] and $C_{p2}Fe(CO)_4$ [24] and vibrational energy relaxation in the azide ion [25].

In addition to detecting light and heat, techniques have been developed to detect photogenerated ions. This usually involves a multi-photon ionization (MPI) scheme. MPI is a popular detection scheme in the gas phase, but has also been applied to solution studies. For example, Braun *et al.* used a novel electric conduction cell to examine kinetic processes in solution [26].

A very powerful detection scheme that is currently being used centers on using one- and two-dimensional arrays of charge-coupled devices (CCD) to record transient spectra. With advances in generating femtosecond white light continuum, this technique offers the possibility of taking a transient absorption spectrum over a few hundred nanometers with a single laser shot, without sacrificing much of the sensitivity. This technique reduces the data collection time by orders of magnitude, thereby increasing the productivity of the spectrometer. The collection of transient spectra is not a new technique. Since the late 1970's, spectra were recorded by combining white-light continuum sources with Vidicon and Reticon detectors. However, CCD detectors are far superior to these earlier array detectors as they substantially increase the signal-to-nosie levels of spectra obtained using continuum sources.

Transient absorption spectroscopy can not only be used to measure population kinetics but also orientational dynamics as well when polarized pump and probe light pulses are employed (see [1], p. 70). The orientational anisotropy, $\Delta I(t)$, defined by Equation (2), can be studied directly by making separate transient absorption measurements with pump and probe beams having parallel and perpendicular polarizations.

$$\Delta I(t) = \frac{I_{||}(t) - I_{\perp}(t)}{I_{||}(t) + 2I_{\perp}(t)}, \tag{2}$$

where $\Delta I(t)$ is the transient absorption polarization anisotropy. The polarization characteristics of the laser beams are usually controlled by calcite polarizers and waveplates, and Pockels cells or photoelastic modulators are used when periodic switching between different polarizations are needed. In cases when population information is desired, the relative polarizations of pump and probe pulses are set at 54.7°, which is the so-called 'magic angle' [1].

2. Optical Sampling Techniques

Even though the two pulse, pump-probe interferometric setup has been the dominant arrangement for transient absorption spectroscopy, the recent advances in Ti:Sapphire technology will help to popularize asynchronous optical sampling techniques. The general principle behind this approach is analogous to the way a sampling oscilloscope works. This method is based on two mode-locked, independently tunable lasers. Due to a small cavity length mismatch, the mode-locking frequencies for these two lasers, v_1 and v_2, are slightly different. If at a given momont, a pair of pulses (one from each laser) overlaps in time, then the next pair of pulses will not. The pulse from the laser with the smaller mode-locking frequency laser will arrive later than the output from the other laser by $\Delta v/v^2$ $(v \approx v_1 \approx v_2)$, where $\Delta v = |v_1 - v_2|$, is the frequency mismatch. This process will continue until the delay time between the two pulses reach $1/\Delta v$, at which time the two pulses overlap again. At this point another sweep of this sampling process begins. If we use the faster laser as the pump laser and monitor the transmittance of the slower laser on an oscilloscope, the trace will reflect the transient absorbance changes induced by the first laser with a period of $1/\Delta v$. The cross-correlation signal between these two lasers can be used to trigger the oscilloscope, thereby synchronizing the scope output to the sampling frequency of the two lasers. The time resolution is determined by a combination of three factors: (1) the pulse durations (i.e., cross-correlation), (2) sampling interval $(\Delta v/v^2)$, and (3) the jitter caused by small perturbations in cavity length during one sweep period $(1/\Delta v)$. Kafka *et al.* reported a 150 fs resolution with such a system composed of two mode-locked Ti:Sapphire lasers [27]. Luo *et al.* recently demonstrated that the excited state lifetime of a laser dye could be determined using such an approach [28].

3. Selected Applications

Transient absorption spectroscopy is currently used in many laboratories around the world to study chemical reaction dynamics. Our laboratory has also utilized transient absorption spectroscopy to study a variety of condensed phase reactive processes. In the remainder of this section we consider four examples of transient absorption spectroscopy: (1) the condensed phase photochemistry of OClO [29], (2) electron transfer in the mixed-valent transition metal dimer [30]. $[(NC)_5Ru^{III}CNRu^{II}(NH_3)_5]^{-1}$ (3) electronic dephasing of the dye molecule HITCI in ethylene glycol [31] and (4) the solvation of photogenerated unsaturated metal carbonyl complexes [32].

A. Condensed Phase Photochemistry of OClO: Chlorine and bromine oxides play an important role in the destruction of stratospheric ozone over Antarctica [33]. Recent reports indicate that more than 20% of the ozone decline is caused by the coupling of chlorine and bromine oxide cycles. Among other products, this coupling results in the formation of ClOO and OClO. While the former is known to destroy ozone through the formation of atomic chlorine and molecular oxygen, the role of OClO in the stratospheric ozone balance is currently a topic of debate. While early photochemical studies of OClO indicated that excitation dissociates the molecule into ClO and O, recent gas phase studies established the existence of a competing parallel photochemical channel, producing Cl and O_2 [34–36].

By taking advantage of the well characterized electronic spectroscopy of OClO, Cl, ClO, and O_2 in aqueous solution, the excited state reaction paths of OClO can be determined. In solution, unlike the gas phase, chlorine atoms are easily detected as they readily form charge transfer (CT) complexes with solvent molecules. This gives rise to a CT absorption, which, for many solvents, occurs in the near-UV to visible spectral region [37]. In Figure 2, the absorption spectra of OClO[33], ClO [38] and Cl [39] in water are presented. The gas phase spectra of OClO is shown [39], as no solution spectra has been reported for this highly reactive molecule. In Figure 3, the dynamics at various probe wavelengths following the photolysis of OClO at 355 nm are shown. A variety of different kinetic processes are revealed by the transient data. The solid lines through the data are calculated signals assuming a kinetic model where OClO undergoes competitive photochemical reactions whereby 90% of the molecules dissociate into ClO and O, and 10% isomerize to ClOO followed by thermal decomposition into Cl and O_2.

Using orbital correlation diagrams as a guide, two allowed reaction channels exist that can produce Cl atoms from the electronically excited $^2B_2$ state of OClO. These are given by Equations (3) and (4).

$$\mathrm{OClO} + h\nu \rightarrow \mathrm{ClOO}(A' \text{ or } A'') \rightarrow \mathrm{Cl}(^3\mathrm{P}) + \mathrm{O_2}(^3\Sigma_g^-) \tag{3}$$

$$\mathrm{OClO} + h\nu \rightarrow \mathrm{Cl}(^3\mathrm{P}) + \mathrm{O_2}(^1\Delta_g) \tag{4}$$

The first mechanism, Equation (3), occurs by photoisomerization of OClO to ClOO, which can be produced either in the ground state, A', of the first electronic excited state, A''. While thermodynamically more stable than OClO, ClOO is kinetically unstable and readily dissociates to form Cl and O_2 ($^3\Sigma_g^-$). The second mechanism involves symmetric dissociation (along a C_{2v} reaction coordinate) producing Cl + O_2 ($^1\Delta_g$). The electronic state of the product oxygen molecule provides an experimental handle by which the relative importance of these two chlorine producing mechanisms can be distinguished.

To determine the relative extent of the two allowed mechanisms for Cl generation, the quantum yield for formation of either O_2 ($^1\Delta_g$) or O_2 ($^3\Sigma_g^-$)

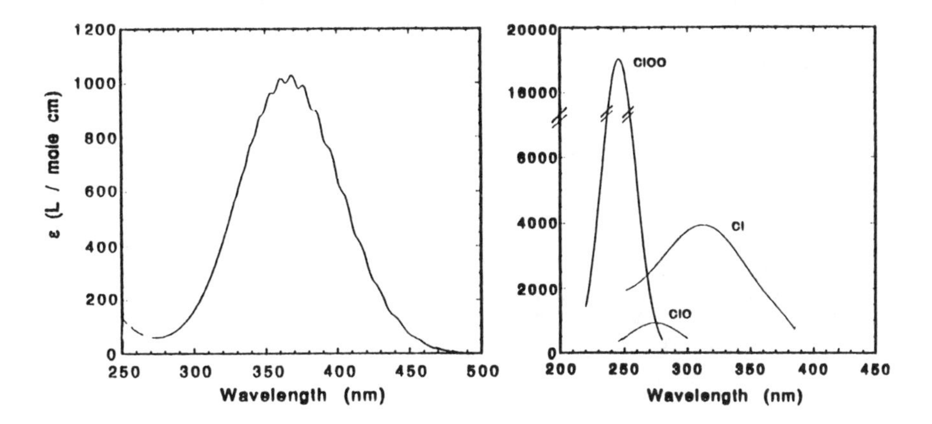

Fig. 2. *Left:* The $^2A_2 \leftarrow ^2 B_1$ absorption spectrum of OClO in water solution. *Right:* The absorption specta of Cl(aq) and ClO(aq) and ClOO(gas). These spectra were reproduced from [38] and [39].

is needed. This was determined by measuring the yields of $^3\Sigma_g^- \leftarrow {}^1\Delta_g$ emission of O_2 following excitation of OClO at 355 nm in deuterated solvents [29b]. By comparing the intensity of this emission (1270 nm) with that of sensitizers for which the yield is known, the quantum yield of O_2 ($^1\Delta_g$) formation from electronically excited OClO in D_2O was found to be 0.005 ± 0.003. Thus, in water solution, 95% of the Cl is produced from the isomerization pathway, Equation (3); the remaining 5% results from the symmetric dissociation, Equation (4). The importance of the symmetric C_{2v} dissociation increases with decreasing solvent polarity. The quantum yield of Cl + O_2 ($^1\Delta_g$) production in C_6D_6 ($E_T(30) = 34.5$ kcal/mole) and carbon tetrachloride ($E_T(30) = 32.5$ kcal/mole) is 0.02 and 0.07, respectively, much larger than in D_2O ($E_T(30) = 63.1$ kcal/mole). The increase in the amount of symmetric C_{2v} dissociation with decreasing solvent polarity suggests that this mechanism may be important in the gas phase chemistry of OClO. This conclusion is consistent with translational energy distributions for the O_2 product in the gas phase photolysis of OClO reported by Davis and Lee [40].

The combination of picosecond transient absorption and time-resolved infrared emission spectroscopy enables one to quantify the photoreactivity of OClO in solution. The overall photochemistry is summarized by Equations (5) to (7).

$$\mathrm{OClO} + h\nu \rightarrow \mathrm{ClOO}(A' \text{ or } A'') \rightarrow \mathrm{Cl}(^3\mathrm{P}) + \mathrm{O_2}(^3\Sigma_g^-) \quad (9.5\%) \tag{5}$$

$$\rightarrow \mathrm{Cl}(^3\mathrm{P}) + \mathrm{O_2}(^1\Delta_g) \quad (0.5\%) \tag{6}$$

$$\rightarrow \mathrm{ClO}(^2\Pi) + \mathrm{O}(^3\mathrm{P}) \quad (90\%) \tag{7}$$

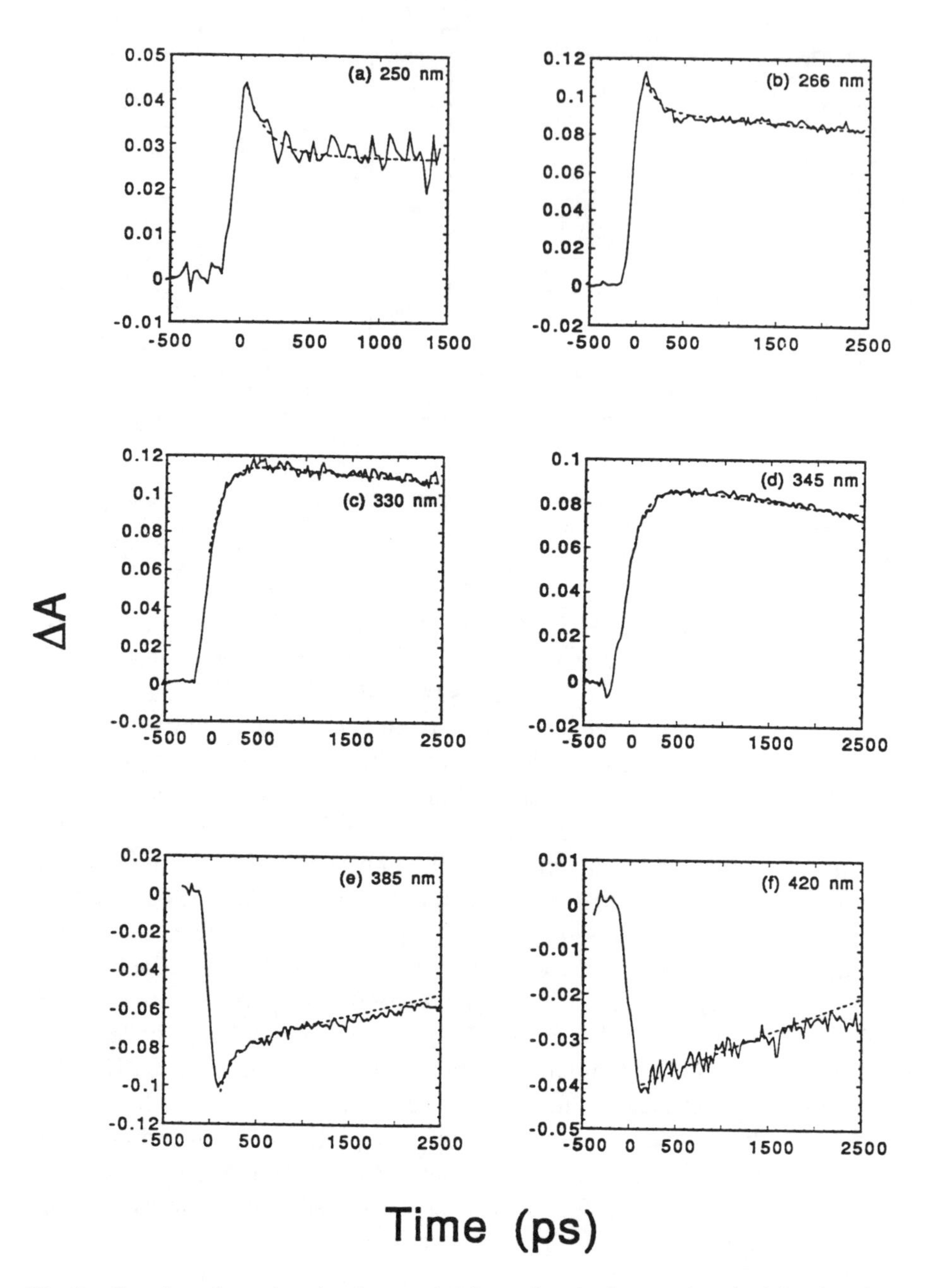

Fig. 3. Transient absorption signals recorded for various probe wavelengths following the photolysis of OClO at 355 nm in water. The dashed lines are calculated absorbance signals where photolysis results in a partitioning of 90% into ClO + O and 10% formation of ClOO. The ClOO subsequently thermally dissociates into Cl + O_2 [29].

B. Photoinitiated Electron Transfer in $[(NC)_5Ru^{III}CNRu^{II}(NH_3)_5]^{-1}$: Electron transfer (ET) reactions occupy a central role in chemistry. In 1992, W. Woodruff and coworkers used picosecond infrared absorption spectroscopy to monitor the electron transfer dynamics of Equation (8) [30].

$$[(NC)_5Ru^{III}CNR^{II}(NH_3)_5]^{-1} \underset{k_{et}}{\overset{h\nu}{\rightleftharpoons}} [(NC_5Ru^{II}CNRu^{III}(NH_3)_5]^{-1} \tag{8}$$

The reaction process was initiated by a picosecond pulse at 600 nm. The sample was then probed by a picosecond infrared laser pulse. Transient decays for three probe wavelengths, 1954, 1994, and 2034 cm^{-1} are shown in Figure 4. Using the band at 2053 cm^{-1}, which corresponds to the terminal RuC≡N stretch in the ground state of $[(NC)_5Ru^{III}CNRu^{II}(NH_3)_5]^{-1}$, the dynamics of back electron transfer were found to occur with a time constant of 6 ± 1 ps. Upon excitation at 600 nm, a new feature was observed at 2110 cm^{-1}, which decayed with a time constant of 0.5 ps. This band was unambiguously assigned to the metal-metal charge transfer state and provides direct evidence that back electron transfer occurs on the femtosecond time scale. The features observed between 1950 cm^{-1} and 2040 cm^{-1}, some of which are shown in Figure 4, are consistent with vibrational cooling of the hot $[(NC)_5Ru^{III}CNRu^{II}(NH_3)_5]^{-1}$ molecule. The data show that the rate constant for vibrational relaxation increases with increasing vibrational quantum number. This study also provides the interesting observation that upon back electron transfer, a large amount of energy (up to 14,000 cm^{-1}) is placed into a single vibrational mode: the terminal RuC≡N stretch. This remarkable selectivity in the vibrational relaxation process could not have been determined from electronic absorption spectroscopy. This study shows the potential power of ultrafast vibrational techniques for studying the detailed dynamics of chemical reactions in solution.

C. Electronic Dephasing of HITCI in Ethylene Glycol: The above examples have focused on measuring population kinetics using absorption spectroscopies. This involves measuring the absorption properties of the chemical systems when the two pulses are well separated in time. When two coherent light pulses overlap in the sample, additional properties of the chemical system can be studied. Such an approach can be used to measure the dynamics of electronic dephasing, a molecular quantity which is usually quantified by the time constant T_2. In principle, several experimental methods can be used to probe the dynamics of electronic dephasing [41–44]. All require that the time resolution of the experiment be competitive with the T_2 time. These studies indicate that the room temperature homogeneous dephasing time of dye molecules is on the order of tens of femtoseconds.

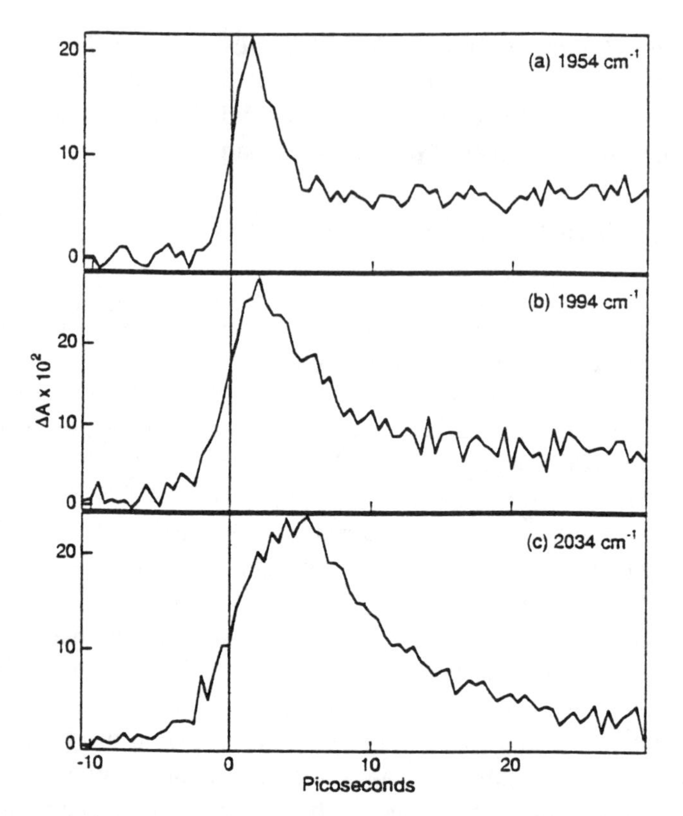

Fig. 4. Transient infrared absorption data following the excitation of the intramolecular charge transfer band of $[(NC)_5Ru^{II}CNRu^{III}(NH_3)_5]^{-1}$ at 600 nm. The dynamics observed at different probe wavelengths reflect the vibrational relaxation of the molecule following back electron transfer [30]. Reprinted with permission.

Our approach to measuring dephasing times is to measure the transient bleaching dynamics of a solute in liquid solution in a degenerate pump-probe arrangement using variable laser pulse-widths [31]. It has been recognized for some time that coherent coupling between the pump and probe laser beams contributes significantly to the transient absorption signal near zero time delay in such experiments. Thus, the transient absorption measurement not only contains the usual population bleaching and recovery terms, but also a coherent coupling spike (sometimes also 'unfairly' referred to as a coherent 'artifact') in the vicinity of time zero, when the pump and probe pulses are derived from the same laser and have identical frequencies. Balk and Fleming were the first to quantify the connection between the shape and amplitude of the coherent spike and the material dephasing time [45]. Based on a density matrix formalism [46] and optical Bloch equations [47], these authors derived a set of equations for the evolution of the probe beam transmittance as a function of delay time between the pump and probe light fields. They also showed via numerical model calculations that the coherent spike is a sensitive function of the ratio of T_2 and laser pulse width. With the exception of T_2, all of the parameters used in their model were experimentally determined. Thus, comparison between experimental data and theoretical calculations can be used to quantify the dephasing time.

Figure 5 (top) displays the pump-probe dynamics of HITCI in room temperature ethylene glycol solution as a function of laser pulse-width. The relative amplitude of the coherent coupling signal to the population bleach is laser pulse-width dependent. For the shortest pulse used, 23 fs, the data primarily reflects population dynamics. Only a small coherent spike is observed around zero delay. However, increasing the laser-pulse width to 42 fs and 54 fs results in an increased amplitude of the coherent signal. The dominant molecular contribution to the relative amplitude of the coherent spike to that of the population dynamics in a pump-probe experiment arises from the ratio between the time resolution of the experiment (Δt) and the electronic dephasing time of the excited state being populated (T_2).

Figure 5 (bottom) shows a comparison between the experimental data obtained with 54 fs laser pulses and theoretical predictions for various T_2 times. The experimental data can be quantitatively reproduced by a dephasing time of 18 ± 2 fs. This study shows that absorption spectroscopy can be used to provide dephasing information in addition to the population dynamics of reactive molecules in solution.

D. Solvation of Photogenerated Unsatured Metal Carbonyl Complexes: Many metal carbonyls have been studied by time resolved electronic and infrared absorption spectroscopy [32, 48]. As an example of this type of chemistry, we consider here the molecule $Cr(CO)_6$. Excitation of $Cr(CO)_6$ leads to dissociation of one of the CO groups, forming the unsaturated intermediate, $Cr(CO)_5$. In solution, a solvent molecule then coordinates to the open site on the metal. Femtosecond absorption studies reported by Nelson and coworkers report the observation of electronically excited $Cr(CO)_6$ following photolysis by a 70 fs laser pulse [48]. The ensueing transient absorption dynamics reveal a time scale of approximately 350 fs for the dissociation into CO and $Cr(CO)_5$. These results indicate that the reactive intermediate is formed on the femtosecond time scale. Once formed, coordination of a solvent molecule to the vacant site on the metal occurs. The time scale associated with this event sheds insight into the dynamics that occur in the first solvation shell. We examined the solvation of the photogenerated $Cr(CO)_5$ fragment in series of hydrocarbon, ether, alcohol, alkyl bromide, and alkyl nitrile solvents [32]. The formation of the solvated complex could be detected by monitoring the temporal evolution of the absorption band characteristic of the six-coordinate molecule $Cr(CO)_5S$

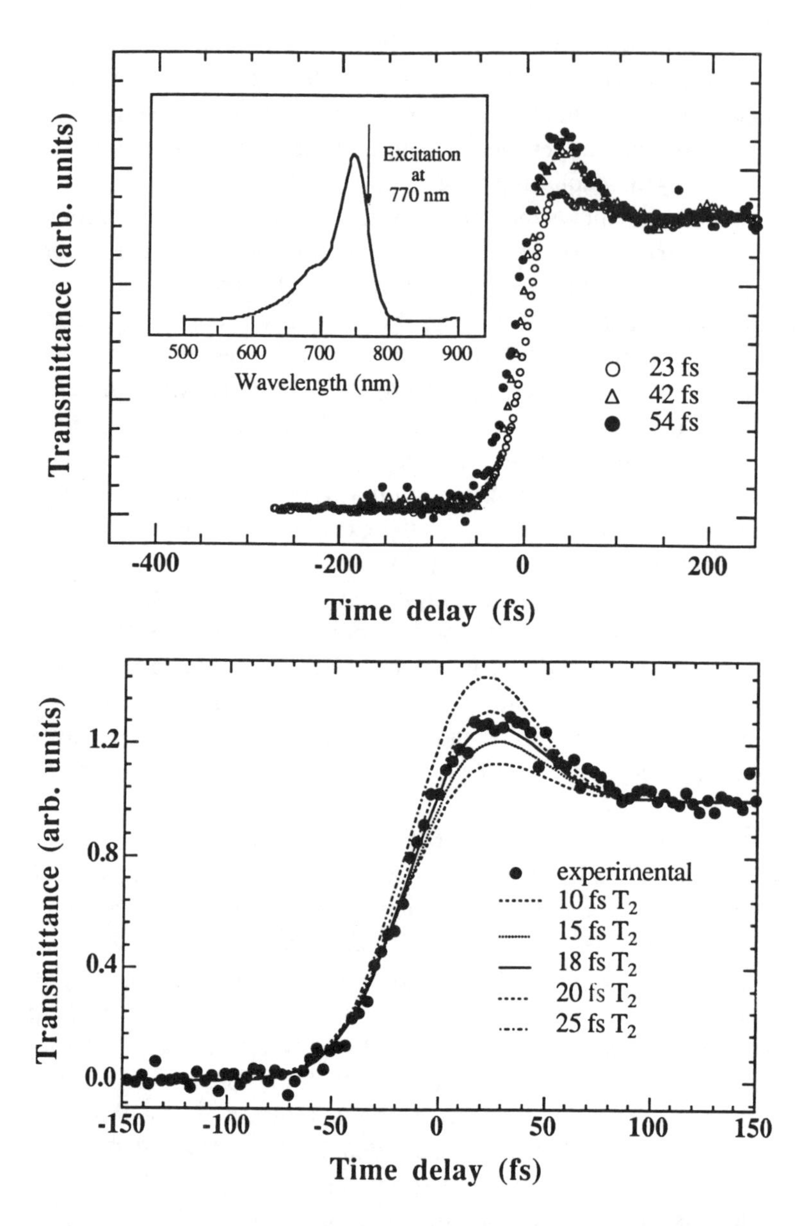

Fig. 5. *Top*: Pump-probe data for HITCI in ethylene glycol. The dynamics plotted were obtained with different laser pulse widths (FWHM): (a) 23 fs, (b) 42 fs, (c) 54 fs. With increasing laser pulse width the coherent spike becomes more evident. These data indicate that the electronic dephasing time of HITCI is comparable to the laser pulse widths being used. The insert displays the absorption spectrum of HITCI in ethylene glycol. The arrow indicates the center of the excitation laser band width. *Bottom*: The experimental data obtained with a 54 fs laser pulse is compared with theoretical predictions. The five calculated curves correspond to different T_2 times: (- - - -) 10 fs, (. . . . .) 15 fs, (——) 18 fs, (– – – –) 20 fs (– · – · – · –) 25 fs. A detuning of 400 cm^{-1} is used, reflecting the difference between the absorption maximum and the central laser frequency [31].

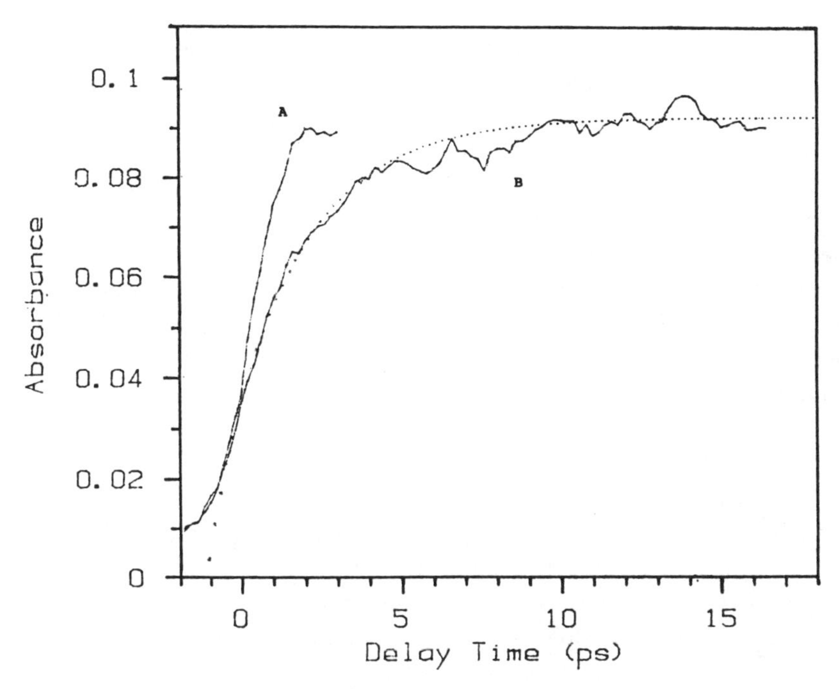

Fig. 6. (A) the $S_1 \leftarrow S_N$ transient absorption of t-stilbene in methanol solution. The experimental absorption signal is a direct measure of the instrument response. (B) The experimentally measured absorbance at 460 nm following the 300 nm photolysis of $Cr(CO)_6$ in methanol solution. The dotted curve is an exponential rise with a time constant of 2.5 ps [32].

(S = solvent molecule). Data for the photolysis of $Cr(CO)_6$ in methanol is shown in Figure 6. The time resolution of this experiment, as limited by the laser pulse width, is $\sim$ 1 ps. As a result, the dynamics of dissociation of CO cannot be resolved. The probe wavelength is 460 nm, the maximum of the absorption band for $Cr(CO)_5MeOH$. For comparison, the rise time of the $S_N \leftarrow S_1$ absorption of t-stilbene, probed at 460 nm, is also plotted. This provides an accurate measure of the instrument response. The data clearly show that the the formation of $Cr(CO)_5MeOH$ occurs on a time scale longer than the instrument response. Fitting the data to an exponential function provides a time scale of 2.5 ps for the solvent coordination process. These dynamics were confirmed by the femtosecond studies reported by Nelson, where the dissociation process as well as the solvation dynamics could be monitored [48].

Studies in longer chain alcohol, ethers, alkyl bromides and alkyl nitriles revealed that a distribution of solvated complexes involving coordination of either an alkane group or the functional group (e.g., nitrile, bromide, hydroxyl) of the solvent molecule were formed following photolysis of the hexacarbonyl. The absorption dynamics revealed that initially formed alkane-coordinated complexes rearranged to form the more stable intermediate in which the functional group of the solvent molecule coordinates to the metal center. A kinetic model for a unimolecular rearrangement mechanism was proposed to account for the entire set of data. Temperature studies on the transient absorption dynamics provided information on the barriers associated with the restructuring of the solvated complexes. The activation energies obtained were similar to those reported for the *syn* rotational barriers in alkanes [32c]. These data were used to develop a molecular mechanism for the solvation dynamics.

B. Emission Spectroscopy

Time-resolved emission spectroscopy is an important and commonly used tool in the study of chemical reaction dynamics in the condensed phase. There are three general approaches to obtaining time resolution in emission spectroscopy: time-correlated single photon counting [49], direct detection by a streak camera [50], and fluorescence up-conversion [51]. Each of these techniques has its limitations and advantages. The best approach often depends on the characteristics of the chemical system being studied.

1. Time-correlated single photon counting (TCSPC)

Usual instruments involved in a TCSPC experiment are a picosecond excitation source (normally a cavity-dumped sync-pumped dye laser operating in the MHz range), a fast photomultiplier, fluorescence collection and dispersion (optional) optics, and counting electronics. The laser power must be sufficiently low such that the probability of producing a detected fluorescence photon from one laser pulse is much less than one.

The basic idea behind the TCSPC technique is to measure the time between stop and start light pulses. In many cases, the start pulse is generated by the excitation light pulse, the stop pulse is generated by a fluorescence photon. This information is processed by converting the time between the start and stop pulse into a voltage using a time-to-amplitude converter. Using this approach, the time-resolved emission is recorded as a histogram of the converted voltage signals. The time resolution of this experiment is determined by the response time of the photomultiplier. The sharper the rising edge of response curve is, the better resolution the time-to-amplitude converter can achieve. Typical TCSPC spectrometer has an instrument response time of $\sim$ 50 ps and a time resolution of $>$20 ps. Since the fluorescence lifetime of many molecules falls in the range of nanoseconds, TCSPC remains a popular technique for lifetime measurements despite its 'limited' time resolution as compared to other techniques.

For many time-resolved optical experiments, the statistical properties of the data obtained are not well understood. TCSPC has the advantage that the signal obeys Poisson statistics. This allows for detailed modelling of the data using correct weighting factors. As a result, the experimental uncertainties for lifetimes obtained from TCSPC experiments can be accurately quantified.

2. Streak camera

Over the years, streak camera detection has been a popular approach for obtaining time resolution better than 100 ps. In some cases, single-shot streak cameras offer a time resolution of better than 1 ps, but the lack of averaging capacity restricts them to strong signals. Synchroscan streak cameras have become available recently which allow signal averaging even with a high repetition rate laser; however, this leads to some deterioration in the time resolution, with 10 ps being a reasonable practical limit for low level signals. Another disadvantage of streak cameras is the spectral response of the photocathode, which typically does not extend beyond 900 nm. In spite of these drawbacks, streak cameras still enjoy wide spread applications for time-resolved fluorescence spectroscopy in the 10 ps time domain.

3. Fluorescence up-conversion

With advances in ultrafast laser technology, sub-100 fs pulses are now routinely generated. However, it has been a challenge to obtain time resolution comparable to the laser pulse width in time-resolved fluorescence spectroscopy. The best technique developed to date relies on optical gating of the fluorescence emission by nonlinear optical crystals, commonly called fluorescence upconversion. This technique is currently used to study a wide variety of chemical phenomena. In the following subsections, this form of spectroscopy is examined in detail. First, the general principles of fluorescence upconversion are introduced. This is followed by a discussion of experimental implementation. Finally, limitations and problems encountered in working on the femtosecond time domain are discussed.

A. Basic concepts: The basic ideas of how optical gating techniques in nonlinear optical crystals are used to obtain time resolution in emission experiments are oultined in Figure 7. The incoherent fluorescence photons excited by an ultrashort (pump) laser pulse are focused into a nonlinear frequency-mixing crystal. Another ultrafast, appropriately delayed (probe) laser beam is overlapped with the fluorescence spot in the crystal. The relative timing sequence

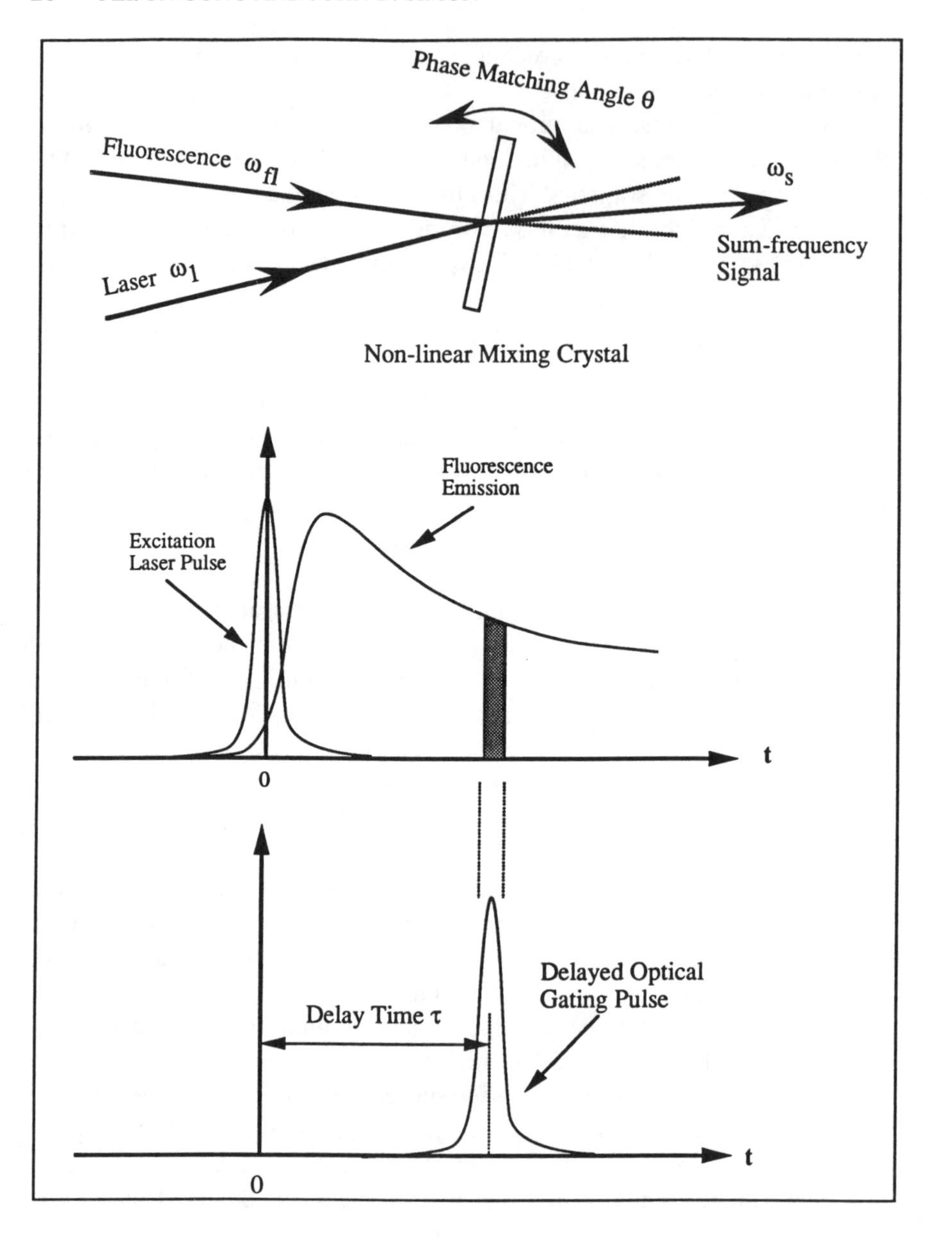

Fig. 7. Principle of fluorescence up-conversion (adapted from [51b]). A sum fequency signal is generated in a non-linear crystal when both of the fluorescence emission and the optical gating pulse are present. The up-converted radiation intensity is proportional to the number of fluorescence photons emitted within the window probed by the gating pulse. Thus, the sum-frequency signal plotted against delay time τ is an accurate representation of the time-dependent fluoresence emission. See text for details.

of the pump, the fluorescence, the probe pulses are illustrated in the lower portions of Figure 7. The crystal is properly oriented to satisfy the phase-matching conditions for the frequency mixing process. This mixing process is normally a sum frequency generation (due to the availability of sensitive detectors in the UV and visible region, i.e., PMTs), however, the discussions also apply to difference frequency mixing, which has been used to time-resolve UV emission. The light generated at the sum frequency is then collected and sent into subsequent light detection and data acquisition devices. The sum frequency signal can be expressed by the following equation.

$$I_{\mathrm{sum}}(\tau) = \int I_{\mathrm{fl}}(\tau) I_{\mathrm{probe}}(t-\tau)\,\mathrm{d}t\ . \tag{9}$$

In Equation (9), I_{sum}, I_{fl}, and I_{probe} are the intensities of the sum, the fluorescence, and the probe beams, respectively. Equation (9) clearly shows that the sum frequency signal is only present when the fluorescence photons and the probe pulse are simultaneously coincident on the crystal. Since the probe pulse is typically much shorter than the time-dependent fluorescence, it acts as an optical gate to the up-conversion signal, analogous to an electronically-gated boxcar integrator. In principle, this 'optical boxcar' has a time resolution comparable to the probe pulse width. The actual time resolution of the measurement also depends on the temporal width of the pump pulse.

B. Experimental implementation: Figure 8 shows a basic setup for an up-conversion experiment. At time $t = 0$, the sample is excited by an ultrashort laser pulse at frequency ω_1. The resulting fluorescence, ω_{fl}, is then collected and focused on a nonlinear crystal. A second laser pulse at frequency ω_2 overlaps the fluorescence spot on the nonlinear crystal. The delay time, τ, between the arrival of the fluorescence and the second laser pulse is typically adjusted by moving a computerized translation stage. The generated sum-frequency light is detected by a photomultiplier after rejection of stray and laser light by a monochromator or prism or both. The signal level of this experiment is usually low due to the small conversion efficiency of the mixing process and the lack of peak intensity of the incoherent fluorescence light. This is especially true for a high-repetition rate, low-power laser system. In these cases, photon counting electronics are often used to improve the signal-to-noise ratio.

The fluorescence up-conversion technique can be used to study orientational dynamics in a similar fashion as transient absorption spectroscopy [52]. The angle between the polarizations of the detected fluorescence and the up-conversion gating pulse is determined by the type of phase matching in the sum-frequncy generation crystal (0° for type I and 90° for type II). Therefore the fluorescence anisotropy can be measured in the same manner as described

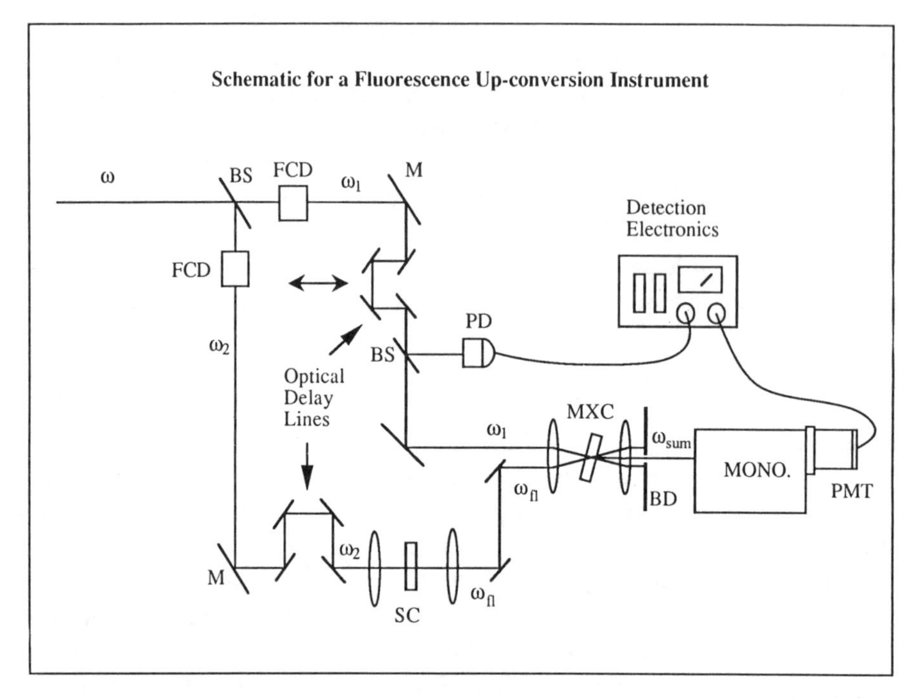

Fig. 8. Schematic for a fluorescence up-conversion instrument. In additon to the abbreviations defined in Figure 1: MXC, mixing crystal; MONO., monochromator; PMT, photomultiplier. The fluoresence light, induced by focusing pump laser beam at ω_2 into a sample cell, is collected by a lens and recombined with the gating pulse at ω_1 in a nonlinear mixing crystal by another focusing lens. The sum frequency radiation is then spatially filtered and selected by a monochromator and detected via a photomulitplier. The photodiode detects a small portion of the gating pulse for normalization purposes. The relative timing between the excitation and the gating pulse is controlled in the same fashion as in transient absorption measurements.

by Equation (2), i.e., measuring separate up-conversion data with parallel and perpendicular polarizations between the excitation and up-conversion gating pulse. When only population information is desired, the polarization angle between the pump and probe lasers should be set at 54.7°. Care must be exercised in selecting fluorescence collection optics that maintain polarization properties.

C. Limiting factors on the time resolution obtainable in up-conversion: In practice, the time resolution attainable in an up-conversion experiment is always longer than the probe laser pulse width. There are two main mechanisms that are responsible for this lengthening of the time resolution: (1) GVD of the emitted fluorescence photons through the sample and collection optics, and (2) group velocity mismatch between the probe pulse and the fluorescence light in the mixing crystal [51]. The first mechanism can be alleviated by using all-reflective (non-dispersive) optics, e.g., a double-Cassegranian, to collect and focus the fluorescence into the up-conversion crystal [51b]. A thin sample cell also helps in reducing the fluorescence dispersion. The second mechanism, however, is unavoidable. The group velocity mismatch physically arises from the fact that the probe laser wavelength and the fluorescence wavelength are different. In order to minimize this effect, a low-dispersion, thin nonlinear crystal is needed. Unfortunately, a low-dispersion crystal often has a low nonlinear susceptibility as well, e.g, KDP, resulting in inefficient mixing. The nonlinear mixing signal scales with the thickness of the crystal, so the desire to obtain the shortest time resolution needs to be balanced with the fact that a reasonable signal level is required to have a useful experiment. Two ways to reduce the group-velocity mismatch are to keep the probe wavelength as close as possible to the fluorescence wavelength and use type I phase matching ($e + e \rightarrow o$; or $o + o \rightarrow e$) whenever possible.

In some applications, the excitation laser pulse is obtained by frequency doubling the fundamental laser light [51a]. In this case, the doubling process may place some severe limits on the shortest pump pulse that can be obtained. This then limits the overall temporal resolution of the up-conversion experiment. To keep the pump pulse broadening to a minimum, one not only has to consider the GVD of the fundamental laser pulse, but more importantly, one has to take into account the GVD of the second harmonic light. For normally-dispersing materials, the refractive index increases with increasing frequency. As GVD scales both with frequency and the change in refractive index with frequency, it will be greatest in the ultraviolet region. Another limiting factor in generating short pulses at the second harmonic frequency is the band-width limit of the crystal as determined by phase-matching conditions. This effect is especially pronounced for sub-50 femtosecond pulses. Recently innovative schemes have been developed to circumvent this limit [52]. The broad spectral profile also complicates the resolution of the fluorescence emission from the possibly overlapping probe laser wavelengths. In general, the fluorescence wavelength must be displaced from the central pump wavelength by at least the FWHM of the pump spectral band width.

4. Selected Examples

In this section we consider some examples of chemical dynamics that have been studied by time resolved emission spectroscopy. The examples recounted below focus on two applications of femtosecond time-resolved emission spectroscopy that utilized the technique of fluorescence up-conversion.

A. Solvation Dynamics: Time-Dependent Stokes Shift: What are the timescales associated with the inertial, rotational and translational times of molecules in liquids? This question has spurred a tremendous experimental and theoretical research effort during the last decade [17]. Understanding the individual and collective dynamics of molecules in liquids is central to developing comprehensive conceptual and quantitative models of chemical reactivity in solution. Prior to using ultrafast spectroscopies to examine the dynamic response of a liquid, time scales for solvent relaxation were determined from dielectric loss measurements coupled to continuum theories for the liquid [54]. More recently, measurement of the time-dependent Stokes shift (TDSS) of an electronically excited solute molecule [17] has been used to elucidate the time-dependent response of a liquid to small perturbations.

TDSS studies involve monitoring the evolution of the emission spectrum of a dye molecule that is electronically excited by an ultrashort light pulse. Formation of the excited state occurs more rapidly than nuclear rearrangement of the solvent molecules. As a result, the excited state molecule finds itself initially in the equilibrium solvation of the ground state. With increasing time, the solvent restructures itself in response to the charge distribution of the excited-state of the probe molecule. This results in a time-dependent stabilization of the energy of the electronically excited molecule, reflected in a corresponding dynamic red-shift in the emission spectrum.

TDSS data have been recorded for many probe molecules in numerous solvents and at varying temperature. A variety of techniques have been used, depending on the time-scale of the measurements. Early measurements by TCSPC or streak camera detection were restricted to a time resolution of greater than 10 ps. These studies reveal that the *average* solvation time is close to that predicted by dielectric continuum models [55]. However, the time dependence of the solvent response is not accounted for by bulk dielectric models. Nonexponential solvent relaxation is observed for liquids that are adequately described by a (single time scale) Debye dielectric response [55].

Computer simulations were the first to reveal a large amplitude (~50%), ultrafast (10's of fs), initial component in the TDSS response [56]. These results were interpreted in terms of librational motion in the case of water and inertial relaxation in a variety of nonprotic solvents. However, before one could invoke such ultrafast relaxation behavior as important in solution phase chemical reactivity, experimental verification was required. Measurement of dynamics on this time-scale is only possible using femtosecond lasers coupled with up-conversion detection. In 1991, Rosenthal *et al.* reported the experimental observation of an ultrafast relaxation component in the TDSS of the dye LDS-750 in acetonitrile [55]. That component accounted for ~80% of the amplitude and could be well described by a decay time of 70 fs. The slower tail, accounting for the remaining 20% of the relaxation dynamics, was exponential with a time constant of 200 fs. The agreement between experimental data and predictions from computer simulations was nearly quantitative. The computer simulations showed that the majority of the relaxation ($\sim$70%) was through interaction with the small number of solvent molecules that sit in the first solvent shell around the solute. The simulations further suggested that the rapid contribution results from independent single molecule motion. This comparison of data obtained from ultrafast emission spectroscopy with molecular dynamics computer simulations facilitated a molecular-based understanding of the dynamics of solvation.

B. Intramolecular Electron Transfer: Many recent experiments focus on the rate constants of excited state intramolecular charge transfer reactions [57]. The molecules studied contain donor and acceptor groups which are covalently linked. Such molecular structures enable the study of electron transfer kinetics in the absence of diffusion. The reaction is initiated by an ultrashort light pulse. The reaction dynamics can generally be followed through emission spectroscopy. We consider here the molecule bianthryl (*BA*) [58]. The excited-state intramolecular charge transfer reaction of *BA* is characterized by low activation energies, $\Delta G^{\ddagger} \leq RT$. One of the aims of the research on this molecule is to determine the relative importance of solvent dynamics and intramolecular vibrational effects on the magnitude of the charge transfer rate. Clearly, assessing the importance of solvent motion on charge transfer processes relies heavily on the time-scales discussed in the previous section.

Considering the results of the TDSS studies, to examine a chemical reaction that occurs on a time-scale comparable to solvent motion requires femtosecond time resolution. The dynamics of the charge transfer reaction of electronically excited *BA* in solution was determined from time-resolved fluorescence dynamics. The emission dynamics revealed the time dependence of the reactant population, thereby providing a direct measure of the electron transfer rate. The reaction dynamics were examined in a variety of polar aprotic solvents. In contrast to many previous reports on related intramolecular charge transfer reactions, Kang *et al.* reported that there was a very poor agreement between the experimentally measured electron transfer rates and the solvent relaxation times as determined from dielectric continuum theory [58]. However, an improved agreement was observed between the electron transfer rates and the inverse of the solvent relaxation times obtained from TDSS studies. This later observation strongly supports the conclusion that the reaction process is solvent controlled. The data indicated that the simple relationship from dielectric continuum ideas may not bc valid.

Kang *et al.* showed that the dynamics of charge transfer in *BA* could be quantitatively accounted for by a theoretical model which incorporated four important features: (1) an Onsager cavity and semi-empirical treatment for the solvent coordinate, (2) an electronically adiabatic description of the mixing between the reactant and product states, (3) a generalized Langevin equation for treating the motion along the reaction coordinate and (4) an empirical solvatochromic/vibronic description for predicting fluorescence and absorption spectra [58]. Excellent agreement was observed between calculated and experimental steady-state absorption and emission spectra and time dependent emission spectra. The observed agreement between theory and experiment provided convincing evidence that polar solvation dynamics control the electron transfer rate for BA.

C. Raman Spectroscopy

Transient Raman spectroscopy can be classified into two classes, incoherent (spontaneous) Raman and coherent Raman, depending on the nature of the Raman scattering process. In this section, we briefly examine spontaneous Raman spectroscopy.

1. Spontaneous Raman Spectroscopy

The use of transient spontaneous Raman spectroscopy to examine chemical dynamics on the picosecond time scale has only become productive during the past five years. The low signal levels of these experiments provided an experimental challenge that required advances in both the laser light sources and optical detection. Due to recent advances in generating stable high repetition rates of ultrashort light pulses and optical detectors, e.g., CCD's, Raman spectroscopy is quickly becoming an important tool in the study of chemical processes. Typical experimental setups are similar to a transient absorption arrangement, except the detected signal is a Raman signal instead of a transmitted laser beam (see Figure 1). Time resolved spontaneous Raman spectroscopy is currently being used by several research groups to measure population kinetics. Raman spectroscopy, in principle, is a more powerful experimental tool than either of the two electronic spectroscopies previously described. In addition to providing information on the dynamics of reactive intermediates, Raman data can be used to extract information on vibrational relaxation processes and molecular strutcure.

2. Examples

In a series of elegantly designed experiments, Mathies and coworkers studied the photo-induced concerted ring-opening reactions of cyclopolyenes in hexanes [59]. From the time evolution of the Raman bands that are characteristic of the ring-opened products, these authors showed unequivocally that ring opening was completed in 8 to 10 ps.

Hopkins and colleagues have performed a series of resonant Raman experiments on a variety of photodissociation reactions, e.g. I_2 [60], $Cr(CO)_6$ [61], carbonmonoxyhemoglobin [62], CH_3I [63]. Their data have been used to extract information on chemical processes such as geminate recombination (caging), and vibrational cooling.

Several workers have used transient Raman spectroscopy to study the excited state isomerization of *t*-stilbene and related molecules [64, 65]. For example, a recent report by Gustafson and coworkers examined the ground and excited-state Raman spectra of *trans*-4,4′-diphenylstilbene [66]. Many of the vibrational modes associated with the biphenyl portion of the molecule exhibited solvent-dependent intensities. These effects were attributed to variation in the Franck-Condon overlap between S_1 and S_N as a result of differences in the planarity of the biphenyl moiety in different solvents. The dynamics suggested that dielectric stabilization of S_1 by the solvent, not bulk viscosity, controlled the conformational dynamics.

D. Polarization Spectroscopy

A time-resolved polarization experiment is similar in many aspects to a transient absorption experiment. Both are two-wave mixing processes. In both cases the probe laser beam intensity is monitored as a function of the relative delay time with respect to the pump pulse. However, in a time-resolved polarization experiment, the polarization properties of the probe beam are monitored as well. The polarization of the light pulse can be affected by both the real and imaginary parts of the refractive index, unlike an absorption experiment in which only the imaginary part of the refractive index is monitored.

Transient polarization experiments can be further separated into two classes, depending on whether linear or circular polarization changes are measured. These two cases are discussed separately below.

1. Transient linear birefringence and dichoism

In a normal linear polarization spectroscopy configuration (see Figure 9), a linearly polarized pump pulse at frequency ω_1 is focused on an isotropic sample. The pump pulse induces an anisotropic polarization, which in principle can bc both birefringent and dichroic. This induced polarization, especially for the case of birefringence, is commonly known as the Optical Kerr Effect (OKE) [67]. Subsequent to the passage of the pump pulse through the sample, a linearly polarized probe pulse, at frequency ω_2 and polarized at 45° from

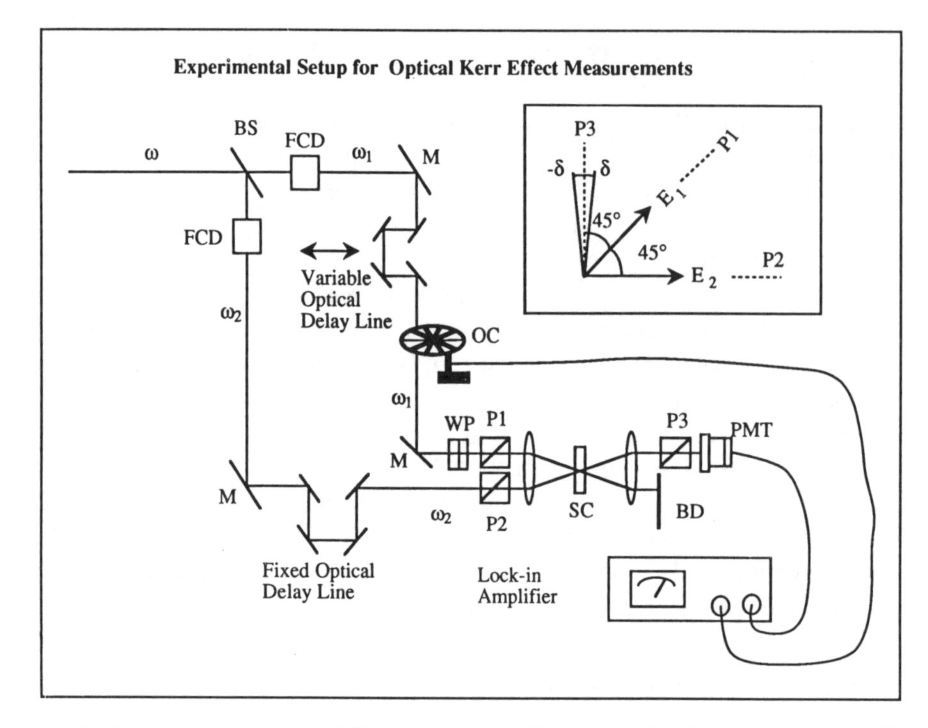

Fig. 9. Experimental setup for OKE measurements. The pump and probe pulses are derived in similar manners as in transient absorption experiments. OC is an optical chopper that modulates the pump beam intensity and provides a reference frequency for the lock-in amplifier. WP and P1 are a waveplate and a polarizer, respectively, that are used to rotate and select the polarization of the pump beam such that it is 45° relative to the probe laser polarization, which is controlled by P2. The detection polarizer P3 in front of the PMT is oriented at $90° \pm \delta°$ relative to P2 (see text for more details). The insert shows the polarization relations among the pump, probe and detection polarizers. The detection scheme is homodyne when $\delta = 0$ and heterodyne when $\delta \neq 0$.

the pump field polarization, is incident on the excited region of the sample. The transmitted probe pulse is detected through an analyzer polarizer oriented at –45°. (This detection scheme is often referred as homodyne detection, in contrast to heterodyne detection, which will be discussed later). If the sample medium were isotropic no light would be transmitted through the analyzer polarizer. However, because of the anisotropic polarization of the medium induced by the pump laser, some of the probe intensity is transmitted through this analyzer. As the sample anisotropy relaxes (by electronic and vibrational dephasing, population decay, rotational relaxation, or other mechanisms) the probe intensity transmitted by the polarizer decreases.

The sensitivity of polarization spectroscopy can be enhanced by exploiting the methods of optical heterodyning [68]. In this detection scheme the probe light field transmitted through the analyzer polarizer (the signal field E_S) is mixed with a second light field (the so-called local oscillator field E_{LO}) at the same frequency. The signal that is detected by the photodetector, I_D, can then be expressed by Equation (9).

$$I_D = I_S + I_{LO} + \frac{nc}{4\pi}\text{Re}[E^*_{LO} \cdot E_S] \,. \tag{10}$$

If $I_{LO} \gg I_S$ the third term in Equation (10) dominates (the I_{LO} term can easily be removed by using a mechanical chopper combined with lock-in detection). Such an approach has the effect of increasing sensitivity and linearizing the signal (the signal is now proportional to the signal field instead of the squared modulus). A detailed signal-to-noise analysis has been carried out by Eesley *et al.* to determine the optimum magnitude of the local oscillator [69]. Depending on whether E_{LO} is chosen as linearly or elliptically polarized, the heterodyne method will selectively detect the dichroic or birefringent contributions to I_S [72].

Polarization spectroscopy is extremely sensitive to the polarization properties of the probing light field. Thus the pair of crossed polarizers in the probe beam must be of high optical quality providing an extinction ratio of better than 10^{-6}. Small birefringences in optical components can distort the signal significantly, especially when the signal is composed primarily of birefringence. Care must be exercised in interpreting the experimental decay curves to avoid being misled by small external birefringences induced by the focusing lens, the beam splitter or other optical elements with poor polarization preserving properties. Waldeck and colleagues developed a procedure in which the dichroic term can be extracted without the interference from the birefringent term by taking the difference of two scans with the same magnitude in-phase local oscillator but of opposite sign [70]. This technique has greatly improved the accuracy of the extracted population information from these heterodyne optical Kerr experiments. Cho et al reported a similar procedure for the birefringent case where the dichroic terms were cancelled out [71].

Since the first time-resolved polarization experiment reported by Shank and Ippen [72], numerous groups have further developed and refined the technique and applied it to a wide variety of problems including, for example, rotational dynamics of dye molecules in fluids [70, 73, 74], molecular reorientation of simple liquids [75], inter- and intra-molecular librational and vibrational dynamics of neat liquids [76], predissociation and vibrational dynamics of a diatomic molecule in solution [77], intramolecular electronic excitation transfer dynamics [78], and the transition-state spectroscopy of O_3 photodissociation [79].

The high sensitivity of OKE measurements as compared with conventional transient absorption experiments arises from the fact that OKE is a background free signal. This high sensitivity is demonstrated by Lotshaw *et al.* for OKE measurements on neat acetonitrile [76]. In their experiments, excellent signal-to-noise ratios were observed for off-resonant excitation of the transparent sample by a 15 mW CW mode-locked dye laser. The transient birefringent response from the neat liquid was attributed to the coherent excitation of intermolecular oscillators, except for an instantaneous electronic contribution which followed the cross-correlation between the pump and the probe pulse. McMorrow and Lotshaw analyzed their OKE data with a Fourier-transform technique and obtained the intermolecular vibrational spectrum of acetonitrile. These authors emphasized that the femtosecond transients probe the role of the intermolecular potential energy surfaces in shaping the short-time vibrational aspects of the intermolecular dynamics in this liquid and discussed possible connections between OKE experiments and solvation dynamics. Recently, Cho *et al.* quantified the connection between OKE experiments and dynamic fluorescence Stokes shift measurements in the same liquid [80]. The latter experiment was carried out with a fluorescence up-conversion apparatus and was discussed above.

Scherer *et al.* reported an elegant transient absorption measurement of the predissociating B state of I_2 in liquid hexane recorded by optical heterodyne detection [77]. From the decay of the amplitude of the vibrational quantum beats originating from the B state, these authors were able to conclude that the predissociation (i.e., curve-crossing) occurs within 200 fs (three vibrational periods). The power of optical heterodyne detection is further demonstrated by Chen *et al.* in their study of the photodissociation dynamics of O_3 in the Chappuis band ($\sim$600 nm) [79]. With the help of the zero-background high-sensitivity detection offered by optical heterodyning and multiplexed broad band probe (over 300 nm), these workers were able to directly observe the wave packet motion on complicated potential energy surfaces during dissociation. What is even more remarkable is that this experiment was carried out in a 50 torr gas cell, where the number density was much lower than that normally encountered in condensed phase experiments.

2. *Transient circular dichroism*

In a circular dichroism experiment the differential absorption of left and right circularly polarized light is measured. In 1988, Xie *et al.* developed a polarization modulation device that enabled one to measure the temporal evolution of circular dichroism signals in the spectral region from 355 nm to 800 nm. The technology developed involved coupling polarization modulation techniques with a high repetition rate tunable picosecond laser. The details of the experiment [81] and the associated optical theory [82] have been published. This technique was used to characterize the evolution of the circular dichroism spectrum of two photoinitiated biological processes: photodissociation of CO from carbonmonoxymyoglobin and carbon-monoxyhemoglobin [83], and electron transfer in photosynthetic bacteria [84]. In principle, this experimental technique can be used to study the evolution of geometric changes of any reacting molecule that is chiral.

E. Coherence Spectroscopies

Many forms of coherence spectroscopy have been used to examine chemical dynamics in solution. Included in this group of techniques are coherent Raman spectroscopy [85], impulsive stimulated scattering [86], transient gratings [87] and photon echo [88]. In addition, it is important to note that new experimental techniques have emerged that enable one to study reaction dynamics of the surfaces of liquids. Researchers in this latter field have exploited the anisotropic nature of a surface to study dynamics using second harmonic and sum frequency generation. Applications of these techniques to a wide variety of chemical processes have occured during the last few years and the work has recently been reviewed [89]. Instead of providing a detailed discussion of the wide range of ultrafast coherence spectroscopies that are currently being used in the study of chemical reaction dynamics, we focus on a new technique that is emerging as a powerful new approach for studying reaction dynamics, phase-locked pulse spectroscopy.

Phase-Locked Optical Spectroscopy: Several experiments designed to carry out pump-probe spectroscopy with phase-locked pulse pairs have been reported. As early as 1978, Salour used a HeNe laser to control the relative phases of two pulses at fixed time delays [90]. These pulses were used to demonstrate a two-photon optical Ramsey fringe spectroscopy. Diels and coworkers examined the effect of phase-stabilized multiple pulse excitation on multiphoton excitation processes [91]. Fayer and co-workers used phase-locked picosecond pulses to measure the free induction decay of sodium vapor [92]. In 1990, Fleming and coworkers reported the first study of ultrafast molecular dynamics in a gas phase sample using phase-locked femtosecond pulses [93].

Controlling the relative phases of two femtosecond pulses (pump and probe) is not a trivial task. When the pulses are temporally overlapped, the interference pattern (or second harmonic signal) can be used to provide information on the relative phases of the two pulses, which without some sort of stabilization will randomly vary from pulse to pulse. However, if one wants to measure the dynamic evolution of a chemical system, it is necessary to control the relative phases of the two pulses at time delays where there will be no overlap. This task was cleverly solved by Scherer *et al.* by taking 10% of the pump and probe beam intensities and linearly chirping them in a monochromator [93]. This process temporally broadens the laser pulses. The linear dispersion of the monochromator preserves the phase relationship between the pump and probe pulse. Thus, setting the phases of the chirped pulses (which do temporally overlap), determines the phase relationship of the original femtosecond pulses (which do not temporally overlap). Using this approach, the relative phase of the two optical pulses can be set for delay times that are nearly 1000 times the pulse duration.

The ability to carry out phase-locked measurements was demonstrated on a gaseous sample of I_2. The wavelength of light used (centered between 608 and 613 nm, and of duration 50–70 fs) is resonant with the $B \leftarrow X$ absorption band of I_2. The experiment involved measuring the fluorescence following excitation of the molecule by a pair of phase-locked pulses that are separated by a variable time delay. In such an experiment a positive signal reflects an increase in the excited state population with respect to the two one-beam-only contributions to the population. This effect arises from the wave packet interference that results from the fact that the phase-relationship between the pulse-pair is controlled. A negative signal indicates that the excited state population is decreased by such wave packet interference. The observation of interferences in the time resolved data was discussed in terms of various theoretical models. Although this spectroscopic technique was demonstrated on a gas-phase sample, the potential applications to condensed phase systems are exciting.

6. Summary

This chapter has presented a discussion of many of the ultrafast experimental techniques that are being used to study the chemistry of condensed phase systems. Most of the remainder of this book deals with the detailed study of chemical systems using these transient spectroscopic techniques. The field of ultrafast chemistry is rapidly growing. With new developments in both the generation and manipulation of ultrashort light pulses and detector technology, new experiments are becoming possible.

Acknowledgements

This work was supported by the National Science Foundation, Division of Experimental Physical Chemistry, the General Medicine Institute of the National Institute of Health and the Materials and Medical Free Electron Laser Program administered by the Office of Naval Research. We thank Dr. Xiaoliang Xie, Dr. Robert Dunn, Dr. Yi Jing Yan, Robert Doolen, Hans Deuel, and Bret Flanders for their contributions to the chemical applications described.

References

1. Fleming, G. R., Chemical Applications of Ultrafast Spectroscopy (Oxford Univ. Press, New York, 1986).
2. (a) Zewail, A. H., *Faraday Discuss. Chem. Soc.* **91**, 207 (1991); (b) Khundkar, L. R. and Zewail, A. H., *Ann. Rev. Phys. Chem.* **41**, 15 (1990); (c) M. Gruebele and A. H. Zewail: *Phys. Today* **43**, No. 5, 24 (1990); (d) A. H. Zewail: *Science* **242**, 1645 (1988).
3. (a) Walkup, R. E., Misewich, J. A., Glownia, J. H., and Sorokin, P. P., *J. Chem. Phys.* **94**, 3389 (1991); (b) Walkup, R. E., Misewich, J. A., Glownia, J. H., and Sorokin, P. P., *Phys. Rev. Lett.* **65** 2366 (1990); (c) Glownia, J. H., Misewich, J. A., and Sorokin, P. P., *J. Chem. Phys.* **92**, 3335 (1990).
4. (a) Simon, J. D., *Rev. Sci. Instrum.* **60**, 3597 (1989); (b) Squir, J. and Mourou, G., *Laser Focus World* **28**, 51 (1992).
5. *Ultrafast Phenomena, I–VIII*, Springer-Verlag Series in Chemical Physics, Berlin, 1976–1992.
6. Fork, R. L., Greene, B. I., and Shank, C. V., *Applied Phys. Lett.* **38**, 671 (1981).
7. Knox, W. H., Downer, M. C., Fork, R. L., and Shank, C. V., *Opt. Lett.* **9**, 552 (1984).
8. Fork, R. L., Shank, C. V., and Yen, R. T., *Appl. Phys. Lett.* **41**, 223 (1982).
9. Spence, D. E., Kean, P. N., and Sibbett, W., *Optics Lett.* **16**, 42 (1991).
10. Huang, C.-P., Kapteyn, H. C., McIntosh, J. W., and Murnane M. M., *Optics Lett.* **17**, 139 (1992).
11. Salin, F., Squier, J., Mourou, G., and Vaillancourt, G., *Optics Lett.* **16**, 1964 (1991).
12. Shen, Y. R., *The Principles of Nonlinear Optics* (Wiley, New York, 1984).
13. (a) Banin, U. and Ruhman, S., *J. Chem. Phys.* **98**, 4391 (1993); (b) Harris, A. L., Brown, J. K., and Harris, C. B., *Annu. Rev. Phys. Chem.* **39**, 341 (1988); (c) Schwartz, B. J., King, J. C., Zhang, J. Z., and Harris, C. B., *Chem. Phys. Lett.* **203**, 503 (1993); (d) Zhang, J. Z. and Harris, C. B., *J. Chem. Phys.* **95**, 4024 (1991).
14. Engh, R. A., Petrich, J. W., and Fleming, G. R., *J. Phys. Chem.* **89**, 618 (1985).
15. (a) Migus, A., Gauduel, Y., Martin, J. L., and Antonetti, A., *Phys. Rev. Lett.* **58**, 1559 (1987); (b) Lu, H., Long, F. H., Bowman, R. M., and Eisenthal, K. B., *J. Phys. Chem.* **93**, 27 (1989).
16. (a) Paige, M. E. and Harris, C. B., *J. Chem. Phys.* **93**, 1481 (1990); (b) Harris, C. B., Smith, D. E., and Russell, D. J., *Chem. Rev.* **90**, 481 (1990); (c) Elsaesser, T. and Kaiser, W., *Ann. Rev. Phys. Chem.* **42**, 83 (1991).
17. (a) Maroncelli, M., MacInnis, J., and Fleming, G. R., *Science* **243**, 1674 (1989); (b) Barbara, P. F. and Jarzeba, W., *Adv. Photochem* **14**, 1 (1990); (c) Barbara, P. F., Walker, G. C., and Smith, T. P., *Science* **256**, 975 (1992); (d) Fleming, G. R. and Wolynes, P. G., *Phys. Today* **43**, 36 (1990); (e). Rossky, P. J. and Simon, J. D., Nature, submitted; (f) Maroncelli, M., *Mol. Liquids*, in press; (g) Bagchi, B., *Ann. Rev. Phys. Chem.* **40**, 115 (1989); (h) Barbara, P. F., Walsh, P. K., and Brus, L. E., *J. Phys. Chem.* **93**, 29 (1989).
18. Xie, X. and Simon, J. D., *Opt. Comm.* **69**, 303 (1987).
19. Rothberg, L. J., Bernstein, M., and Peters, K. S., *J. Chem. Phys.* **79**, 2569 (1983).
20. Hochstrasser, R. M., Anfinrud, P. A., Diller, R., Han, C., Iannone, M., and Lian, T., in *Ultrafast Phenomena VII*, page 429, Harris, C. B., Ippen, E. P., Mourou, G. A., and Zewail, A. H., Eds. (Springer, Berlin, 1990).
21. Iannone, M., Cowen, B. R., Diller, R., Maiti, M., and Hochstrasser, R. M., *Appl. Opt.* **30**, 5247 (1991).
22. Locke, B., Lian, T., and Hochstrasser, R. M., *Chem. Phys.* **158**, 409 (1991).
23. Anfinrud, P. A., Han, C., and Hochstrasser, R. M., *Proc. Natl. Acad. Sci.* **86**, 8387 (1989).
24. Anfinrud, P. A., Han, C. H., Lian, T. Q., and Hochstrasser, R. M., *J. Phys. Chem.* **95**, 574 (1991).
25. Owrutsky, J. C., Kim, Y. R., Li, M., Sarisky, M. J. and Hochstrasser, R. H., *Chem. Phys. Lett.* **184**, 368 (1991).

26. Braun, C. L., Smirnov, S. N., Brown, S. S., and Scott, T. W., *J. Phys. Chem.* **95**, 5529 (1991).
27. Kafka, J. D., Pieterse, J. W., and Watts, M. L., *Opt. Lett.* **17**, 1286 (1992).
28. Luo, N. and Robinson, G. W., private communication.
29. (a) Dunn, R. C., Simon, J. D., *J. Am. Chem. Soc.* **114**, 4869 (1992); (b) Dunn, R. C., Anderson, J., Foote, C. S., Simon, J. D., *J. Am. Chem. Soc.*, in press; (c) Simon, J. D., Vaida, V., *Science*, in preparation; (d) Dunn, R. C., Simon, J. D., *J. Am. Chem. Soc.*, submitted; (e) Dunn, R. C., Flanders, B. N., Vaida, V., Simon, J. D., *Spectrochim Acta* **48A**, 1293 (1992); (f) Dunn, R. C., Richard, E. C., Vaida, V., Simon, J. D., *J. Phys. Chem.* **95**, 6060 (1991).
30. Doorn, S. K., Stoutland, P. O., Dyer, R. B., Woodruff, W. H., *J. Am. Chem. Soc.* **114**, 3133 (1992).
31. (a) Cong, P., Deuel, H. P., Simon, J. D., *Chem. Phys. Lett.* **212**, 367 (1993); (b) Note added in proof: We have developed a non-Markovian treatment for the electronic dephasing dynamics in room-temperature liquids, see Cong, P., Yan, Y. J., Deuel, H. P., Simon, J. D., *J. Chem. Phys.*, submitted.
32. (a) Simon, J. D., Xie, X., *J. Phys. Chem.* **90**, 6751 (1986); (b) Xie, X., Simon, J. D., *J. Am. Chem. Soc.* **112**, 1130 (1990); (c) O'Driscoll, E., Simon, J. D., *J. Am. Chem. Soc.* **112**, 6580 (1990).
33. (a) Molina, M. J. and Rolan, F. S., *Nature* **249**, 819 (1974) (b) Solomon, S., Mount, G. H., Sanders, R. W., Schmeltekopf, A. L., *J. Geophys. Res.* **92**, 8329 (1987); (c) Solomon, S., *Rev. Geophys.* **26**, 131 (1988); (d) Anderson, J. G., Brune, W. R., Chan, R. J., *J. Geophys. Res.* **94**, 11480 (1989); (e) Brune, W. H., Anderson, J. G., Toohey, D. W., Fahey, D. W., Kawa, S. R., Jones, R. L., McKenna, D. S., and Poole, L. R., *Science*, **252**, 1260 (1991).
34. (a) Vaida, V. and Simon, J. D., *Science*, in prepation; (b) Vaida, V., Solomon, S., Richard, E. C., Ruhl, E., and Jefferson, A., *Nature*, **342**, 405 (1989).
35. (a) Richard, E. C. and Vaida, V., *J. Chem. Phys.* **94**, 153 (1991); (b) Richard, E. C. and Vaida, V., *J. Chem. Phys.* **94**, 163 (1991)
36. Bishenden, E., Hancock, J., and Donaldson, D. J., *J. Phys. Chem.* **95**, 2113 (1991).
37. Buhler, R. E., *Rad. Res. Rev.* **4**, 233 (1972).
38. Klaning, U. K. and Wolff, T., *Ber. Bunsen-Ges. Phys. Chem.* **89**, 243 (1985).
39. Mauldin, R. L., III, Burkholder, J. B., Ravishankara, A. R., *J. Phys. Chem.* **96**, 2582 (1992).
40. Davis, H. F. and Lee, Y. T., *J. Phys. Chem.* **96**, 5681 (1992).
41. Nelson, K. A. and Ippen, E. P., *Adv. Chem. Phys.* **75**, 1 (1989) and references therein.
42. (a) Nibbering, E. T. J., Wiersma, D. A., and Duppen, K., *Phys. Rev. Lett.* **68**, 2464 (1991); (b) Nibbering, E. T. J., Duppen, K., and Wiersma, D. A., *J. Chem. Phys.* **93**, 5477 (1990).
43. (a) Bardeen, C. J. and Shank, C. V., *Chem. Phys. Lett.* **203**, 535 (1993); (b) Brito Cruz, C. H., Fork, R. L., Knox, W. H., and Shank, C. V., *Chem. Phys. Lett.* **132**, 341 (1986); (c) Becker, P. C., Fragnito, H. L., Bigot, J-Y., BritoCruz, C. H., Fork, R. L., and Shank, C. V., *Phys. Rev. Lett* **63**, 505 (1989); (d) Bigot, J-Y., Portella, M. T., Schoenlein, R. W., Bardeen, C. J., Migus, A., and Shank, C. V., *Phys. Rev. Lett.* **66**, 1138 (1991).
44. Lawless, M. K. and Mathies, R. A., *J. Chem. Phys.* **96**, 8037 (1992).
45. Balk, M. W. and Fleming, G. R., *J. Chem. Phys.* **83**, 4300 (1985).
46. (a) Mukamel, S., *Phys. Rev A* **28**, 3480 (1983); (b) Mukamel, S., *Phys. Rep.* **93**, 1 (1982); (c) Boyd, R. W. and Mukamel, S., *Phys. Rev. A* **29**, 1793 (1984).
47. Allen, L. and Eberly, J. H., *Optical Resonance and Two-Level Atoms* (Dover, New York, 1975).
48. (a) Joly, A. G. and Nelson, K. A., *J. Phys. Chem.* **93**, 2976 (1989); (b) Lee M. and Harris C. B., *J. Am. Chem. Soc.* **111**, 8963 (1989); (c) Wang, L., Zhu, X., Spears, K. G., *J. Am. Chem. Soc.* **110**, 8695 (1988).
49. O'Connor, D. V. and Phillips, D., Time-correlated single photon counting (Academic Press, London, 1984).
50. See, for example, Schmidt, J. A. and Hilinski, E. F., *Rev. Sci. Instrum.* **60**, 2902 (1989).

51. (a) Kahlow, M. A., Jarzeba, W., DuPruil, T. P., and Barbara, P. F., *Rev. Sci. Instrum.* **59**, 1098 (1988); (b) Shah, J., *IEEE J. of Quantum Electronics*, **24**, 276 (1988).
52. For a recent example of polarization-resolved up-conversion experiments, see: Xie, X., Du, M., Mets, L., and Fleming, G. R., SPIE conference proceedings, (1992).
53. Hofmann, T., Mossavi, K., Tittel, F. K., and Szabo, G., *Opt. Lett.* **17**, 1691 (1992).
54. (a) Frohlich, H., *Theory of Dielectrics* (Oxford Univ. Press, Oxford, 1949); (b) Davies, M., in *Dielectric Properties and Molecular Behavior*, eds. Hill, N. E., Vaughan, W. E., Price, A. H., and Davies, M. (Van Nostrand, London, 1969); (c) Kivelson, D. and Madden, P. A., *Ann. Rev. Phys. Chem.* **93**, 839 (1984).
55. Rosenthal, S. J., Xie, X., Du, M., and Fleming, G. R., *J. Chem. Phys.* **95**, 4715 (1991).
56. (a) Maroncelli, M., *J. Chem. Phys.* **94**, 2084 (1991); (b) Carter, E. A. and Hynes, J. T., *J. Chem. Phys.* **94**, 5961 (1991); (c) Fonseca, T., Ladanyi, B. M., *J. Phys. Chem.* **95**, 2116 (1991); (d) Maroncelli, M. and Fleming, G. R., *J. Chem. Phys.* **89**, 5044 (1988).
57. (a) Lippert, E., Rettig, W., Bonacic-Koutecky, V., Heisel, F., and Miehe, J. A., Adv. Chem. Phys. **68**, 1 (1987); (b) Doolen, R. and Simon, J. D., *J. Am. Chem. Soc.* **114**, 4861 (1992) and references cited therein.
58. (a) Kang, T. J., Kahlow, M. A., Giser, D., Swallen, S., Nagarajan, V., Jarzeba, W., Barbara, P. F., *J. Phys. Chem.* **92**, 6800 (1988); (b) Kahlow, M. A., Kang, T. J., Barbara, P. F., *J. Phys. Chem.* **91**, 6452 (1987); (c) Kang, T. J., Jarzeba, W., Barbara, P. F., Fonseca, T., *Chem. Phys.* **149**, 81 (1990).
59. Reid, P. J., Wickham, S. D., and Mathies, R. A., *J. Phys. Chem.* **96**, 5720 (1992).
60. Lingle, R., Xu, X. B., Yu, S.-C., Zhu, H. P., and Hopkins, J. B., *J. Chem. Phys.* **93**, 5667 (1990); (b) Lingle, R., Xu, X. B., Yu, S.-C., Chang, Y. J., and Hopkins, J. B., *J. Chem. Phys.* **92**, 4628 (1990); (c) Xu, X., Yu, S.-C., Lingle, R., Zhu, H., and Hopkins, J. B., *J. Chem. Phys.* **95**, 2445 (1991).
61. Yu, S-C., Xu, X. B., Lingle, R., Jr, Hopkins, J. B., *J. Am. Chem. Soc.* **112**, 3668 (1990)
62. (a) Lingle, R., Xu, X. B., Zhu, H. P., Yu, S.-C., and Hopkins, J. B., *J. Am. Chem. Soc.* **113**, 3992 (1991); (b) Lingle, R., Xu, X. B., Zhu, H. P., Yu, S.-C., and Hopkins, J. B., *J. Phys. Chem.* **95**, 9320 (1991).
63. Hopkins, J. B., personal communication.
64. (a) Weaver, W. L., Huston, L. A., Iwata, K., and Gustafson, T. L., *J. Phys. Chem.* **96**, 8956 (1992); (b) Kwata, K. and Hamaguchi, H., *Chem. Phys. Lett.* **196**, 462 (1992); (c) Langkilde, F. W., Wilbrandt, R., Negri, F., and Orlandi, G., *Chem. Phys. Lett.* **165**, 66 (1990);
65. Ci, X. P. and Myers, A. B., *Chem. Phys. Lett.* **158**, 263 (1989).
66. Butler, R. M., Lynn, M. A., Gustafson, T. L., *J. Phys. Chem.* **97**, 2609 (1993).
67. Hecht, E., *Optics*, 2nd Edition (Addison-Wesley, Reading, Mass., 1987) page 611.
68. Eesley, G. L., Levenson, M. D., and Tolles, W. M., *IEEE J. of Quantum Electronics*, **QE–14**, 45 (1978).
69. Eesley, G. L., Levenson, M. D., and Tolles, W. M., *Appl. Phys.* **19**, 1 (1979).
70. Alavi, D. S., Hartman, R. S., and Waldeck, D. H., *J. Chem. Phys.* **94**, 4509 (1991).
71. Cho, M., Du, M., Scherer, N. F., Fleming, G. R., and Mukamel, S., submitted to *J. Chem. Phys.*
72. (a) Ippen, E. P. and Shank, C. V., in *Ultrafast Light Pulses*, ed. Shapiro, S. L. (Springer-Verlag, Berlin, 1983); (b) Shank, C. V. and Ippen, E. P., *Appl. Phys. Lett.* **26**, 62 (1975).
73. Waldeck, D., Cross, A. J., Jr., McDonald, D. B., and Fleming, G. R., *J. Chem. Phys.* **74**, 3381 (1981).
74. Alavi, D. S., Hartman, R. S., and Waldeck, D. H., *J. Chem. Phys.* **94**, 4055 (1991).
75. McMorrow, D., Lotshaw, W. T., and Kenney-Wallace, G. A., *IEEE J. of Quantum Electronics*, **24**, 443 (1988).
76. McMorrow, D. and Lotshaw, W. T., *J. Phys. Chem.* **95**, 10395, (1991), and references therein.
77. Scherer, N. F., Ziegler, L. D., and Fleming, G. R., *J. Chem. Phys.* **96**, 5544 (1992).
78. Zhu, F., Galli, C., and Hochstrasser, R. M., *J. Chem. Phys.* **98**, 1042 (1993).

79. Chen, Y., Hunziker, L., Ludowise, P., and Morgen, M., *J. Chem. Phys.* **97**, 2149 (1992).
80. Cho, M., Rosenthal, S. J., Scherer, N. F., Ziegler, L. D., and Fleming, G. R., *J. Chem. Phys.* **96**, 5033 (1992).
81. (a) Xie, X. and Simon, J. D., *Rev. Sci. Instrumen.* **60**, 2614 (1988); (b) Lewis, J. W., Goldbeck, R. A., Kliger, D., Xie, X., Dunn, R. C., and Simon, J. D., *J. Phys. Chem.* **96**, 5243 (1992); (c) Dunn, R. C., Xie, X., and Simon, J. D., *Methods in Enzym.: Mettalobiochemistry C*, Riordam, J. F., and Vallee, B. L., Eds., in press.
82. Xie, X. and Simon, J. D., *J. Opt. Soc. B* **7**, 1675 (1990).
83. (a) Xie, X. and Simon, J. D., *Biochem.* **30**, 3682 (1991); (b) Xie, X. and Simon, J. D., *Proc. SPIE* **1203**, 66 (199); (c) Xie, X. and Simon, J. D., *J. Am. Chem. Soc.* **112**, 7802 (1990).
84. (a) Xie, X. and Simon, J. D., *Biochim. Biophys. Acta.* **1057**, 131 (1991); (b) Simon, J. D., Xie, X., and Dunn, R. C., *Proc. SPIE* **1432**, 211 (1991).
85. (a) Okamoto, H., Inaba, R., Yoshihara, K., and Tasumi, M., Chem. Phys. Lett. **202**, 161 (1993); (b) Laubereau, A. and Purucker, H. G., *Nuovo. Cimento. Della Soc. Ital. de Fisica D* **10**, 979 (1992); (c) Chronister, E. L. and Crowell, R. A., *Chem. Phys. Lett.* **182**, 27 (1991).
86. (a) Yan, Y.-X., Gamble, E. B., Jr., and Nelson, K. A., *J. Chem. Phys.* **83**, 5391 (1985); (b) Ruhman, S., Joly, A. G., and Nelson, K. A., *IEEE J. Quant. Electron.* **QE–24**, 460 (1988).
87. Fourkas, J. T. and Fayer, M. D., *Acc. Chem. Res.* **25**, 227 (1992).
88. See for example (a) Cho, M. H. and Fleming, G. R., *I. Chem. Phys.* **98**, 2848 (1993); (b) Meijers, H. C. and Wiersma, D. A., *J. Lumines.* **53**, 80 (1992); (c) Chronister, E. L. and Crowell, R. A., *Mol. Cryst. & Liq. Cryst.* **211**, 361 (1992).
89. Eisenthal, K. B., *Ann. Rev. Phys. Chem.* **43**, 627 (1992).
90. Salour, M. M., *Rev. Mod. Phys.* **50**, 667 (1978).
91. (a) Diels, J. C. and Stone, J., *Phys. Rev. A* **31**, 2397 (1985); (b) Diels, J. C. and Besnainou, S., *J. Chem. Phys.* **85**, 6347 (1986).
92. Fourkas, J. T., Wilson, W. L., Wackerle, G., Frost, A. E., and Fayer, M. D., *J. Opt. Soc. Am.* **9**, 1905 (1989).
93. (a) Scherer, N. F., Ruggiero, A. J., Du, M., and Fleming, G. R., *J. Chem. Phys.* **93**, 856 (1990); (b) Scherer, N. F., Carlson, R. J., Matro, A., Du, M., Ruggiero, A. J., Romero-Rochin, V., Cina, J. A., Fleming, G. R., and Rice, S. A., *J. Chem. Phys.* **95**, 1487 (1991).

2. Electron Transfer in Solution: Theory and Experiment

M. D. FAYER, L. SONG, S. F. SWALLEN, R. C. DORFMAN and K. WEIDEMAIER
Department of Chemistry, Stanford University, Stanford, CA 94305, U.S.A.

1. Introduction

Photoinduced electron transfer is a fundamental chemical process. Following the transfer of an electron from an electronically excited donor to an acceptor, the resulting radical ions can go on to do useful chemistry. Electron back transfer (geminate recombination), however, quenches the ions and prevents further chemistry from occurring. The initial steps of photosynthesis involve excitation of a donor followed by electron transfer. Electron back transfer to the primary donor would stop the photosynthetic process. However, a specialized spatial array of consecutive acceptors eliminates the back transfer problem and is responsible for the efficiency of photosynthesis [1–3]. In systems of randomly distributed donors and acceptors (liquid or solid solutions) geminate recombination can be very rapid [4]. In liquid solutions, geminate recombination competes with diffusional separation of the photoinduced ions and limits chemical yields. Therefore, understanding phenomena which influence back transfer is not only an important basic problem, but is a problem of considerable practical significance.

In a system of dilute donors and more concentrated acceptors, the initial electron transfer is into a random distribution of acceptors. In solid solution, the positions of the donors and acceptors are fixed. The rate of forward electron transfer depends on the distance-dependent transfer rate that is determined by the intermolecular interactions between the donor and acceptor [5–10]. The transfer rate falls off very rapidly with the separation between donor and acceptor; therefore transfer occurs to acceptors that are relatively close by (a few tens of Å at most). However, contact between the donor and acceptor is not required for transfer to occur.

The back transfer problem is more complex than forward transfer. The distribution of radical-ion pair separations formed by forward transfer is not random, but rather determined by the details of the forward transfer distance-dependent transfer rate and the acceptor concentration. This leads to a complex averaging problem in the theoretical analysis of the combined forward

J.D. Simon (ed.), Ultrafast Dynamics of Chemical Systems, 37–80.

and back transfer dynamics. At short time, donors with nearby acceptors will transfer an electron. Because of the resulting small separation of the ions, back transfer will also be rapid. At longer times, donors with acceptors somewhat further away will transfer an electron. Because of the larger separation between the ions that are generated, back transfer will be slower. By the time ion pairs are formed with large separation, recombination of ions formed with small separations will have occurred. Over time, ion pairs at smaller separations are formed and have time to recombine, while ions at greater separations are only beginning to be formed. This process continues as time increases, and the result is that the probability distribution of separations of ions in pairs moves out as a wave.

In liquid solution, prior to electron transfer, neutral donors and acceptors will undergo spatial diffusion. The rate of diffusion is determined by the viscosity of the liquid and the size of the molecules. The motion of the particles increases the complexity of the dynamics of both forward transfer and geminate recombination. Initially, an excited donor may be far from any acceptor. In solid solution, either electron transfer would be very slow or no transfer would occur. However, in a liquid, diffusive motion can move an acceptor close to an excited donor. If the viscosity of the liquid is reasonably low, the rate of diffusion can be rapid, and diffusion will have a profound effect on forward electron transfer. The rate of transfer and the yield of electron transfer is greatly increased by significant diffusion of the particles. If the donor and acceptor are initially neutral, electron transfer generates oppositely charged particles. The charged particles will be attracted to each other by their Coulomb interaction. Since the forward transfer is relatively short range, the Coulomb attraction can be very strong. This modifies the random diffusive motion. The particles are drawn toward each other, enhancing geminate recombination. However, the diffusive motion makes it possible for the ions to avoid geminate recombination, although the probability for escape may be very small.

It is also possible for the particles to repel each other. If the donor is initially a doubly negative ion and the acceptor is neutral, then following electron transfer both particles will have single negative changes. Now the Coulomb interaction pushes the ions apart, enhancing the probability for escape. In liquid solutions, back transfer competes with diffusional separation of the radical-ions and may have a non-obvious influence on chemical reaction yields [4, 11, 12].

As discussed briefly above and demonstrated by the results of many experimental investigations on a wide range of systems, there is more to photoinduced electron transfer than electron hopping [5, 6, 13–19] between molecules in intimate contact. It has been shown that electron transfer depends not only on the distance between donors and acceptors [5–10] but also on factors such as exothermicity [20–22], reorganization energy [2, 13], temperature [14, 15, 23], solvent polarity [24], molecular spacer (for intra-molecular electron transfer)[7, 13], and electric field [3, 25]. This has motivated theoretical investigations into the dependence of the electron transfer rate on each of these factors [7, 20, 23, 26–29]. The purpose of this article is to give experimental and theoretical insights into the statistical mechanical aspects of the dynamics of photo-induced electron transfer and geminate recombination in liquid solution. The standard quantum mechanical model for the distance-dependent transfer rate between a donor acceptor pair (neutrals or ions) is used. This transfer rate falls off exponentially with distance [5, 6, 8, 10]. In a system of donors and acceptors that are initially randomly distributed and are undergoing diffusive motion, properly performing the spatial averages over configurations is complex. This problem has been treated successfully theoretically [30–34]. However, the difficulty of the problem is greatly magnified when geminate recombination is included. The experimental and theoretical descriptions, given below, of photoinduced electron transfer and geminate recombination in solid and liquid solutions provide a detailed explication of the current understanding of this field.

In Section 2 a model is developed for a system of low concentration donors and high concentration acceptors in solid solution. The donors and the acceptors are in fixed positions. This is the simplest situation, and the theoretical model can be solved exactly. Considerable insight into the nature of the problem is obtained by examining the problem in the absence of diffusion. The time dependent ion pair populations, the spatial distribution of the separations of ions in the pairs, and experimental observables are calculated. In order to put this theory into perspective, it is compared with other formulations of the problem, and the ranges of applicability of each are discussed.

In Section 3, the theory is extended to include diffusion of the particles in liquid solutions. The lifetime of photo-generated ions in solution depends on the forward and back electron transfer parameters [1], the concentration of acceptors, the interaction potential between the ions, and on the diffusion [35–39] constants of the donors and the acceptors. The theory provides a comprehensive description of the competition between electron back transfer and separation by diffusion. When the particles are allowed to diffuse following electron transfer, the Coulomb interaction can greatly influence the dynamics. To illustrate the important situations, three different donor (D) and acceptor (A) combinations will be explored that result in an attractive, a repulsive, or no Coulomb interaction. In addition to other properties, the reactive state survival fraction, *i.e.*, the fraction of ions that escape geminate recombination by diffusing apart, is calculated. The results are compared to previous models that considered only the fast or slow diffusion limits [30–34, 40–43]and electron transfer only at contact [44–47].

In Section 4, the models presented in Sections 2 and 3 will be compared to experiments. Steady state fluorescence yield and time resolved fluorescence measurements are used to determine the population of the donor's excited state, and pump-probe experiments are used to measure the population of ion pairs produced as a result of electron transfer. Since electron transfer quenches the fluorescence from the excited state, fluorescence provides a direct observable for the yield and time dependence of the forward electron transfer process. The pump-probe observable is proportional to the population that is not in the ground state of the system. In the systems studied here, there are only two other states involved, the excited state and the reactive state (the ion pair produced as a result of forward electron transfer). Since the time-dependent excited state population is obtained from the fluorescence measurements, the time-dependent reactive state population is determined by combining the results of the pump-probe and time-resolved fluorescence experiments. An important aspect of this work is that the experiments were performed in both solid and liquid solutions on the same donor-acceptor pair. The solid and liquid solvents were chosen to have very similar dielectric properties. Therefore, within the context of the theoretical model, the only difference between the solid and liquid is the rate of diffusion.

The comparison between solid and liquid solutions provides an important control for judging the applicability of the statistical mechanics theoretical treatment. For a solid solution (fixed donor and acceptor positions), the necessary spatial averages can be performed exactly [36, 37]. Measurements as a function of the acceptor concentration in solid solution yield the forward and back electron transfer parameters. Then, in principle, experimental results from liquid solution can be analyzed using the same transfer parameters and the theory that now includes diffusion. In the experiments presented below, the forward transfer data, obtained from time resolved and steady state fluorescence, are in remarkable agreement with theory. The forward transfer parameters obtained from liquid solution are virtually identical to those found in solid solution. The back transfer dynamics in solid solution, obtained from pump-probe experiments as a function of acceptor concentration, show very good agreement with theory. When the solid solution back transfer parameters are used in the liquid solution theory, very good agreement with experiment is again found without adjustable parameters if the high frequency relative permittivity is used. The use of the low frequency relative permittivity results in very poor agreement between theory and experiment, regardless of how the back transfer parameters are varied. The use of the high frequency relative permittivity is discussed in terms of the time and distance scales involved in the electron transfer dynamics.

2. Solid Solutions

In a system in which there are neutral donors (low concentration) and neutral acceptors (high concentration) randomly distributed in a solid solution, optical excitation of a donor can be followed by transfer of an electron to an acceptor [1]. Once electron transfer has occurred, there exists a ground state radical cation (D^+) near a ground state radical anion (A^-) (this pair represents the reactive state). Since the thermodynamically stable state is the neutral ground state D and A, back transfer will occur. In liquid solution, back transfer competes with separation by diffusion. Here the focus is on a system of donors and acceptors which are in fixed positions. This permits the ensemble averaged dynamics of the coupled forward and back transfer processes to be isolated from the influence of molecular diffusion.

A great deal of effort has been devoted to modeling the microscopic electron transfer rate [1, 2, 20, 26, 48, 49]. It has been shown [5, 6, 8, 10] that a transfer rate which is exponentially dependent on distance works well for electron transfer over a considerable range of distances. This form of the microscopic electron transfer rate is employed to describe both the forward and back electron transfer for a particular donor-acceptor pair.

The forward transfer process is relatively straightforward to model [5, 6, 10, 50–53]. The forward transfer process involves the interaction of a donor with acceptors which are randomly distributed in space. For a donor-acceptor electron transfer rate which falls off exponentially with distance, Inokuti and Hirayama [50] have developed a statistical mechanics theory describing the time dependence of the ensemble averaged forward transfer for the case of point particles.

The back transfer problem is more complex. The distribution of distances between the ions D^+ and A^- is not random. It is determined by the details of the forward transfer process. The distribution will be strongly biased toward small separations. After electron transfer, the system consists of a cation near an anion. Transfer from the anion to a neutral acceptor is not included since there is no net driving force for the transfer [2]. Transfer from the anion to a cation which was not the original source of the electron is not included because the concentration of the donors is assumed to be low and the concentration of the donor cations is even lower. The statistical mechanical theory presented below is an exact solution to the model problem outlined above. The theory calculates the time-dependent probabilities that the donor is excited, that it is a cation (the reactive state), or is a neutral in its ground electronic state. From these results, the average ion-pair separation as a function of time and the average ion-pair (reactive state) existence time are calculated. The effect a particular acceptor has on the reactive state probability, with the influence of the other acceptors properly accounted for, as a function of time and distance is also calculated. The results provide an excellent picture of the dynamics of an ensemble of electron transfer systems. The results of this model have been compared to other approximate solutions of this problem.

A. Theoretical Development

In this section, the donor excited state population function, $\langle P_{ex}(t) \rangle$, and the donor reactive state (ion-pair) population function, $\langle P_{re}(t) \rangle$, are derived. In the model, low concentration donors and high concentration acceptors are randomly distributed and held fixed in a rigid matrix. It is assumed that the donor has only one accessible electronic excited state, and the acceptor has only one acceptor state. All states are singlets. The concentration of donor molecules is low enough that donor-donor electronic excitation transfer does not occur.

At time $t = 0$ an ensemble of dilute donors is optically excited. In the absence of acceptors, the probability of finding the donor still excited at time t, $\langle P_{ex}(t) \rangle$, decays exponentially with excited state lifetime τ, $\langle P_{ex}(t) \rangle = \exp(-t/\tau)$. When acceptors are present the probability decreases more rapidly due to the addition of the electron transfer pathway for quenching the electronic excited state. Electron transfer creates a ground state radical cation (D^+) near a ground state radical anion (A-). Since the thermodynamically stable state consists of the neutral ground state of D and A, electron back transfer will occur. A schematic illustration of these processes is shown in Figure 1.

The rate constants for excited state decay (k), forward electron transfer (k_f), and electron back transfer (k_b) are:

$$k = \frac{1}{\tau}, \tag{1}$$

$$k_f(R) = \frac{1}{\tau} \exp\left(\frac{R_f - R}{a_f}\right), \tag{2}$$

$$k_b(R) = \frac{1}{\tau} \exp\left(\frac{R_b - R}{a_b}\right), \tag{3}$$

where R is the donor-acceptor separation. R_f and R_b are used to parametrize the distance scales of forward and back electron transfer [5, 10, 50]. a_f and a_b characterize the fall off of the electronic wave function overlap of the neutral donor and acceptor states and of the ionic states, respectively [5, 10, 50]. τ is the donor fluorescence lifetime. While the lifetime τ is not inherently related to the forward or backward electron transfer processes (Equations (2) and (3)), this factor is used as a relative clock by which to follow electron transfer temporally.

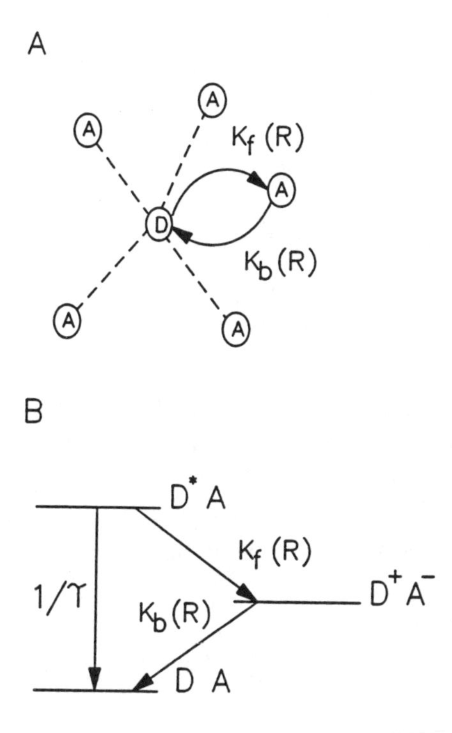

Fig. 1. (A) A diagrammatic representation of forward and backward (geminate) electron transfer. This model depicts the situation studied here, i.e., a neutral excited donor surrounded by a higher concentration of neutral acceptors. (B) A level diagram showing the three states: the ground (*DA*), excited ($D^* A$), and reactive ($D^+ A^-$) states. The three rate processes are represented by their rates, τ, $k_f(R)$, and $k_b(R)$, which are the fluorescence lifetime of the donor, the forward, and the back transfer rates, respectively.

The differential equations describing the processes for a donor and n acceptors having a fixed configuration of donor-acceptor separations given by the set of distances $R_1, \ldots, R_n$ ($\bar{R}$), in a volume *V* are:

$$\frac{\partial}{\partial t} P_{\text{ex}}(\bar{R}, t) = -\left[\frac{1}{\tau} + \sum_{i=1}^{n} k_f(R_i)\right] P_{\text{ex}}(\bar{R}, t), \tag{4}$$

$$\frac{\partial}{\partial t} P_{\text{re}}^i(\bar{R}, t) = k_f(R_i) P_{\text{ex}}(\bar{R}, t) - k_b(R_i) P_{\text{re}}^i(\bar{R}, t), \quad i = 1, \ldots, n \tag{5}$$

where each R_i is the distance from the donor to the ith acceptor. $P_{\text{ex}}(\bar{R}, t)$ is the probability of finding the donor in its excited state. $P_{\text{re}}^i(\bar{R}, t)$ is the probability of finding the donor in its reactive (cation) state with the ith acceptor in its anion state. Summing the solution of Equation (5) over all i gives the total probability of finding the donor in its cation state. The terms multiplying $P_{\text{ex}}(\bar{R}, t)$ in Equation (4) account for processes which remove the donor molecule from its excited state. In Equation (5) the factor $k_f(R_i)P^i(\bar{R}, t)$ describes electron transfer which takes the donor to its cationic state. This is also the rate of ion formation on the ith acceptor. Similarly the term $-k_b(R_i)P^i_{\mathrm{re}}(\bar{R}, t)$ accounts for electron back transfer from the ith acceptor (anion) to the donor (cation), returning the donor cation to its neutral ground state.

The solution of Equation (4) is straightforward:

$$P_{\mathrm{ex}}(\bar{R}, t) = \exp(-t/\tau) \exp\left(-\sum_{i=1}^{n} k_f(R_i)t\right). \tag{6}$$

The decay described by $P_{\mathrm{ex}}(\bar{R}, t)$ depends on the particular acceptor configuration considered. The ensemble average over all possible configurations is $\langle P_{\mathrm{ex}}(t)\rangle$. At low concentrations $\langle P_{\mathrm{ex}}(t)\rangle$ is given by Inokuti and Hirayama (IH) [50]. The IH theory does not include excluded volume effects. These will be included below. Initially the molecules are taken to be point particles. The IH result is

$$\langle P_{\mathrm{ex}}(t)\rangle = \exp(-t/\tau) \exp\left(-4\pi C \int_0^\infty [1 - \exp(-k_f(R)t]R^2 \, \mathrm{d}R\right), \tag{7}$$

where C is the acceptor concentration.

Instead of directly solving Equation (5) for $P^i_{\mathrm{re}}(\bar{R}, t)$, then performing the ensemble average and passing to the thermodynamic limit, the average over all possible spatial configurations of n - 1 acceptors in a volume V is performed for each term in Equation (5):

$$\langle \frac{\partial}{\partial t} P^i_{\mathrm{re}}(R, t)\rangle_{n-1} = \langle k_f(R)P_{\mathrm{ex}}(R, t)\rangle_{n-1} - \langle k_b(R)P^i_{\mathrm{re}}(R, t)\rangle_{n-1}, \tag{8}$$

where $\langle\rangle_{n-1}$ denotes an average over all spatial coordinates except the ith spatial coordinate (the subscript on R has been dropped). $\langle P^i_{\mathrm{re}}(R, t)\rangle_{n-1}$ is the probability of finding the donor in its reactive (cation) state at time t, with an acceptor at R in its anion state. Since the spatial distribution of acceptors at different points is uncorrelated and the ensemble averaging procedure is independent of the time derivative, Equation (8) can be rewritten as

$$\frac{\partial}{\partial t}\langle P^i_{\mathrm{re}}(R, t)\rangle_{n-1} = k_f(R)\langle P_{\mathrm{ex}}(R, t)\rangle_{n-1} - k_b(R)\langle P^i_{\mathrm{re}}(R, t)\rangle_{n-1}. \tag{9}$$

Casting the problem in the form of Equation (9) has an important advantage. It reduces the many particle problem in Equation (5) to a two particle problem. This is the key step which makes the exact solution of this problem tractable.

The solution of Equation (9) is now straightforward:

$$\langle P^i_{\text{re}}(R,t)\rangle_{n-1} = \int_0^t k_f(R)\langle P_{\text{ex}}(R,t')\rangle_{n-1} \times \exp(-k_b(R)(t-t'))\,\mathrm{d}t'. \quad (10)$$

Equation (10) is an exact expression for the probability of a donor molecule being a cation with an anion at position *R*. $k_f(R)\langle P_{\text{ex}}(R,t')\rangle_{n-1}$ (under the integration sign in Equation (10) is the rate of an excited donor transferring an electron at time t' (the rate of reactive state formation), and the exponential term $\exp(-k_b(R)(t-t'))$ is the reactive state survival probability at time *t* of an ion-pair that was created at time t'. Hence the total probability of finding the system in the reactive state at time *t* with the anion at *R* is the product of the rate of reactive state formation with the reactive state survival probability integrated over all times from zero to *t*. Equation (10) was obtained for the initial condition that at $t=0$ all the molecules are neutral, $\langle P^i_{\text{re}}(\bar{R},0)\rangle_{n-1} = 0$.

The explicit expression for $\langle P^i_{\text{re}}(\bar{R},t')\rangle_{n-1}$ is obtained by first averaging Equation (6) over all acceptor coordinates except the ith acceptor coordinate. The result is:

$$\langle P_{\text{ex}}(R,t')\rangle_{n-1} = \exp(-t'/\tau)\exp(-k_f(R)t') \times \left[1 - \frac{4\pi}{V}\int_0^{R_V}[1-\exp(-k_f(R)t')]R^2\,\mathrm{d}R\right]^{n-1}, \quad (11)$$

where $V = 4\pi R_v^3/3$ is the volume in which the acceptors are distributed. Equation (11) substituted into Equation (10) gives:

$$\langle P^i_{\text{re}}(R,t)\rangle_{n-1} = \int_0^t k_f(R)\exp(-t'/\tau)\exp(-k_f(R)t')\exp(-k_b(R)(t-t')) \times \left[1 - \frac{4\pi}{V}\int_0^{R_V}[1-\exp(-k_f(R)t')]R^2\,\mathrm{d}R\right]^{n-1}\mathrm{d}t'. \quad (12)$$

The total time-dependent probability of a donor molecule being a cation is found by averaging over the last spatial position *R*, and summing over all *n* acceptors:

$$\langle P_{\text{re}}(t)\rangle = \sum_{i=1}^{n}\frac{4\pi}{V}\int_0^{R_V}\langle P^i_{\text{re}}(R,t)\rangle_{n-1}R^2\,\mathrm{d}R. \quad (13)$$

Each term in the sum is identical thus

$$\langle P_{\mathrm{re}}(t)\rangle = \frac{4\pi n}{V}\int_0^{R_V} \langle P^i_{\mathrm{re}}(R,t)\rangle_{n-1}\, R^2\, \mathrm{d}R. \tag{14}$$

At this point the thermodynamic limit is taken (the limit as the number of acceptors *n* and the volume *V* go to infinity while keeping their ratio constant). In this limit, *n*/*V* becomes the concentration of acceptors *C*. The result gives the total probability that the donor is a cation:

$$\begin{aligned}\langle P_{\mathrm{re}}(t)\rangle &= 4\pi C \int_0^{\infty}\int_0^{t} k_f(R)\, \exp(-k_f(R)t')\langle P_{\mathrm{ex}}(t')\rangle \\ &\quad \times\, \exp(-k_b(R)(t-t'))\, \mathrm{d}t' R^2\, \mathrm{d}R. \end{aligned} \tag{15}$$

Equation (15), which involves a double integral over space and time, can be readily evaluated numerically.

The theory developed above uses a model of point particles distributed randomly in a solid solution. In a real molecular system, the donor and the acceptor molecules occupy finite volumes. When the results given above are modified to include donor-acceptor excluded volume and acceptor-acceptor excluded volume [40, 54, 55], Equations (7) and (15) become, respectively,

$$\begin{aligned}&\langle P_{\mathrm{ex}}(t)\rangle = \\ &\quad \exp(-t/\tau)\, \exp\left(\frac{4\pi}{d^3}\int_{R_m}^{\infty} \ln[1-p+p\, \exp(-k_f(R)t]R^2\, \mathrm{d}R\right), \end{aligned} \tag{16}$$

$$\begin{aligned}\langle P_{\mathrm{re}}(t)\rangle &= 4\pi C \int_{R_m}^{\infty}\int_0^{t} \frac{k_f(R)\, \exp(-k_f(R)t')}{1-p+p\, \exp(-k_f(R)t')} \\ &\quad \times \langle P_{\mathrm{ex}}(t')\rangle\, \exp(-k_b(R)(t-t'))\, \mathrm{d}t'\, R^2\, \mathrm{d}R, \end{aligned} \tag{17}$$

where $R_m = R_D + R_A$, where R_D is the radius of the donor, R_A is the radius of the acceptor. R_m is the distance at which the donor and acceptor are in contact. $p = Cd^3$ and *d* is the diameter of the acceptor excluded volume. The electron transfer rates in Equations (2) and (3) depend exponentially on distance. Therefore electron transfer is a relatively short range process. The distance scale is on the order of R_f and R_b. The short range nature of the transfer rates has led some to believe that only the single nearest acceptor is important in determining the dynamics of electron transfer and recombination [6, 52, 53]. In this sense, the problem would reduce to a one acceptor calculation. Using the results given above, it is possible to address the question of the many particle nature of this problem.

Prior to passing to the thermodynamic limit, the problem is cast in terms of a finite number of acceptors. Equation (15) gives the cation probability for an ensemble having an infinite number of acceptors and concentration *C*. Equation (14) can be used to obtain the same quantity for any finite number of acceptors *n* in a volume *V* such that their ratio is *C*. Thus Equation (14) gives the probability that the donor is a cation as a function of time and the number of acceptors. It has been shown [11, 56] that for even moderate concentration the convergence is quite slow. Thus, an accurate calculation requires the thermodynamic limit.

The results given in Equations (16) and (17) will be used below to calculate observables and analyze experimental data. However, it is interesting to examine the reactive state probability in another way that is not experimentally accessible. For a system of randomly distributed donors and acceptors, it is possible to look at the influence of a particular acceptor on the reactive state probability as a function of time and the donor-acceptor separation. To investigate the effect of the ith acceptor, it is necessary to average over the positions of all other acceptors, since they in part determine the rate of electron transfer to the ith acceptor when it is at location *R*. The expression for this probability, $\langle P^i_{\text{re}}(R,t)\rangle_{n-1}$ (the reactive state distribution function), for a finite number of point particles is given by Equation (12). In the thermodynamic limit the expression is

$$\langle P^i_{\text{re}}(R,t)\rangle_{n-1} = \int_0^t k_f(R)\,\exp(-k_f(R)t')\langle P_{\text{ex}}(t')\rangle\,\exp(-k_b(R)(t-t'))\,\mathrm{d}t'. \quad (18)$$

$\langle P^i_{\text{re}}(R,t)\rangle_{n-1}$ given in Equation (18) is the time dependent probability that the ith acceptor, located at *R*, is an ion. $\langle P^i_{\text{re}}(R,t)\rangle_{n-1}$ vs. distance for a unit volume element about *R* is displayed in Figure 2 for the time, *t*, varying from 0.01 ns to 100 ns. The electron transfer parameters are a_f = 1.0 Å, R_f = 10.0 Å, a_b = 1.0 Å, and R_b = 10.0 Å. The concentration, *C*, of the acceptors is 0.1 M, and the excited state lifetime is 16 ns. For a given time, the curves show the probability of having ion pairs with various ion separation distances. Consider one of the curves for a particular time, *t*. If each point on the curve is multiplied by $4\pi CR^2$ and the curve is integrated, the resulting value is the reactive state probability at the time *t*.

In Figure 2, for each time, there is a most probable cation-anion separation, and this distance increases as *t* increases. At short times, most ion pairs that are created have very small ion separations. These pairs are created quickly,

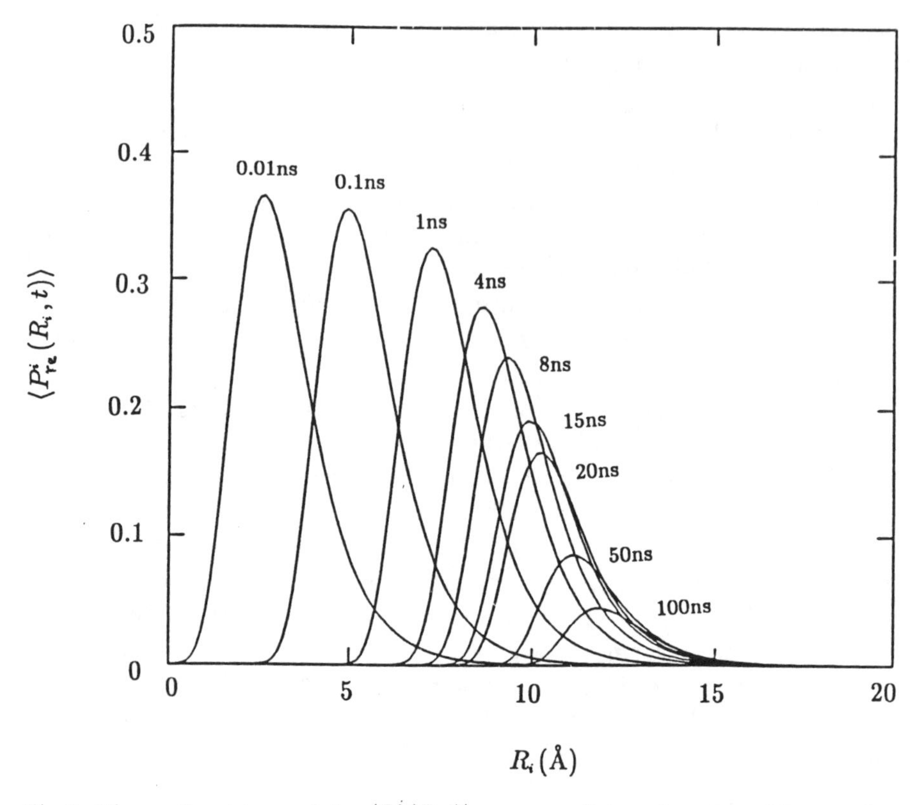

Fig. 2. The reactive state population $\langle P^i_{\rm re}(R,t)\rangle_{n-1}$ versus distance for values of time varying between 0.01 ns to 100 ns. The electron transfer parameters are $a_f = a_b = 1.0$ Å, $R_f = R_b = 10.0$ Å. The concentration of acceptors $C = 0.1$ M, and the excited state lifetime τ 16 ns. Each curve shows the probability of an ion pair existing at the given time at a separation distance R. As time increases, ion pairs formed at small separations have time to recombine, while ion pairs at larger distances are slowly beginning to be formed. The result is that the spatial distribution of ion pairs becomes weighted to greater distances at larger times.

but because of the small separations, recombination is very rapid. As time increases, it becomes more probable that ions will be created which are further apart, and the distribution of ion separation distances becomes larger. As can be seen from the figure, it is as if the distribution of separations moves out as a wave. Pairs with small separations are created and vanish. Then pairs with larger separations are created and vanish. It can also be seen from the figure that for a given set of parameters, there is an effective maximum separation. This arises because of the excited state lifetime which acts to cut off very slow, long range transfer events.

B. Comparisons with Other Methods

In this section, three other methods that appear in the literature for calculating the radical-ion probability for a donor surrounded by a random distribution of acceptors in solid solution will be compared to the exact result given in II.A. Details of the comparisons and the methods are given in reference [57]. The methods are the one acceptor approximation (*OA*) [56], the nearest neighbor approximation (*NN*) [52, 53, 58], and the one acceptor series approximation (*OAS*) [59]. The exact result is labeled *ER*. High (1.0M), medium (0.1M), and low (0.01M) concentrations of acceptors were used to find where each theory is useful by comparing it to the exact results. For the purposes of comparison, a reasonable set of forward and back transfer parameters were used. To simplify computations, the forward and back transfer parameters are identical, $k_f(R) = k_b(R)$. The parameters used for the rates are $\tau = 20$ ns, $R_f = R_b = 10.0$ Å, $a_f = a_b = 0.5$ Å, and the excluded volumes are zero ($R_m = d = 0$).

Figure 3 shows the reactive state population as a function of time and concentration. For Figure 3A, the concentration is very low, $C = 0.01$M. The *OA* and *OAS* results are the same and lie slightly above the *NN* and *ER* results. The *OA* method and the *OAS* method are virtually the same except that the *OAS* is a power series in the *OA* result. The *OA* and *OAS* results overcount the number of radical-ions. Since the time-dependent excited state probability is the same for all the curves at this concentration [57], the differences between the methods at low concentration is only the description of the recombination dynamics. The *NN* and *ER* results are the same.

At moderate concentration ($C = 0.1$M, Figure 3B) we see that all the curves are different. The *OA* and *OAS* results are above the *NN* and *ER* results, and the differences are now large. At high concentration ($C = 1.0$M, Figure 3C) we see something different. The *OA* result rises and falls within 1 ns. At this high concentration, the *OA* method considers a volume that is too small to catch all the possible electron transfer events and so undercounts the number of radical-ions. It is possible to improve the *OA* result by modifying it to include an infinite volume. This is shown as a fifth curve labeled OA^{∞}. The OA^{∞} curve is similar to the *OAS* curve. Both of these curves are profoundly in error as can be seen by comparing them to the exact result, curve *ER*. On the other hand, the NN curve is a reasonable approximation to the exact result for this set of transfer parameters.

The *OA* and *OAS* methods are only reasonable at extremely low concentrations where there is virtually no electron transfer. In contrast, the *NN* result is a reasonable approximation, but its validity for a given set of transfer parameters must be tested against the exact result.

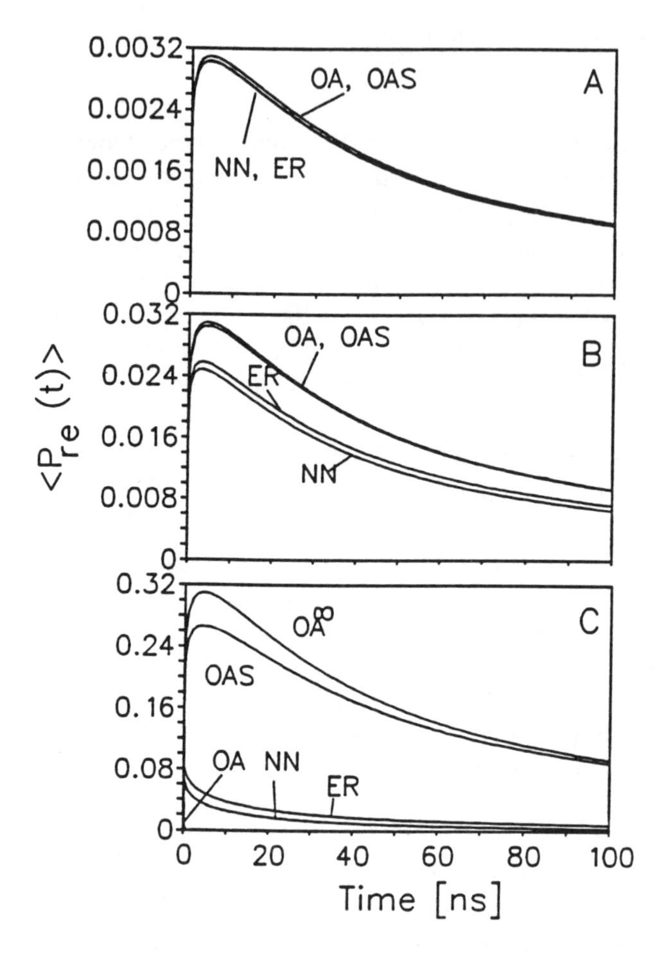

Fig. 3. The reactive state population as a function of time, $P_{re}(t)$. The calculations were performed at three concentrations: *A* is 0.01 M, *B* is 0.1 M, and *C* is 1.0 M. The electron transfer parameters used are $R_f = R_b = 10.0$ Å, $a_f = a_b = 0.5$ Å, and $\tau = 20.0$ ns. *OA* is the one-acceptor approximation, *OAS* is the one acceptor series result, *NN* is the nearest neighbor result, and *ER* is the exact result (Equation (17)). In *C*, *OA*∞ is the *OA* result with the upper limit of integration extended to infinity.

3. Liquid Solution

In this section, the theory of photoinduced electron transfer and geminate recombination in solid solutions [56] is extended to include diffusion of the donors and acceptors. This makes the theory applicable to liquids and provides a comprehensive description of the competition between geminate recombination and separation by diffusion.

The nature and role of diffusion in chemical systems has been studied for many years [58, 60, 61]. More recently, a considerable amount of work has been done on the influence of translational and rotational diffusion on excitation transport among molecules in liquids [30–34, 40–43]. In these treatments, the assumption of a slow or fast diffusion process relative to the transfer time is assumed to make the mathematics more tractable. The influence of diffusion on electron transfer has also been studied both experimentally [4, 46, 62, 63] and theoretically [12, 44–47, 64–68]. In these studies various assumptions have limited the general applicability of the theoretical results. In one study, the limit of low acceptor concentration was used to obtain information only on forward electron transfer [12]. In other studies, transfer is allowed only at contact between the donor and acceptor [44–47, 66].

Here, the problem of photoinduced electron transfer and geminate recombination is treated for the full range of diffusion constants and concentrations and for any combination of forward and back electron transfer parameters. The model encompasses the three situations that illustrate the full range of possibilities. Case 1 has as its excited state a neutral excited D^* molecule with a neutral A molecule. Its reactive state is an ion pair of opposite charges (D^+A^-), while the ground state is the neutral D and A. Case 2 has an excited state composed of a neutral donor and an acceptor A^{++} with a +2 charge, a reactive state with like charges on the molecules (D^+A^+), and a ground state which has the neutral donor D and dication acceptor A^{++}. In Case 3 the model has an excited state comprised of an excited neutral D^* with a positively charged A^+. The reactive state has a neutral A with a positively charged D^+, and the ground state has a neutral D with a positively charged A^+. The cases were chosen because they cover an attractive, a repulsive, and no Coulomb interaction, respectively, following electron transfer, and there is no Coulomb interaction between the initial species. Therefore, it is reasonable to take the initial distribution of donors and acceptors as random. The material presented below can be applied to non-random initial conditions by performing the appropriate spatial average.

A. Theoretical Development

The excited state, $\langle P_{\mathrm{ex}}(t)\rangle$, and reactive state, $\langle P_{\mathrm{re}}(t)\rangle$, populations will be derived. The probabilities of finding the donor molecule in its excited or reactive state for a system containing only two molecules, the donor D and the acceptor A, in solution is solved first. These results are then extended to a system that has many acceptors, and the ensemble average is performed over all acceptor positions.

At $t = 0$, the donor is optically excited. In the absence of the acceptor, the probability of finding the donor excited decays exponentially with the excited state life time, τ, where $\langle P_{\mathrm{ex}}(t)\rangle = \exp(-t/\tau)$. When an acceptor is present, the probability decreases more rapidly due to the addition of the electron transfer pathway.

The survival probabilities of the excited, $S_{\text{ex}}(R,t)$, and reactive states, $S_{\text{re}}(R,t)$, for solid solution [56, 65, 69] are given by the following equations for one acceptor in the absence of fluorescence:

$$S_{\text{ex}}(R,t) = \exp(-k_f(R)t), \tag{19}$$

$$S_{\text{re}}(R,t) = \exp(-k_b(R)t). \tag{20}$$

These are the probabilities that the state (excited or reactive) will still survive after a time t with a donor-acceptor separation of R in the absence of fluorescence. To include fluorescence one multiplies the survival probability by $\exp(-t/\tau)$.

In liquid solution, the situation is more complicated because the positions of the particles are not static. The donor and acceptor molecules undergo diffusive motion characterized by their corresponding diffusion constants D_D and D_A. It is convenient to describe the position of the acceptor in a reference frame whose origin coincides at any instant with the center of mass of the donor. In this reference frame, the acceptor undergoes diffusive motion relative to a stationary donor characterized by the diffusion constant $D = D_D + D_A$ [34, 39, 46, 66]. For the three cases introduced above, there are no significant intermolecular forces between the donor and acceptor prior to electron transfer. This means that the donors and the acceptors are randomly distributed and are freely diffusing. The survival probabilities for the excited and reactive states in liquid solution are [46, 47]:

$$\frac{\partial}{\partial t}S_{\text{ex}}(t \mid R_0) = D\nabla^2_{R_0}S_{\text{ex}}(t \mid R_0) - k_f(R_0)S_{\text{ex}}(t \mid R_0), \tag{21}$$

$$S_{\text{ex}}(0 \mid R_0) = 1, \tag{22}$$

$$4\pi R_m^2 D\frac{\partial}{\partial R_0}S_{\text{ex}}(t,\ R_0)\,|_{R_0=R_m} = 0, \tag{23}$$

$$\frac{\partial}{\partial t}S_{\text{re}}(t \mid R_0) = L^+_{R_0}S_{\text{re}}(t \mid R_0) - k_b(R_0)S_{\text{re}}(t \mid R_0), \tag{24}$$

$$S_{\text{re}}(0 \mid R_0) = 1, \tag{25}$$

$$4\pi R_m^2 D\frac{\partial}{\partial R_0}S_{\text{re}}(t \mid R_0)|_{R_0=R_m} = 0, \tag{26}$$

Where $\nabla^2_{R_0}$ for the spherically symmetric case considered here is

$$\nabla^2_{R_0} = \frac{2}{R_0}\frac{\partial}{\partial R_0} + \frac{\partial^2}{\partial R_0^2} \tag{27}$$

$V(R_o)$ is the interaction potential between the donor and the acceptor in the reactive state divided by K_BT. Here $V(R_0)$ is a Coulomb potential and is given by:

$$V(R_0) = \left[\frac{Z_D Z_A e^2}{4\pi\varepsilon_0\varepsilon_r K_B T}\right]\frac{1}{R}, \tag{28}$$

where Z_D and Z_A are the numbers and signs of the charges on the donor and the acceptor, respectively, *e* is the charge of the electron, ε_O is the permittivity of free space, ε_r is the relative permittivity of the solvent, K_B is Boltzmann's constant, and *T* is the temperature. $L^+_{R_0}$ is given by [46, 47]:

$$L^+_{R_0} = \frac{1}{R_0^2}\exp(V(R_0))\frac{\partial}{\partial R_0} D R_0^2 \exp(-V(R_0))\frac{\partial}{\partial R_0}. \tag{29}$$

Equations (21) and (24) for the excited and reactive state survival probabilities cannot be solved analytically. There are, however, analytical solutions for very small [67] and very large diffusion constants [68] and for transfer only at contact [46, 47]. Here, the equations for the survival probabilities will be solved numerically since the diffusion constants explored in this work are intermediate between the small and large diffusion limits, and electron transfer occurs over a range of distances with a distance-dependent rate.

The partial differential equations describing electron transfer with diffusion for a donor and *N* acceptors having an initial configuration of R_{0j} $(j = 1, \ldots, N)$ and a configuration of R_j $(j = 1, \ldots, N)$ at a time *t* are given below for an infinitely long excited state fluorescence lifetime:

$$\begin{aligned}\frac{\partial}{\partial t}&P_{\text{ex}}(R_1, \cdots, R_n, t \mid R_{01}, \cdots, R_{0N})\\ &= \sum_{j=1}^{N}[D\nabla_j^2 - k_f(R_j)]\\ &\quad\times P_{\text{ex}}(R_1, \cdots, R_N, t \mid R_{01}, \cdots, R_{0N}),\end{aligned} \tag{30}$$

$$P_{\text{ex}}(R_1, \cdots, R_N, 0 \mid R_{01}, \cdots, R_{0N}) = \prod_{j-1}^{N} \frac{\delta(R_j - R_{0j})}{4\pi R_{0j}^2}, \tag{31}$$

$$4\pi R_m^2 D \frac{\partial}{\partial R_j} P_{\text{re}}(R_1, \cdots, R_N, t \mid R_{01}, \cdots, R_{0N})\,|_{R_j = R_m} = 0, \quad (j = 1, \cdots, N), \tag{32}$$

$$\frac{\partial}{\partial t} P_{\rm re}^{i}(R_1, \cdots, R_N, t \mid R_{01}, \cdots, R_{0N})$$
$$= \sum_{j=1}^{n} L_{R_j} P_{\rm re}^{i}(R_1, \cdots, R_N, t \mid R_{01}, \cdots, R_{0N})$$
$$-k_b(R_i) P_{\rm re}^{i}(R_1, \cdots, R_N, t \mid R_{01}, \cdots, R_{0N})$$
$$+k_f(R_i) P_{\rm ex}(R_1, \cdots, R_N, t \mid R_{01}, \cdots, R_{0N})$$
$$(i = 1, \cdots, N), \tag{33}$$

$$P_{\rm re}^{i}(R_1, \cdots, R_N, 0 \mid R_{01}, \cdots, R_{0N}) = 0, \quad (i = 1, \cdots, N) \tag{34}$$

$$4\pi R_m^2 D \; \exp(-V(R_j)) \frac{\partial}{\partial R_j} \exp(V(R_j))$$
$$\times P_{\rm re}(R_1, \cdots, R_N, t \mid R_{01}, \cdots, R_{0N}) \, |_{R_j = R_m} = 0,$$
$$(j = 1, \cdots, N), \tag{35}$$

where $P_{\rm ex}(R_1, \ldots, R_N, t \mid R_{01}, \ldots R_{0N})$ is the excited state probability with the initial condition given by Equation (31) and boundary condition given by Equation (32). $P_{\rm re}^{i}(R_1, \ldots, R_N, t \mid R_{01}, \ldots, R_{ON})$ is the probability that the system is in the reactive state and the ith acceptor has the electron. L_{Rj} is the Smoluchowski operator [46, 47]:

$$L_{R_j} = \frac{1}{R_j^2} \frac{\partial}{\partial R_j} D R_j^2 \; \exp(-V(R_j)) \frac{\partial}{\partial R_j} \; \exp(V(R_j)). \tag{36}$$

The initial condition is given by Equation (34), and the boundary condition is given by Equation (35). This is only exact in the limit of $D_D = 0$, but is a very accurate approximation in three dimensional systems [66].

The solutions to the above equations for the state populations in the thermodynamic limit are:

$$\langle P_{\rm ex}(t) \rangle =$$
$$\exp(-t/\tau) \; \exp\left(\frac{4\pi}{d^3} \int_{R_m}^{\infty} \ln[1 - p + pS_{\rm ex}(t \mid R_0)] R_0^2 \, {\rm d}R_0 \right), \tag{37}$$

where d is the acceptor diameter, $p = Cd^3$, and C is the concentration of acceptors in number density units.

$$\langle P_{\rm re}(t) \rangle \; = \; 4\pi C \int_{R_m}^{\infty} \int_{0}^{t} \frac{S_{\rm re}(t - t' \mid R_0) \langle P_{\rm ex}(t') \rangle}{1 - p + pS_{\rm ex}(t' \mid R_0)}$$
$$\times \left[- \frac{\partial S_{\rm ex}(t' \mid R_0)}{\partial t'} \right] \; {\rm d}t' \; R_0^2 \, {\rm d}R_0, \tag{38}$$

where $\langle P_{ex}(t)\rangle$ is given by Equation (37).

The probability that the system is found in any one of the three states, ground ($\langle P_{gr}(t)\rangle$), excited ($\langle P_{ex}(t)\rangle$), or the reactive ($\langle P_{re}(t)\rangle$) state is unity. Therefore, the ground state population is:

$$\langle P_{gr}(t)\rangle = 1 - \langle P_{ex}(t)\rangle - \langle P_{re}(t)\rangle. \tag{39}$$

B. Model Calculations

In this section, the excited state (Equation (37)) and reactive state (Equation (38)) population functions will be displayed for a variety of parameters. The reactive state survival fraction will also be calculated. The partial differential equations for the survival probabilities were solved numerically. The formal treatment presented in the last section performed some of the necessary averaging analytically [70]. It also formulated the problem in a manner amenable to numerical analysis. The Crank-Nicholson method [71] was used to solve these differential equations, and Gaussian quadrature [71] was used for numerical integration. In each case, great care was exercised to ensure the accuracy of the numerical procedures [70]. It was necessary to choose small enough distance and time steps to give stable and accurate solutions. The reactive state probability (probability of finding ion pairs) is plotted in Figure 4 as a function of time and diffusion constant for each of the three cases defined above. The back transfer parameters are equal to the forward transfer parameters: The parameters are $a_f = a_b = 1$ Å, $R_f = R_b = 12$ Å, $\tau =$ 20 ns, $R_m = 9$ Å, $d = 7.2$ Å, and $C = 0.3$ M. All calculations were performed using a temperature of 298 K. Case I has an attractive Coulomb potential, case II has a repulsive Coulomb potential and case III has no potential. The relative permittivity, ε_r, is 78.5, the value for water. Following $t = 0$, forward electron transfer creates ion pairs, and back electron transfer destroys them. As the diffusion constant increases, the probability of finding an ion pair at long time increases for this set of transfer parameters. Even for $D = 1$ Å^2/ns, the results are significantly different from the solid solution ($D = 0$). $D = 100$ Å^2/ns approaches infinitely fast diffusion. For $D = 0$ and $D = \infty$, the three cases of Coulomb interactions give the same results. In solid solution the particles are fixed. In the case of infinite diffusion constant, the particle motion is so fast that the reactive state once created moves far away from the donor and does not back transfer. For diffusion constants of 1, 10, and 100 Å^2/ns, the three cases are different. Faster back transfer results in a lower reactive state probability at any time. The repulsive case gives the highest reactive state probability, while the attractive case gives the least. The differences among the curves are not great because of the high value of the relative permittivity used in the calculations.

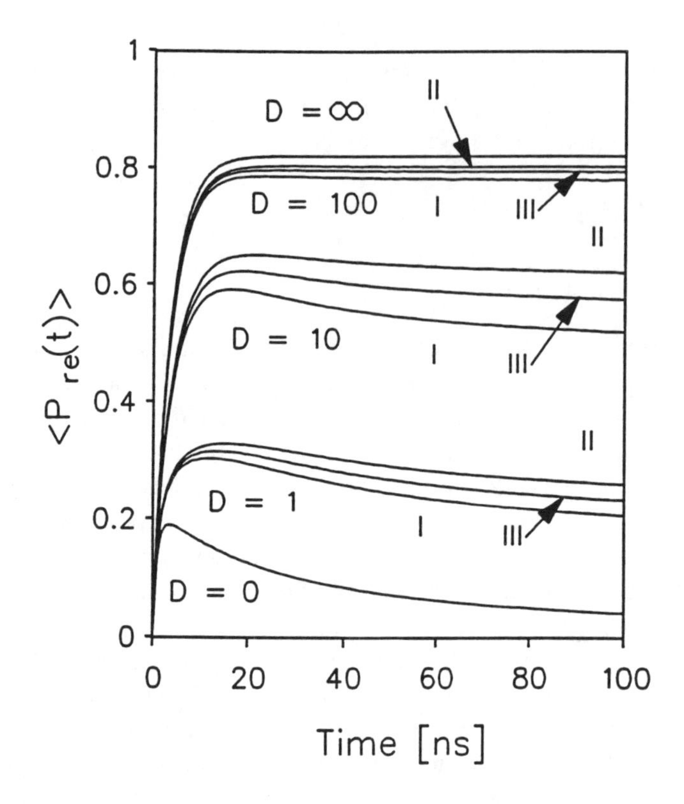

Fig. 4. The reactive state probability as a function of time. The back transfer parameters are equal to the forward transfer parameters: The parameters are $a_f = a_b = 1.0$ Å, $R_f = R_b = 12.0$ Å, $\tau = 20.0$ ns, $R_m = 9.0$ Å, $d = 7.2$ Å, and $C = 0.3$ M. Case I has an attractive Coulomb potential, case II has a repulsive Coulomb potential and case III has no potential. $\varepsilon_r = 78.5$, the value for water. As the diffusion constant, D, increases so does the reactive state probability.

In Figure 5 the reactive state survival fraction, $f_s(t)$, is plotted as a function of diffusion constant. The parameters are the same as those used for Figure 4 except $\varepsilon_r = 24.3$, the value for ethanol. The survival fraction is defined as the fraction of all reactive states formed that still survive at a time t. The reactive state probability is obtained from Equation (38). The total number of reactive states formed by time t is given by Equation (38) in the limit that the back transfer rate goes to zero, $\langle P_{\text{re}}(t)\rangle_{kb} = 0$. The survival fraction is given by:

$$f_s(t) = \frac{\langle P_{\text{re}}(t)\rangle}{\langle P_{\text{re}}(t)\rangle_{k_b=0}}. \tag{40}$$

In the limit of long time, this gives the reactive state escape probability. This is the fraction of all reactive states that did not undergo geminate recombination. In the figure, the survival fraction at $t = 100$ ns is plotted as a function of diffusion constant for the three cases. In all three cases, the survival

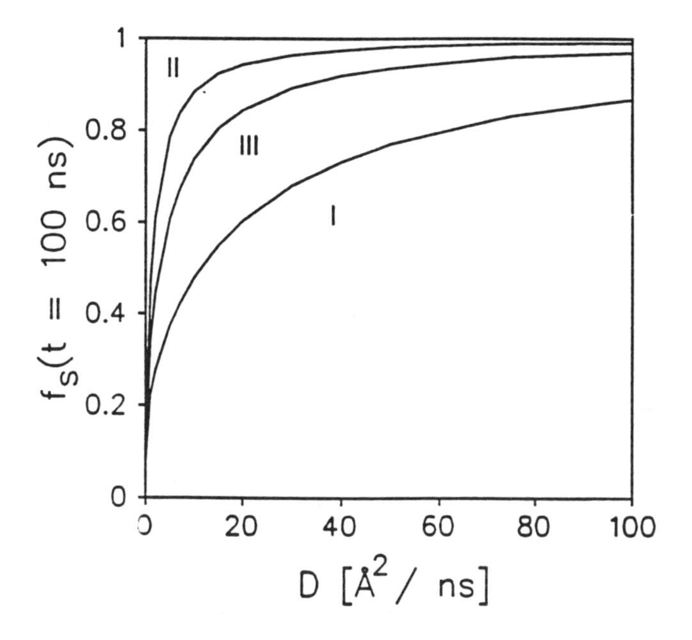

Fig. 5. The reactive state survival fraction at $t = 100$ ns as a function of diffusion constant. The parameters are the same used in Figure 4 except with $\varepsilon_r = 24.3$, the value for ethanol. Case I has an attractive Coulomb potential, case II has a repulsive Coulomb potential and case III has no potential. For all cases, as the diffusion constant increases, the survival fraction also increases.

fraction increases rapidly between $D = 0$ and 20 Å^2/ns. At larger diffusion constants, the survival fraction increases more slowly. Thus, not only are more reactive states generated with larger diffusion constants, but a greater percentage of them survive. Case I, which has the attractive potential, has the smallest survival fraction, while case II, which has the repulsive potential, has the largest survival fraction. At higher relative permittivities, the curves calculated for the three cases would be closer together, while at smaller relative permittivities they would be further apart.

In Figure 6 the survival fraction is plotted as a function of relative permittivity for a diffusion constant of $D = 10$ Å^2/ns with the other parameters the same as Figures 4 and 5. As can be seen, the no potential case (III) is independent of ε_r and lies between the repulsive (II) and attractive (I) cases. At sufficiently large relative permittivities, the three cases should converge to the same result since the Coulomb interaction is effectively screened. As seen in the figure, the three cases get closer together and more symmetric about the no potential case at large relative permittivities. At smaller relative permittivities the cases become quite different. For the attractive case, the survival fraction falls to zero, while the survival fraction for the repulsive case goes almost to one as the relative permittivity goes to zero. This dramat-

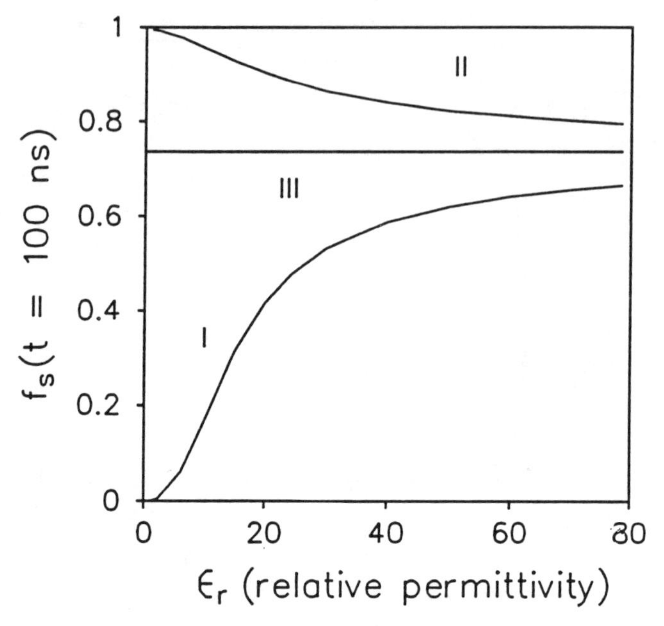

Fig. 6. The reactive state survival fraction at $t = 100$ ns as a function of relative permittivity. The parameters are the the same as in Figure 4 except $D = 10$ Å^2/ns. As the relative permittivity increases the three cases get closer together while at small relative permittivities they diverge.

ically illustrates the importance of the Coulomb interaction in determining the extent of geminate recombination.

Figure 7 presents a comparison between the theory for electron transfer and geminate recombination presented here and other methods that have been employed. In Figure 7, the reactive state probability given by Equation (38) is plotted (no potential, case III), along with two approximate models for the reactive state survival probability. In all curves, the excited state survival probability, $S_{ex}(t \mid R_0)$, and the excited state probability, $\langle P_{ex}(t) \rangle$, were calculated using Equations (21) and (37), respectively. Thus the differences will be associated with the manner in which geminate recombination is handled. The parameters used in the calculations are the same as those used for Figure 4 except that $D = 1$ Å^2/ns. In Curve A, $S_{re}(t \mid R_0)$ (used in Equation (38)) is calculated using the Collins and Kimball model [46]:

$$S_{re}(t \mid R_0) = 1 - \frac{R_m}{R_0}\left[\frac{k}{k + 4\pi R_m D}\right] \times \left[\operatorname{erfc}\left(\frac{R_0 - R_m}{\sqrt{4Dt}}\right) - \exp(\alpha^2 Dt + \alpha[R_0 - R_m]) \times \operatorname{erfc}\left(\alpha\sqrt{Dt} + \frac{R_0 - R_m}{\sqrt{4Dt}}\right)\right], \tag{41}$$

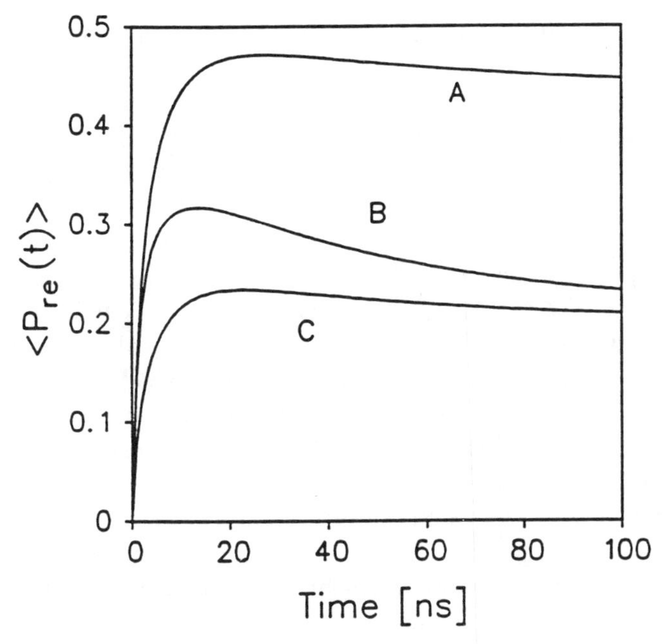

Fig. 7. The reactive state probability as a function of time for various models for the reactive state survival probability, $S_{re}(t \mid R_0)$ for the no potential case (III). The rate of reactive state formation is given by the model developed here. Curve *A* uses the Collins and Kimball model for $S_{re}(t \mid R_0)$, curve *B* has $S_{re}(t \mid R_0)$ given by Equation (24), and curve *C* uses the Smoluchowski model for $S_{re}(t \mid R_0)$. The parameters are the same as used in Figure 4 except $D = 1.0$ Å^2/ns and $k = 1275$ Å^3/ns for the Collins and Kimball model and was calculated using Equation (43). The Smoluchowski model underestimates the reactive state probability while the Collins and Kimball model overestimates it.

where

$$\alpha = \frac{4\pi D R_m + k}{4\pi D R_m^2}. \tag{42}$$

It is necessary to choose a value for *k* consistent with the detailed theory. *k* is given by:

$$k = 4\pi \int_{R_m}^{\infty} k_f(R) R^2 \, \mathrm{d}R \tag{43}$$

which, for the parameters used here, is $k = 1275$ Å^3/ns. Curve *B* uses the theory presented here (Equation 24), and curve *C* uses the Smoluchowski model in which:

$$S_{\mathrm{re}}(t \mid R_0) = 1 - \frac{R_m}{R_0} \operatorname{erfc}\left(\frac{R_0 - R_m}{\sqrt{4Dt}}\right). \tag{44}$$

As can be seen in the figure, the Smoluchowski model underestimates the reactive state probability, while the Collins and Kimball model overestimates it. The Collins and Kimball model, although further from curve B than the Smoluchowski model in these calculations, is better at higher diffusion constants. As the diffusion constant becomes smaller, the Collins and Kimball model will increasingly underestimate the amount of back electron transfer. This is because all the electron transfer happens at contact in the Collins and Kimball model, while the model presented here allows transfer at separations other than contact. Also, the disagreement between the curves in Figure 7 would be greater had the rate of ion formation been calculated using the Smoluchowski model in C and the Collins and Kimball model in A. The agreement was improved by using a more accurate description of the forward transfer. Furthermore, if data taken as a function of diffusion constant were fit with either the Collins and Kimball or Smoluchowski model, the electron transfer parameters arising from the fits would not be consistent from one diffusion constant to another. It is interesting to note that the Collins and Kimball model is better at short time, while the Smoluchoski model is better at long time.

4. Comparison of Theory and Experiment

In this section, an experimental study is presented on the influence of molecular diffusion on photoinduced electron transfer and the experimental results are compared to the theory presented above. Steady state fluorescence yield and time resolved fluorescence measurements were used to measure the population of the donor's excited state, and pump-probe experiments were used to measure the population of ion pairs produced as a result of electron transfer. Since electron transfer quenches the fluorescence from the excited state, fluorescence provides a direct observable for the yield and time dependence of the forward electron transfer process. The pump-probe observable is proportional to the population that is not in the ground state of the system. Since there are only two other states involved, the excited state and the reactive state (the ion pair produced as a result of forward electron transfer) and since the excited state population is obtained from the fluorescence measurements, the reactive state population is obtained by combining the results of the pump-probe and time resolved fluorescence experiments.

Experiments were performed in both solid and liquid solutions on the same donor/acceptor pairs. In solid solution, the viscosity is infinite; there is no molecular diffusion. Thus, the data will be analyzed using the state populations derived in Section 2: In liquid solution, the molecules are diffusing and the state populations derived in Section 3 will be used. The results of Section 2 are the limiting case of the Section 3 results. The solid and liquid solvents were chosen to have very similar dielectric properties. Therefore, within the context of the theoretical model, the only differences between the solid and liquid is the rate of diffusion. Thus, although they have vastly different diffusion constants, the solid and liquid solutions should yield similar electron transfer parameters from the data fits. The experiments provide a detailed test of the theory and concepts presented above.

A. EXPERIMENTAL PROCEDURES

1. Samples

The samples are composed of rubrene [*RU*], the donor, and duroquinone [*DQ*], the acceptor, dissolved in either sucrose octaacetate [*SOA*], or diethyl sebacate [*DES*]. *SOA* is a glass at room temperature while *DES* is a liquid. In the presence of light and oxygen, *RU* in solution will irreversibly oxidize. The presence of dust particles in samples increases the amount of scattered light and noise in the data. Concentration inhomogeneities in the samples would lead to inconsistent results. These problems shaped the sample preparation technique [72].

For solid solution, the glass, *SOA*, was twice recrystallized in ethanol. *DQ* was twice sublimated. *RU* is difficult to sublimate or recrystallize, and instead, a small amount of *RU* was dissolved in a degassed (with argon) solution of *SOA* in spectral grade acetone in the dark. This solution was filtered through a 0.2 μm filter into a 1 mm path length optical cell. The cell was connected to a vacuum line and the pressure in the cell was gradually lowered to evaporate the acetone. The sample was then melted (under vacuum, $\approx 10^{-6}$ torr). The cell was removed from the vacuum line, and *DQ* was added. The sample was placed back on the vacuum line and sealed off. The sample was melted to dissolve the *DQ*. While molten, the sample was shaken. This step was repeated several times to ensure a homogeneous distribution of *DQ*. The concentrations of *DQ* and *SOA* were determined spectroscopically for the solid solution.

The liquid samples were made using serial dilution to obtain the desired concentrations for *RU* and *DQ*. Typically the *RU* concentration was 10^{-4} M, corresponding to an optical density (*OD*) of ≈ 0.2 at 532 nm. The *DQ* concentration ranged from 0 to 0.5 M. The solvent diethyl sebacate was filtered. Each sample was placed into a 1 mm optical cuvette. The samples were freeze-pump-thawed (3 to 5 cycles) to remove oxygen to prevent *RU* decomposition and sealed under vacuum. The *RU* concentration (10^{-4} M) was chosen to eliminate dimer formation, reabsorption, and energy transfer [72].

The viscosities of pure *DES* and *DQ/DES* solutions were measured with an Ubbelohde viscometer. At the highest concentrations of *DQ*, the viscosities of the solutions differed from the pure solvent by a small amount. The measured viscosities were used.

2. Fluorescence Measurements

The reduction in the *RU* fluorescence quantum yield as a function of the *DQ* acceptor concentration was measured in the following manner. Single pulses at a 1.0 kHz repetition rate from a *CW* pumped acousto-optically mode-locked and Q-switched Nd:YAG laser were doubled to 532 nm. Green single pulses (fwhm $\approx$ 100 ps) were used for sample excitation. A sample holder was constructed to ensure that each sample was reproducibly illuminated with the same amount of light and the same solid angle of fluorescence collected. The fluorescence was collected by a lens, and the light went through a set of 532 nm cutoff filters to eliminate scattered excitation light. The broad band fluorescence was detected by a phototube, and the signal was sent to a lock-in amplifier. To obtain the relative yield accurately, all the samples were positioned the same way relative to the excitation-detection setup. For each sample, a corresponding yield was measured for a sample that just contained *RU* and solvent (*SOA* or *DES*). Dividing the fluorescence intensities of the *RU/DQ*/solvent samples by the plain *RU*/solvent sample after correcting for small *RU OD* differences gave the relative fluorescence yields.

The time resolved fluorescence decays of the *RU/DQ/DES* solutions were measured by time-correlated single photon counting. The laser system and the single photon counting instrument have been described in detail [73]. Briefly, the frequency-doubled (532 nm) output from an acousto-optically mode-locked *YAG* laser is used to synchronously pump a cavity dumped dye laser. The excitation wavelength was 555 nm. The fluorescence is detected in a 60 nm window centered at 590 nm. Scattered light is blocked from entering the detector (a multichannel plate). The broad spectral bandwidth of this arrangement essentially eliminates the influence of solvent relaxation on the time dependence of the emission. The instrument response of the system is $\approx$ 70 ps. The output of the instrument is transferred to a computer for data analysis. The detection polarization was set to the magic angle from the excitation polarization to remove the influence of rotational relaxation of the donor molecules on the time dependence of the fluorescence.

The time resolved fluorescence decays of the *RU/DQ/SOA* solutions were measured on a different apparatus (before the single photon counting setup was available). However, the same monochromator and multichannel plate setup was used, and fluorescence was detected at the magic angle. Excitation pulses (532 nm) came from the laser described above for the fluorescence yield measurements. The output of the multichannel plate went to a boxcar averager. The sampling window of the boxcar was positioned in time by a 10 V ramp, giving a time range of 100 ns. The digital output of the boxcar was added to the data from previous shots by computer until an adequate signal to noise ratio was obtained. The overall time response of the system (1.2 ns) was measured by observing the excitation pulse. The system impulse response was recorded and used for convolution with theoretical calculations to permit accurate comparison to the data.

3. Pump-Probe Experiments

The laser used in the fluorescence yield measurements was also used for the pump-probe experiments. For the pump-probe experiment a single green pulse (532 nm, fwhm = 100 ps) is selected and split. One pulse is the excitation pulse and it passes through a halfwave-plate and polarizer before passing through the sample. The half-wave-plate and polarizer permit adjustment of the intensity and polarization of the pump beam. The probe pulse passes down a mechanical delay line then through a half-wave-plate and polarizer before it impinges on the sample. The probe is set to the magic angle relative to the pump beam. The signal is detected through another polarizer set to the same angle as the probe beam. The spot sizes were 200 and 150 μm for the pump and probe, respectively. The angle between the beams was about 3 degrees. The signal is detected by a large area photodiode. A second probe beam passes through a different spot in the sample. This probe, which is not affected by the pump pulse is detected by a second, identical photodiode. This second probe beam is constantly measured to remove any fluctuations in the signal due to temporal variations in laser intensity. The outputs of the two photodiodes are inputs into a differential amplifier. The output of the differential amplifier is sent to a lock-in amplifier. The pump beam is chopped so that the signal recorded is the difference between the transmitted intensity of the probe with and without the pump. As the ground state recovers, less intensity is transmitted and the signal decreases. The intensities of the pump and probe beams were chosen such that there were no intensity artifacts. This was achieved by lowering the laser power until no further change in the shape of the signal was observed. The intensity of the pump beam was typically $\geq$ 20 times that of the probe.

4. Physical Parameters

The dynamics of electron transfer and back transfer in solid and liquid solutions are determined by ten parameters. In addition to the *RU* excited state lifetime, τ, there are four other rate parameters; the forward electron transfer rate parameters, R_f and a_f, and the back electron transfer parameters, R_b and a_b. There are two molecular sizes: the radius of the donor, R_D, and the radius of the acceptor, R_A. The contact distance is given by $R_m = R_D + R_A$ and the acceptor diameter, $d = 2R_A$. There are three other parameters: the sum of the donor and the acceptor molecular diffusion constants in *DES*, *D*, the relative permittivity of *DES*, ε_r, and the concentration of *DQ*, *C*, in the various samples. There are other parameters involved in constructing the observables such as absorption and stimulated emission cross-sections for the *RU* ground and excited states; these will be discussed further below. The electron transfer parameters will be obtained from fits to the data. All of the other parameters can be obtained from independent measurements.

The fluorescence lifetimes of *RU* in *SOA* and *DES* were measured using time resolved fluorescence with samples that contained only *RU* and solvent. The diffusion constants were obtained from measuring the viscosity of *DES* and using the Stokes-Einstein equation. The relative permittivity of DES was obtained by capacitance measurements, and the concentrations of the samples were obtained spectroscopically.

The molecular sizes were obtained from X-ray crystallographic data [74]. The values used for the analysis are the donor-acceptor contact distance R_m = 9 Å and the acceptor diameter d = 7.2 Å. These were calculated by dividing the volume of the crystallographic unit cell by the number of molecules per unit cell for each molecule. This procedure gave the molecular volumes. Space filling models of the molecules were also constructed, and the molecular volumes agreed with the numbers obtained using the X-ray crystallographic data. The molecules are taken to be spheres.

The viscosity of *DES* was measured as a function of *DQ* concentration. For the concentrations of *DQ* used, the following diffusion constants ($D = D_A + D_D$ were obtained from the viscosities and the molecular volumes: C = 0.031 M, D = 18.3 Å^2/ns; C = 0.110 M, D = 18.4 Å^2/ns; C = 0.210 M, D = 18.6 Å^2/ns; and C = 0.330 M, D = 18.8 Å^2/ns. These diffusion constants are on the same order as those obtained for similar molecules [46, 75].

The relative permittivity of *DES* was measured using a capacitance bridge and an air gap capacitor [76]. The measured ε_r was 5. This was also performed as a function of *DQ* concentration and no variation was found.

The measured life times of *RU* in *SOA* and *DES* are 15 and 14.4 ns, respectively. These numbers fall in the range of values measured by others [75]. At room temperature *RU* has virtually unit quantum yield for fluorescence [75, 77]. Thus, there is no significant triplet formation.

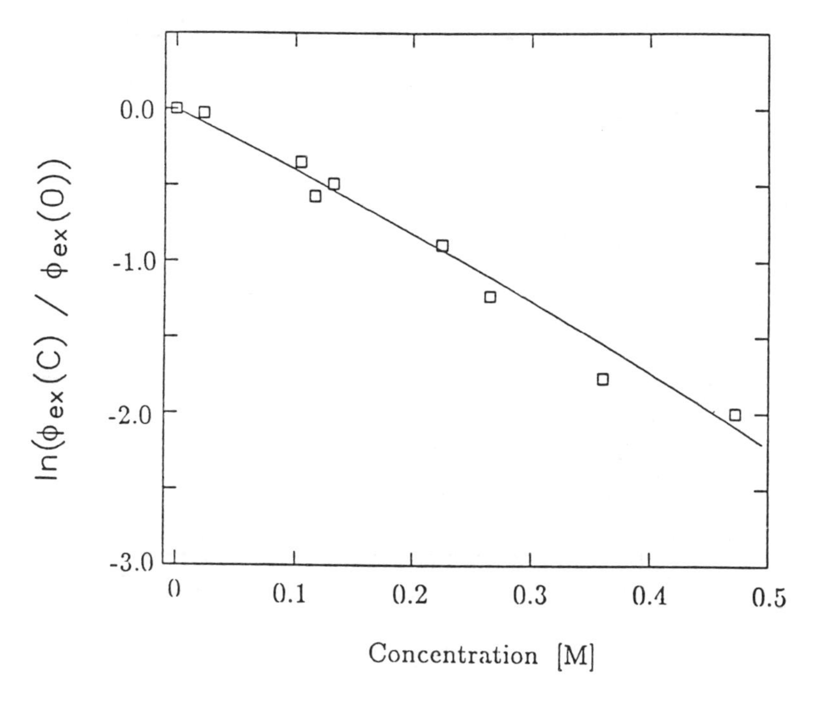

Fig. 8. The natural log of the relative fluorescence yield plotted as a function of the acceptor concentration for the solid solution. From this plot, one of the two forward transfer parameters is determined. $R_f = 13.1$ Å.

B. Results

There are three basic issues to address. (1) Can the statistical mechanical model describe the data, i.e., can the shape and concentration dependence be reproduced with a single set of fitting parameters? (2) Is there a consistency between the parameters obtained for the solid and liquid solutions? (3) How do the dynamics in the solid and liquid solutions compare?

1. Fluorescence Experiments

Figures 8 and 9 display the fluorescence yield data taken on solid and liquid solutions respectively. As the acceptor concentration increases, the amount of electron transfer increases, decreasing the fluorescence yield. The relative fluorescence yield was calculated using:

$$\phi_{\mathrm{ex}} = \frac{1}{\tau} \int_0^\infty \langle P_{\mathrm{ex}}(t) \rangle \, \mathrm{d}t. \tag{45}$$

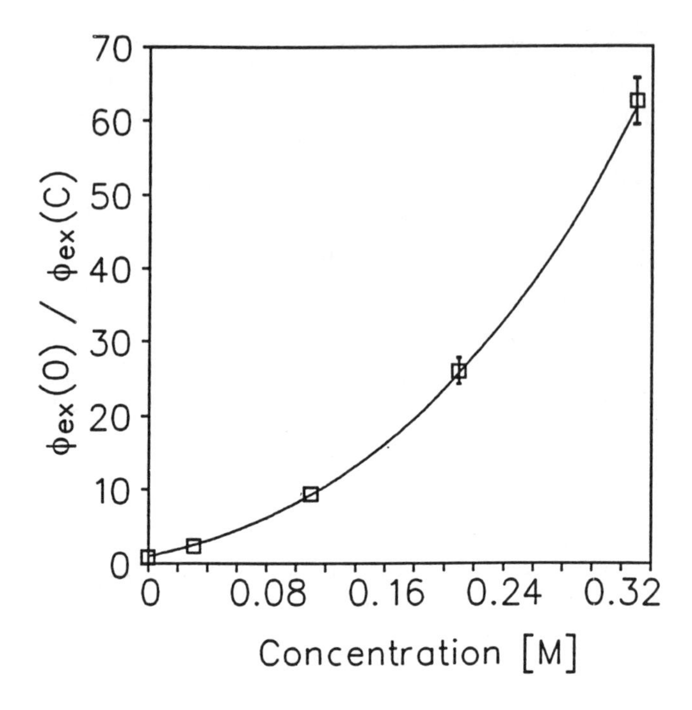

Fig. 9. A Stern-Volmer plot of the relative fluorescence yield data for liquid solution. Error bars are shown for two points. For the other data points, the error bars are smaller than the data squares. The best fit (solid line) gave $a_f = 0.22$ Å and $R_f = 12.8$ Å.

The solid solution data is displayed as a Perrin plot [1, 46, 75]. It is a plot of the natural log of the relative fluorescence yield vs. concentration of the acceptor. The data from the liquid solution is a Stern-Volmer plot [1, 46, 75]. It is a plot of the inverse fluorescence yield vs. concentration of the acceptor.

Inokuti and Hirayama [50] found that the Perrin plot is essentially linear and its slope is related to the R_f parameter. In the data analysis here, Equation (45) was used to fit the data, using Equation (16) for the excited state population. In principle, Equation (45) depends on the two forward electron transfer parameters, R_f and a_f, and the concentration of acceptors, C. However, Inokuti and Hirayama [50] have found that ϕ_{ex} is not sensitive to large changes in a_f for the calculations without excluded volume. Calculations with excluded volume show that this is still true. Therefore, by comparing steady state fluorescence yield data to ϕ_{ex} obtained from Equation (45), the R_f value for solid solution is uniquely determined. The solid line through the data in Figure 8 is the best fit. This gives a value of $R_f = 13.1$ Å for the solid solution. This is a single parameter fit.

The liquid solution data (Figure 9) is given as a Stern-Volmer plot. In the limit that diffusion is infinitely fast, the data would fall on a straight line. In the limit that the reaction is diffusion limited, the initial slope is related to the rate parameters [1, 46, 75]. The experimental results are between these two limits. Equation (45) was used to fit the data using Equation (37) for the excited state population. For each concentration, the measured diffusion constant (given above) was employed. In Figure 9, the solid line is the best fit; the symbols are the data. Error bars are given for the two highest concentrations. For the lower concentrations, the error bars are smaller than the symbols. The fit is highly sensitive to the values of both a_f and R_f. It is not possible to pick an arbitrary value of one of the parameters and obtain a fit by varying the other. As can be seen, the curve goes through the middle of the error bars. The fit is obtained when $a_f = 0.22$ Å and $R_f = 12.8$ Å.

The time resolved fluorescence quenching data in the solid and liquid solutions are presented in Figures 10 and 11, respectively. The calculated excited state population $P_{\text{ex}}(t)$ given by Equation (16) for solid solution or Equation (37) for liquid solution is convolved with the instrument response function and then compared to the data. The appropriate equation is

$$S(t) = \int_0^\infty IR(t - t')\langle P_{\text{ex}}(t')\rangle \, \mathrm{d}t', \tag{46}$$

where $IR(t)$ is the measured instrument response. For the solid solution, only one parameter, a_f, was adjusted to fit the data. An $a_f = 0.22$ Å was obtained. In Figure 10, the data and the theory agree quite well. (At long time there is a small bump in the data caused by an instrumental problem).

For the liquid solutions, either a_f or R_f or both can be adjusted to fit the experimental data (Figure 11). All of these procedures were used to analyze the data. The fits are shown in Figure 11 as the solid lines through the data. The data and fits are in such close agreement that it is difficult to distinguish them in some of the curves. The fits were very sensitive to the values of both a_f and R_f. Because of the high quality of the data and the sensitivity of the fits to the parameters, it is possible to determine them accurately. The best fits yield $a_f = 0.22$ Å and $R_f = 12.8$ Å. These are identical to the parameters obtained from the fit to the fluorescence yield data.

Comparing the solid (no diffusion) and liquid (diffusion) results shows that the a_f values are identical and the R_f values differ by $\approx 2\%$. a_f and R_f are molecular parameters describing the forward transfer dynamics. Since the electron transfer dynamics are affected by the solvent reorganization energy [1], it is not surprising that these parameters are slightly different from one solvent to another due to small differences in the solvents' dielectric properties. The fact that almost identical electron transfer parameters are obtained for liquid diethyl sebacate ($a_f = 0.22$ Å and $R_f = 12.8$ Å) and rigid sucrose octaacetate ($a_f = 0.22$ Å and $R_f = 13.1$ Å) solutions demonstrates

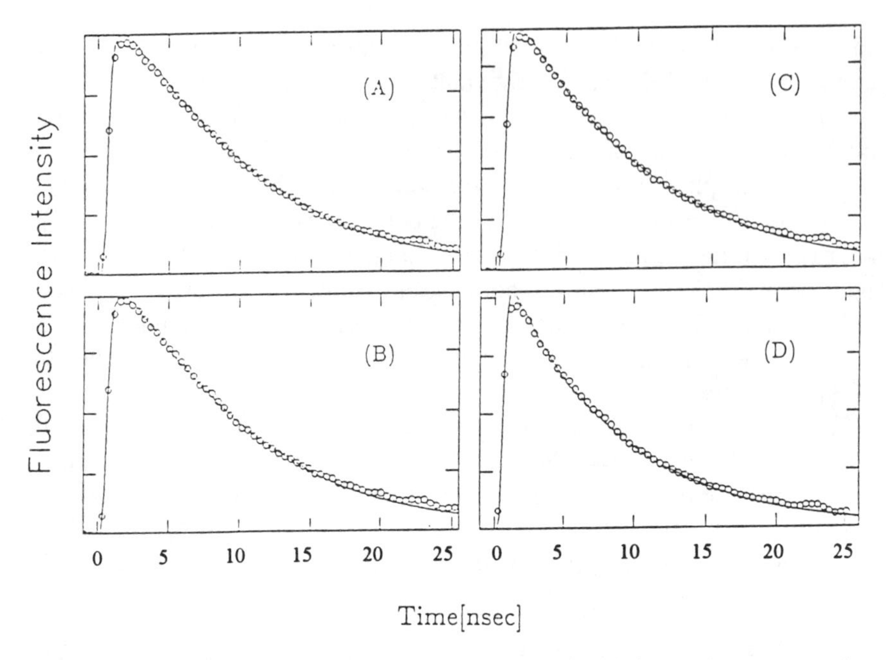

Fig. 10. Time-resolved fluorescence data and theory for solid solution for four concentrations. The circles are the experimental data, and the lines are the theoretical curves. The concentrations for plots *A, B, C, D* are 0.105 M, 0.134 M, 0.224 M, and 0.470 M, respectively. The best fit is obtained with a_f = 0.22 Å and R_f = 13.1 Å. a_f was the only parameter adjusted. The bump in the data at very long times is due to instrumental error.

that a_f and R_f are not arbitrary fitting parameters, but rather they are good molecular parameters that characterize the forward transfer process.

It can be seen by comparing Figures 10 and 11 that the excited state populations decay much faster because of electron transfer in the liquid solutions than in the solid solutions. Figure 12 shows a calculation of the time-dependent donor quenching by electron transfer for the solid (*A*) and liquid (*B*) solutions. The excited state decay is not included so that the differences in forward electron transfer can be seen clearly. The parameters used are taken from the experiments and are given in the figure caption. The difference is dramatic. At short time (< 5 ps) the diffusion of the molecules in the liquid is negligible and the curves are identical. At longer time, diffusion leads to more electron transfer events and a faster depletion of the excited state. Diffusion of the molecules has a profound influence on the transfer process in spite of the fact that the transfer parameters, a_f and R_f, are virtually identical in the two systems. This can be understood qualitatively. In a solid, acceptors that are too far away from the excited donor to receive an electron within the donor lifetime do not participate in the transfer dynamics. In a liquid, an acceptor too far from a donor at $t = 0$ can move in at later times and receive

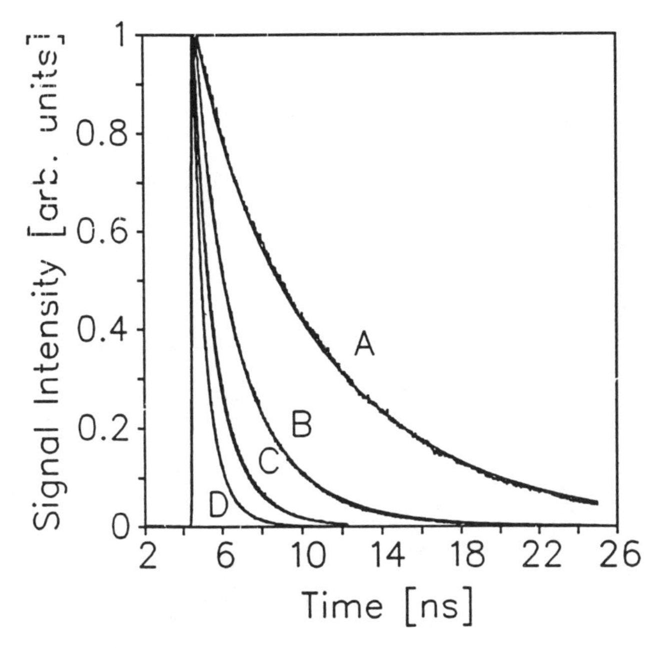

Fig. 11. Time-resolved fluorescence data and theory for liquid solution for four concentrations. The concentrations for plots *A, B, C, D* are 0.031 M, 0.11 M, 0.21 M, and 0.33 M, respectively. The best fit is obtained with a_f = 0.22 Å and R_f = 12.8 Å.

an electron. Because of this motion, more acceptors pass through the range within which transfer is likely (during the life time of the excited state) than would normally be there in solid solution. The net result is that, all other things being equal, depletion of the excited state will increase with diffusion rate.

2. *Pump-Probe Experiments*

In the simplest situation, the pump-probe observable is proportional to the population that is not in the ground state. This is the situation if the only absorbing state is the ground state and there is no stimulated emission or excited state-excited state (ES-ES) absorption. The pump-probe observable is then given by:

$$S(t) = A[\langle P_{\mathrm{ex}}(t)\rangle + \langle P_{\mathrm{re}}(t)\rangle], \tag{47}$$

where A is an arbitrary constant. Equation (47) comes from conservation of probability (see Equation (39)).

Stimulated emission will occur if the probe wavelength is in the spectral region of the fluorescence of the donor. Because of stimulated emission, there is amplification of the probe intensity as it passes through the sample. This

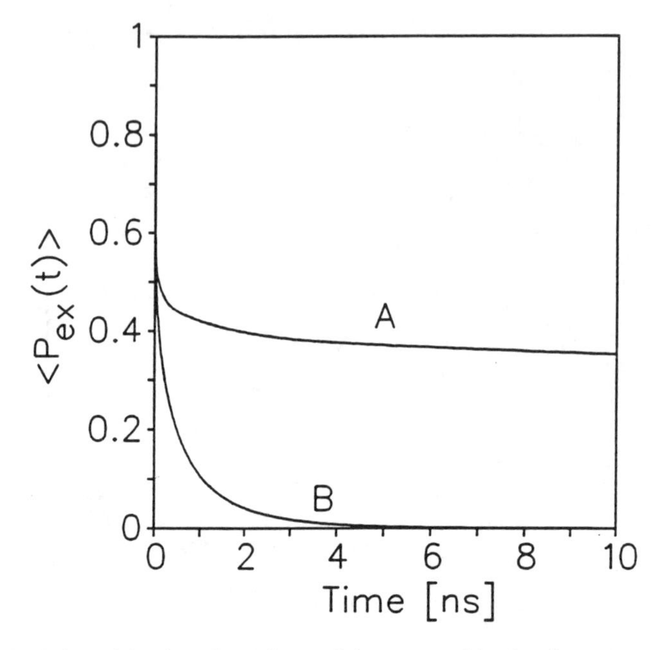

Fig. 12. Calculation of the time dependence of donor quenching by electron transfer for the solid (curve *A*) and the liquid (curve *B*) solutions. Lifetime decay of the excited state is not included so that electron transfer can more directly be compared. The parameters are $C = 0.27$ M, $a_f = 0.22$ Å, $R_f = 13.0$ Å and for liquid solution $D = 18.6$ Å^2/ns. Diffusion results in a large increase in electron transfer.

extra signal is proportional to the excited state population and the stimulated emission cross-section. The contribution to the signal from removal of population from the ground state is proportional to the ground state absorption coefficient (cross-section). The inclusion of stimulated emission in the analysis is accomplished by adding the ratio of the stimulated emission cross-section to the ground state absorption cross-section to the coefficient in front of the exited state population [78]. This gives:

$$S(t) = A[(1.0 + \beta)\langle P_{\text{ex}}(t)\rangle + \langle P_{\text{re}}(t)\rangle], \tag{48}$$

where β is the ratio of the stimulation emission cross-section to the absorption cross-section at the wavelength of interest.

Following optical excitation, ground state depletion and stimulated emission increase the detected probe intensity. Conversely, excited state-excited state (ES-ES) absorption decreases the probe intensity. This effect can be included in the same manner as stimulated emission. The ratio of the absorption cross-section for the ES-ES absorption to the ground state absorption cross-section is subtracted from the coefficient in front of the excited state population. The observable is:

$$S(t) = A[(1.0 + \beta - \gamma)\langle P_{\text{ex}}(t)\rangle + \langle P_{\text{re}}(t)\rangle], \tag{49}$$

where γ is the ratio of the ES-ES absorption cross-section to the ground state absorption cross-section.

In both *SOA* and *DES* at 532 nm, excited *RU* has a significant amount of stimulated emission and ES-ES absorption [77]. Although these two effects work in opposite directions on the signal, they do not completely cancel. The β's were measured in both *SOA* and *DES* at 532 nm [72]. The β's obtained for *SOA* and *DES* are 0.5 ± 0.1 and 0.3 ± 0.1, respectively. These are similar to those measured by others [77, 79]. A recent study [77] was able to measure the ES-ES absorption of *RU* at 532, yielding a $\gamma = 1.0 \pm 0.2$. The RU^+ and DQ^- ions do not absorb at 532 nm [80] and therefore do not need to be included in the analysis.

For accurate comparison to the data $S(t)$ is convolved with the appropriate pulse shape functions of the pump and the probe. The signal, $I(t)$ is given by the convolution:

$$I(t) = \int_{-\infty}^{\infty} G_{\text{probe}}(t - t') \int_{-\infty}^{t'} G_{\text{pump}}(t'')S(t' - t'')\, \mathrm{d}t''\, \mathrm{d}t', \tag{50}$$

where $G_{\text{probe}}(t')$ and $G_{\text{pump}}(t'')$ are Gaussians whose full width at half maxima are determined by autocorrelation.

Since there is a significant uncertainty in the coefficient in front of the excited state population in Equation (49), the procedure used to fit the data involved adjustment of this coefficient within the error bars of β and γ.

Figure 13 displays pump-probe data taken on solid solutions for four acceptor concentrations. At short time the data is dominated by a coherence artifact. Measurements of the instrument response show that after 0.5 ns the data is not influenced by the coherence spike. For this reason, the data fitting was done at times greater than 0.5 ns. As can be seen in Figure 13, the theory fits the solid data very well using a single choice of the back transfer parameters. The only noticeable deviation is a slight mis-match at intermediate time in the highest concentration (*D*). The fits yielded the back electron transfer parameters: $R_b = 12.3$ Å and $a_b = 0.9$ Å. Spanning the region within the error bars of β and γ, the best fit was obtained with a $\beta = 0.4$ and a $\gamma = 1.2$. Within the error bars of β and γ, the best fit R_b varied by less then 2% and a_b did not vary. Although the parameters did not vary significantly, the quality of the fits at other values of β and γ were clearly inferior. Thus the choices of the β and γ parameters do not degrade the reliability of the electron transfer parameters.

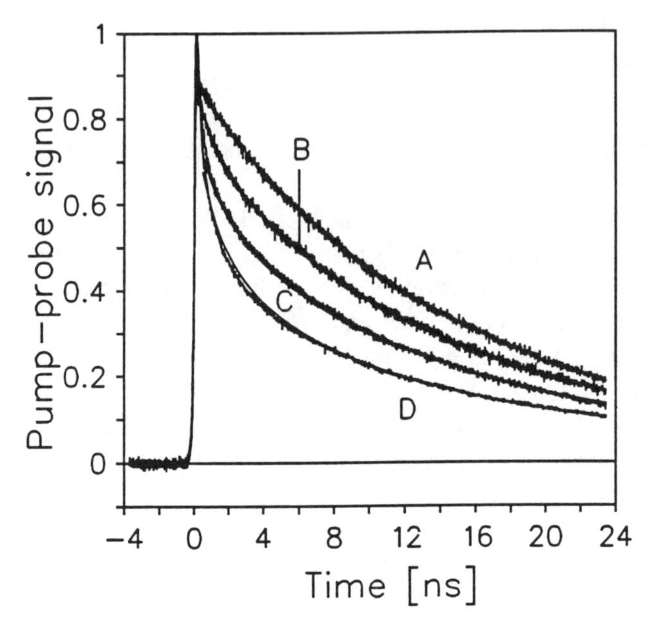

Fig. 13. The pump-probe data for the solid solution. The curves labeled *A, B, C,* and *D* have the following concentrations: 0 M, 0.051 M, 0.098 M, and 0.23 M, respectively. The peak at short time is due to a coherence artifact. The parameters that fit the data are R_b = 12.3 Å and a_b = 0.9 Å.

In fitting the forward transfer data in the solid and liquid solutions, it was determined that the forward transfer parameters were virtually identical in the two media. In fact, if the solid transfer parameters had been used without modification, the agreement with the liquid data would be virtually the same as that displayed in Figure 11. The excellent agreement between theory and data shows that it is possible to handle the diffusion of the particles accurately. Comparing the back transfer in the solid and liquid is fundamentally different because of the Coulomb interaction between the ions. In the solid solution, the ions are immobile, so the fact that the particles are charged does not change the ability of the statistical mechanical theory to do the averaging over the non-random spatial distribution exactly. In the liquid, it is not only necessary to account for diffusive motion, but also the non-diffusive motion caused by the attractive Coulomb interaction. This interaction is mediated by the dielectric properties of the solvent (see Equation (24)). In the theory, the dielectric properties come in through a single parameter, the relative permittivity.

The data for liquid solutions are presented in Figure 14. It shows the pump-probe observable for several concentrations, 0 M, 0.04 M, 0.13 M, and 0.22 M (Curves A-D, respectively). The spike around t = 0 is a coherence

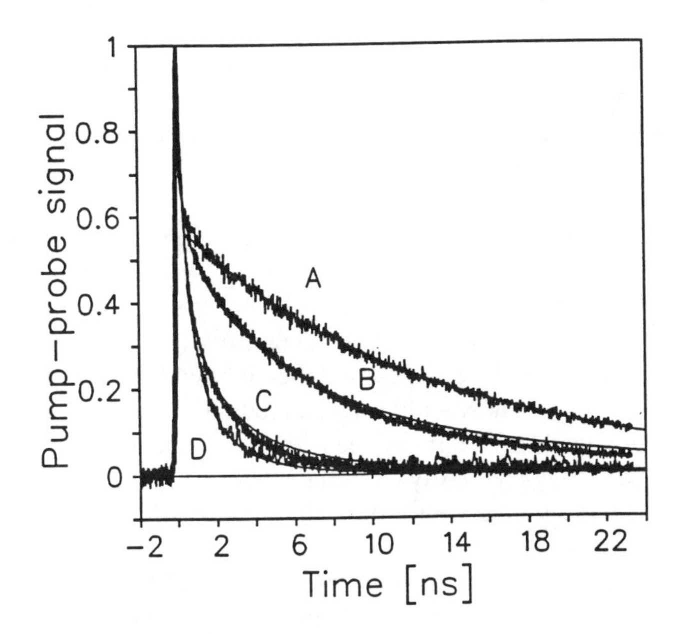

Fig. 14. The pump-probe data for liquid solution. The curves labeled *A, B, C*, and *D* have the following concentrations: 0 M, 0.04 M, 0.13 M, and 0.22 M, respectively. The peak at short time is due to a coherence artifact. The smooth lines are the fits using the back transfer parameters obtained for solid solution, R_b = 12.3 Å and a_b = 0.9 Å, and ε_r = 2.

artifact. As the acceptor concentration is increased, the signal decays much faster. The signal depends both on the forward and back transfer rates. At higher concentration the rate of forward transfer is increased (Figure 11). However, due to the distance dependence of the transfer parameters, the ion spatial distribution changes as well. With larger acceptor concentration, the ion density is shifted toward smaller ion pair separations. This larger concentration results in an increase of both forward and back transfer, causing the ions, on average, to be created closer together. In other words, the ratio of number of ions being formed at small separations to the number formed at large distances varies directly with the concentration of acceptors.

The solid lines shown in Figure 14 are calculations using the back transfer parameters measured in the solid solution, i.e., R_b = 12.3 Å and a_b = 0.9 Å. The ratios of the stimulated emission cross-section and the ES-ES cross-section to the ground state absorption cross-section were adjusted within their error bars to obtain the best fit. Like the solid solution, this improved the agreement between the data and the calculated curves, but as discussed below, did not change the back transfer parameters that gave the best agreement with the data. The values are β = 0.4 and γ = 0.8. The relative permittivity used in the calculation is ε_r = 2. This is the high frequency value [81]. As can be seen from the figure, the agreement between theory and experiment is quite good. This is particularly true considering the fact that there are no adjustable parameters in the calculation. The forward transfer parameters were obtained from the fluorescence measurements, and the back transfer parameters are those measured from the solid solution. However, the high frequency relative permittivity was employed.

Figure 15 again shows the data but with calculations using the low frequency relative permittivity [81], $\varepsilon_r = 5$; all other parameters are identical. The agreement between the data and the calculations is poor. Extensive fitting of the data using the low frequency relative permittivity was performed. R_b and a_b were varied, both independently and together, so that their ratio remained constant, over an extremely wide range, including values that are unphysical. The agreement could not be improved and became increasingly worse as the parameters were adjusted away from the values measured in solid solution. The calculated curves have fundamentally the wrong shape. Clearly the use of the low frequency relative permittivity underestimates the strength of the Coulomb interaction. When the high frequency relative permittivity is used, the calculated curves have close to the correct functional form and are able to reproduce the concentration dependence observed in the data. This suggests that the use of the high frequency relative permittivity is appropriate. Intermediate values of the relative permittivity were also used, and it was found that the data were more and more closely fitted by the calculated curves as the high frequency value was approached.

The relative permittivity has two contributions. Ultrafast polarization of the solvent electrons as well as certain restricted motions of molecular substituents will occur much faster than the time scale of the electron transfer being observed in these experiments. The second contribution involves reorientation of the solvent molecules to align their dipoles with the fields generated by the sudden formation of the ions. This will occur with essentially the rate of orientational relaxation of the solvent molecules. The orientational relaxation time, τ_r, was calculated approximately by making a molecular model, determining its volume, V, and using the Debye-Stokes-Einstein equation,

$$\tau_r = V\eta/KT, \tag{51}$$

where η is the viscosity and T is the absolute temperature. *DES* can assume a wide variety of conformations. The various conformations have different volumes, and therefore different τ_r's. From models, the volumes of various conformations were determined. The slowest τ_r is 600 psec, while the fastest is 150 psec. Typical values are in the range of 400 to 500 psec.

The pump-probe data shown in Figure 14 decays more slowly than τ_r, particularly at low acceptor concentration. However, the long decays are in part due to the rate of forward transfer. The correct comparison of time

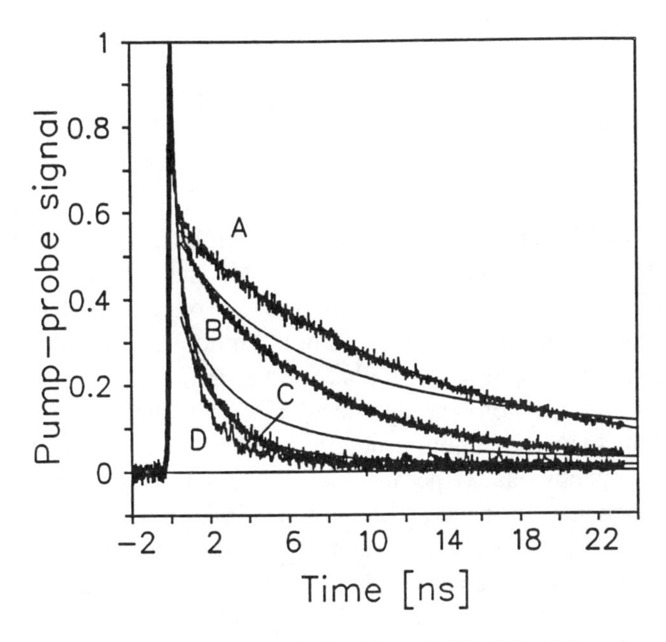

Fig. 15. The pump-probe data for liquid solution shown in Fig. **14** and fits using ε_r = 5. All the other parameters are the same as those used for the fits in Figure 14. The curves labeled *A*, *B*, *C*, and *D* have the following concentrations: 0 M, 0.04 M, 0.13 M, and 0.22 M, respectively. The smooth lines are the fits using the same back transfer parameters as those obtained for solid solution, R_b = 12.3 Å and a_b = 0.9 Å.

scales is between τ_r and the life time of ion pairs once they are created. This comparison can be made by calculating the average survival time using the excited state (Equation (21)) and reactive state survival probabilities (Equation (24)). The average survival time is obtained by calculating the probability that the ions exist at time t:

$$\langle P(t)\rangle = \frac{\int_{R_m}^{\infty} S_{\mathrm{re}}(R \mid t) 4\pi C k_f(R) \int_0^{\infty} S_{\mathrm{ex}}(R \mid t')\langle P_{\mathrm{ex}}(t')\rangle \,\mathrm{d}t'\, R^2 \,\mathrm{d}R}{\int_{R_m}^{\infty} 4\pi C k_f(R) \int_0^{\infty} S_{\mathrm{ex}}(R \mid t')\langle P_{\mathrm{ex}}(t')\rangle \,\mathrm{d}t'\, R^2 \,\mathrm{d}R}. \quad (52)$$

This probability decay curve can be defined as having a characteristic lifetime given by the time at which it has fallen to e^{-1} times its original value. This lifetime is taken to be the average survival time of the reactive state pairs. Given that a reactive state pair is formed at time t = 0, the average length of time which this pair lives before geminately recombining is the e^{-1} point of the $\langle P(t)\rangle$ curve. The value obtained from the calculation of $\langle P(t)\rangle$ is approximately 500 ps. This value is somewhat skewed to long time due to the small fraction of ions which escape recombination. This escape acts to increase the survival probability, thus increasing the lifetime of $\langle P(t) \rangle$.

This characteristic lifetime of the ions is on the same order as the orientational relaxation time τ_r. While this result is not conclusive, it suggests that the high frequency relative permittivity may give an appropriate description of the Coulomb interaction between the reactive ions in solution. The agreement between the data and calculations using the high frequency relative permittivity with *no adjustable parameters* indicates that additional aspects of the overall system should be taken into account. The orientational relaxation time, τ_r, that was calculated is for *DES* as a bulk liquid. However, the orientational relaxation time of interest is actually that of *DES* in close proximity to the *RU* cation and the *DQ* anion. These molecular ions will make a significant perturbation of the local DES liquid structure. An increase in τ_r locally would make the high frequency relative permittivity appropriate. Another consideration is the distance scale associated with the ion pair Coulomb interaction. Typical distances for the initial ion pair separations are 12 to 13 Å center-to-center. This is only 3 to 4 Å edge-to-edge. While the electric field is not strictly directed along the line connecting the molecular centers, there are very few solvent molecules participating in screening the Coulomb interaction. There will only be one or two solvent molecules between the ions. In contrast, the relative permittivity is a bulk property. For the short distances that are important in electron transfer, the molecularity of the solvent can be important [4]. The experimental results presented here are consistent with a solvent that mediates the Coulomb interaction in a manner that can be characterized by the bulk high frequency relative permittivity, most likely because of a combination of the time scale for solvent reorientation and the short distances involved.

The forward and back electron transfer parameters obtained by fitting the data yield theoretical curves that are consistent with the experimental results over a range of concentrations in both solid and liquid solutions. This demonstrates the ability of the statistical mechanical model to handle the influence of diffusion on electron transfer. The accurate description of back transfer, which involves diffusion modified by the Coulomb interaction with non-random initial conditions, is significant. While any form of the distance dependence of the transfer rate can be used in the theory, the exponential form employed here provides a reasonable description for this system as demonstrated by the agreement between theory and experiment. However, the very shortest range transfer events, which occur faster than the experimental time scale (50 to 100 psec), are not probed.

5. Concluding remarks

In this paper we have examined photoinduced forward electron transfer and back transfer (geminate recombination) in both liquid and solid solution. The focus has been on the comparison of experiments with a statistical mechanical model [69, 70] of the role of donor and acceptor diffusion on electron transfer dynamics. In this regard, the contrast between liquid and solid solutions is particularly important. Previously, theoretical treatments of the combined forward and back transfer problem have been given for solid solution [52, 53, 56, 58, 59, 70] (no particle diffusion), for diffusion in the limiting cases of very fast diffusion and very slow diffusion [30–34, 40–43, 67, 68], and for diffusion with electron transfer only at contact between donor and acceptor [44–47]. Here, the experiments and theoretical calculations addressed the more general situation of distance-dependent electron transfer with an intermediate rate of diffusion.

The forward transfer data and theory agree extremely well. There is excellent agreement between the data and theory for a variety of acceptor concentrations in liquid solution. In addition, the results in liquid solution produced almost the identical electron transfer parameters as those obtained from the solid solution experiments. The spatial averages performed to describe the solid solution system are exact. The agreement between the electron transfer parameters obtained in solid and liquid suggests that these are not arbitrary fitting parameters and that the approximations used in doing the spatial averaging with diffusion are accurate.

In the type of solid solution considered here (no acceptor self-exchange), all forward electron transfer is followed by geminate recombination since there is no possibility of ion escape by diffusion. The forward transfer parameters obtained from fluorescence measurements are used in the calculations of the pump-probe experimental data. The pump-probe data then yields the back transfer parameters. The agreement between the calculations and the data for several acceptor concentrations is very good, although not quite as good as in the case of forward transfer.

The most important aspect of this work is the investigation of geminate recombination in liquid solution following photoinduced forward electron transfer. The forward electron transfer produces a radical cation and a radical anion. Since the separation of the ions at the time of formation is small, the Coulomb interaction is significant, even in solvents with large relative permittivities [69]. In the system studied here, the solvent has a low relative permittivity, and the Coulomb attraction dominates the motion of the particles. Simple diffusion describes the particle motions for the neutral donors and acceptors involved in the forward transfer, but the diffusive motion is significantly modified by the Coulomb attraction of the ions undergoing back transfer.

Comparison between pump-probe data in liquid solution and the theoretical model, using no adjustable parameters, was found to be good when the high frequency relative permittivity was used in the calculation. The back transfer parameters were obtained from experiments on solid solutions. In the theoretical model [69], the properties of the solvent come in through the relative permittivity. The proper choice of the value to be used for the relative permittivity is complex. It is necessary to compare the solvent's orientational relaxation time to the ion survival time. In the system studied here, it was found that the average ion survival time is approximately equal to the orientational relaxation time. This indicates that the low frequency relative permittivity is not appropriate. Perturbations of the local liquid structure and the molecularity of the solvent on the small distance scale involved in electron transfer may move the appropriate value of the relative permittivity further toward the high frequency value.

The agreement between experiments and calculations and the consistency of the results obtained from liquid and solid solution data strongly support the validity of the statistical mechanical model. Therefore, experiments combined with the theory can provide considerable insight into the competition between geminate recombination and separation of ions by diffusion. In the future, examination of a variety of donor/acceptor systems and solvents will elucidate the molecular and solvent properties that control geminate recombination.

Acknowledgment

This work was supported by the Department of Energy, Office of Basic Energy Sciences (DE-FG03–84ER13251). K. W. acknowledges support from an NSF pre-doctoral fellowship.

References

1. Guarr, T. and McLendon, G., *Coordination Chem. Rev.* **68**, 1 (1985).
2. Devault, D. Q., *Rev. Biophys.* **13**, 387 (1980).
3. Popovic, Z. D., Kovacs, G. J., and Vincett, P. S., *Chem. Phys. Lett.* **116**, 405 (1985).
4. Eads, D. D., Dismer, B. G., and Fleming, G. R., *J. Chem. Phys.* **93**, 1136 (1990).
5. Domingue, R. P. and Fayer, M. D., *J. Chem. Phys* **83**, 2242 (1985).
6. Huddleston, R. K. and Miller, J. R., *J. Phys. Chem.* **86**, 200 (1982).
7. Beratan, D. N. J., *Am. Chem. Soc.* **108**, 4321 (1986).
8. Siders, P., Cave, R. J., and Marcus, R. A., *J. Chem. Phys.* **81**, 5613 (1984).
9. Zamaraev, K. I., Khairutdinov, R. F., and Miller, J. R., *Chem. Phys. Lett.* **57**, 311 (1978).
10. Strauch, S., McLendon, G., McGuire, M., and Guarr, T., *J. Phys. Chem.* **87**, 3579 (1983).
11. Dorfman, R. C. and Fayer, M. D., *J. Chem. Phys.*, (1992).
12. Najbar, J., *J. Chem. Phys.* **120**, 367 (1988).

13. Taube, H., *Electron Transfer Reactions of Complex Ions in Solution*, (Academic Press: New York, 1970).
14. Fleming, G. R., Martin, J. L., and Breton, J., *Nature* **333**, 190 (1988).
15. Kemnitz, K., Nakashima, N., and Yoshihara, K., *J. Phys. Chem.* **92**, 3915 (1988).
16. Simon, J. D. and Su, S., *J. Chem. Phys.* **87**, 7016 (1987).
17. McGuire, M. and McLendon, G., *J. Phys. Chem.* **90**, 2549 (1986).
18. Huppert, D., Ittah, V., Masad, A., and Kosower, E. M., *Chem. Phys. Lett.* **150**, 349 (1988).
19. Kemnitz, K., *Chem. Phys. Lett.* **152**, 305 (1988).
20. Siders, P. and Marcus, R. A., *J. Am. Chem. Soc.* **103**, 748 (1981).
21. Mataga, N., Kanda, Y., and Okada, T., *J. Phys. Chem.* **90**, 3880 (1986).
22. Chen, P. and Danielson, E., *J. Phys. Chem.* **92**, 3708 (1988).
23. Brunschwig, B. S., Ehrenson, S., and Sutin, N., *J. Am. Chem. Soc.* **106**, 6858 (1984).
24. Kosower, E. M. and Huppert, D., *Ann. Rev. Phys. Chem.* **37**, 127 (1986).
25. Lockhart, D. J., Goldstein, R. F., and Boxer, S. G., *J. Chem. Phys.* **89**, 1408 (1988).
26. Kestner, N. R., Logan, J., and Jortner, J., *J. Phys. Chem.* **78**, 2148 (1974).
27. Yan, Y. J., Sparpaglione, M., and Mukamel, S., *J. Phys. Chem.* **92**, 4842 (1988).
28. Sparglione, M. and Mukamel, S., *J. Chem. Phys.* **88**, 3263 (1988).
29. McConnell, H., *J. Chem. Phys.* **35**, 508 (1961).
30. Agrest, M. M., Kilin, S. F., Rikenglaz, M. M., and Rozman, I. M., *Opt. Spectrosc.* **27**, 514 (1969).
31. Kusba, J. and Sipp, B., *Chem. Phys.* **124**, 223 (1988).
32. Sipp, B. and Voltz, R., *J. Chem. Phys* **79**, 434 (1983).
33. Sipp, B. and Voltz, R., *J. Chem. Phys* **83**, 157 (1985).
34. Steinberg, I. Z. and Katchalski, E., *J. Chem. Phys.* **48**, 2404 (1968).
35. Gordon, R. G., *J. Chem. Phys* **44**, 1830 (1965).
36. Uhlenbeck, G. E. and Ornstein, L. S., *Phys. Rev.* **36**, 823 (1930).
37. Cichocki, B. and Felderhof, B. U., *J. Chem. Phys.* **89**, 1049 (1988).
38. Smoluchowski, M. V., *Z. Physik. Chem. (Leipzig)* (1918), **92** 129.
39. Crank, J., *The Mathematics of Diffusion*, (Oxford University Press: London, 1956).
40. Baumann, J. and Fayer, M. D., *J. Chem. Phys.* **85**, 4087 (1986).
41. Yokota, M. and Tanimoto, O., *J. Phys. Soc. Japan* **22**, 779 (1967).
42. Allinger, K. and Blumen, A., *J. Chem. Phys.* **72**, 4608 (1980).
43. Allinger, K. and Blumen, A., *J. Chem. Phys.* **75**, 2762 (1981).
44. Schulten, Z. and Schulten, K., *J. Chem. Phys.* **66**, 4616 (1977).
45. Werner, H. J., Schulten, Z., and Schulten, K., *J. Chem. Phys.* **67**, 646 (1977).
46. Rice, S. A., *Diffusion-Limited Reactions*, (Elsevier: Amsterdam, 1985).
47. Agmon, N. and Szabo, A., *J. Chem. Phys.* **92**, 5270 (1990).
48. Onuchic, J. N., Beratan, D. N., and Hopfield, J. J., *J. Phys. Chem.* **90**, 3707 (1986).
49. Sutin, N., *Acc. Chem. Res.* **15**, 275 (1982).
50. Inokuti, M. and Hirayama, F., *J. Chem. Phys.* **43**, 1978 (1965).
51. Miller, J. R., Beitz, J. V., and Huddleston, R. K., *J. Am. Chem. Soc.* **106**, 5057 (1984).
52. Tachiya, M. and Mozumder, A., *Chem. Phys. Lett.* **28**, 87 (1974).
53. Tachiya, M., *J. Chem. Soc., Faraday Trans. II* **75**, 271 (1979).
54. Blumen, A. and Manz, J., *J. Chem. Phys.* **71**, 4694 (1979).
55. Blumen, A., *J. Chem. Phys.* **72**, 2632 (1980).
56. Lin, Y., Dorfman, R. C., and Fayer, M. D., *J. Chem. Phys.* **90**, 159 (1989).
57. Dorfman, R. C., *Doctoral Thesis, Department of Chemistry*, Stanford University, 1992.
58. Chandrasekhar, S., *Rev. Mod. Phys.* **15**, 1 (1943).
59. Mikhelashvili, M. S., Feitelson, J., and Dodu, M., *Chem. Phys. Lett.* **171**, 575 (1990).
60. Onsager, L., *Phys. Rev.* **54**, 554 (1938).
61. Debeye, P., *J. Electrochem. Soc.* **82**, 265 (1942).
62. Song, L., Dorfman, R. C., Swallen, S. F., and Fayer, M. D., *J. Phys. Chem.* **95**, 3454 (1991).

63. Chatterjee, P., Kamioka, K., Batteas, J. D., and Webber, S. E., *J. Phys. Chem.* **95**, 960 (1991).
64. Najbar, J. and Turek, A. M., *Chem. Phys.* **142**, 35 (1990).
65. Tachiya, M., *Radiat. Phys. Chem.* **21**, 167 (1983).
66. Szabo, A., Zwanzig, R., and Agmon, N., *Phys. Rev. Lett.* **61**, 2496 (1988).
67. Rabinovich, S. and Agmon, N., *Chem. Phys.* **148**, 11 (1990).
68. Agmon, N., *J. Chem. Phys.* **90**, 3765 (1989).
69. Dorfman, R. C., Tachiya, M., and Fayer, M. D., *Chem. Phys. Lett.* **179**, 152 (1991).
70. Dorfman, R. C., Lin, Y., and Fayer, M. D., *J. Phys. Chem.* **94**, 8007 (1990).
71. Press, W. H., Flannery, B. P., Teukolsky, S. A., and Vetterling, W. T., *Numerical Recipes in C*, (Cambridge University Press: Cambridge, 1988).
72. Song, L., Swallen, S. F., Dorfman, R. C., Weidemaier, K., and Fayer, M. D., *J. Phys. Chem.* **97**, 1374 (1993).
73. Stein, A., Peterson, K. A., and Fayer, M. D., *J. Chem. Phys.* **92**, 5622 (1990).
74. Kennard, O., Watson, D. G., and Rodgers, J. R., *Crystal Data Determinative Tables, 3rd ed.*, (U. S. Department of Commerce, National Bureau of Standards, and the JCPDS-International Center for Diffraction Data: 1978).
75. Birks, J. B., *Photophysics of Aromatic Molecules*, (Wiley-Interscience: London, 1970).
76. Shoemaker, D. P., Garland, C. W., Steinfeld, J. I., and Nibler, J. W., *Experiments in Physical Chemistry*, 4th edition, (McGraw-Hill Book Company: New York, 1981).
77. Lohmannsroben, H. G., *App. Phys. B.* **47**, 195 (1988).
78. Song, L. and Fayer, M. D., *J. Luminescence* **50**, 75 (1990).
79. Yee, W. A., Kuzmin, V. A., Kliger, D. S., Hammond, G. S., and Twarowski, A. J., *J. Am. Chem. Soc.* **101**, 5104 (1979).
80. Shida, T., *Electronic Absorption Spectra of Radical Ions*, (Elsevier: Amsterdam, 1988).
81. The high-frequency dielectric constant was obtained by calculating the square of the high frequency index of refraction. The index of refraction was obtained from the Aldrich Chemicals catalog as well as measured using an Abee refractometer. The two values agreed to within error. The low frequency dielectric constant was measured using a capacitance bridge at 1 KHz and 120 Hz. It was found to be constant within this range.

3. Ultrafast Electron and Proton Reactivity in Molecular Liquids

Y. GAUDUEL
Laboratoire d'Optique Appliquée, CNRS URA 1406, INSERM U275, Ecole Polytechnique, ENS Techniques Avancées, 91120, Palaiseau, France

General Introduction

The understanding of chemical reactions at the microscopic level correspond to fundamental challenges for the physical-chemistry and the biophysics. The elementary steps of a reaction often correspond to very short lived transitional states between reactants and products [1–8]. At the molecular level, the lifetime of transient steps ($t \leq 10^{-12}$ s) correspond to angström or subangström displacements.

Reaction dynamics in dissipative condensed phase media (rare gas solutions and molecular liquids) may be observable on the time scale of molecular motions i.e. on a subpicosecond time scale. In liquids, the time scale for reactions can occur through several order of magnitude and ultrafast intramolecular electron transfer can be determinant for the behavior of intra or intermolecular energy transfers at longer time. At the molecular level, vibrational and rotational motions correspond to the fundamental parameters which influence reaction dynamics through chemical bonds. One of the main fundamental question on chemical reaction in molecular environment of polar on non polar liquids is to understand and explain how environment assist or impede a reaction. A key point to understand concerns the role of the microscopic dynamics of the solvent during charge transfer reactions. For instance, regarding redox reactions in molecular liquids there has been a great deal of discussion over the meaning of frequency factor and the relative role of the collisional and dielectric components solvent friction [9–14]. The role of solvent frictions in the definition of frequency factors of reactions encompass the development of high time resolution photochemical studies and computational solution chemistry.

The understanding of reaction dynamics require to observe in real time the nature of transitions states by using high-time resolution technics. Ultrashort optical pulses offer the opportunity of a direct discrimination of the short lived nonequilibrium configurations using different spectroscopic methods [15, 16]. The recent technological advances in the generation, measurement

J.D. Simon (ed.), Ultrafast Dynamics of Chemical Systems, 81–136.

of ultrashort laser pulses and high-time resolution spectroscopy have allow to push investigations of chemical reaction down to the picosecond and even the femtosecond time scale [17–20]. Consequently, direct informations can be obtained on the dynamics of ultrafast radical reactions linked to charge transfer, intramolecular and intra-ionic processes. The experimental works performed in the femtosecond time domain can mixe different spectroscopic technics involving the frequency domain (fluorescence, hole burning, four wave mixing, polarization) Raman spectroscopy, or different versions of the excited-probe configurations. The best time resolutions are in the range 30–100 fs.

During the last ten years, significant advances in the understanding of chemical dynamic have been obtained. These advances encompass simultaneous progresses in experimental works and computational chemistry. Time-resolved measurements permit to obtain detailed informations on chemical reaction dynamics. The comparison of experimental data with the predictions of suitable classical, semi-classical or quantum mechanical theories become possible and constructive. The performances of different generations of computer with vector or parallel architectures are not stange to the development of theoretical chemistry in condensed phases including Monte Carlo (MC) or Molecular Dynamics (MD), quantum path integral MC (QPIMC), quantum path integral MD (QPIMD) or splitting operator method (SOM). The theoretical works involve structural computations of liquid or dynamical comportment of solvent molecules during solvation reactions [21–28].

The objective of this chapter is to focus on recent experimental works of chemical reactions in liquid phase. We restrict our considerations to chemical reaction dynamics in molecular liquids and more particularly in polar protic liquids. The description of primary events involving ultrafast coupling between single charges and molecular liquids can help to extend our knowledge on (i) physical picture of transition states, (ii) chemical reactions at the microscopic level, (iii) the relationship between nonequilibrium configurations and the statistical physics of protic solvents.

The manuscript is organized as follow: the first section deals with considerations on recent developments of time-resolved electron relaxation and solvation dynamics in polar liquids. The section II will cover investigation of ultrafast electron transfers (electron attachment and reactivity of nonequilibrium electronic states) considering the influence of solvation dynamics. In the section III we will center our attention on ultrafast reactions which involve concerted electron-proton transfer reactions. The last part of this chapter will be devoted to the conclusions.

1. Electron Relaxation in Molecular Liquids

1.A. Solvation Dynamics and Chemical Reactions

The understanding of solvent effects on charge transfer reactions require to obtain detailed informations on the solvation dynamics of single charge or excited states of molecules. The importance of dynamical solvent effect on the rate of charge transfer reactions is particularly evident for small activation barrier reactions: activationless, solvent-controlled fast intramolecular electron transfer for which the free energy (ΔG) of the reaction is small compared to kT [8, 27]. The solvent part is dominant in the contribution of the free activation energy. When the coupling between reactants and products is weak, the energy profiles for the electronic wave functions of the initial and final states $(\Psi_i(\Psi_f))$ cross in single zone. That means that the equalization of the energies for the reactants and the products are occasional and due to solvent fluctuations (non-adiabatic coupling). In this limit case, it is now well established that there is no involvment of solvent dynamics in the rate determining step. The other limit corresponds to the adiabatic case for which the coupling between the reactants and products are strong; the reaction zone is large and defines a single potential energy surface. Most of the classical theories predict that the rate constant k_{et} is proportional to the inverse of the longitudinal time (T_L^{-1}) or the experimental solvation time (T_{obs}^{-1}) [9, 14, 29–31].

Electron transfer kinetics in solutions have often been analyzed and interpreted in the frame work of the general adiabatic theory of Marcus [3]. Although electron-transfer dynamics are not always characterized by a classical rate constant [32], a general formulation of the chemical reaction concerns the rate constant k which can be expressed as:

$$k = \nu_{\text{eff}} K_{\text{el}} \Gamma_n \exp(-\Delta G/k_B t), \tag{1}$$

where ν_{eff} is the effective frequency for motion along the reaction coordinates, K_{el} the electronic transmission factor, Γ_n the nuclear tunelling factor, ΔG the free energy of activation, k_B the Boltzman constant and T the temperature [8].

The dynamical solvent effects involve friction parameters (collisional or dielectric origin) which appear in the three frequency prefactors of equation. The electronic transmission coefficient K_{el} can be expressed from the probability for a transition from the initial state to the final state diabatic energy surface through the crossing region:

$$K_{\text{el}} = 2P_0/(1 + P_0) \tag{2}$$

with $P_0 = 1 - \exp(-\pi\gamma)$: the surface hopping probability. The non-adiabatic limit is characterized by the following expression of K_{el} [3, 27]:

$$k_{\text{el}} = 2 \mid H_{\text{if}} \mid^2 \; \pi^{3/2}/h\nu_{\text{eff}} k_B T E_\lambda^{1/2} \tag{3}$$

in which H_{if} is the transfer integral and E_{γ} the reorganization energy.

The investigations on solvent effects during chemical reactions and especially charge transfer processes have been mainly directed toward the study of the time-dependent response of a polar protic or aprotic solvents to a transient change of charge distribution of solute or elementary charges. The discrimination of time-resolved emission spectrum of a molecular probe permits to analyze the relaxation of non-equilibrium configurations of solute/solvent mixtures [9, 13, 18, 33, 34]. The studies on the time dependent fluorescence shift (TDFS) of an excited molecular probe, give significant informations on the dynamics of solvent cage formation. Molecular probes are chosen when the dipolar moment of the excited state is higher than the fundamental state S_0. Consequently, during the solvation process of the solute, the molecules of solvent reorganize around the new dipole moment. The emission spectrum of the probe shifts to higher wavelengths during this polarization induced solvent cage reorganization.

The first classical models used to analyze the experimental correlation function $C(t)$ have been developed considering the macroscopic properties of the solvent. In the classical models, the solute is equivalent to a point dipole and the solvent is characterized by a dielectric continuum with the frequency dependent complex dielectric constant [35]. The potential energy linked to polar interactions $E(t)$ is function of the time dependent reaction field $R(t)$ and the dipole moment of the solute $\mu(t)$:

$$E(t) = R(t)\mu(t). \qquad (4)$$

The nonequilibrium transition between the ground and excited states of the solute correspond to measurable fluorescence Stokes shift dynamics. The determination of $R(t)$ is obtained by Fourier transform of the frequency domain reaction field $R(\omega)$.

The simple continuum model predicts an analogy between the longitudinal relaxation time of the solvent and the correlation function $C(t)$ [9, 18]. However, the predictive analogy between the correlation function $C(t)$ and the monoexponential relaxation time of the solvent is not often confirmed. Many experimental data have demonstrated that the dynamics of solvation of molecular probes (T_{obs}) is more complex and can deviate significantly from the longitudinal time T_L [10, 18, 29]. The discrepancy between macroscopic predictions of the continuum theory and picosecond or femtosecond experimental data may be due to several factors: the inhomogeneity of the electronic excitation of the probe, the molecular shape of the probe (molecular factor); the local dielectric saturation of the solvent [13, 36]; the cooperative effects of the solvation response, the specific solute-solvent interactions involving the local hydrogen bonding lattice.

More refined approaches on the solvation dynamics of molecular probes have been developed taking into account some molecular aspects of the Debye solvent and particularly the inhomogeneous character of the medium. In this way, the generalization of the continuum theory by the Mean Spherical Approximation treatment (MSA), have been developed by Wolynes and predicts the existence of a hierarchy of solvation times [37–39]. This hierarchy of solvation times is due to the influence of the electric field in the vicinity of the molecular probe on the dynamical response of solvent molecules. In the vicinity of the molecular probe, the solvent cannot be seen as a continuum because the relaxation is more relevant of a single particule time (T_D). The relaxation process of the solvent my be due to more complexes motions such as translational motion of the solvent dipoles [36], solvent viscoelasticity or non markovian effects [39]. This later case has been considered when the ratio 'P' ($P = (D \cdot T_D/a^2)$) involving the solvent self diffusion coefficient (D), the solute radius 'a' and the Debye time (T_D) is greater than 1.

To analyse solvation effects on charge transfer reactions, computational Molecular Dynamics simulations have been performed for polar solvents (water, alcohols) and apolar media (acetonitrile) [40–42]. Most of these MD simulations consider the linear response theory (LR): the non equilibrium response of the solvent around the molecular probe which undergoes a large variation of its charge distribution can be predicted from equilibrium dynamics of the solvent in the vicinity of the ground state of the probe. This linear response breaks down when specific molecular motions are linked to the early solvation step. For instance, in alcohols, the MD simulation of molecular probes solvation need to consider complex solvent responses such as the predominant rôle of O-H motions.

1.B. Electron Relaxation in Neat Liquid Water

Let us now focus on the fundamental chemical problem which concerns the transfer of elementary charges in liquids. From a more general point of view, the femtosecond spectroscopy of non-equilibrium states of elementary charges (electron, proton) in polar protic liquids permit to investigate primary steps of charge transfer: formation of the hydration cage around an electron, encounter pair formation, ion-molecule reaction, electron attachment to solvent molecule, early electron-ion pair recombination.

Dynamical comportment of excess subexcitation electron in polar liquids (alcohols, water), is of particular interest because this elementary charge implicates fascinating issues on its coupling with the solvent molecule (Figure 1). Bound states exist only through the interaction with the surrounding medium. Since the discovery of the hydrated electron by Hart and Boag [43],

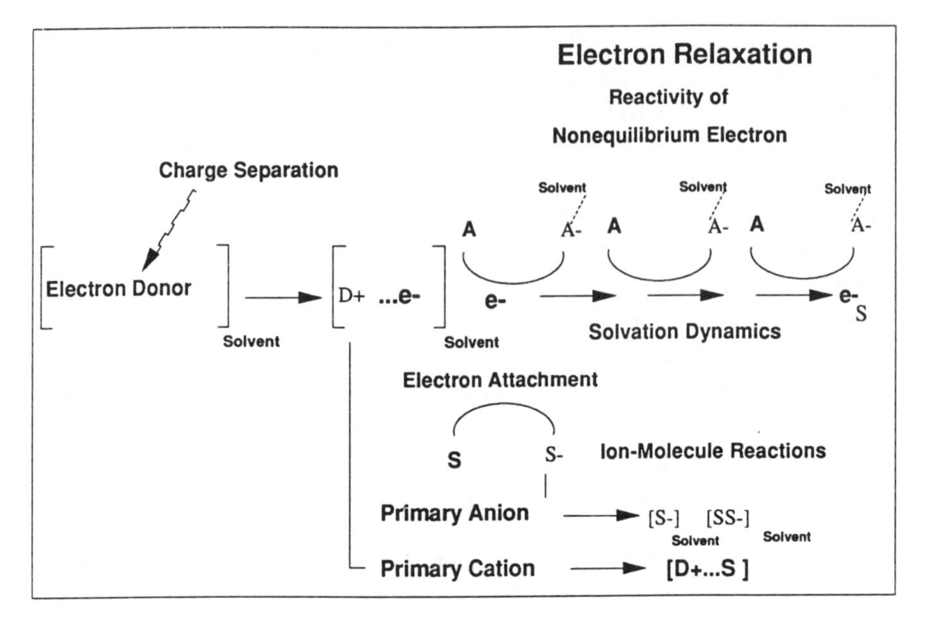

Fig. 1. Sequential events of ultrafast one-electron transfer processes in molecular liquids.

issues concerning the mechanisms involved in the electron-water molecules interactions have remained unresolved during two decades [44–48].

In polar and nopolar liquids, the electron can be used as a microprobe to test the local structure of the environment through investigation of dynamical electron-molecules couplings [49]. Indeed, when an excess electron is injected in a polar solvent, the following early steps (thermalization, localization, trapping and solvation) are expected to include contributions for charge induced polarization and/or preexisting configurational order of the liquid. Concomitantly to the experiments on solvation in water, structural models of molecular liquids have been developed by statistical mechanisms or molecular dynamics computer simulation. Numerous theoretical works on electron localization and solvation in polar bulk have used dielectric continuum or semicontinuum models and more recently semi-quantum approaches [50–55].

The rôle of water molecules in charge transfer reactions can be linked to the microdynamic properties of this polar protic liquid: change in the hydrogen bound networks, existence of clathrate-like holes. In this way, the chemical and physical properties of liquid water would have a determinant influence on charge transfer processes and the quantum nature of the hydrated electron is of great interest to investigate electron transfer reaction at the submolecular level [56–58].

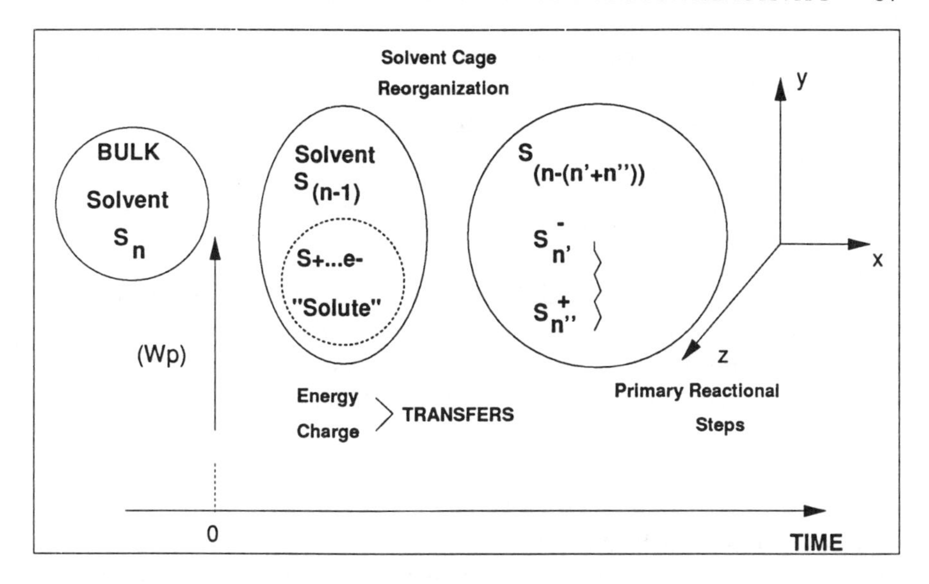

Fig. 2. Representation of spatio-temporal events linked to energy transfers and charge-solvent interactions in neat polar or non polar liquids.

The femtosecond spectroscopic investigations of the early events in liquid water (physical and chemical stages) permit to obtain unique informations on ultrafast reactions which occur at the temporal shell of molecular motions. The discrimination of time-dependence of electron-water molecules coupling offer the opportunity to analyze the transient fluctuations of the liquid configurations around the spatial distribution of charge (Figure 2). In this way, the single charge can be used as a microprobe of the local structures which fluctuate in the librational or vibrational time domain.

The time scale of physical and chemical events involved in the absorption of energy by water molecules extends from 10^{-16} s (formation of excited or ionized state of water) to 10^{-12} s (thermal orientation of water molecules) and to 10^{-7} s (formation of molecular products). Considering the specific case of water molecules, the energy exchanges which follow the initial energy deposition result in the formation of several intermediate states including quasi-free or dry electron (e^-_{qt}), thermalized electron (e^-_{th}), localized electron (e^-_{loc} or e^-_{trap}), hydrated electron (e^-_{hyd}) and prototropic species (H_3O^+, OH). Several non-equilibrium states of electron has been recently identified in the femtosecond regim.

The direct photoionization of water molecules by short laser pulses is possible in specific conditions because light can be absorbed through a nonlinear process. Indeed, neutral water molecules can absorb high intensity light pulses ($I > 10^{12}$ W cm^{-2}) by two photons process [59]. In the UV spectral region (190–310 nm) where water does not absorb a photon, the two absorption coefficient becomes non negligible when dealing with multigigawatt peak power pulses: 4×10^{-13} m W^{-1} at 310 nm. In the case of a wave plane propagation through the sample, a two photons absorption process can be expressed by the equation:

$$\frac{\partial I}{\partial x} + \left(\frac{1}{\nu}\right) \frac{\partial I}{\partial t} = -BI^2 \tag{5}$$

with I the radiation intensity, B the two photons absorption coefficient and ν the light velocity in the medium. The absorbed energy – incident energy ratio (E_{abs}/E_0) permit to determine the two photons absorption coefficient. For instance, in the case of a gaussian shape and for a small optical path length, B is expressed as follow:

$$E_{\text{abs}}/E_0 = BE_0 \cdot l/2\pi^{(1/2)}\tau_p S. \tag{6}$$

In pure liquid water, regarding the physical reasons for the existence of complex photochemical channels we must consider the photophysics of water molecules. Although several uncertainties remain about the estimate of the ionization potential of water molecules in liquid phase $I_{\text{Liq}} = IP_g + P^+ + V_0$ with IP_g the ionization potential in the gaseous phase, P^+ the adiabatic electronic polarization of the medium by the positive ion (H_2O^+) and V_0 the conduction band edge energy of the solvent [60]. The existence of a low dissociation channel for H_2O molecules has been suspected to compete with the direct ionization of the $A^{\sim}$ state ($1b_1 \rightarrow 3sa_1$ for instance) [61, 62]. Theoretically, the estimate of the energy required to generate electron through vertical ionization (Born-Oppenheimer) is around 8.5 eV within the uncertainty about the value of V_0 [60, 63]. Experimentally several works have suggested that the threshold energy for the formation of photogenerated hydrated electron through one or two photons will be around 6.5 eV [59, 61]. In our femtosecond photochemical experiments which use ultraviolet pulse (E = 4eV) and high energy density, the two photons absorption is above the estimate of the ionization potential. This leads to the generation of subexcitation electron with excess energy of about 1.5 eV. Of course, the important point to discuss concerns the ionization process which can occur at energy lower than the vertical ionization threshold.

During the interaction of ionizing radiation with an aqueous phase, the absorption of energy initiates the ionization process. This step is followed by energy exchanges between excess electron and the molecules of solvent [64]. The energy exchange inside a local region leads to complex couplings between the elementary charge or water cation and the protic solvent (Figures 2 and 3).

The ultraviolet multiphotonic ionization of water molecules leads to the formation of several transient species including prototropic and electronic

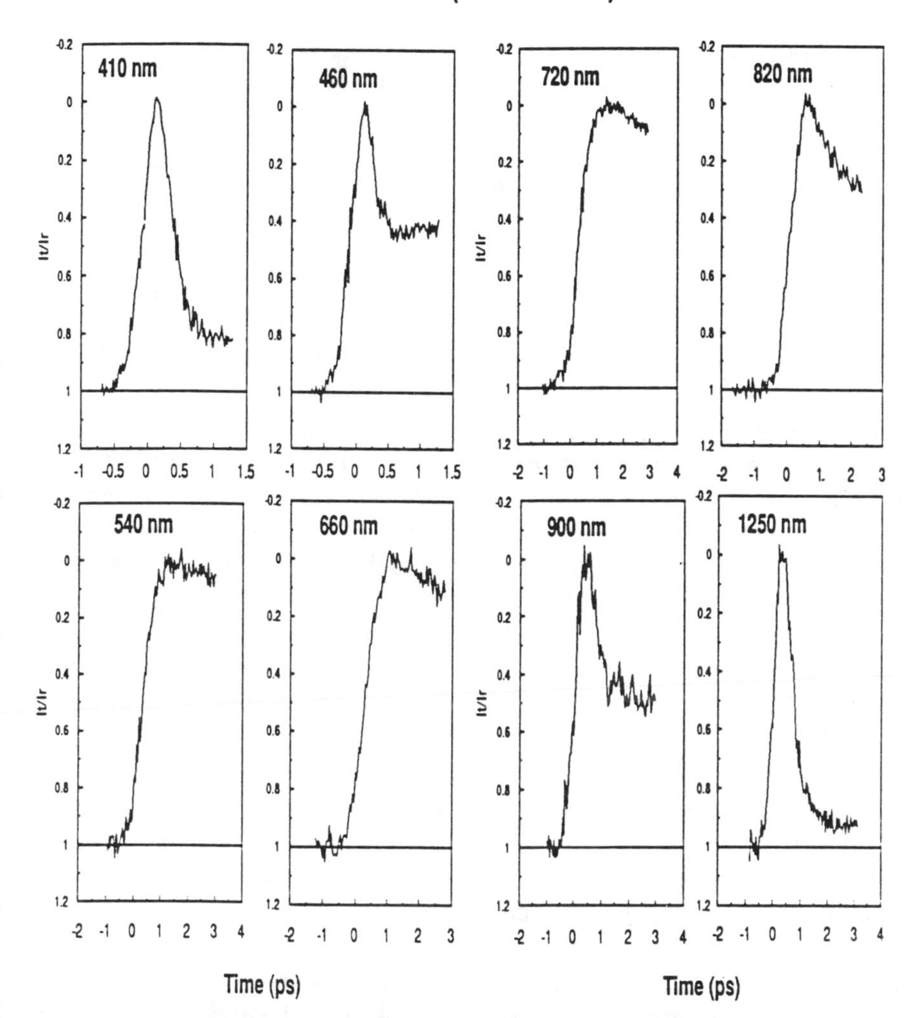

Fig. 3. Set of time-resolved absorption data for different test wavelengths following the femtosecond photoionization of water molecules by ultraviolet pulses ($E_{Excit.}$ 4 eV, 310 nm) at 294 K. In the red spectral region the absorption signal is due to the fully hydrated state of the electron. The transient infrared signal represents the non equilibrium state of localized electron (e_{ir}). In the near ultraviolet (410 nm), the transient short lived component is assigned to the relaxation of the water cation (X_2O^+, X = H,D) through an ion molecule reaction (reaction 8).

entities (reactions 7–9).

$$X_2O + (h\nu\ 310\ \text{nm}) - k_1 \rightarrow [X_2O]^* \rightarrow X_2O^+ + e^-_{qf}$$
$$X = \text{H, D}\ (k_1 = 1/T_1) \qquad (7)$$

$$X_2O^+ + X_2O - k_2 \rightarrow X_3O^+ + OX \quad (k_2 = 1/T_2) \qquad (8)$$

$$e^-_{qf}n + X_2O - k_3 \rightarrow e^-_{IR} - k_4 \rightarrow e^-_{VIS} \quad (k_3 = 1/T_3; k_4 = 1/T_4). \quad (9)$$

Femtosecond spectroscopic investigations in the visible and infrared have permitted to demonstrate that electron hydration in pure liquid water proceeds through at least one intermediate state (localized or prehydrated electron) whose the lifetime is in the femtosecond regime [65]. One non-equilibrium electronic configuration is characterized by an infrared absorption band extending above 1250 nm. This infrared band is build up with a rate constant of 9×10^{12} s^{-1} in H_2O. Its relaxation follows a monoexponential law whose the rate constant ($K_4 = 1/T_4 = 4 \times 10^{12}$ s^{-1}) is similar to the appearance of an absorption band in the visible spectral region (Figure 4). In agreement with previous pulse radiolysis experiments, this long-lived state exhibits a maximum optical transition at 1.7 eV and a 0.8 eV bandwith [66a]. The very fast appearance of e^-_{prehyd} is at a time comparable to any nuclear motion, solvent dipole orientation or thermal motion of water molecules and its implies that efficient mechanisms involved in the localization process do not require large molecular and dynamical reorganization. The ultra-short lived infrared tail has been assigned to the existence of a prehydrated electronic state [65] and its relaxation to an internal transition (non adiabatic transition) of the excited hydrated electron toward a fully relaxed state [67–69]. Moreover, the absence of a significant continuous shift between the infrared and the visible bands demonstrates that the relaxation of water molecules in the vicinity of excess electron involves extremely small water motions.

The relaxation of the transient infrared electronic state would involved limited molecular reorganization such as librational and/or OH vibrational motions. We have proposed that this precursor is a state where the electron is still spatially extended [65]. Is it already a thermalized state? This would imply that the excess kinetics energy of the electron has been transfered to the water molecules in less than the risetime T_1 of the infrared spectrum. Such ultrafast thermalization is quite possible if we take into account the existing numerous mode of vibration of the solvent and particularly the most energetic vibrational mode OH (antisymetric stretch). Previous theoretical works have yeld thermalization time of electron in liquid water to be in the range $T_{th} = 2.4–4 \times 10^{-14}$ s [70]. The measured initial electron trapping time (110 fs) is longer than these estimates. This would suggest that if the early energy loss rate of electron in water is about 10^{13} eV s^{-1}, electron would get thermalized before being localized. In the appearance of a prehydrated state (e^-_{prehyd}), the efficient role of shallow traps i.e. ($V_0 = 0.58$ eV) for a cross section around 20 Å^2 would correspond to a spatially extended electron-trap of about 4 Å. This size is larger than the radius of the fully hydrated electron (r = 2.3–2.8 Å) [64]. These first femtosecond experiments cannot permit to establish if during the trapping time electron creates its own trapping site

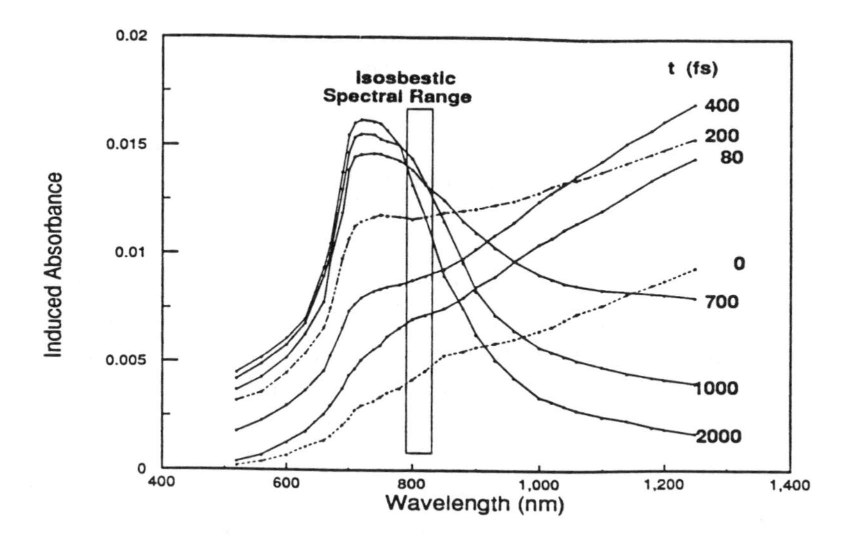

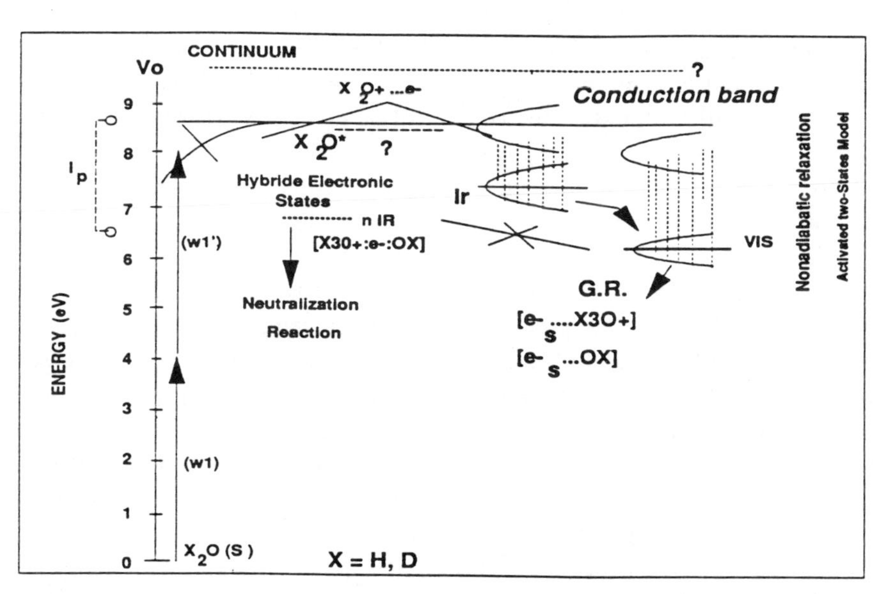

Fig. 4. A: Time dependence of absorption spectra of different configurations of excess electron in pure liquid water at 294 K. An *isosbestic spectral range* can be defined from these data. B: Energy diagram showing the different photochemical channels involved the photoejection of low energy electron following the femtosecond ultraviolet excitation of a neat aqueous phase.

(self trapping mechanism) or searchs for pre-existing shallow traps identified as small structural fluctuation or miniclusters. The important question was to precise whether localized electron are self trapped entities or their localization is favoured by pre-existing trapping sites present in the liquid.

Additional spectroscopic investigations performed in heavy water ($D_2O >$ 99.95%) have permitted to observe that the trapping time ratio between D_2O and H_2O ($T_{1(D_2O)}T_{1(H_2O)}$) does not exceed 1.09 [71]. This value is lower than (i) the ratio of energetic vibrational mode (OH vs OD) which is $\sqrt{2}$ times greater in H_2O than in D_2O [72], (ii) the estimate of the ratio of the energy loss rate [50]. If the rate of energy loss for the photoejected election during the thermalization and trapping steps is dependent on the coupling with the most energetic vibrational mode of the OX bond (antisymetric stretch), the time necessary for complete energy dissipation during localization would increase in the same proportion as the ratio of energetic vibrational mode (OH/OD). As the H/D isotope effect on trapping time is lower than this value ($\sqrt{2}$), we have concluded that the thermalization step in D_2O remains very fast and is not the limiting factor in the prehydration step. The monoexponential relaxation dynamics of the infrared electronic state remains largely independent of the physical properties of the polar solvent. Moreover, there is a discrepancy between the prediction of the Debye theory and the experimental data [65, 71]. The absence (i) of significant H/D effect on the relaxation dynamics of the infrared electronic state, (ii) of a continuous shift between the long wavelength tail (infrared trapped electron: e_{prehyd}) and the broad visible band due to fully hydrated state (e_{hyd}) support the conclusion that the epithermal electron reaches its hydration state without a dominant dielectric response of the water molecules (Figure 4). Accompanying femtosecond investigations on electron solvation dynamics, there have been intense computed simulations of excess electron in clusters and bulk for which the excess electron is described as a quantum particle and water molecules are treated classically with rigid or flexible molecular models: SPC, ST_2, MCY, T_1P_4P, RWK2 models [54, 73–75]. The electron-water interaction is modelized by a pseudopotential which includes Coulomb polarization, exclusion and exchange contributions [53–55, 74, 76]. In liquid water, quantum statistical studies have permitted considerable progress on nonequilibrium electronic states, calculation of eigenstates, eigenwaves, transient optical spectroscopy and relaxation dynamics of electron in aqueous phases. In neat liquid water (bulk), computed simulations predict the existence of initial pre-existing trapping sites which are arising from statistical fluctuations and molecular clustering. In particular, these computer simulations predict a repartition of electronic eigenstates for an excess electron in the absence of solvent reorganization. The density of shallow traps is more favorable for the initial excess electron localization in liquid water. These sites are due to fluctuations of the electronic density or solvent molecule orientation and correspond to a range of potential well depth extending from 0 to –32.2 $Kcal.mol^{-1}$ [53, 74]. The distribution of pre-existing sites behave monotonically in energy down to –1.4 eV. If we take into account the estimates on the distribution of statistical fluctuations from molecular dynamics simulation, this does not mean that all the solvent configuration intervene in an ultrafast electron localization.

Computational transient spectroscopy of electronic eigenstates in the infrared are compatible with experimental femtosecond spectral data at short time ($t < 200$ fs). However, the adiabatic quantum simulation of the electron solvation in pure liquid water (ground state dynamics via adiabatic transitions) predict solvation times which are faster than experimental data: an initial localization with energy drop (~ 2.5 eV) occurs in 30 fs through continuous spectral shift and narrowing for initial step; A second component would occur in 200 fs and correspond to heat dissipation and translational reordering of the medium [74]. Similar computed dynamical characteristic are obtained for adiabatic electron solvation in water clusters [75]. Finally, the theoretical adiabatic simulations of electron trapping and solvation in pure water suggest that different populations of non-equilibrium electronic configurations (prehydrated states, excited solvated state, trapped electron by pre-existing deep traps) can contribute to the infrared signal before the electron gets its equilibrium state (hydrated state).

In pure liquid water, the physical view of the electron solvation as an internal conversion process (non-adiabatic transition of an excited hydrated state) has been considered by developing computational quantum simulations of nonadiabatic electronic transient relaxations. Direct comparisons between experimental data and theoretical predictions would permit to obtain detailed understandings of fast electron-water molecules couplings in liquid phase. Femtosecond photochemical investigations performed in pure liquid water at room temperature have permitted to discriminate the existence of additionnal non-equilibrium electronic configurations during the electron solvation steps (Figure 4). These experimental works concern the detailed mechanisms of an electron relaxation in a polar protic solvent and more particularly the influence of energy transfer processes in the vicinity of prototropic species. This important aspect of electron relaxation in polar protic liquid and its physical meaning at the microscopic level will be discussed at length in Section 3.

1.C. Solvation Dynamics in Ionic Solutions

1. Charge Separation and Ionic Strength Effects

The investigation of ionic aqueous solutions by flash photolysis have allowed the understanding of excess electron solvation in an ionic environment and the influence of ionic strength on the bimolecular rate constant of reactions [66b, 77].

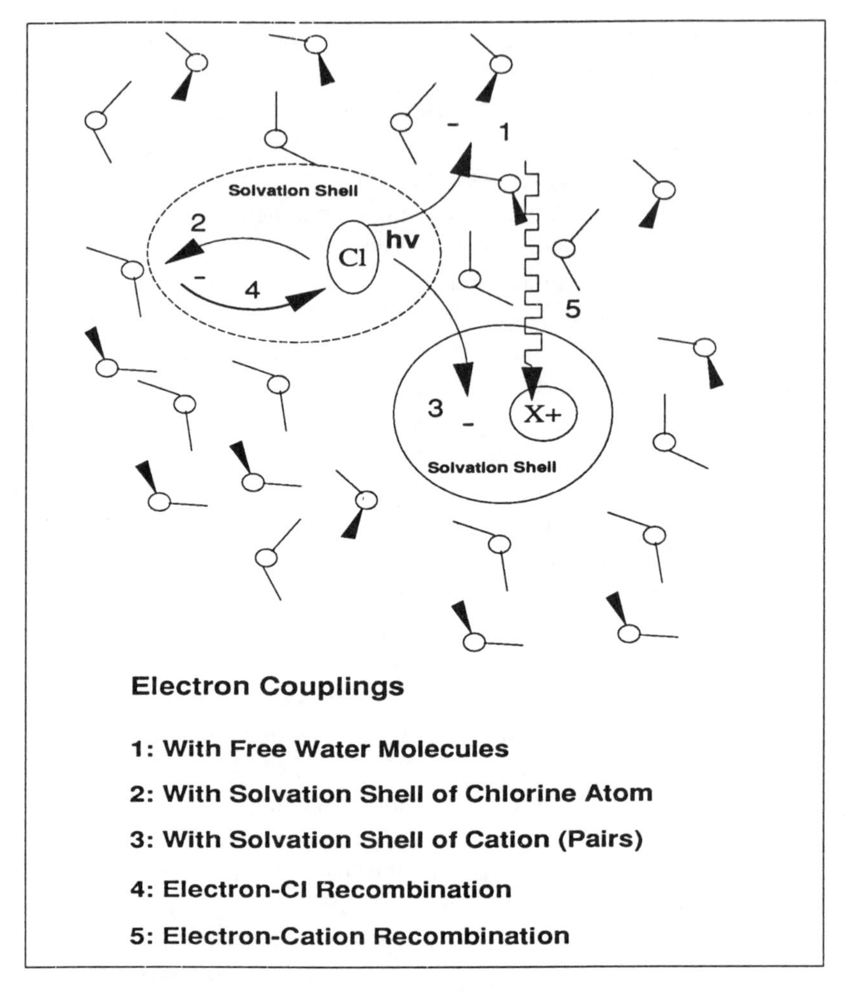

Fig. 5. Representation of different ultrafast electron relaxation channels and early recombination reactions in ionic aqueous solution of XCl (X = Na, Li, H). For explanations see the text.

The femtosecond photochemistry of ionic aqueous solutions have been recently developed to investigate charge transfer dynamics in function of the ionic strength and the nature of counterions [46, 78, 79]. The photodetachment of subexcitation electron from a halide, can been achieved by pumping with ultraviolet pulses via a charge transfer to solvent (CTTS) i.e. through a highly excited state of the anion. This transient state dissociates to give an epithermal electron and a chlorine atom. Considering the transient absorption spectra obtained on the subpicosecond time scale, different channels of electron coupling with water molecules or ions have been discriminated (Figure 5).

The primary steps which follow the energy deposition in a solution correspond to the photoejection, localization and solvation of excess electron or to an early deactivation of nondissociative excited states.

$$\begin{aligned} Cl^- + h\nu \to (Cl^-) \;\to\; Cl + e^-_{qf} \to (Cl \cdots e^-)^*_{nH_2O} \\ \to\; (Cl \cdots e^-)_{nH_2O} - k_r \to (Cl^-)_{nH_2O} \end{aligned} \qquad (10)$$

$$e^-_{qf} + n(H_2O) \to \text{Localization}\,(T_1) \to \text{Solvation}\,(T_2) \to e^-_{hyd}. \qquad (11)$$

The femtosecond photochemical investigations on ionic aqueous solutions (XCl) have permitted to obtain fundamental informations on dynamics of electron trapping and hydration and early electron-atom recombination.

In dilute ionic aqueous solutions (NaCl = 0.4M), the photodetachment of an electron from Cl^- has been performed with ultraviolet femtosecond pulse and the subsequent hydration process of excess electron studied by femtosecond absorption spectroscopy. At ambient temperature, electron relaxation proceeds through at least one intermediate state. This transient species absorbes in the infrared and rises with a time constant of 120 fs. This transient species (e^-_{prehyd}) relaxes then towards the fully hydrated electron following a pseudo first order [71]. The best computed fit of the experimental curves taking into account the convolution of the apparatus time with the infrared electron population evolution leads to conclude that we see the transition of a single state towards a fully hydrated state. The lifetime of this transitional state equals 250 fs in dilute solution but increases in more concentrated solutions of NaCl (Figure 6). Similarly to pure liquid water, the IR band relaxation is concomitant to the risetime of a broad visible band assigned to a fully hydrated state. This state is characterized by a binding energy around 1.7 eV.

The spectroscopic investigations of time-dependent absorption spectra of subexcitation electron in ionic aqueous media exhibit a significant effect of the ionic strength on the high energy tail band. In concentrated aqueous solutions of XCl ($X = Li^+, Na^+$) the amplitude of the infrared signal assigned to a non fully hydrated electron decreases [80]. Indeed, when the number of water molecules approaches the limit for hydration of Na^+ counterions and chloride ions (NaCl 6M at 294 K), the solvation shells of Na^+ and Cl^- are characterized by high activation energies for the sodium or chlorine relaxation process and the rotational correlation time of water molecules. In this favorable distribution of ionic clusters, there is a disruption of the usual hydrogen-bonded structure (a regular tetrahedral structure) that is generally observed in liquid water [81, 82]. Femtosecond infrared spectroscopy demonstrates, that in a such concentrated aqueous solution, electron relaxation dynamics is longer than in non saturated sodium chloride solutions.

Figure 7 shows the influence of the ionic strength (0–11 M) on electron trapping and hydration dynamics and on transient spectra of non-equilibrium

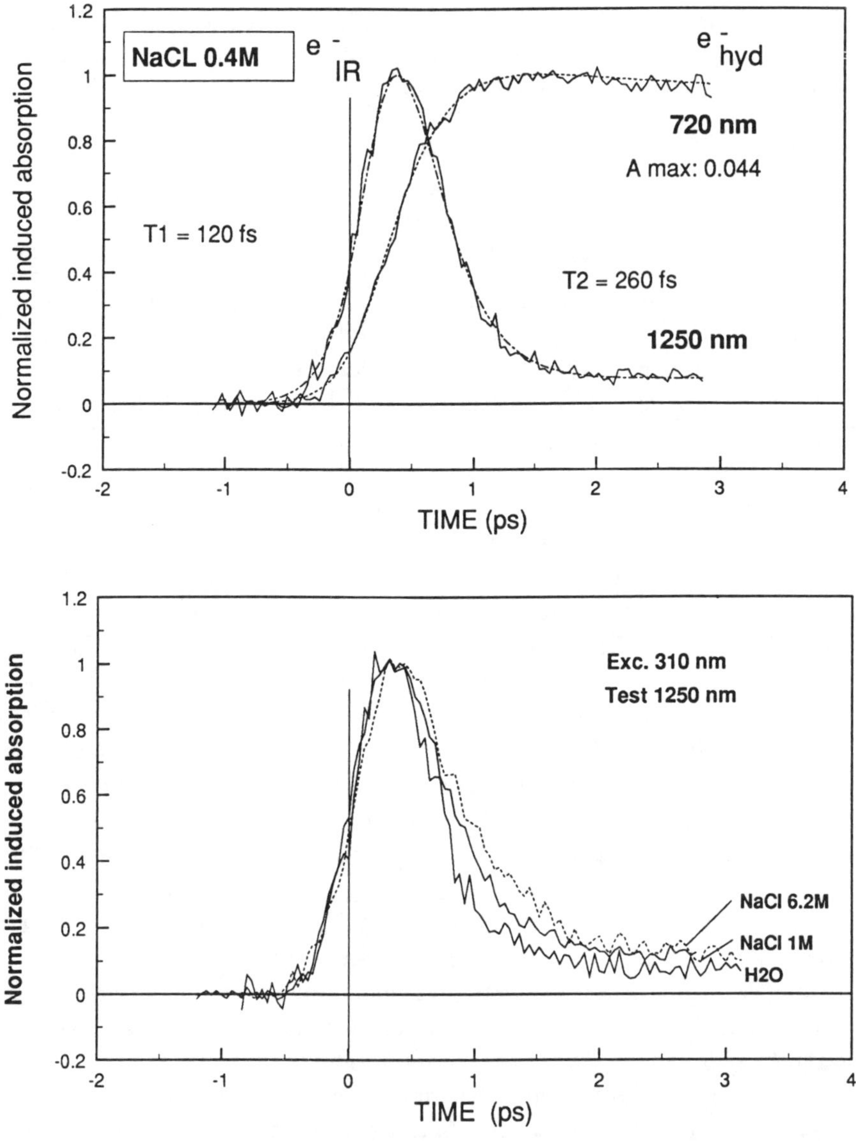

Fig. 6. Upper part: Time dependence of induced absorbance in the near infrared (1250 nm) and visible (720 nm) following the femtosecond ultraviolet photoionization of sodium chloride aqueous solution. The dotted lines represent the best computed fits of experimental data and give T_1 = 120 fs and T_2 = 260 fs. Lower part: Effect of ionic concentration (NaCl 0–6M) on the infrared electron relaxation at 294 K.

electronic states. The electron localization step dynamics (T_1) remains independent on the nature and the concentration of the ionic species. The most significant ionic strength effects are observed on the second step of electron hydration dynamics (T_2) and the nature of the transient electronic states. The spectral characteristic of the hydrated electron at short times (600 fs) are dependent on the ionic strength. When the hydration shell around Li^+ or Cl^- correspond to thigh bound water, $[H_2O]/[LiCl] = 5$, an important band peaking in the red spectral region can be observed. Compared to the previous spectrum of e^-_{hyd} in pure water, a blue shift of about 0.2 eV is observed in concentrated lithium chloride solution. This shift is attributed to a change of the electron hydration energy induced by the presence of the cation [83]. Transient spectral data performed in the visible and infrared permit to assume that in such concentrated ionic aqueous solutions the density and configurational fluctuations provide potential well into which the excess epithermal electron may be directly hydrated. The favorable spatial distribution of deep traps created by the presence of stable counter ion (Na^+ or Li^+) would represent a specific order of the liquid for direct electron capture and subsequent stabilization.

At high ionic strength, the rotational correlation time (T_c) obtained by ^{1}H NMR is longer than in bulk water [81] whereas the ^{1}H NMR relaxation time T is significantly reduced by slow motions of water molecules. Femtosecond data demonstrate that the electron solvation dynamics remains independent on the relaxation time of the ionic atmosphere but is largely faster than the formation time of ionic atmosphere all around the hydrated electron (T_r), as estimated by the Equation (12) [84]:

$$T_r = [3.55 \times 10^{-9} \sum_j Z_j]/[\mu \sum_j \Lambda_j]. \qquad (12)$$

In this expression, T_r is expressed in picoseconds and Λ_j represents the equivalent conductance of each type of ion in the solution.

These femtosecond spectroscopic data tend also to indicate that in concentrated ionic aqueous media (Na^+ or Li^+), the electron hydration dynamics remains largely independent on macrospopic physical properties such as the dielectric constant, the viscosity and the kinematic viscosity but is likely influenced by the trapping sites configurations linked to the presence of ionic entities.

Recent experimental and theoretical studies on femtosecond molecular motions in liquid water have proposed that the librational motion of water, representing the configuration disorder of the 'frozen' liquid at the 10^{-14} time scale, will be involved in the initial electron localization process. The effects of ionic species on the configurational disorder of molecular liquids remain to be clarified, in particular the role of local ionic strenght and electronic field on the fluctuations due to translational and reorientational motions of the water molecules and on the oscillator strength of the non fully solvated electron.

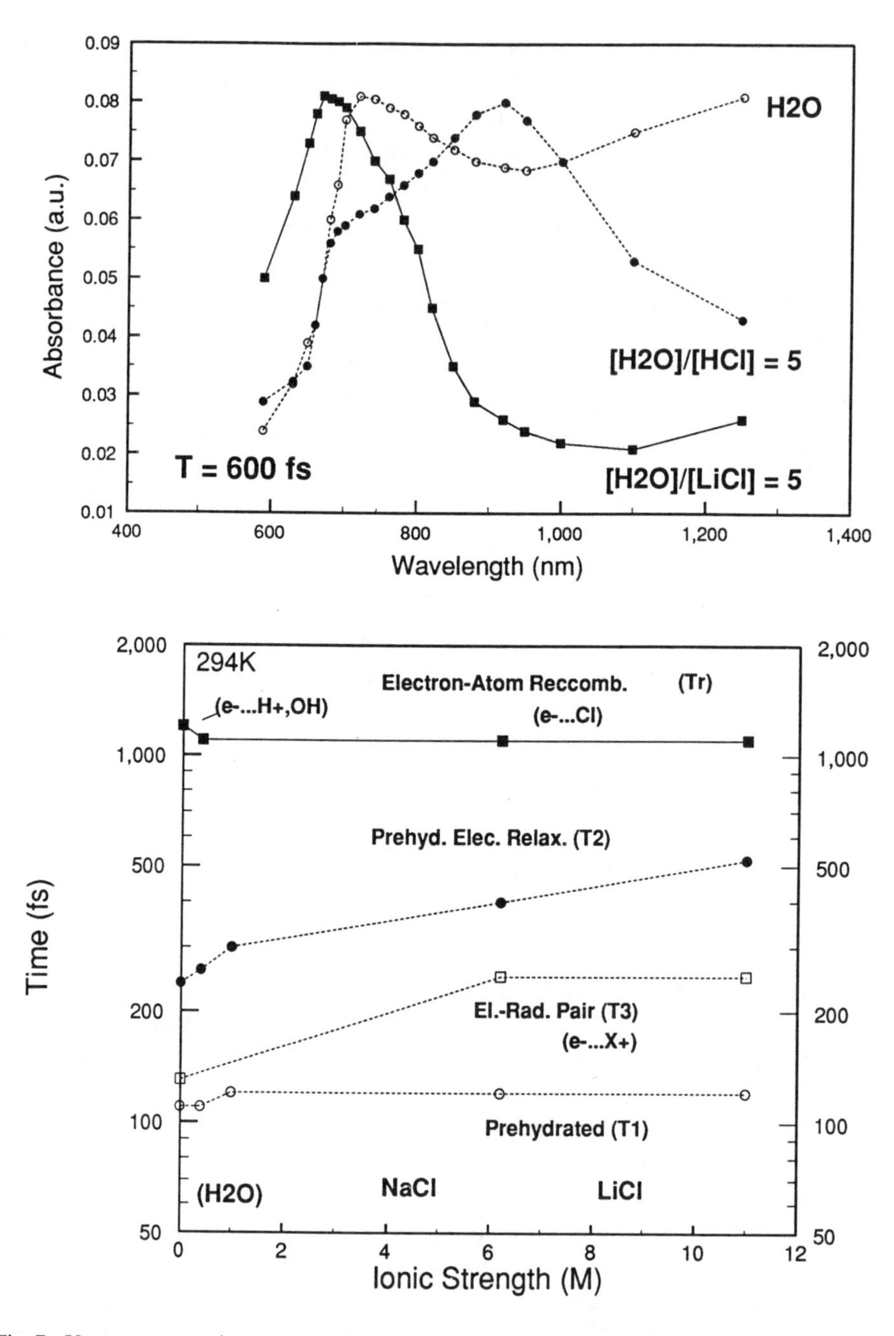

Fig. 7. Upper part: transient spectra of non-equilibrium electronic configurations in pure water and concentrated ionic aqueous solution (HCl and LiCl). The points reported on the transient spectra have been determined from the experimental kinetics obtained in the visible and the near infrared spectral region. Lower part: influence of the ionic strength on the dynamics of electron localization (T_1), solvation (T_2), electron-ion pair formation (T_3) and early geminate recombination (T_r) in aqueous solutions.

The investigation of early steps following the femtosecond photoionization of halide (Cl^-) in ionic aqueous solution has permitted to demonstrate the existence of specific photochemical channels which are dependent on the ionic concentration and on the nature of the counter ion. These channels are represented in Figure 5 and the different characteristic times can be summarized as follow:

1 Electron Localization and Relaxation: T_1, T_2
2 Electron Localization and Relaxation: T_1', T_2'
3 Electron-Radical Pairs Formation: T_3:
(Long lived Pairs: $X^+ = Na^+, Li^+$)
(Short lived Pairs: $X^+ = H+$: T_{EP}
4 Early Electron-Chlorine Atom Recombination (1D): T_r

$$A(t) \sim \text{erf}\,(T_r/t)^{1/2}$$

5 Ultrafast Reaction of hydrated electron (3D): T_5

$$(X^+ = H^+)$$

$$A(t) \sim \exp(-t/T_5)^{1/2}$$

2. Direct identification of electron-ion pairs

In concentrated ionic aqueous solutions, it has been suggested that excess electron can react with counterions via the formation of electron-ion or radical pairs. An ion of particular interest in the domain of chemical reactivity is the hydronium ion or hydrated proton.

In water the reaction of hydrated electron with the hydronium ion (reaction 13) exhibits a rate constant of $2.3 \times 10^{10}\ M^{-1}\ s^{-1}$ [85, 86]. This value is about 30% of the estimate for a diffusion controlled reaction using a reaction radius of 5 Å, and diffusion coefficients of $9 \times 10^{-5}\ cm^2\ s^{-1}$ and $4.7 \times 10^{-5}\ cm^2\ s^{-1}$ for hydronium ion and hydrated electron respectively [85, 87]. This reaction will be not diffusion controlled but influenced by the efficiency of reaction during electron-radical pairs formation.

$$H_3O^+ \ + \ e^-_{hyd} \leftrightarrow [\text{Encounter Pair}]? \rightarrow H + H_2O$$
$$(k_{Exp} \sim 2.3 \times 10^{10}\ M^{-1}\ s^{-1}). \qquad (13)$$

In acid media, the reaction rate between the hydrated electron and proton (hydronium ion, H_3O^+) is also smaller than the calculated bimolecular rate defined as the diffusion controlled limit process [86]. Furthermore, the hydrated proton does not reduce the initial yield of hydrated electron [88]. This phenomena represents an important exception to the empiric relationship between C_{37} and $k(e^- + H_3O^+)$ and will be attributed to the formation of an encounter pair $(H_3O^+ : e^-)_{hyd}$ [88].

Several years ago, Czapski and Peled [89] have suggested the existence of a relationship between the lifetime of suspected encounter pair ($e_s^- \ldots H_3O^+$) and the rate constant of the reaction 13. The lifetime of this encounter pair would be one of the limiting factor of the electron-proton reaction.

The investigation of the primary steps of a fast photoinduced single electron transfer in aqueous solution containing high concentration of hydronium ion (HCl 11 M) have permitted to identify an encounter pair ($e^-_{hyd} \ldots H_3O^+)_{hyd}$ at ambient temperature [90]. When the number of water molecules involved in the solvation of ionic species is very low ($[H_2O]/[H_3O^+] = 5$), the transient absorption spectra exhibit three characteristic bands ($\lambda^1_{max} < 1$eV, $\lambda^2_{max} = 1.35$ eV, $\lambda^3_{max} = 1.72$ eV) which are assigned to an a non fully hydrated electron (e^-_{prehyd}), an encounter electron-ion pair ($H_3O^+ : e^-)_{hyd}$ and an electron stabilized in solvent cage $(e^-)_{hyd}$ respectively (Figure 7).

The important structureless spectral distribution clearly observed between 800 and 1000 nm has been assigned to the fact that a significant fraction of excess electron can be localized within the solvation shell of the cation $[(H_3O+)_{nH_2O})]$. The formation of the encounter pair $(e^-:H_3O^+)_{hyd}$ is found to follow a rate constant of 4×10^{12} s^{-1} and the one exponential law relaxation (time constant of 850 fs) corresponds to the deactivation of this transient encounter pair; the cleavage rate constant $[(H_3O^+ : e^-)_{nH_2O} \rightarrow H_2O + H]$ equals 1.17×10^{12} s^{-1} at ambient temperature. Indeed, the encounter pair deactivation dynamics is faster than the average lifetime of H_3O^+ but remains comparable to the H-bond time scale [90–94]. These data clearly demonstrate that the single electron transfer: e^-_{qf} + n H_2O + $H_3O^+ \rightarrow (e^- \ldots H_3O^+)_{hyd}$ occurs prior the electron hydration phenomenon and probably corresponds to a localization of the electron in shallow traps within the solvation shell of the cation $[(H_3O^+)_{nH_2O})]$ (channel 3 of Figure 5).

In these conditions, the probability P' that an excess electron becomes localized inside the encounter volume is not negligible when $[H_2O]/[H_3O^+]$ = 5:

$$P' = 1 - \exp(-4\pi r^3_{\mathrm{eff}}[\mathrm{S}]/3 \times 10^3). \tag{14}$$

Let us check this point: the effective reaction radius has been calculated from the expression related to the C_{37} values [88]:

$$r_{\mathrm{eff}} = 7.35\ \mathrm{C}_{37}^{-1/3} \tag{15}$$

In water, the C_{37} value of H_3O^+ equals 11 M and the estimate of r_{eff} is 3.41 Å. Reporting this value in expression 14, we obtain $P' = 0.84$ for the probability of an encounter pair formation in aqueous hydrochloric acid solution ($[H_2O]/[HCl] = 5$). Consequently the channel corresponding to electron localization and solvation outside the hydration shell of H_3O^+ represents about 16% of the single electron transfer [90].

The three modes of electron couplings in concentrated hydrochloride solutions can be summarized as follows:

$$[Cl^-, H_3O^+]_{5(H_2O)} + h\nu \rightarrow (Cl : H_3O^+)_{hyd} + e^-_{qf} - -k_1, k_2 \rightarrow (e^-)_{hyd} + Cl \tag{16}$$

$$(H_3O^+)_{hyd} + e^-_{qf} - -k_3 \rightarrow [(H_3O^+ : e^-)_{hyd} \leftrightarrow (H_3O)^+_{hyd}] - -k_{EP} \rightarrow H + H_2O \tag{17}$$

with k_1, k_2, k_3, $k_{EP} = 1/T_1$, $1/T_2$, $1/T_3$ and $1/k_{EP}$ respectively.

The understanding of the couplings between electron-hydronium ion pair and water molecules needs to define whether the relaxation of these transient states are triggered by intracomplex structural changes (geometrical perturbations of the hydration cage) owing to the fact that pairs relaxation occurs at a very short time i.e. before that diffusion of proton becomes appreciable. In acid solutions, the torsionnal vibration of the water molecules changes the potential well of the hydrated proton and the polarization of the hydrogen bond. The reorientational relaxation of water molecules close to the hydronium seems to be faster than in bulk water. It is interesting to notice that the cleavage rate constant of the encounter pair occurs at a similar time scale that the H-bond mean lifetime or average lifetime of the hydronium ion in liquid water [91–94]. We have suggested that the limiting factor of the deactivation dynamics of the encounter pair correspond to the activation energy of the radical-ion bond cleavage reaction including either a proton migration from a hydronium ion to neighbour water molecules or a local polarization effect on H bonds [90]. In this way, the initial reactivity of excess electron with hydrated hydronium ion $(H_3O^+ + e^-)_{hyd}$ would depend both on (i) the local structure of this cation in the vicinity of water molecules, (ii) the initial electron-hole pair distributions, (iii) the H bond dynamics between H_3O^+ and water molecules or on proton migration from hydronium to neighbouring water molecules. In this last case, the relaxation of the encounter pair can be compatible with vibrational modes of water molecules in the femtosecond range: vibrational OH bonds, librational and translational modes of the hydronium ion.

2. Electron Attachment and Solvent Effects

2.A. Formation and Reactivity of Primary Anion in Molecular Liquids

Femtosecond optical techniques allowing the generation of intense optical pulses from the near UV to the near IR can be used to perform the investigations of ultrafast photophysical and photochemical reactions in non polar media. The interaction of excess energy electron with molecular liquids may lead to solvation phenomena via polarization processes. The preceeding section deals with solvation dynamics in molecular liquids. When the solvent molecules exhibit a high electron affinity, electron attachment to solvent can correspond to the main coupling process for which an irreversible reaction of a primary anion yields a secondary anion (Figure 1). Numerous examples in both groups of liquids have been studied [95]. A particular type of electron-liquid interactions seems to exist involving both a solvation process and a direct electron attachment to solvent molecules. Indeed non polar solvent containing sulfure atom exhibit a high electron affinity which can lead to a direct electron attachment to the solvent molecules although this molecular liquid exers a non negligible capacity to solvate subexcitation electron. In liquid dimethylsulfide (DMS: CH_3SCH_3), solvated electron and secondary anion ($CH_3SSCH_3^-$) have been observed at the nanosecond and picosecond regimes [96]. Up to now, it was suggested that the early formation of the primary anion ($CH_3SCH_3^-$) and the localized electron (e^-_{Loc}) result to the existence of two embranchments involving competitive reactions with a very shortlived common precursor. The density fluctuations of this liquid would influence the dynamical comportement of the single charge transfer through solvation or electron attachment. This implies also that during the reactivity of the primary anion with surrounding solvent molecules, the SS bond formation is as fast as the electron solvation itself and would occur at the femtosecond range [97].

Femtosecond photochemistry of DMS permit to clarify several points about the chemical reactions for which sulfide atom or disulfide bridge are involved. The primary events due to the femtosecond photoionization of DMS by ultraviolet pulse are given in the reactions [18]–[21]. The ionization potential of DMS in the gas phase being $IP = 8.685$ eV, the ultraviolet pulse of 100 fs duration, centered around 310 nm ($E = 4$ eV) has been used to initiate the photodetachment of a subexcitation electron by two photons phenomena (reaction [18].

Electron Photodetachment:

$$(CH_3)_2S + 2h\nu \rightarrow (CH_3)_2S^+ + e^-_{qf} \tag{18}$$

Electron Localization:

$$e^-_{qf} + n(CH_3)_2S \rightarrow e^-_{Loc} \qquad (19)$$

Electron Attachment (Formation of primary anion):

$$e^- + (CH_3)_2S \rightarrow \{\text{Intermediate States}\} \rightarrow (CH_3)_2S^- \qquad (20)$$

Ion-Molecule Reaction (Chemical bond formation with umpaired electron):

$$(CH_3)_2S^- + n(CH_3)_2S \leftrightarrow (CH_3SCH_3)^{-*}_{\cdots(n)(CH_3)_2S}$$
$$\rightarrow C_2H_6 + (CH_3SSCH_3)^-_{(n-1)(CH_3)_2S}. \qquad (21)$$

Subsequently to electron photodetachment, the primary cation ($(CH_3)_2S^+$) and the subexcitation electron undergo interactions with surrounding solvent molecules. In this paragraph, we will not discuss the reactivity of primary cation by complexation with a second molecule of solvent to yield the secondary cation ($(CH_3)_2SS(CH_3)_2^+$) with two-centers-three-electron bonds ($2\sigma/1\sigma^*$) [98]. Experimental works are in progress to understand the mechanism of cation-molecule reactions in non polar liquids.

The first infrared spectroscopy experiments performed in neat DMS at ambient temperature and at the femtosecond time scale have permitted to establish the existence of an induced absorption in the near infrared (Figure 8). This absorption exhibits a long lifetime i.e. over several hundred of picoseconds at 1330 nm. The best computed fits of the appearance time of the infrared band (1330 nm) give an electron solvation time of $T_{solvation} = 120$ fs at 294 K. When an electron scavenger like biphenyl (Ph_2: 0.26 M) is added to DMS, the infrared signal follows a monoexponential decay (pseudo first order kinetics) with a time constant of 25 ps $\pm$ 1 ps at 294 K (Figure 9). The complete decay of the infrared signal at long time ($t \sim 100$ ps) permits to suggest that the transient infrared component is due to a unique species. Considering previous results obtained in the nanosecond time scale, it is reasonable to assign the existence of the infrared signal to a localized state of the electron. In presence of biphenyl, the fast scavenging process can be defined by the reactions [22], [23].

$$DMS + 2h\nu \rightarrow DMS^+ + e^-_{IR} \qquad (22)$$

$$e^-_{IR} + Ph_2 \rightarrow Ph_2^- \qquad (23)$$

The bimolecular rate constant of reaction 23 can be determined from infrared experimental data and equals $1.5 \pm 0.2 \times 10^{11}$ M^{-1} s^{-1}. The high

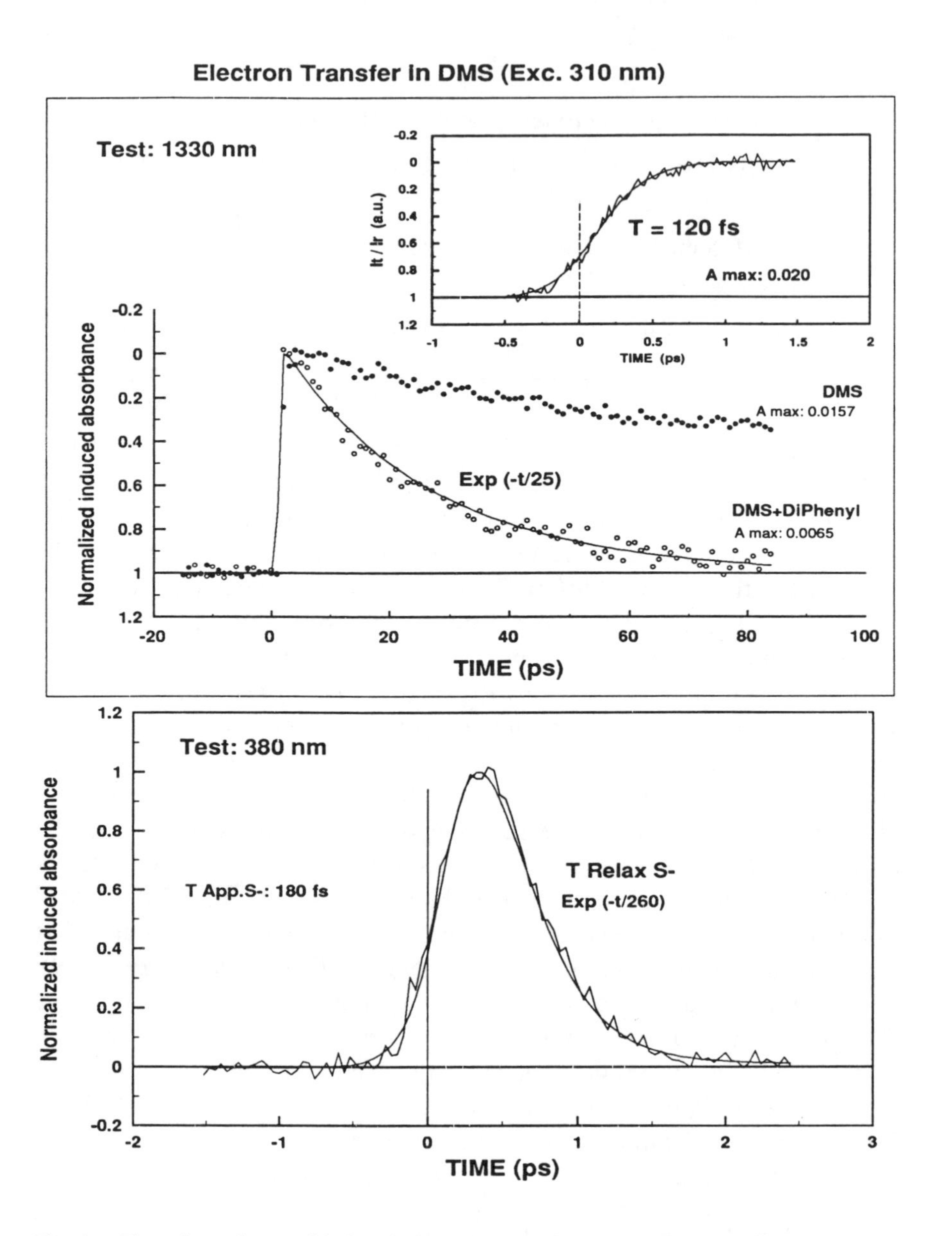

Fig. 8. Time dependence of induced absorption of neat Dimethylsulfide (DMS) at 294 K following photoexcitation with 100 fs laser pulses. The normalized kinetics are shown for two test wavelengths (1330 nm and 380 nm). The infrared component is assigned to the existence of a localized state of excess electron. The best computed fit gives a solvation time of 120 fs. The near ultraviolet absorption exhibits a very short lifetime component which is due to the reactivity of the primary anion (CH_3SCH_3).

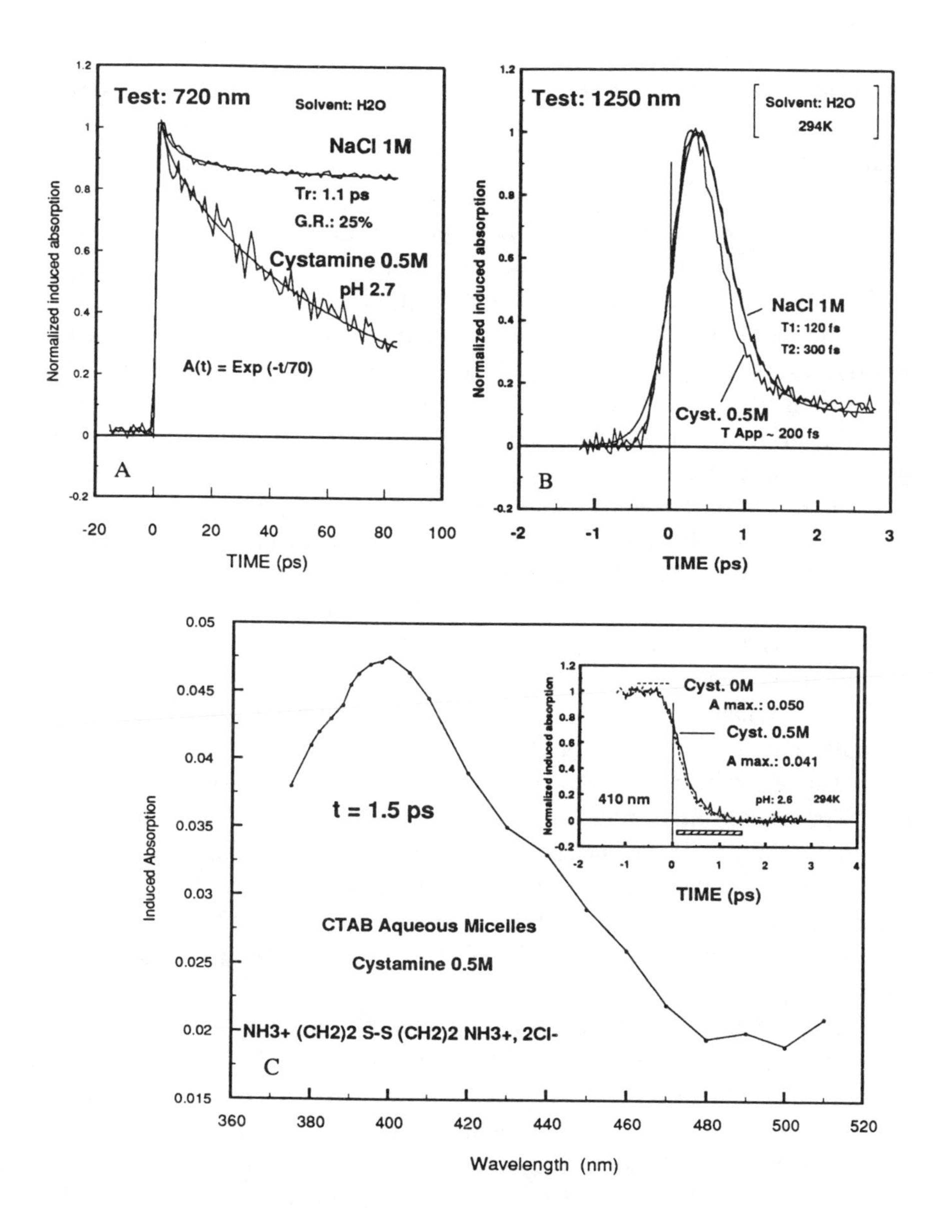

Fig. 9. A, B: Set of time-resolved induced absorption data in aqueous solutions containing sodium chloride or dihydrochlorhydrate of cystamine. The influence of the counterion (Na^+ or NH_3^+ $(CH_2)_2$ SS $(CH_2)_2$ NH_3^+) on the lifetime of infrared prehydrated electrons (1250 nm) is shown. Part C of the figure represents the transient spectrum of SS radical in anionic aqueous biomicelles containing cystamine (0.5M, pH 2.3).

bimolecular rate constant is due to the great mobility of solvated electron in such media ($\mu_{e_s^-} = 1.4 \times 10^{-2}$ cm^2 V^{-1} s^{-1}) and is in agreement with previous experimental data [96]. A quantitative comparison of the signal obtained in absence and in presence of Ph_2 also reveals that an early capture of excess electrons by biphenyl can occur before this charge gets its solvated state. The fact that the maximum absorbance is lower in presence of Ph_2 would indicate that even though the cross section of Ph_2 would be very high, the scavenging efficiency of Ph_2 on the precursors of e_s is strong enough to deplete the initial absorbance by more than 50%.

Femtosecond spectroscopy data obtained with neat DMS solution show that in the near ultraviolet, the time-dependence of induced absorption significantly differs from those obtained in the infrared (Figure 8). The maximum of absorbance is particularly intense, almost twice more than in the infrared. The ultrashort lived component has been tentatively assigned to a precursor of the secondary anion CH_3-SS-CH_3^- [97]. This precursor will be generated through an ultrafast electron attachment with a DMS molecule and would correspond to a solvated or complexed form of the primary anion (CH_3-S-$CH3^-$). The risetime of the induced component at 380 nm is not short as the pulse duration ($t < 100$ fs). The analysis of time-resolved data in the spectral range (380–420 nm) are conducting with a kinetical model which takes into account the convolution of the pump-probe temporal profile and the expected signal rise dynamics of transient species. At 380 nm, the transient species exhibits an appareance time of 180 fs and a lifetime of 260 fs $\pm$ 10 fs. This second characteristic time is in agreement with the value obtained at 420 nm [97]. The pseudo first order kinetics of absorption decay at 380 nm can be reasonably assigned to the last step of an ion-molecule reaction (reaction 21). This reaction would involve a primary negative adduct such as the primary anion ($(CH_3)_2S^-$) or its complexed or solvated states and lead to the formation of a secondary anion $(CH_3SSCH_3)^-$ characterized by an umpaired electron.

Additional experimental investigations on the early reactivity of the primary anion in neat DMS are of interest to discriminate whether:

- Ultrafast formation of the secondary anion may involve pre-structurated configurations of the solvent such as dimers with disulfide bridge.
- Early reaction of the primary anion with electron scavenger (Ph_2) can occur concomittantly to the formation of a secondary anion.

At this stage, the first femtosecond studies obtained in neat DMS address the problem of dual behavior of excess electron through localization step and/or attachment to solvent molecule (reactions 19–21).

In neat DMS, we have seen that a very short lived species is tentatively assigned to a precursor of the secondary anion and will be generated via an ultrafast electron reaction with solvent molecules (Figure 8). The femtosecond spectroscopy of sulfide compounds in neat liquid DMS or in aqueous solutions (cystamine) permit to discriminate the influence of a molecular response during the formation and the relaxation of secondary anion (SS^-) and to extend our knowledge on the reactivity of nonequilibrium electronic states and primary anions [98].

2.B. Reactivity of Non-equilibrium States of Electron

It is now well established that an electron becomes localized or solvated in a polar liquid involving several nonequilibrium states. An important point regarding the domain of ultrafast reaction dynamics concerns the existence of single charge transfers with non fully relaxed electron and the rôle of solvent dynamics during ultrafast redox reactions (Figure 1).

Femtosecond studies on photochemical initiation of ultrafast monoelectronic transfer have been recently performed in different aqueous media (homogeneous solutions and organized assemblies) [99]. The Figure 9 illustrates some of more significant results obtained with a biomolecule containing a disulfide bridge (cystamine). In aqueous solution the femtosecond photoionization of cystamine dihydrochloride (RSSR, $\mathbf{R = (CH_2)_2\ NH_3^+, Cl}$) initiates generation of non equilibrium electronic states before complete hydration of excess electron (reactions 10, 11). Fully hydrated electron reacts with cystamine through bimolecular reaction:

$$e_{hyd}{}^- + (RSSR)_{nH_2O} \rightarrow (RSSR^-)_{nH_2O}. \qquad (24)$$

The part A of Figure 9 represents picosecond study of a univalent reduction reaction (reaction 24 which is limited by the diffusion of reactants. Experimental bimolecular rate constant determined by infrared spectroscopy at the picosecond time scale equals $2.9 \times 10^{10}\ M^{-1}\ s^{-1}$ at 294 K and pH 2.7. Femtosecond data obtained in the same spectral region (part B of Figure 9) demonstrate the existence of an ultrafast charge transfer reaction (reaction 25).

$$[RSSR]_{nH_2O} + e^-_{\text{non fully relaxed}} \leftrightarrow [RSSR^-]_{nH_2O} \rightarrow [RSSR^-]_{nH_2O}. \quad (25)$$

These data establish that the apparent lifetime of the localized electron is influenced by the presence of a SS bridge. In aqueous solution of cystamine, an ultrafast reactivity of infrared electron with the biomolecule initiates the formation of an anionic radical (RSSR). At very short time ($t \sim 1.5$ ps), the transient spectrum of this radical has been discriminated in biomolecular systems (Figure 9). The ultrafast formation of the anionic radical cannot be explained by a diffusion-controlled reaction between solvated electron and the molecular accpetor. More precisely, this electron transfer process occurs within the solvation time of electron ($t < 500$ fs) and would be linked to an ultrafast reactivity of non equilibrium states of excess electron (reaction 25) [100].

A comparison of data on electron attachment to sulfide molecule (monomer) and ultrafast reactivity of non equilibirum electronic states with disulfide bridge (dimer) permits to underline the rôle of molecular response of a solvent during single electron transfer. Significant computed results are reported in the Figure 10. In pure DMS, the electron capture on a solvent monomer and the resulting formation of the primary anionic radical (S^-) occurs within a temporal window ($t < 1$ ps) which is similar to those of electron localization. Moreover, in aqueous solution of cystamine, the SS^- radical can be identified within the first ps and is in agreement with an ultrafast capture of non-equilibrium electronic states by the biomolecule. In these two situations, the formation of an anionic radical with sulfides is linked to a single charge transfer reaction and likely corresponds to an electronic response without significant molecular reorganization of solvent in the vicinity of newly charge distribution. When the anionic radical SS^- is generated by the reaction of a primary anion (S^-) with surrounding molecules ($S^- + S \rightarrow SS^-$) femtosecond spectroscopic studies permit to establish that the appareance time of SS^- is delayed in comparison with that happen when non relaxed electron reacts with pre-existing SS bridges (cystamine). This time delay ($\sim$ 500 fs) correspond to a specific molecular response of solvent during primary anion-molecule reaction (reaction 21).

The femtosecond investigations of ultrafast reactions of excess electron with sulfide compounds would permit (i) to more completely understand the role of electronic and molecular responses during the formation and solvation of primary and secondary anion, (ii) to understand the influence of solvent dynamics and density fluctuations on the reactivity of primary anion, (iii) to define the nature of the limiting factors during the formation of chemical bond with umpaired electron (2c, $3e^-$).

3. Reaction Dynamics and Polar Protic Solvent Effects

The understanding of solvent effects on charge transfer reactions require to obtain detailed informations on the time-dependent response of solvent molecules to a change of change distribution. The rôle of water molecules in charge transfer reactions is linked to the microdynamic properties of this liquid. Its structure is fluctuating through change in the hydrogen bond networks and the existence of clathrate-like holes [56].

The femtosecond photophysics of water molecules is a powerful tool for the investigation of ultrafast primary reactions that occur after an initial energy deposition and charge separation. During the interaction of ionizing radiation with an aqueous phase, the absorption of energy initiates several ionization

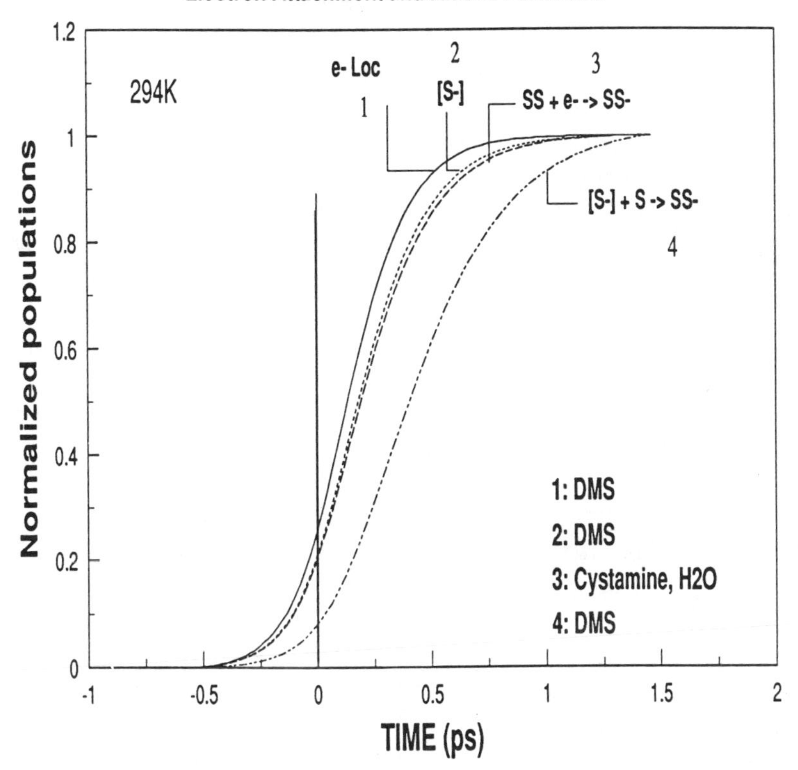

Fig. 10. Ultrafast charge transfer with sulfides including electronic and molecular events: computed appareance time of electronic populations: localized electron (e^-_{loc}), primary anion (S^-), secondary anion (SS^-) in solutions of organic sulfur compounds at 294 K.

processes. The initial energy deposition in the bulk phase is followed by the formation of transient electronic states and a water cation. In this way, the femtosecond spectroscopy of non-equilibrium states of elementary charges (electron, proton) in liquid water permit to investigate primary steps of charge transfer reactions in a polar protic solvent: formation of the hydration cage around an electron, encounter pair formation, ion-molecule reaction, electron attachment to solvent molecule, early electron-ion pair recombination and to obtain unique informations on ultrafast reactions which occur at the temporal shell of molecular motions.

The H/D isotope substitution represents a powerful tool which can be used to discriminate, at the molecular level, the influence of the energy vibrational mode (OH, OD) on the early steps of short lived non-equilibrium electronic states in the near infrared spectral region. Although some molecular and dielectric properties of light and heavy water are very similar [101] significant differences in the microscopic structure of these two solvents can be mentioned: deuterated water exhibits stronger hydrogen bonds than in normal water [102], the energetic vibrational mode (OD vs OH) is $2^{1/2}$ times lower in D_2O than in H_2O [103]. Moreover, the lifetime of protropic species is about two times longer in D_2O [91]. These differences support the conclusion than deuterated water corresponds to a more ordered liquid than light water [104]. The Figure 11 illustrates the possibility that, in neat liquid water, very short pulses can test different mechanism of early coupling between electron and prototropic species through short or long range interactions. It is of particular interest to understand how the initial spatial distribution of the electron-prototropic species pairs can involve hybride transition states and/or proton motions.

3.A. Early Recombination Reactions

The femtosecond studies of primary events occuring in pure liquid water permits to investigate the mechanism of early couplings between fully relaxed electron and prototropic species and offer the opportunity to better understand the influence of the protic solvent on reaction dynamics. Following the two photons excitation of water molecules, the initial spatial distribution of the electron-ion pair and the high rate constants of the recombination of e^-_{hyd} with both the hydronium ion or the hydroxyl radical (reactions 26, 27 do influence the early reactivity of the hydrated electron with these primary prototropic species [71, 105, 106].

$$H_3O^+ + e^-_{hyd} \rightarrow H_2O + H \quad (k = 2.3 \times 10^{10}\ M^{-1}\ s^{-1}) \tag{26}$$

$$OH + e^-_{hyd} \rightarrow OH^- \quad (k = 3 \times 10^{10}\ M^{-1}\ s^{-1}). \tag{27}$$

Figure 12 shows that within the first picoseconds after the photoionization of water molecules, a non negligible fraction of the hydrated electron population (55%) reacts rapidly with the two possible nearest neighbours (H_3O and OH) produced by an ultrafast ion molecule reaction (Equation (28). Considering the early electron-hydronium or electron-hydroxyl radical couplings, it is fundamental to determine whether the ultrafast recombination process can be analyzed by a classical theory. In the Onsager's theory, it is assumed that the charge carriers are brownian particules and their motions under the action of their mutual Coulomb field are diffusive obeying the Smoluchoswki equation [107].

$$nH_2O + (H_2O^+) \rightarrow (H_3O^+ + OH)_{n-1(H_2O)}. \tag{28}$$

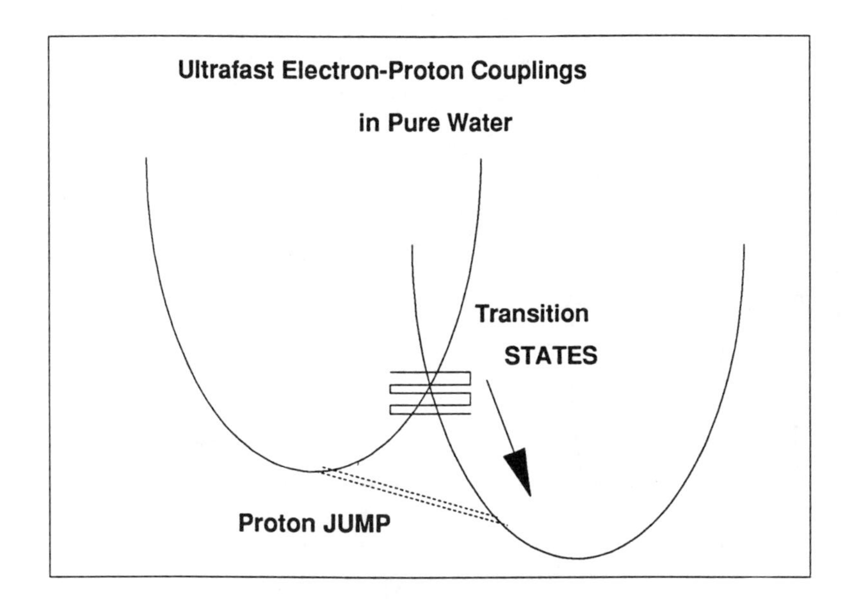

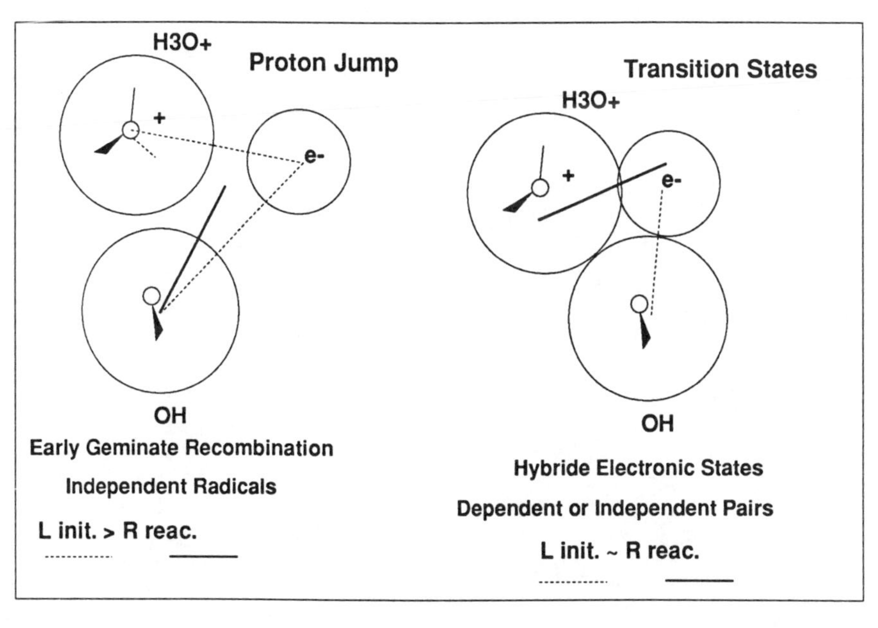

Fig. 11. Influence of the initial pair distributions on the early couplings between excess electron and prototropic species (H_3O^+, OH) in pure liquid water. Two limit cases are considered. 1) Independent pairs for which the initial radius is longer than the reaction radius ($L_{\text{Init}} > R_{\text{reac}}$); an early geminate recombination occurs through proton jump. 2) initial pair distribution favor the existence of hybride electronic states for which L_{Init} equals the reaction radius (R_{reac}).

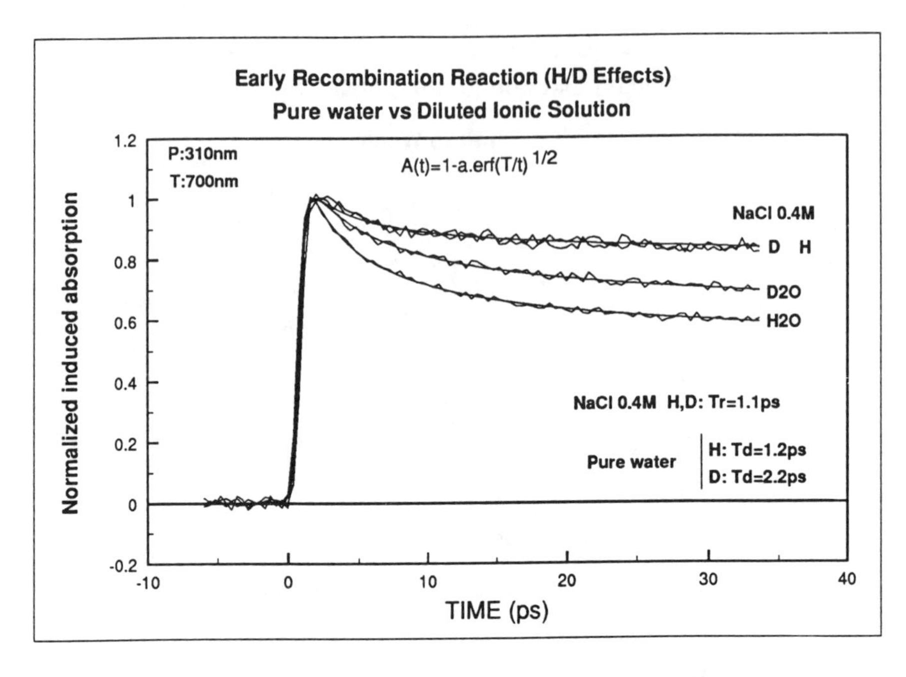

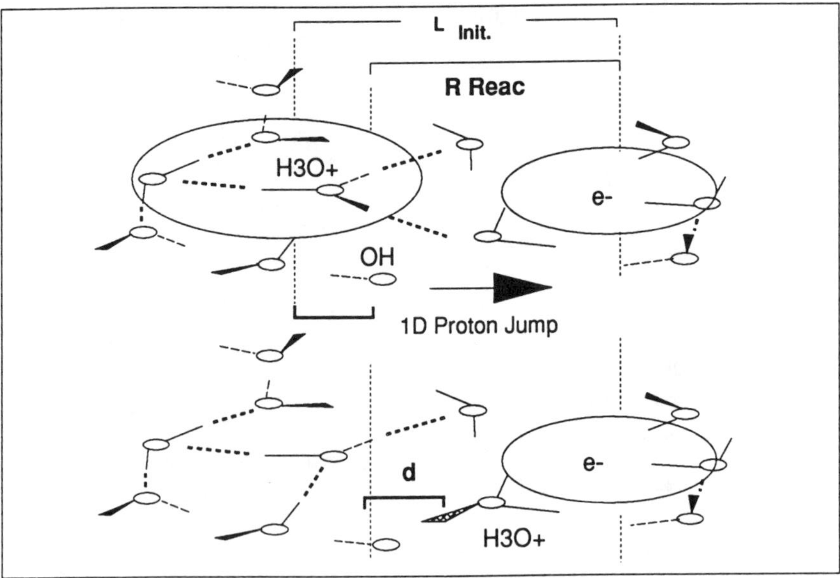

Fig. 12. Early behavior of the hydrated electron following the femtosecond ultraviolet photoionization of ionic solute (NaCl) or solvent molecules. A comparison of the H/D isotope substitution effect is also reported. The smooth lines represent the theoretical best fits of the data. The lower part of the figure represents the proton jump process (1D proton jump) assigned to the ultrafast recombination reaction in pure water (reactions 26, 27).

Within the first picosecond following the photoionization of water molecules, the prototropic species (H_3O^+) and OH can be separated by only few Å i.e. few molecular radius [108]. Consequently, the macroscopic parameters of the reaction (radius reaction or coefficient diffusion) cannot describe the microcopic structure of the reaction area. The analysis of geminate recombination at very short time by a classical diffusive model seems to exhibit few incertities in regard to the microscopic properties of the solvent: the short lifetime of the prototropic species (H_3O^+ and OH), the high mobility of the hydronium ion, the nature of interactions between different pairs [109]. Recent theoretical developments have shown that the Smoluchowski equation breaks down when the reaction occurs within the Onsager radius (7 Å in water) [110].

Time-resolved spectroscopic data on the non-homogeneous recombination process in pure water have been analyzed considering the microscopic structure of the reaction area (Figure 11). In this kinetical approach, we have suggested that the electron-radical pairs ($H_3O^+ \ldots e^-_{hyd}$ or $OH \ldots e^-_{hyd}$) executes a onedimensional (1 D) walk before undergoing an ultrafast geminate recombination [71, 106]. The lower part of the Figure 12 illustrates the finite process (proton jump) we have used to analyse the ultrafast recombination reaction. The analytical solution of a recombination controlled by a finite process (1 D diffusion) can be expressed by the following expression:

$$C_{e^-_{hyd}}(t) = \int_{-\infty}^{+t} \frac{dNe^-_{hyd}(t)}{dt} \left(1 - \gamma \, \mathrm{erf}\sqrt{\frac{T_d}{(t-t')}}\right) dt' \tag{29}$$

with:

$$N_{e-hyd}(t) = N_0[[1-1/(T_2-T_1)]\cdot[T_2 \exp(-t/T_2) - T_1 \exp(-t/T_1)]]. \tag{30}$$

In the expression (29), $1/T_d$ corresponds to the jump rate of the recombination process: $T_d \times 1.2$ ps (jump rate of $0.83 \times 10^{12}\ s^{-1}$ in light water (Figure 12). The random walk law of the early decay explains by itself the kinetics for all the times and satisfies the time-dependence $(1/(t)^{1/2})$ of the relaxation observed at longer time. The jump rate of the neutralization processes is found to be significantly influenced by an H/D isotope substitution ($0.45 \times 10^{12}\ s^{-1}$ in D_2O and $0.83 \times 10^{12}\ s^{-1}$ in H_2O) [71, 106].

Let us consider the analysis of the early recombination reaction occurring between hydrated electron and prototropic species (OH, H_3O^+) in the framework of a proton jump characterized by a frequency λ over a distance 'd' (diffusive motion by a finite process). For instance, taking into account the high mobility of the hydroniumion (H_3O^+) compared to e_{-hyd} ($D^+_{H_3O} = 9 \times 10^{-5}\ cm^2\ s^{-1}$ vs $D^-_{e_{hyd}} = 4.75 \times 10^{-5}\ cm^2\ s^{-1}$ [108, 111], the proton jump approach emphasizes the rôle of the dynamical structure of the electron-ion pair recombination and would involve the time dependence of hydrogen bonds polarization. We can consider a limit case of the ultrafast neutralization reaction for which the diffusion coefficient of the hydrated proton (H_3O^+) is defined by the expression $D = (\lambda d^2)/6$ [108]. In this hypothesis, for a mean H_3O^+ lifetime of 10^{-12} S and a reaction radius (R_{reac}) equal to 5 Å we obtain a proton jump distance δ of 2.19 Å. That means that an electron-ion pair which have an initial separation length $L_{\text{Init}} = R_{\text{Reac}} + \delta$ of about 7 Å can relax through a single proton jump (*unidirectional diffusive motion by a finite process*). This initial length (7 Å) is also equivalent to the Onsager distance over which structural characteristic of the solvent must be remembered [107]. The proton jump rate we suspect to be linked to the neutralization process (geminate recombination) is found to be significantly influenced by an H/D isotope substitution [71]. In agreement with previous pulse radiolysis experiments, this would demonstrate that the initial spatial distribution of photogenerated hydrated electrons and prototropic species is slightly broader in D_2O than in H_2O [112]. The initial spatial distribution of electron and prototropic species can be influenced by the rate of energy deposition through coupling with the vibrational modes of the polar protic solvent.

Femtosecond investigations of ionic aqueous solutions underline that ultrafast recombination of hydrated electrons with prototropic species are specifically governed by specific molecular motions of the protic solvent around the electron and newly created radical species (Figure 12). Indeed, an H/D isotope substitution is only observed when the prototropic species are directly involved in the electron-radical pair recombination. This is not the case in diluted ionic aqueous solutions for which the photodetachment of electron occurs from the halide. Indeed, the limiting factor in the early geminate recombination would correspond to the activation energy of the hydrated electron-cation bond cleavage reaction, including a proton jump with a local polarization effect of the electric field on H bonds.

It is interesting to notice that in aqueous solutions with various concentration of hydronium ion, the initial electron ion-pair distribution influences significantly the early behavior of the hydrated electron population. The Figure 13 illustrates this fact with three limit cases: pure liquid water, intermediate concentration of H_3O^+ ($[H_2O]/[HCl] = 7$) and very concentrated hydrochloride solution ($[H_2O]/[HCl] = 5$). In pure water, as previously discussed, the computed best fits of the early decay assumes the existence of a 1D random walk law. At intermediate concentration ($[H_2O]/[HCl] = 7$), the absorption decay follows a 3D recombination and the reaction occurs in an isotropic media. An isotope effect is observed as in pure water. The last case ($[H_2O]/[HCl] = 5$) corresponds to the deactivation of a transition state (encounter pair $(H_3O^+ : e^-)_{\text{hyd}}$) according to a monoexponential law [90].

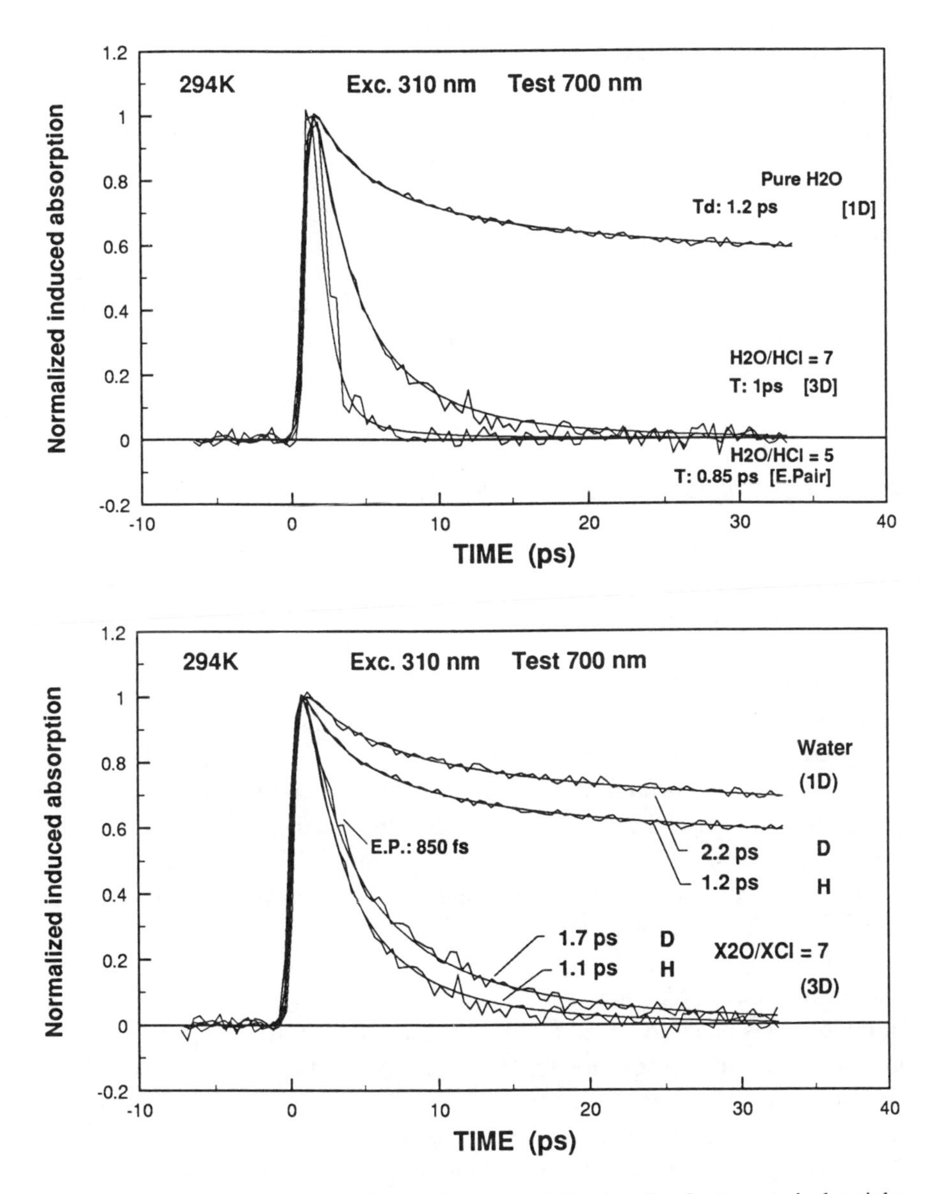

Fig. 13. Early behavior of the hydrated electron following the femtosecond ultraviolet ionization of pure water and aqueous solutions of HCl ([H2O]/[HCl] = 7 and 5). The lower part of the figure represents the effect of an H/D isotope substitution on the early behavior of the hydrated electron.

From NMR studies on characteristic times of prototropic species [91], comparisons can be made with the present results. T_d is similar to the short lifetime of hydronium ion. The jump rate ratio $(T_d)^{-1}_{D_2O}/(T_d)^{-1}_{H_2O}$ is in the range of values defined by the prototropic lifetimes ratio [106]. The existence of H/D isotope effects on both the dynamics of recombination and the lifetimes of the prototropic species suggest that one of the limiting step in the recombination process of the hydrated electron with prototropic species would be the lifetime of hydrated X_3O^+ or OX. This interpretation supports the analytical model for which the jump distance equals the initial recombination length. The model implies that within a $(X_3O^+ : OX)_{hyd} \ldots (e^-)_{hyd}$ pair a proton transfer from X_3O^+ executes a 1D walk in the vicinity of the hydrated electron before undergoing recombination (Figure 11). Ultrafast recombination between electron and prototropic species is governed by specific molecular motions of the protic solvent around the electron and radical species.

3.B. Ultrafast Electron-Proton Couplings in Water

1. The Ion-Molecule Reaction: $nH_2O + (H_2O) \rightarrow (H_3O^+ + OH)_{n-1(H_2O)}$

In the past many attemps have been made to identify the water cation in gas-phase and successful investigations have permitted to observe its absorption and emission spectra in the range 350–660 nm [113]. Using transient absorption spectroscopy in the femtosecond time scale, recent experiments have been conducted in the spectral range 410–460 nm in order to determine if an induced absorption characterized by an ultrashort lifetime can be tentatively assigned to the existence of the water cation H_2O^+ (dry positive hole). The dry hole reacts with an adjoining water molecule and through an ultrafast proton transfer gives the cationic ion H_3O^+ (hydrated proton) and the radical OH (reaction 28) [114]. This reaction is likely one of the fastest which occur in polar solvent and represents an ideal case to learn more about ultrafast proton transfer in a protic liquid.

During the initial energy deposition due to femtosecond ultraviolet ionization of water molecules, an instantaneous species absorbing in the near ultraviolet has been observed to appear in less than 100 fs (Figure 14). This ultrashort transient absorption has been assigned to the water cation (H_2O^+) and its relaxation would then correspond to the ion-molecule reaction (reaction 28) for which the measured cleavage rate constant equals 10^{-13} s^{-1} at 294 K [114]. It is important to note that the precursor of the hydronium ion and the hydroxyl radical would relax following an ultrafast proton transfer whose rate constant is faster than the final relaxation step of the trapped electron. This means that a favorable structured environment $(H_3O^+ \ldots OH)_{hyd}$ can be created before than an electron gets its final equilibrium state. This

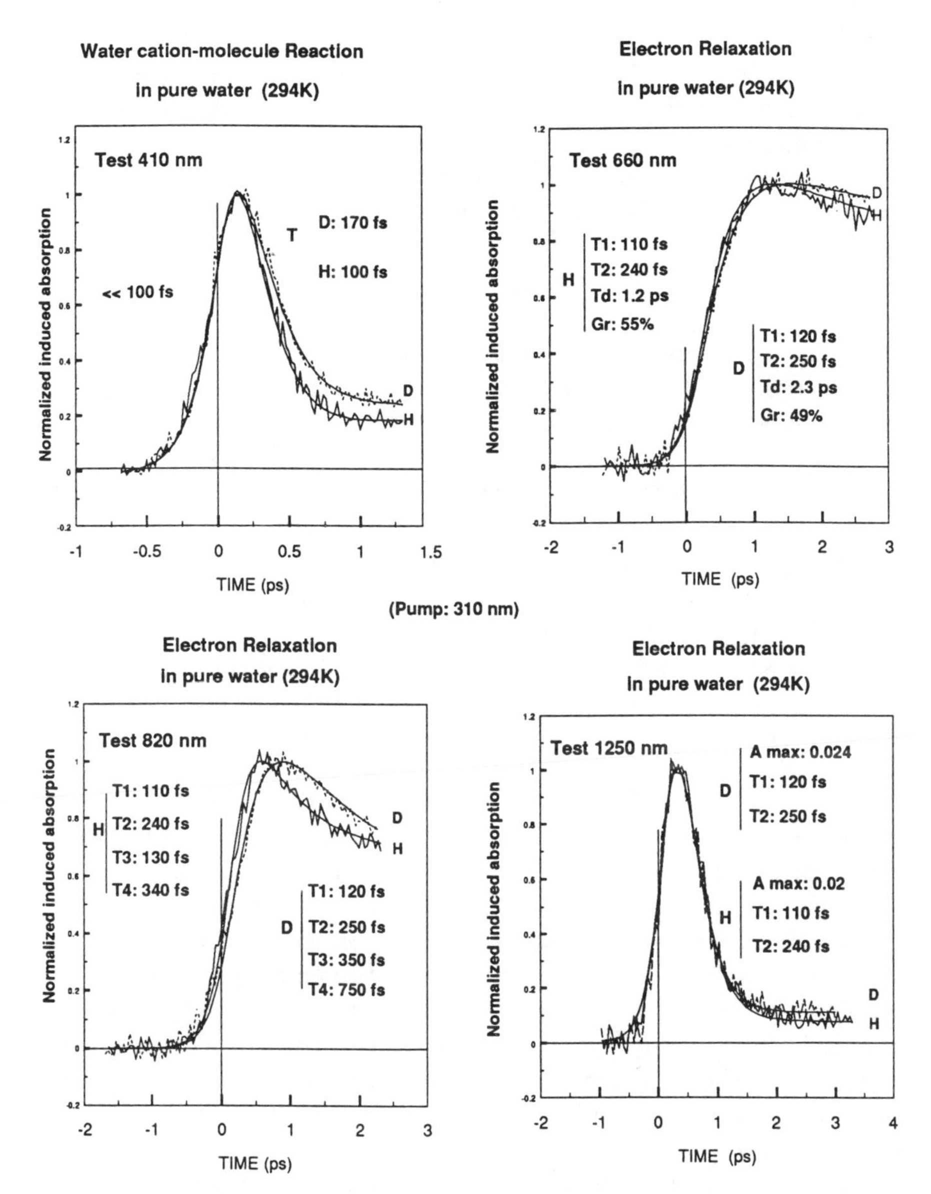

Fig. 14. Set of time-resolved spectroscopy data in pure light and heavy water following photoionization by ultraviolet femtosecond pulses (λ: 310 nm, E: 4 eV).

important point concerns the ultrafast electron-hole coupling in liquid will be discussed at length in the following paragraphs. One important point clarified by femtosecond spectroscopy is that the ion-molecule reaction exhibits a significant H/D isotope effect on the relaxation time of the fastest component. The effects shown in the Figure 14 demonstrate that the ultrafast proton transfer from the water cation to water molecule is likely dependent on specific properties of the protic solvent such as the vibrational energy of antisymetric stretch. Concerning the physical meaning of a positive hole migration in a polar protic solvent, we should wonder whether this process involves or not some important structural changes including geometrical perturbation of the hydration cage around the water cation or an excess electron.

2. Positive Hole Migration and Non-Equilibrium Electronic States

It is interesting to discuss more at length the relationship that exists between an H/D isotope effect on the ion-molecule reaction and the formation time of nonequilibrium electronic state in pure liquid water. The water cation-molecule reaction observed in the near ultraviolet occurs with a transition probability of about 10^{13} s^{-1} i.e. few vibrational X—OX periods (X = H,D) [111, 115]. The limiting factor for the cleavage rate constant of the water cation X_2O^+ would be due to the dynamics of hydrogen bonds formation and the weak activation energy of the ultrafast proton transfer with adjacent water molecules along the hydrogen bond.

Previous experimental investigations have permitted to estimate that the migration frequency of the hole was about 21 times greater that the ion-molecule reaction (reaction 28) [116]. That means that a resonant proton transfer ($X_2O^+ + X_2O \rightarrow X_2O + X_2O^+$) would occurs with a frequency of $k_1/21$ i.e. $\sim 2.1 \times 10^{14}$ s^{-1} in H_2O and 1.23×10^{14} s^{-1} in D_2O. In light water, this value remains very comparable to the estimates on the vibration frequency of the H—OH bond $\nu_{H-OH} = 1.15 \times 10^{14}$ s^{-1} [117].

In the femtosecond spectroscopic studies of primary events in water, several key points concern the role of neutral or charged neoformed prototropic species (i) in the initiation of ultrashort lived favourable traps for fast electron localization, (ii) in a solvent cage effect which can limit the probability of the excess electron to escape far of the primary positive hole (X_2O^+). These two aspects can be linked to the influence of the behavior of water cation on the time dependence of the early electron- positive hole or electron-radical pair couplings in water.

One important question concerns the detailed mechanisms of an electron relaxation in this polar protic solvent and more particularly the energy transfer processes that can occur in the coulombic field of H_2O^+ or H_3O^+ or within the solvation shell of OH. During the solvation step, the profile of the energy loss can be influenced by an initial electron-pair distribution and the solvent relaxation phenomena around prototropic species. Computer simulations of hydrated hydronium ion at 300K have shown that the relaxation of water molecules around a new hydronium ion would be faster than the mean lifetime of the ion [118]. Ab-initio calculations on the structural characters of water molecules linked to the solvation shells of the hydronium ion in dilute solutions have demonstrated that the radial distribution functions $g_{OO}(r)$ and $g_{OH}(r)$ exhibit a maximum at 2.48 Å and 3.2 Å respectively [118]. These values are in agreement with structural informations obtained by X-ray and neutron diffraction [119]. Indeed, over a 7 Å distance range around the hydronium ion two hydration shells can be considered. If an excess electron is directly trapped in the structured hydration shell of H_3O^+ or OH, the initial charge separation distance would be shorter than the Onsager radius ($r_c = 7$ Å in water) but remains very similar to the estimates of reaction radius: e^- ... H_3O^+: 5 Å, e^- ... OH: 6 Å [85, 87].

Femtosecond photochemical investigations performed in pure light and heavy water at room temperature have permitted to discriminate the existence of a non-equilibrium electronic configuration in the near infrared spectral region (Figures 14, 15) and to precise whether the infrared electron (prehydrated state) and the neoformed electron-radical pair exhibit a common precursor. The discrimination of a short lived nonequilibrium electronic state in the near infrared (around 820 nm) have need to perform a carefull analysis of kinetical data at very short time ($t \sim$ 2ps) and longer time ($t \sim 10$ ps) including the existence of a broad visible band, the long tail of infrared electrons and the spectral component due to hybride electronic state [120].

Time-resolved data are characterized by very low signal values ($\Delta A_{\max} \sim 0.05$). Their analysis have been conducted considering a kinetical model for the expected signal rise dynamics which includes all the different transient species and the convolution of the pump-probe pulses temporal profile (Equations (31)–(38)). More precisely the induced absorbance at a specific test wavelength (λ) and for a time delay (τ) between the excitation and the probe beams is defined as the sum of the contribution $A_i^\lambda(\tau)$ of all the species i.e. the contributions of (i) the infrared prehydrated electrons, (ii) the fully hydrated electrons populations whose a fraction rapidly recombines with prototropic species (X_3O^+ or OX), (iii) a non equilibrium electronic state assigned to a neoformed electron-radical or electron-ion pairs [$X_3O^+ : e^-$:OX, with X = H,D]. This last transient species would correspond to a transition state for which the photogenerated electrons would be initially localized in the vicinity of prototropic species (water cation: X_2O^+ or its derivates: hydronium ion: X_3O^+ and hydroxyl radical: OX).

$$\Delta A^\lambda(\tau) = \sum A_i^\lambda(\tau) = \sum C_i \cdot \varepsilon_i \cdot l. \tag{31}$$

The transient signal triggered by a pump beam (I_{PU}), through a non linear phenomena (n order), is defined by the expression:

$$S(t) = \int_{-\infty}^{+\infty} A(t' - t) I_{PU}^n \, \mathrm{d}t'. \tag{32}$$

This signal can be probed by test pulse with ρ the temporal delay between the pump (I_{PU}) and probe (I_{PR}) pulses. The time-resolved discrimination of the physical phenomena $A(t)$ follows the relationship:

$$S'(t) = \int_{-\infty}^{+\infty} I_{PR}(t+\tau) \int_{-\infty}^{+\infty} A(t'-t) I_{PU}^{n} \, \mathrm{d}t'. \tag{33}$$

With the variable change $t = t - (\rho + t')$, $S'(t)$ will be defined by the expression:

$$S'(t) = \int_{-\infty}^{+\infty} A(t) I_{PR}(t+\tau) \int_{-\infty}^{+\infty} \mathrm{d}t' I_{PU}^{n}(t') I_{PR}(t+\tau'). \tag{34}$$

The first term represents the response of the molecular system and the second one to the correlation function between the pump and the probe pulses. The computed fits of the experimental traces are obtained considering symmetrical biexponential pump and probe pulses. The time broadening factor occuring when the pump and the probe wavelengths overlap with different group velocities is measured on sample for which an "instantaneous" response can be discriminated:

$$I_{PU}(t) = I_{PR}(t) = \exp - \mid t/\tau_{imp} \mid . \tag{35}$$

The time-resolved data are analyzed taking into account the determination of the zero time delay and the refactive effect index for the different samples. This last point is of particular importance at very short time when we consider the shirp of the zero time delay that occurs between two samples (H_2O, D_2O) having refractive index $n_1(\omega)$ and $n_2(\omega)$. The temporal delay between the two samples of optical length 'l' will be:

$$\begin{aligned} \Delta\tau &= T_{PR} - T_{PU} \\ &= \frac{1}{C}\left(\left[\frac{n_2(\omega_{PU}) - n_2(\omega_{PR})}{n_2(\omega_{PU}) \cdot n_2(\omega_{PR})}\right] - \left[\frac{n_1(\omega_{PU}) - n_1(\omega_{PR})}{n_1(\omega_{PU}) \cdot n_1(\omega_{PR})}\right]\right). \end{aligned} \tag{36}$$

In the particular case of small signal the time dependence of the different populations during the pumping and probing follows the relationship:

$$\Delta A_i^{\lambda}(\tau) = \varepsilon_i^{\lambda} \cdot l \int_{-\infty}^{+\infty} C_i(\tau - \tau') \cdot C^{\lambda}(\tau') \, \mathrm{d}\tau' \tag{37}$$

with:

$$C^{\lambda}(\tau) = \int_{-\infty}^{+\infty} I_{PR}(t+\tau) I_{PU}^{2}(t) \, \mathrm{d}t \, \mathrm{d}\tau'. \tag{38}$$

In these expressions, I is the interaction length, $C^{\lambda}(\tau')$ the normalized correlation between the probe and the pump pulse, ε^{λ} and C_i the molar extinction coefficient and the concentration of species i respectively.

Figure 14 permits to fully compare the influence of an H/D isotope substitution on the risetime and the early behavior of the induced absorption in the red and near infrared spectral regions. At 720 nm the early relaxation phenomena has been assigned to the existence of an ultrafast geminate recombination between hydrated electron and prototropic entities. At 820 nm, the data show that a non negligible fraction of the signal disappears faster than at 720 nm. This fact demonstrates that the relaxation phenomena observed in the isosbestic spectral range of transient spectra (Figure 4) cannot be only assigned to the existence of an early geminate recombination between fully hydrated electrons and prototropic species [67, 120]. The total relaxation process corresponds to a complex non exponential decay and includes all the transient components. A short-lived non equilibrium electronic state has been identified and assigned to the existence of a second photochemical channel: reaction (39).

$$
\begin{aligned}
n(\mathrm{X_2O}) + h\nu &\rightarrow (\mathrm{X_3O^+} \cdots \mathrm{OX})_{\mathrm{hyd}} + \mathrm{e^-_{qf}} - -k_5 \\
&\rightarrow [(\mathrm{X_3O} \cdots \mathrm{OX})^*_{\mathrm{hyd}}] \leftrightarrow [(\mathrm{X_3O^+} : \mathrm{e^-} : \mathrm{OX})_{\mathrm{hyd}}] \\
&\qquad (k_5 = 1/T_5) \\
[(\mathrm{X_3O..OX})^*_{\mathrm{hyd}}] &\leftrightarrow [(\mathrm{X_3O^+} : \mathrm{e^-} : \mathrm{OX})_{\mathrm{hyd}}] - - - k_6 \\
&\rightarrow \mathrm{X} + \mathrm{X_2O} \quad (k_6 = 1/T_6) \\
&\rightarrow \mathrm{X_3O^+} \ldots \mathrm{OX}? \\
&\rightarrow \mathrm{e^-_{hyd}} + (\mathrm{X_3O^+} \ldots \mathrm{OX})? \\
&\rightarrow (\mathrm{X_2O})_2\mathrm{e^-}? \qquad (39)
\end{aligned}
$$

In light water, the appearance time of a near infrared induced aborption (T_5) is very close to that of the infrared electron (T_3) and its mean lifetime (T_6) to the presolvated electron one (T_4). The carefull analysis of the H/D substitution effects on the risetime of the signal at 820 nm permits to discreminate the existence of a significant H/D isotope substitution effect on the formation time ($T_5 = 1/k_5$) and the mean lifetime ($T_6 = 1/k_6$) of the nonequilibrium electronic configurations (X_3O^+ : e^-:OX, with X = H,D). More precisely, the transient state assigned to an hybride electronic state appears with a time constant of 130$\pm$20 fs in light water and 320$\pm$20 fs in D_2O. The deactivation rate of the hybride electronic states, defined as $k_6 = 1/T_6$, is lower in deutered water than in light water: 1.36×10^{12} s^{-1} against 3×10^{12} s^{-1} respectively [121].

The initial yield of fully or non fully hydrated electrons generated by femtosecond ultraviolet excitation of water molecules can be dependent on the rate of energy scattering with the vibrational modes of neutral water molecules or the transient solvent configurations due to prototropic entities. The computed simulations of the transient populations of electrons (prehydrated states, electron-radical pair and fully relaxed hydrated states) reported in the Figure 15 permit to compare the effects of an H/D isotope substitution on the time dependence of the different populations of prototropic and electronic species in pure water. At very short time (t < 4ps) this figure shows that in H_2O and D_2O the ultrashort lived infrared electrons population (prehydrated state) exhibits a maximum at $t \sim 300$ fs. An H/D isotope substitution does not modified neither the temporal dependence of this non-equilibrium electronic population nor its maximum value ($C_{e^-\text{prehydmax}}$). It is interesting to notice that the population of fully relaxed hydrated electrons is never at equilibrium. As previously discussed, this situation is due to the influence of prototropics species through an early 1D walk geminate recombination (Equations (26), (27)). Computed data of Figure 15 demonstrate that within the first 2 ps, the spectroscopic contribution of hydrated electron detected in H_2O (D_2O) corresponds to 90% and (95%) of a saturated population ($C_{e^-\text{hydmax}}$). In other word, H/D isotope substitution modifies the early yield of photogenerated hydrated electrons in liquid phase and 10% (5%) of hydrated electron, cannot be detected at early time. Indeed, the direct observation of an isosbestic point is hindered [65, 120]. Additional data reported in the Figure 5 exhibit significant influence of an isotopic substitution on the population of hybride electronic state ($[(X_3O^+ : e^-:OX)_{hyd}]$) in the near infrared.

It is interesting to discuss more at length the relationship that exists between an H/D effect on the ion-molecule reaction (reaction 8) and the formation time of nonequilibrium electronic state in the near infrared (reaction 39). The water cation-molecule reaction observed in the near ultraviolet occurs with a transition probability of about 10^{13} s^{-1} i.e. few vibrational X—OX period [114]. The limiting factor for the cleavage rate constant of the water cation X_2O^+ would be due to the dynamics of hydrogen bonds formation and the weak activation energy of the ultrafast proton transfer with adjacent water molecules along the hydrogen bond. The influence of an H/D isotope substitution on T_5/T_2 ratio permits to precise that the subexcitation electron would probably test several configurations in the vicinity of prototropic species before to get an hybride electronic state (Figure 4). A limitation step in the formation of neoformed encounter pairs would correspond to proton motions during the ion-molecule reaction. Femtosecond data of Figures 14 and 15 show that the migration of water cation in D_2O is about 30% larger than in H_2O. Similar H/D isotope effect is obtained on the T_5/T_2 ratio [121]. This result suggests that a positive hole migration can be, at a microscopic level, a determinant factor for the existence of transient coupling between electron and prototropic species. This does not exclude that the elec-

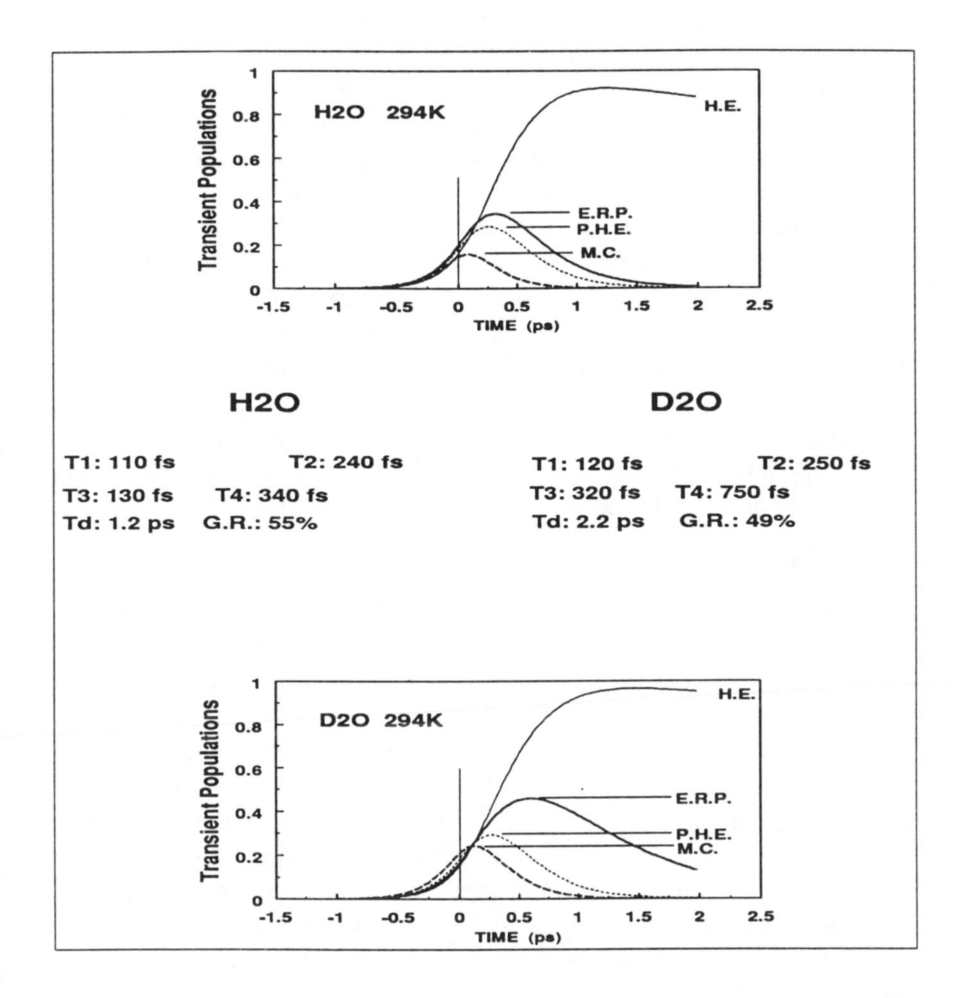

Fig. 15. Computed simulation of time-dependent non-equilibrium populations in pure H_2O and D_2O: H. E.: hydrated electron; E.R.P.: electron-radical pair; P.H.E.: prehydrated electron; M.C.: molecular cation (X_2O^+, X=H, D).

tron can contribute to influence the local environment of prototropic species with polarization of hydrogen bonds through proton dispersion forces. Monte Carlo simulations have shown that polarization of hydrogen bonds are three times more than of water molecules [122].

Numerous discussions have been mentioned concerning the photon energy required to perform vertical Born-Oppenheimer ionization of liquid water at ambient temperature [59–61, 63]. Using picosecond pulses, the production of hydrated electrons can be observed around 6.5 eV i.e. 2 eV lower than the estimate of ionization potential (8.76 eV) [60]. These experimental data have suggested that ultrafast internal energy exchange and cage effects, which are not considered in the macroscopic ionization threshold, can affect the early charge separation and photoionization quantum yield. It is interesting to notice that the value 6.5 eV is very near the energy of the first electronically excited state (A^1B_1) whose the photodecomposition gives OH (XII) and $H(^2S)$ [123].

Experimental studies on liquid water have suggested the existence of a low energy channel under the energy band gap for which the excitation energy required to produced hydrated electron and prototropic species equals 5.78 eV [61, 62]. This charge transfer process can compete with the direct dissociation of the $\sim A$ state ($1b_1 \rightarrow 3sa_1$), and involve transitions from favorable site geometry to permit the thermodynamic cycle of an autoionization process (low energy photochemical channel). The cooperative effect of two water molecules would enhance the photogeneration of electron on short distance i.e. the trapping of the excess electron would occur in the first two solvation shells of the prototropic species.

$$2\,H_2O \rightarrow (H_3O^+)_{hyd} + (OH)_{hyd} + e^-_{hyd}. \qquad (40)$$

Femtosecond studies on H/D isotopic substitution effects clearly demonstrates the existence of a low energy photochemical channel. Femtosecond ultraviolet excitation of water molecules through two-photons process ($2\times$ 4 eV) can represent sufficient energy to initiate via lattice vibrations, ultrashort-lived configurations (electron-radical pair). The kinetical data obtained in the red and near infrared confirms that the femtosecond ultraviolet excitation of water molecules initiates the formation of early electron-radical pairs (e^- ... X_3O^+; e^- ... $OX)_{nX_2O}$, X = H,D) through transient couplings between electron and prototropic species (Figures 4, 11). In light and heavy liquid water, the early charge transfer linked to the ion-molecule reaction (Equation 28) is fastest that the formation of electron-radical pairs (reaction 39 and Figures 14, 15). This means that very transient favourable structured environments for electron localization can be triggered by the presence of neoformed prototropic species (X_2O^+ or its derivates). In this hypothesis, excess electron would get localized or trapped state after energy scattering with energetic vibrational mode and vibration frequency (H—OH vs D—OD). Ultrafast nuclear motions such as proton transfer can favor the existence of local solvent cage configurations whose the microscopic structure would be dependent on a concerted phenomena occuring between electron, hydrogen bonds lattice and prototropic species. Table I and Figure 16 represent concerted and non concerted electron-proton transfers which can be discriminated at early time in pure water [71, 114, 120, 121].

The significant change observed on T_5 (reaction 39 between light and heavy water underlines the role of a concerted mechanism in the existence

TABLE I

Primary events linked to concerted and nonconcerted electron-proton transfers in pure water. Photoexcitation of water molecules is initiated by ultraviolet femtosecond pulses. Steps are numbered as in Figure 16

n X_2O (X = H,D) + x$h\nu$ → Primary Reactional STEPS	
STEPS	
1	$X_2O^+ + n(X_2O) \rightarrow (X_3O^+)_{n_1X_2O} + (OX)_{n2X_2O}$
2, 4	$e^- + n'(X_2O) \rightarrow e^-_{IR} \rightarrow e^-_{Vis}$
3	$n''(X_2O) \rightarrow [OX : e^- : X_3O^+]_{n''-2(X_2O)}$
5	$[OX : e^- : X_3O^+]_{n''-2(X_2O)} \rightarrow OX^-, X, n''X_2O$
6	$e^-_{Vis} + (OX, X_3O^+)_{(n_1+n_2)(X_2O)} \rightarrow OX^-, X, n''''X_2O$

of a low photochemical channel for which a proton transfer and an electron trapping would be involved to give an hybride electronic state (Rydberg state or neoformed encounter pair). It is interesting to notice that the formation time ratio $(T_5(D_2O)/T_5\ (H_2O) \sim 2.4$ is very similar to the ratio of predissociation rate of Rydberg states which has been shown to be dependent on rotational sub level and exhibit significant isotope effect: $T(C\ \sim)_{D_2O}T(C\ \sim)_{H_2O}$ = 2.4 [124]. As previously suggested short-lived configurations of solvent molecules resulting of electronic excitation would be equivalent to Rydberg states whose adjacent water molecules can influence their behavior. Initial configurations involving several water molecules would correspond to favorable local trap for which cooperative effects on electron ejection made be probable under the energy band gap of water. If the energy distribution of pre-existing traps is defined by fluctuations density of the solvent, then ultrafast structural liquid reorganization initiated by short lived prototropic species would favor electron localization by self trapping phenomena.

An important point to be considered on ultrafast electron transfer in a polar protic solvent concerns the comparison between the dynamics of formation (T_3) and relaxation (T_4) of infrared electron through a two states activated model with those of neoformed electron-radical pairs (T_5, T_6). In pure light water at ambient temperature (Figure 15), femtosecond spectroscopic data show that the formation time of early electron-pair (T_5) remains similar to the dynamics of electron localization (T_3): T_5(H)/T_3(H) = 1.18. Consequently, it could be suggested the existence of a common precursor for the infrared electron (prehydrated electron) and the neoformed encounter pair [120]. However, the study with H/D substitution permits to clarify this interpretation and to

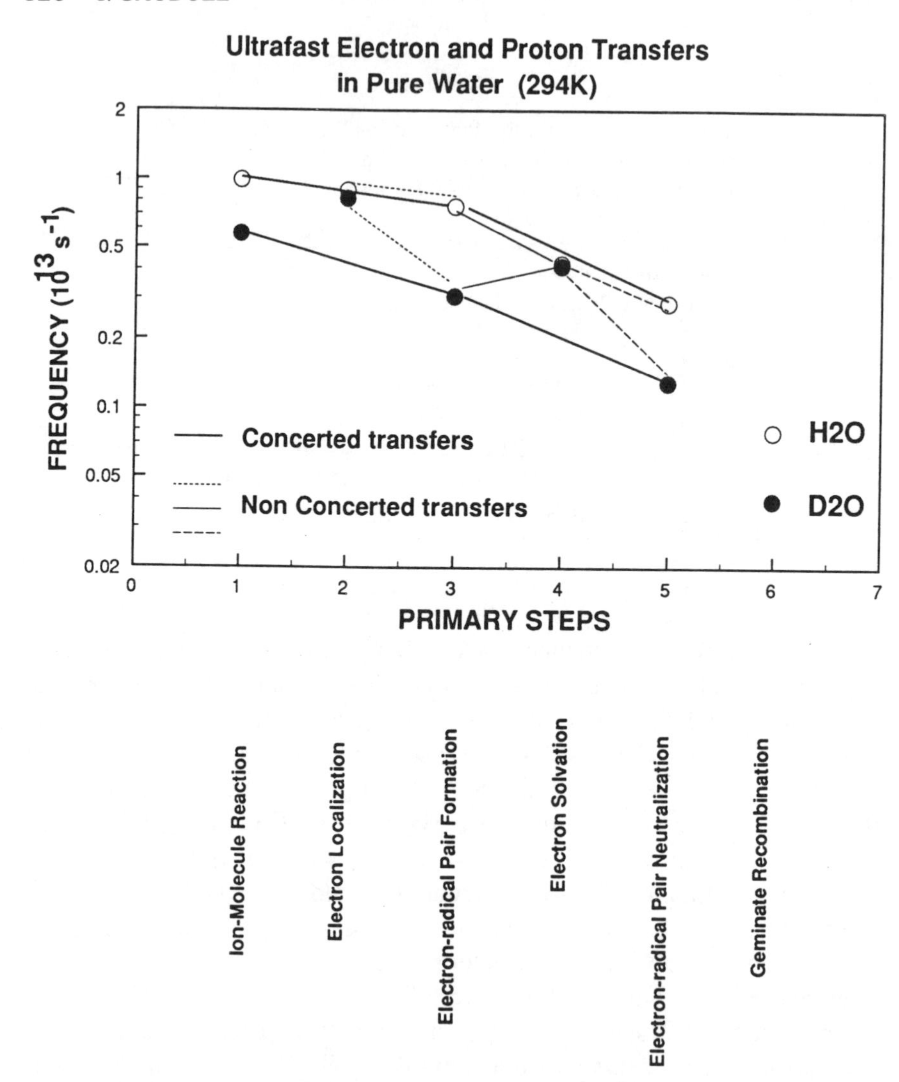

Fig. 16. Synthetical representation of electron and proton transfer frequencies in pure liquid water (H_2O, D_2O) following femtosecond ultraviolet photoionization. Primary reactionnal steps are studied by femtosecond absorption spectroscopy. Concerted and non concerted charge transfers can be clearly discriminated.

conclude that the existence of a competitive process from a commom precursor of presolvated electron and electron-radical pairs is unlikely [121]. The presolvation of excess electron and the formation of electron-radical pairs do not involved similar responses of the water molecules.

3. *Transition States and Charge Transfer Dynamics*

Femtosecond spectroscopic investigations of charge transfer in pure deutered water permit us (i) to precise whether several photochemical channels can lead to electron solvation, (ii) to extend the knowledge on primary reactions between electron and prototropic species. Figure 16 shows the existence of two well defined relaxation processes for electronic states in the infrared and near infrared. The electronic relaxation does not occur in the same way when excess electron is localized far or in the vicinity of neoformed prototropic species. The experimental data obtained in the near infrared show the existence of a short-lived electronic component for which the frequency of neutralization process (Table I, step 5) is significantly dependent on the protic character of the solvent: $k_6 = 1/T_6 = 0.29 \times 10^{13}$ s^{-1} in H_2O and 0.13×10^{13} s^{-1} in D_2O. This means that the probability to obtain hydrated electron from neutralization reaction of electron-radical pairs (step 4 vs step 5 of Figure 16 and Table I) is very low. This figure underlines the existence of a discrepancy between a non adiabatic relaxation of infrared electron and an electron-radical pair deactivation.

It is interesting to compare the relaxation frequency of hybride electronic states assigned to electron-radical pair $[(X_3O^+ : e^- {:} OX)_{\rm hyd}]$ with the estimate of the lifetime of excited molecules produced by ultrafast neutralization of unrelaxed electron with primary water cation: $e^- + X_2O^+ \rightarrow [X_2O^*]_{nX_2O} \rightarrow 2X_2O$ (Figure 17). In absence of surrounding water molecules, the deactivation of $[X_2O^*]$ can be estimated by formula of Henly and Johnson [125]:

$$\tau^{-1} = \nu_{\rm H-OH}((E^* - E_{\rm dis})E^*)^2 \tag{41}$$

for which E^* equals the exciting energy of neutral molecules ($\sim$ 6.25 eV) and $E_{\rm dis} \sim 5.11$ eV. The H-OH decay channel occurs with a time constant of 3×10^{-13} S i.e. a frequency of 0.33×10^{13} s^{-1}. The Figure 17 shows that the deactivation frequency of electron-radical pairs in H_2O is similar to this estimate ($k_6 = 1/T_6 \sim 0.29 \times 10^{13}$ s^{-1} at 294 K). This means that an early rearrangment of water molecules during the deactivation of encounter pair (step 6 of Figure 16 and Table I) can be equivalent to a relaxation of an excited state of water molecule $[H_2O^*]_{nH_2O}$. Considering that we cannot obtain direct spectroscopic informations on the ultrafast quasi free electron-water cation reaction ($H_2O^+ + e^- \rightarrow H_2O^*$, $t \sim 10^{-14}$ S), it is not easy to determine the quantum yield of electron solvation from virtual excited states of water molecules. This point is more complex if we consider the existence of different statistical configurations of hybride electronic states in the near infrared at very short time (Figure 4). Recent theoretical works on electron solvation in bulk have suggested that different populations of non equilibrium electronic states (prehydrated states, excited solvated state, trapped electron

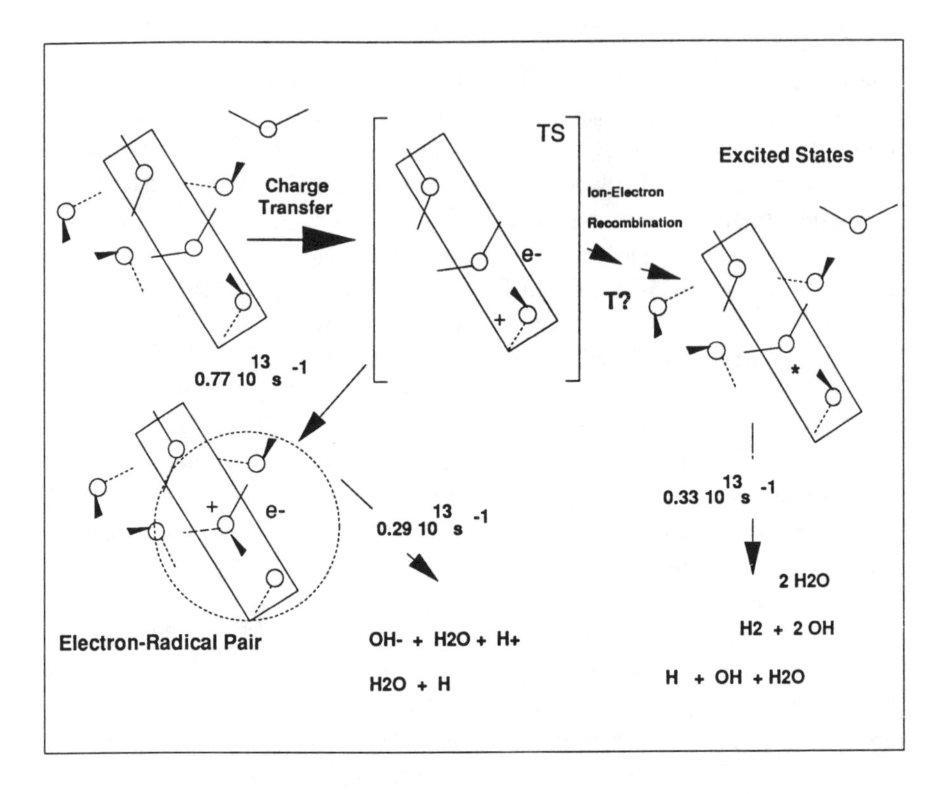

Fig. 17. Comparison between the relaxation frequency of hybride electronic states linked to an electron-radical pair (H_3O^+:e^-:OH) or to an excited water molecule produced by ultrafast recombination ($e^- + H_2O^+ \rightarrow (H_2O)^*_{nH_2O} \rightarrow 2H_2O$.

by pre-existing deep traps) can contribute to the transient infrared signal [126].

Concerning the behavior of non-equilibrium electron in the framework of a low photochemical channel, the relaxation process correspond to ultrafast neutralization between localized states and prototropic species without the intervention of a significant diffusion process (Figure 17). The existence of neoformed encounter pairs (e^-:X_3O^+:OX) can be dependent on dynamics of water molecules rearrangments in the vicinity of prototropic entities. The reorganization of water molecules may require breaking of the hydrogen bonds for which important parameters such as the OH strech mode will be concerned. The cleavage rate constant of electron-radical pair occurs at a similar time scale that the H-bond mean lifetime or average lifetime of the hydrated proton [93, 94]. Let us discuss the relationship that can exist between the lifetime of transition states and the electron-proton reaction efficiency (reaction 42). As previously shon in Section 2, the experimental bimolecular rate constant is about 30% of the estimate for a diffusion controlled reaction [85, 87]. Reaction (42) will not be diffusion controlled but influenced by the activation rate constant K_{act}. The lifetime of the encounter pair would be one of the main important parameter of the electron-proton reaction (reaction 42).

$$\begin{aligned} e_{\text{hyd}}^{-} + (H_3O^+)_{nH_2O} - - - k_D \rightarrow [\text{T. States}] \leftarrow k_{\text{Act}} \\ \rightarrow \gg H + (H_2O)_{nH_2O}. \end{aligned} \quad (42)$$

Femtosecond photochemical studies performed in pure liquid water emphase the existence of nonequilibrium electronic states which have been assigned to neoformed encounter pairs between electrons and hydrated hydronium ion (X_3O^+) or hydroxyl radical (OX), X = H, D. The mean lifetime of these transient electronic states in pure water at 294 K (H_2O and D_2O) is less than 1 ps (Figures 14, 16, 17). For a given concentration of hydronium ion, if we consider that the probability of reaction (42) is dependent on the lifetime of encounter pair ($e^- \ldots H_3O^+$)nH_2O, then the ratio of the pseudo-first order constant ($k_{\text{Exp.}}/k_{\text{Theor}}$) can be expressed as follow:

$$\begin{aligned} k_{\text{Exp.}}/k_{\text{Theor.}} &\sim 0.3 \sim \nu_{\text{Eff}} EP/\nu_{\text{TS}} \sim (1/T_{\text{Eff}} EP)/(1/T_{\text{TS}}) \\ &\sim T_{\text{TS}}/T_{\text{Eff}} EP. \end{aligned} \quad (43)$$

In this expression, $v_{\text{Eff}} EP = 1/T_{\text{Eff}} EP$ represents the frequency of formation of *efficient encounter pair*. The limit case would correspond to those for which the lifetime of the encounter pair is limited by the mean lifetime of hydrated proton $(H_3O^+)_{\text{hyd}}$. In this hypothesis, the transient states of the reaction (43) can be characterized either by an activation barrier or by a tunneling process. The second parameter $v_{TS} = 1/T_{TS}$ represents the mean formation frequency of efficient and unefficient transition states. In a first reasonable approximation, this frequency can be defined from the experimental lifetime of encounter pair (e^-:H_3O^+): $T_{TS} \sim T_{\text{MeanEP}} = T_6$: 340 fs and 750 fs in H_2O and D_2O respectively [121]. Consequently, the theoretical ratio $T_{TS}/T_{\text{Eff}} EP$ defined by the Equation (43) equals 0.40 in H_2O and 0.53 in D_2O (Table II). These two values are estimated by defect owing to the fact that in our experimental conditions, up to now, we cannot decriminate the respective contributions of X_3O^+ and OX in the ultrafast neutralization process of electron-radical pairs. Consequently, the mean lifetime of electron-radical pairs we measure at 820 nm represents the average lifetime of two populations of pairs (e^-:H_3O^+ and e^-:OH) whose the formation probability would be dependent on transient microscopic structures of solvent cage.

NMR studies have shown that the lifetime of OH (OD) radical is 2 (3.5) times longer than the H_3O^+ (D_3O^+) ion [91]. If we take $T_{\text{Eff}}^{M} EP$ equivalent to the average lifetime of the two prototropic species $(T_{X_3O^+} + T_{\text{OX}})/2$ i.e. 1.12 ps in H_2O and 3.35 ps in D_2O, then the $T_{TS}/T_{\text{Eff}} EP$ ratios equals 0.26

TABLE II

Comparison between the macroscopic rate constant of the electron-proton reaction in water (reaction 26) and the characteristic times of ultrashort lived states identified by femtosecond infrared spectroscopy

SOLVENT	T^*_{TS}	$T_{\text{EFF}}(EP)$ (X_3O^+)**	$T_{TS}/T_{\text{EFF}}(EP)$	$T^M_{\text{EFF}}(EP)$ (X_3O^+, OX)**	$T_{TS}/T^M_{\text{EFF}}(EP)$
H_2O	340 fs	850 fs	**0.40**	1.12 ps	**0.30**
D_2O	750 fs	1.4 ps	**0.53**	3.35 ps	**0.23**

* Experimental lifetimes of transition state (encounter pairs) measured in the infrared [120, 121].

** Lifetime of prototropic species, X = H, D (from [91]).

in H_2O and 0.24 in D_2O (Table II). These calculations on the probability of reaction (43) via the existence of non reactive electron-radical pairs underline that (i) the lifetime of hybride states of electron in water are shortest than those of isolated prototropic species in very dilute solution, (ii) the reaction between electron and proton is not dependent on the lifetime of hydrated hydronium ion but rather short-lived electron-ion pair.

In future research, we should wonder whether the relaxation of nonequilibrium electronic states involves intra complex structural changes for which proton mobility, H bond mean lifetime or average lifetime of prototropic species will be determinant factors for activation energy of the reaction (42). Moreover, in a polar protic solvent the importance of many-body effects must be considered (i) the influence of the Coulomb attraction between electron and hydronium upon the electron: OH couplings; (ii) dynamical couplings between proton with neutral water molecules; (iii) the rôle of cooperative effects between water molecules [127, 128]. These aspects need to obtain better understanding of ultrafast reactions considering the microscopic structure of solvent cage around excess electron or prototropic species and the physical meaning of solvent coordinates at very short time (Figure 18).

Conclusions

Several key points can be underlined in conclusion of this chapter. within the physico-chemical stages (10^{-15}–10^{-12} s). In molecular liquids, ultrashort laser pulses allow to initiate selective photochemical processes (photoexcitation of molecular probes, photoejection of charge) to investigate (i) primary events within the physico-chemical stages (10^{-15}–10^{-12} s); (ii) early steps of single charge transfers in connection to solvation dynamics.

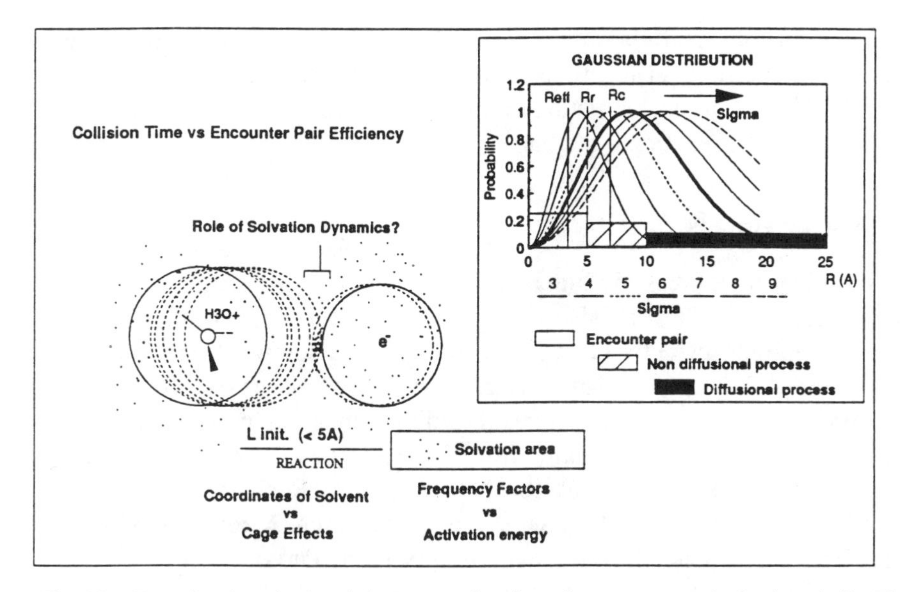

Fig. 18. Reactional regim involving an overlapping of two reactants hydration shells. The efficiency of the reaction ($e^-_{hyd} + H^+_{hyd} \rightarrow H_{hyd}$) is discussed in the framework of transient states (encounter pair) for which solvent coordinates equal $L_{Init} \sim R_{eff} > 5$ Å. The insert represents a Gaussian distribution of electron-prototropic species for different standard deviations (σ). The significant parameters are R_c: onsager radius in water, R: mean reaction radius of hydrated electron with hydronium ion or hydroxyl radical; R_{eff}: effective radius of the hydronium ion.

Experimental investigations of non-equilibrium electronic states in polar protic solvents permit to get unique informations on the dynamics of primary steps of elementary charge transfer reactions: formation of the hydration cage around an excess electron, encounter pair formation, ion-molecule reaction, electron attachment to solvent molecule, early electron-proton recombination. In liquid water, transient local configurations linked to the protic character of this solvent excer major effects on solvent cage reorganization and reaction dynamics within the non diffusional regime: dry positive hole reactivity; electron-hole pair neutralization. These cage effects are dependent on energetic vibrational mode of the solvent and statistical density fluctuations. An improvement in the knowledge of concerted charges transfers would consist to better describe reaction coordinates i.e. electronic clouds displacements during complexation mechanism. Spectral identification of ultrashort lived intermediates during concerted electron-proton transfer in water provide guidance for future developments on transition states theories in condensed matter.

The recent progress observed in experimental and theoretical works on ultrafast single charge transfer in liquid phases would lead to a better understanding of solvent effects on reaction dynamics. One of the most exciting challenge in physical chemistry and biology will be to extend our knowledge of charge transfer reactions at the molecular level keeping in mind the rôles of the microscopic structure of a solvent and the electronic or molecular responses of the reactional media.

Acknowledgments

I would like to express my indebtedness to Drs. A. Antonetti, A. Migus, S. Pommeret (X-ENSTA, Palaiseau), Drs. J. Belloni, J. L. Marignier (Orsay), Dr. S. Pimblott (USA) for helpful discussions and acknowledge the technical assistance of J. Bottu, G. Hamoniaux, N. Yamada for laser spectroscopy and data processing. Several experimental studies mentioned in this chapter have been supported by grants in aids from Direction des Recherches et Etudes Techniques (DRET, Paris), Centre National de la Recherche Scientifique (CNRS), Institut National de la Santé et de la Recherche Médicale (INSERM) and Groupe de Recherches Internationales Servier (Paris).

References

1. Kramers, H. A., *Physica VII* **4**, 284 (1940).
2. Noyes, R., *Prog. React. Kinetics* **1**, 129 (1961).
3. Marcus, R. A., *J. Chem. Phys.* **24**, 966 (1956); Marcus, R. A., *Ann. Rev. Phys. Chem.* **15**, 155 (1964); Sumi, H. and Marcus, R. A., *J. Chem. Phys.* **84**, 4894 (1986).
4. Burshtein, A. I., Khudyakov, I. V., and Yakabson, B. I., *Prog. Reaction Kinetics* **13**, 221 (1984).
5. Smith, I. W. M., in *'Topics in Current Physics', Nonequilibrium vibrational kinetics*, M. Capitelli Ed., Vol. 39, Springer-Verlag, 112 (1986).
6. Wiley, N. Y., *Kinetics of Nonhomogeneous Processes*, G. R. Freeman Ed., (1987).
7. *Chemical Reactivity in Liquids; Fundamental Aspects'*, M. Moreau, P. Turcq Eds., Plenum Press (1988).
8. Hynes, J. T., *Ann. Rev., Phys. Chem.* **36**, 573 (1985); Hynes, J. T., *J. Phys. Chem.* **90**, 3701 (1986); For recent reviews see 'The Theory of Chemical Reactions Dynamics', M. Baer Ed., Chemical Rubber, Boca Raton, FL (1986).
9. Castner, E. W., Maroncelli, M., and Fleming, G. R., *J. Chem. Phys.* **86**, 1090 (1987); Castner, E. W., Fleming, G. R., Bagchi, B., and Maroncelli, M., *J. Chem. Phys.* **89**, 3519 (1988).
10. Bagchi, B., *Ann. Rev. Chem.* **40**, 115 (1989) and references therein.
11. Zwan, G. V. and Hynes, J. T., *J. Chem. Phys.* **78**, 4174 (1983).
12. Mc Manis, G. E., Golovin, M. N., and Weaver, M. J., *J. Phys. Chem.* **90**, 6563 (1986); Mc Manis, G. E. and Weaver, M. J., *J. Chem. Phys.* **90**, 1720 (1989).
13. Simon, J. D. and Su, S., *J. Chem. Phys.* **87**, 7016 (1987); Su, S. and Simon, J. D., *J. Phys. Chem.* **93**, 753 (1989).
14. For recent molecular theories, see: Rips, I., Klafter, J., and Jortner, J., *J. Chem. Phys.* **88**, 3246, 89, 4288 (1988) and references therein.
15. 'Picosecond Phenomena', C. V. Shank, E. P. Ippen, S. L. Shapiro Eds., Springer Verlag, Berlin Heidelberg (1978); Shank, C. V., Greene, B. I. (1982); Gauduel, Y., Migus, A., Martin, J. L., Lecarpentier, Y., Antonetti, A., *Ber. Bunsen.-Ges. Phys. Chem.* **89**, 218 (1985); Migus, A., Antonetti, A., Etchepare, J., Hulin, D., Orszag, A., *J. Opt. Soc. Am. B* **2**, 584 (1985); Kosower, E. M., Huppert, D., *Ann. Rev. Phys. Chem.* **37**, 127 (1986).

16. 'Applications of time-resolved optical spectroscopy', Brückner, V. Feller, K. H. Grummt, U. W. Eds., Elsevier (1990).
17. Photoinduced Electron Transfer, Fox M. A. and M. C. Eds., Parts A-D, Elsevier (1988); Huppert, D., Ittah, V., Kosower, E. M., *Chem. Phys. Lett.* **144**, 15 (1988).
18. See the recent review of Barbara, P. F., Jarzeba, W., In '*Advances in Photochemistry*', Volman, Hammond, D. H., Gollnick, G. S., K. Eds., Wiley-Interscience, **15**, 1 (1989); Mataga, N., Hirata, Y., In '*Advances in multiphoton processes and spectroscopy*', Lin, S.H. Ed., Vol. 5, World scientific, **175** (1989).
19. 'Ultrafast Phenomena VII', Harris, C.B. Ippen, E.P. Mourou, G.A. Zewail A.H. Eds., *Springer Verlag*, Berlin (1990).
20. Dantus, M., Bowman, R.M., Gruebele, M., Zewail, A.H., *J. Chem. Phys.* **91**, 7437 (1989); Rose, T.S., Rosker, M.J., Zewail, A.H., *J. Chem. Phys.* **91**, 7415 (1989); Dantus, M., Bowman, R.M., Zewail, A.H., *Nature* **343**, 737 (1990). For general reviews see also: Zewail, A.H., Bernstein, R.B., *In Chemical and Engineering News* **66**, 24–43 (1988); Khundkar, L.R., Zewail, A.H., *Ann. Rev. Phys. Chem.* **41**, 15 (1990).
21. Yu, H., Karplus, M., *J. Chem. Phys.* **89**, 2366 (1988); Bader, J.S., Chandler, D., *Chem. Phys. Lett.* **157**, 501 (1989).
22. Sprik, M., Klein, M.L., *J. Chem. Phys.* **89**, 1592 (1988).
23. Yang, D.Y., Cukier, R.I., *J. Chem. Phys.* **91**, 281 (1989).
24. Benjamin, I., Wilson, K.R., *J. Chem. Phys.* **90**, 4176 (1989) and references therein; Lee, L.L., Li, Y.S., Li Wilson, K.R., *J. Chem. Phys.* (1991).
25. Kuharski, R.A., Bader, J.S., Chandler, D., Sprik, M., Klein, M.L., and Impey, R.W., *J. Chem. Phys.* **89**, 3248 (1988).
26. Smith, J.M., Lakshminarayan, and Knee, J.L., *J. Chem. Phys.* **93**, 4475 (1990).
27. Newton, M.D., *Chem. Rev.* **91**, 767 (1991).
28. Clementi, E., Corongiu, G., Bahattacharya, D., Feuston, B., Frye, D., Preiskorn, A., Rizzo, A., and Xue, W., *Chem. Rev.* **91**, 679 (1991).
29. Maroncelli, M., Castner, E.W., Bagchi, B., and Fleming, G.R., *Faraday Discuss. Chem. Soc.* **85**, 199 (1988).
30. Calef, D.F., Wolynes, P.G., *J. Phys. Chem.* **87**, 3387 (1983).
31. Cole, R.H., *Ann. Rev. Phys. Chem.* **40**, 1 (1989) and references therein.
32. Joachim, C., *Chem. Phys.* **116**, 339 (1987); Reimers, J.R.; Hush, N.S., *Chem. Phys.* **146**, 89 (1990) and references therein; Kosloff, R., Ratner, M.A., *Israel J. Chem.* **30**, 45 (1990).
33. Kahlow, M.A., Kang, T.J., Barbara, P.F., *J. Phys. Chem.* **91**, 6452 (1987); Nagarajan, V., Brearley, A.M., Kang, T., Barbara, P.F., *J. Chem. Phys.* **86**, 3183 (1987); Jarzeba, W., Walker, G.C., Johnson, A.E., Kahlow, M.A., Barbara, P.F., *J. Phys. Chem.* **92**, 7039 (1988). Kahlow, M.A., Jarzeba, W., Kang T.J., Barbara, P.F., *J. Chem. Phys.* **90**, 151 (1989). Weaver, M.J., McMannis, G.E., Jarzeba, W., Barbara, P.F., *J. Phys. Chem.* **94**, 1715 (1990).
34. Declemy, A., Rulliere, C., Kottis, P., *Laser Chem.* **10**, 413 (1990).
35. Frolich, H., Theory of dielectrics, Clavendon, Oxford (1950); Davidson, D.W., Cole, R.H., *J. Phys. Chem.* **19**, 1984 (1951).
36. Zwan, G., Hynes, J.T., *J. Phys. Chem.* **89**, 4181 (1985); Loring, R.F., Mukamel, S., *J. Phys. Chem.* **87**, 1272 (1988); Tachiya, M., *Chem. Phys. Lett.* **159**, 505 (1989).
37. Wolynes, P.G., *J. Chem. Phys.* **86**, 5133 (1987); Nichols III, A.L., Calef, D.F., *J. Chem. Phys.* **89**, 3783 (1988).
38. Rips, I., Klafter, J., Jortner, J., *J. Chem. Phys.* **88**, 3246 (1988); Stell, G., Zhou, Y., *J. Chem. Phys.* **91**, 4869, 4879, 4885 (1989).
39. Chandra, A., Bagchi, B., *Chem. Phys. Lett.* **151**, 47 (1988); idb *J. Chem. Phys.* **92**, 6833 (1990); *Chem. Phys.* **156**, 323 (1991).
40. Karim, O.A., Haymet, A.D., Banet, M.J., and Simon, J.D., *J. Phys. Chem.* **92**, 3391 (1988).
41. Fonseca, T. and Ladanyi, B., *J. Phys. Chem.* **95**, 2116 (1991).

42. Maroncelli, M., *J. Chem. Phys.* **94**, 2084 (1991).
43. Hart, E.J. and Boag, J.W., *J. Am. Chem. Soc.* **84**, 4090 (1962)
44. Rentzepis P. M., Jones R. P., and Jortner J. J., *J. Chem. Phys.* **59**, 766 (1973); Chase W. J. and Hunt J. W., *J. Phys. Chem.* **79**, 2835 (1975).
45. Baxendale J. H., *Can. J. Chem.* **78**, 1996 (1977); Jonah C.D., Matheson M. S., Miller J. R., and Hart E. J., *J. Phys. Chem.* **80**, 1267 (1976).
46. Wiesenfeld J. M. and Ippen E. P., *Chem. Phys. Lett.* **73**, 47 (1980).
47. Webster B. C. and Howat G., *Rad. Res. Rev.* **4**, 259 (1972); Dainton F. S., *Chem. Soc. Rev.* **4**, 323 (1975); Brodsky A. M. and Tsarevsky A. V., *Ad. Chem. Phys.* **44**, 483 (1980).
48. Kenney-Wallace G. A. and Jonah C. D., *J. Phys. Chem.* (1982) **86**, 2572; Lewis M. A. and Jonah C. D., *J. Phys. Chem.* **90**, 5367 (1986); Hirata Y. and Mataga N., **94**, 8503 (1990).
49. *The Chemical Physics of Solvation*, Parts A, B, C. In *Studies in physical and theoretical chemistry* **38**, Edited by Dogonadze R. R., Kalman E., Kornyshev A. A., Ulstrup J., Elsevier (1988); Bartczak W. M., Hummel A., *Radiat. Phys. Chem.* **27**, 71 (1986); Tachiya M., Schmidt W. F., *J. Chem. Phys.* **90**, 2471 (1989) and references therein.
50. Popkie H., Kistenmacher H., and Clementi E., *J. Chem. Phys.* **59**, 1325 (1973); Reimers J. R., Watts R. O., and Klein M. L., *Chem. Phys.* **64**, 95 (1982); Jorgensen W. L., Chandrasekhar J., Madura J.D., Impey R. W., and Klein M. L.; *J. Chem. Phys.* **79**, 926 (1983); Speedy R., Madura J., and Jorgensen W. L., *J. Phys. Chem.* **91**, 909 (1987).
51. Schnitker J., Rossky P. J., and Kenney-Wallace P. J., *J. Chem. Phys.* **85**, 2926 (1986).
52. Computer Simulations of Liquids", Allen M. P., Tidesly D. J., *Oxford Science Publications* (1990).
53. Rossky P. J., *J. Opt. Soc. Am.* B **7**, 1727 (1990) and references therein.
54. Jonah C. D., Romero C., and Rahman A., *Chem. Phys. Lett.* **123**, 209 (1986); Romero C. and Jonah C. D., *J. Chem. Phys.* **90**, 1877 (1989).
55. Wallqvist A., Martyna G., and Berne B. J., *J. Phys. Chem.* **92**, 1721 (1988).
56. *Water Science Reviews*, Felix Franks Ed., Cambridge University Press, NY., 1985.
57. Dore J. C. in "Water Science Reviews 1", Felix Franks Ed., Cambridge University Press, NY., 2 (1985); Toukan K., Ricci M. A., Chen S., Loong C., Price D. L., and Teixeira J., *Physical Reveiw A* **37**, 2580 (1988).
58. Robinson G. W., Thistlethwaite P. J., and Lee J., *J. Phys. Chem.* **90**, 4224 (1986); Hameka H. F., Robinson G. W., and Marsden J., *J. Phys. Chem.* **91**, 3150 (1987); Guissani V., Guillot B., and Bratos S., *J. Chem. Phys.* **88**, 5850 (1988).
59. Nikogosyan D. N., Oraevsky A. O., and Rupaso V. I., *Chem. Phys.* **77**, 131 (1983).
60. Grand D., Bernas A., and Amouyal E., *Chem. Phys.* **44**, 73 (1979); Bernas A., Ferradini C., and Jay-Gerin J. P., *Chem. Phys. Lett.* **170**, 492 (1990).
61. Sokolov V. and Stein G., *J. Chem. Phys.* **44**, 3329 (1966); Boyle J. W., Ghormley J. A., Hochanadel C. J., Riley J. F., *J. Phys. Chem.*, **73**, 2886 (1969).
62. Han P. and Bartels D. M., *J. Phys. Chem.* **94**, 5824 (1990).
63. Delahay P., *Acc. Chem. Res.* **15**, 40 (1982).
64. Klassen, N. V. "In Radiation Chemistry", Farhataziz and Rodgers Eds., VCH, p 29–64 (1987) and references therein.
65. Gauduel, Y., Martin, J. L., Migus, A., and Antonetti, A. In Ultrafast Phenomena V, Fleming and Siegman Eds, Springer Verlag, New York, (1986) 308. Migus, A., Gauduel, Y., Martin, J. L., and Antonetti, A., *Phys. Rev. Lett.* **58**, 1559 (1987): Gauduel, Y., Migus, A., Martin, J. L., Antonetti, A., *Chem. Phys. Lett.* **108**, 318 (1984); *Rad. Phys. Chem.* **34**, 5 (1989).
66. (a) Hart, E. J. and Anbar, M., "The Hydrated Electron", Wiley, New York, 1970; (b) Okazaki, K. and Freeman, G., *Can. J. Chem.* **56**, 2313 (1978); Kondo, Y., Aikawa, M., Sumiyoshi, T., Katayama, M., and Kroh, J., *J. Phys. Chem.*, **80**, 2544 (1984).
67. Lu, H., Long, F. H., Bowman, R. M., and Eisenthal, K. B., *Phys. Rev. Lett.* **64**, 1469 (1990).

68. Webster, F., Schnitker, J., Friedrichs, M. S., Friesner, R. A., and Rossky, P. J., *Phys. Rev. Lett.* **66**, 3172 (1991); Webster, F. A., Rossky, P. J., and Friesner, R. A., *Computer Physics Comm.* **63**, 494 (1991).
69. Neria, E., Nitzan, A., Barnett, R. N., and Landman, U., *Phys. Rev. Lett.*, In press (1991).
70. Mozumder, A., *Radiat. Phys. Chem.* **34**, 1 (1989) and references therein; Rips, I., Silbey, R. J., *J. Chem. Phys.*, in press (1991) and reference therein.
71. Gauduel, Y., Pommeret, S., Yamada, N., Migus, A., and Antonetti, A., *J. Phys. Chem.* **95**, 533 (1991).
72. Konovalov, V. V., Raitsiimring, A. M., Tsvetkov, Y. D., *Radiat. Phys. Chem.* **32**, 623 (1988).
73. Hilczer, M., Bartczak, W., Kroh, J., *J. Chem. Phys.* **89**, 2286 (1988); Hilczer, M., Bartczak, W., *J. Phys. Chem.* **94**, 6165 (1990); Pommeret, S., Gauduel, Y., *J. Phys. Chem.* **95**, 4126 (1991).
74. Motakabbir, K. A., Rossky, P. J., *Chem. Phys.* **129**, 253 (1989); Schnitker, J., Rossky, P. J., *J. Phys. Chem.* **93**, 6965 (1989).
75. Barnett, R. N., Landman, U., Cleveland, C. L., and Jortner, J., *J. Chem. Phys.* **88**, 4421 (1988); 88, 4429 (1988); Barnett, R. B., Landman, U., and Nitzan, A., *J. Chem. Phys.* **90** (1989); 91, 5567 (1989); Batnett, R. N., Landman, U., and Nitzan, A., *J. Chem. Phys.* **93**, 6535 (1990).
76. Wallquist, A., Thirumalai, D., and Berne, B. J., *J. Chem. Phys.* **85**, 1583 (1986).
77. Matheson, M. S., Mulac, W. A., and Rabani, J., *J. Phys. Chem.* **67**, 2613 (1963); Logan, S. R., *Trans. Farad. Soc.* **63**, 3004 (1967).
78. Gauduel, Y., Migus, A., Chambaret, J. P., and Antonetti, A., *Rev. Phys. Appl.* **22**, 1755 (1987); Gauduel, Y., Pommeret, S., and Antonetti, A., *J. Phys. Cond. Mett.* **2**, SA171 (1990).
79. Long, F. H., Lu, H., Eisenthal, K. B., *J. Chem. Phys.* **91**, 4193 (1989); Long. H., Lu, H., Shi, X., Eisenthal, K. B., *Chem. Phys. Lett.* **169**, 165 (1990).
80. Gauduel, Y. et al. Manuscript in preparation.
81. Water and Aqueous solutions, G. W. Neilson and J. E. Enderby, Adam Higer Ed. (1985).
82. Impey, R. M., Madden, P. A., and McDonald, I. R., *J. Phys. Chem.* **87**, 5071 (1983).
83. Kreitus, I., *J. Phys. Chem.* **89**, 1987 (1985).
84. Coyle, P. J., Dainton, F. S., and Logan, S. R., *Proc. Chem. Soc. London* **73**, 219 (1964).
85. Buxton, G. V., Greenstock, C. L., Helman, W. P., and Ross, A. B., *J. Phys. Chem.* **17**, 513 (1988); Tumer, J. E., Hamm, R. N., Wright, H. A., Ritchie, R. H., Magee, J. L., Chatterjee, A., and Bolch, W. E., *Radiat. Phys. Chem.* **32**, 503 (1988).
86. Wolff, R. K., Bronskill, M. J., and Hunt, J. W., *J. Chem. Phys.* **53**, 4211 (1970); Peled, E. and Czapski, G., *J. Phys. Chem.* **75**, 3626 (1971); Dorfman, L. M. and Taub, I. A., *J. Am. Chem. Soc.* **85**, 2370 (1963).
87. Biondi, C. and Bellugi, L., *Chem. Phys.* **62**, 145 (1981) and reference therein; Baker, G. C., Fowles, P., Sammon, D. C., and Stringer, B., *Trans. Faraday Soc.* **66**, 1498 (1970).
88. Lam, K. Y. and Hunt, J. W., *Int. J. Radiat. Phys. Chem.*, **7**, 317 (1975); Razem, D. and Hamill, W. H., *J. Phys. Chem.* **81**, 1625 (1977).
89. Czapski, G., Peled, E., *J. Phys. Chem.* **73**, 593 (1973).
90. Gauduel, Y., Pommeret, S., Migus, A., Yamada, N., and Antonetti, A., *J. Am. Chem. Soc.* **112**, 2925 (1990).
91. Halle, B. and Karlstrom, G., *J. Chem. Soc. Faraday Trans.* **70**, 1047 (1983).
92. Pernoll I. and Maier U., *J. Chem. Soc. Farad. Trans.* **2**, 71, 201 (1975); Hayd A., Weidemann E. G., and Zundel G., *J. Chem. Phys.* **70**, 86 (1979): *Kanno H., Hiraishi J., Chem. Phys. Lett.* **107**, 438 (1984); Janoschek R., Weidemann E. G., Pfeifer H., and Zundel G., *J. Am. Chem. Soc.* **94**, 2387 (1972).
93. Glick R. E. and Tewari K. C., *J. Chem. Phys.* **44**, 546 (1966); Rabideau S. W. and Hetch H. G., *J. Chem. Phys.* **47**, 544 (1967).
94. Conde O. and Teixera J., *J. Physique* **44**, 525 (1983).

95. Chaudri A. A., Göbl, M., Freyholdt T., and Asmus K. D., *J. Am. Chem. Soc.* **106**, 5988 (1984); Belloni J. and Marignier J. L., *Rad. Phys. Chem.* **34**, 157 (1989) and references therein.
96. Marignier J. L. and Belloni J., *J. Phys. Chem.* **85**, 3100 (1981); Belloni J., Marignier J. L., Katsumura Y., and Tabata Y., *J. Phys. Chem.* **90**, 4014 (1986).
97. Gauduel Y., Pommeret S., Antonetti A., Belloni S., and Marignier J. L., *J. Phys. Supl. V* **C5–161** (1991).
98. Asmus K. D., In *'Sulfur-centered Reactive Intermediates in Chemistry and Biology'*, Ed. C. Chatgilialoglu, Asmus K. D., *Plenum Press*, NY, 155 (1990) and reference therein.
99. Gauduel Y., Berrod S., Migus A., Yamada N., and Antonetti A., *Biochem.* **27**, 2509 (1988); Gauduel Y., Pommeret S., Yamada N., Migus A., and Antonetti A., *J. Am. Chem. Soc.* **111**, 4974 (1989): *J. Opt. Soc. Am. B* **7**, 1528 (1990).
100. Wollf R. K., Aldrich J. E., Penner T. L., and Hunt J. W., *J. Phys. Chem.* **79**, 210 (1975).
101. Collie C. H., Hasted J. B., and Riston D. M., *Proc. Phys. Soc.* **60**, 145 (1948)
102. Nemethy G. and Sheraga H. A., *J. Chem. Phys.* **41**, 680 (1964).
103. Anbar M. and Meyerstein D., *Trans. Farad. Soc.* **62**, 2121 (1966).
104. Kuharski R. A. and Rossky P. J., *J. Chem. Phys.* **82**, 5164 (1985).
105. Lu H., Long F. H., Bowman R. M., and Eisenthal K. B., *J. Phys. Chem.* **93**, 27 (1989).
106. Gauduel Y., Pommeret S., Migus A., and Antonetti A., *J. Phys. Chem.* **93**, 3380 (1989).
107. Onsager L., *Phys. Rev.* **34**, (1938); Hubbard J. and Onsager, L., *J. Chem. Phys.* **67**, 4850 (1977).
108. Turner J. E., Magee J. L., Weight H. A., Chatterjee A., Hamm R. N., and Ritchie R. H., *Radiat. Res.* **96**, 437 (1983); Chernovitz A. C. and Jonah C. D., *J. Phys. Chem.* **92**, 5946 (1988).
109. Pimblott S., *J. Phys. Chem.* **95**, 6946 (1991); Goulet T. and Jay-Gerin J. P., *J. Chem. Phys.* **96**, 5076 (1992).
110. Tachiya M., *J. Chem. Phys.* **87**, 4108 (1987).
111. Buxton G. V., *'In Radiation Chemistry'*, Farhataziz and Rodgers Eds., VCH, p 321–349 (1987).
112. Jonah C. D. and Chernovitz A. C., *Can. J. Chem.* **68**, 935 (1990).
113. Lew H. and Heiber I., *J. Chem. Phys.* **58**, 1246 (1972); Dutuit O., Tabche-Fouhaile A., Nenner I., Frolich H., and Guyon P. M., *J. Chem. Phys.* **85**, 584 (1985).
114. Gauduel Y., Pommeret S., Migus A., and Antonetti A., *Chem. Phys.* **149**, 1 (1990).
115. Zaider M. and Brenner D. J., *Radiat. Res.*, **100** 245 (1984); Schmidt K. H. and Buck W. L., *Science* **151**, 70 (1966).
116. Ogura H. and Hamill W. H., *J. Phys. Chem.* **77**, 2952 (1973).
117. Kaplan I. G., Miterev A. M., and Sukhonosov Y., *Rad. Phys. Chem.* **36**, 493 (1990).
118. Newton M. D. and Ehrenson S., *J. Am. Chem. Soc.* **93**, 4971 (1971); Newton M. D., *J. Chem. Phys.* **67**, 5535 (1977); Fornili S. L., Migliore M., and Palazzo M. A., *Chem. Phys. Lett.* **125**, 419 (1986).
119. O'Ferral R. A., Koeppl G. W., and Kresge A., *J. Am. Chem. Soc.* **93**, 1 (1971); Giguere P. A. and Guillot J. C., *J. Phys. Chem.* **86**, 3231 (1982).
120. Pommeret S., Antonetti A., and Gauduel Y., *J. Am. Chem. Soc.* **113**, 9105 (1991).
121. Gauduel Y., Pommeret S., and Antonetti A., *J. Phys. Chem.* **97**, 134 (1993).
122. Janoschek R., Weidemann E. G., Pfeifer H., and Zundel G., *J. Am. Chem. Soc.* **94**, 2387 (1972); Pernoll I. and Maier U., *J. Chem. Soc. Farad. Trans.* **2**, 71, 201 (1975); Hayd A., Weidemann E. G., and Zundel G. *J. Chem. Phys.* **70**, 86 (1979): Kanno H. and Hiraishi J., *Chem. Phys. Lett.* **107**, 438 (1984).
123. Engel V., Meijer G., Bath A., Andresen P., and Schinke R., *J. Chem. Phys.* **87**, 4310 (1987).
124. Docker M. P., Hodgson A., and Simons J. P., *Mol. Phys.* **57**, 129 (1986).
125. Henly E. and Johnson E., *The chemistry and Physics of high energy reactions*, University Press (1969).
126. Messmer M. C. and Simon J. D., *J. Phys. Chem.* **94**, 1220 (1990).
127. Hameka H. F., Robinson G. W., and Marsden J., *J. Phys. Chem.* **91**, 3150 (1987).
128. Tachikawa H. and Ogasawara M., *J. Phys. Chem.* **94**, 1746 (1990).

4. The Dynamics of Anion Solvation in Alcohols*

Y. LIN* and C. D. JONAH*
Chemistry Division, Argonne National Laboratory Argonne, IL 60439, U. S. A.

1. Introduction

The importance of the solvent in chemical reactivity has been long recognized,[1] albeit the quantification of the various roles that the solvent can play was not possible. These roles include: supplying or dissipating energy by acting as a heat bath, confining reactants, and solvation of ions. Solvation will be important for any reactions in which charge is created, destroyed or transferred, including intra-molecular charge redistribution, which, for example, leads to giant dipole formation in excited states.

Some of the earlier measurements of solvation have included electron solvation in liquids and glasses [2–9]. These results exposed the complexity that can occur in solvation kinetics. For example both the kinetic behavior and the rate of solvation processes have been found as a function of temperature [2–5]. These temperature-dependent studies clearly show that the solvation occurs by multiple processes with different energetics.

Electron solvation may not be representative of the solvation processes that are critical in many electron transfer reactions. In electron transfer reactions, there exists a site within the solution that provides a local minimum for the electron (an orbital on a molecular framework). The strength of this attractive potential is modified by the solvation of an ion in the solution but this modification is not the dominant term.

Dipole solvation studies have taken place using both time-dependent and time-independent approaches [10–23]. In the time-independent approaches [20–23], the shift of the absorption and emission spectrum of molecules is observed as a function of solvent for probing the solvation energy. However little or no kinetic information can be obtained from these studies. Only the equilibrium solvation of the ground state is measured. The shifts of fluorescence spectra as a function of time have been recently used to measure the time dependence of the solvation [10–19]. These experiments have pro-

* Work performed under the auspices of the Office of Basic Energy Sciences, Division of Chemical Science, US-DOE under contract number W-31-109-ENG-38.

J.D. Simon (ed.), Ultrafast Dynamics of Chemical Systems, 137–162.

vided considerable information on solvation dynamics and such studies are presently an active field of research.

Little work has been done measuring ion solvation, except for the study of electron solvation, which, as discussed above, may not be closely related to molecular ions. Recently, in a review of solvation studies, it was suggested that information on solvation process can also be extracted from the electron transfer rates [24]. However because the electron transfer rates depend on many factors, and not all of these can be well quantified, direct measurement of the solvation process of ions is clearly the better approach.

In this article we will review both experimental and theoretical studies of the solvation dynamics in dipolar liquids. Our emphasis will be on our measurements of aromatic ion solvation using pulse radiolysis. The specific issues that we will address are: (1) The structure and energetics of the stable charged species involved in solvation. (2) The dynamical pathways between the unsolvated species and the solvated species. and (3) The development of computational model that describes the solvated state.

The arrangement of the article will be: Section 2 will contain a short review of the literature of the different types of solvation measurements. Section 3 will contain a short summary of the theoretical development in the field. It is given so that the reader may be able to evaluate the experiments with the expected results. No quantitative comparisons will be attempted. Section 4 will contain a description of the experimental techniques that will be used for our measurements. Section 5 will contain a summary of our experimental data, with the emphasis on the amount of the spectral shift. Section 6 contains a description of a simple Monte Carlo model that we use to describe our results. Section 7 will contain a summary of the experimental kinetics of solvation. Finally, Section 8 will summarize the article. The major portion of our work will be described in Sections 5–7.

2. Review of the Literature

A number of reviews have appeared recently on electron solvation [25], experimental studies of dipole solvation [26–28], and intramolecular electron and proton transfer dynamics [29, 30]. Bagchi has recently reviewed the microscopic theories of solvation and solvent dynamics [31].

The dynamics of solvation are measured by creating a species that is not in equilibrium with the solvent on a time scale that is short compared to the solvent relaxation processes. These species could be an ion, a giant dipole or an electron. Such a species will create a strong local field on its environment, forcing the surrounding solvent molecules to rearrange themselves to form a configuration that will solvate the newly created species. As these solvent rearrangements take place, the solvent will perturb the electronic states of the molecule, and will cause shifts in the solute absorption or emission spectrum.

If an excited state with a large dipole is created, the emission spectrum of the excited state will change as the solvent relaxes around the excited molecule. The measurement must be made in competition with the decay of the fluorescent process. Such excited dipole can be formed by optical absorption or by intramolecular proton transfer or electron transfer reactions. If an ion is created, its transient absorption spectrum can be measured. These measurements must be done in competition with the reaction of the ion. Ions can be created by photoionization or by reaction with electrons produced by radiolysis.

The results of the dipole solvation measurements to date may be summarized as follows: (i) The observed dynamics are mainly dependent on the properties of the solvent, although there is evidence that the measured dynamics are not independent of the probe molecule [28]. Below we will discuss the effects on solvation dynamics as the probe changes from an electron to an anion or to a molecular dipole. (ii) The relaxation process is generally a non-exponential process and the time scale for solvation is between τ_L, the longitudinal relaxation time and τ_D, the Debye relaxation and usually closer to τ_L [26]. The value of τ_L reflects the time scale for the collective reorientation of the solvent molecules while τ_D reflects the reorientation of the individual molecules that are interacting with the probe molecules. Only for a few cases, involving solvents with very high dielectric constants, are the solvation times much longer than τ_L [13, 17]. This leads finally to (iii): There appears to be a correlation between the deviation from τ_L and the static dielectric constant ε_0.

3. Summary of Theoretical Treatments

Theoretical treatments of the dynamics of solvation in polar liquids can be divided by the approaches used. In a continuum model, no cognizance is taken of the fluid structure. These models make use of well-defined and well-measured physical parameters [32–35]. Another class of models makes explicit use of the molecular structure of the fluid and of the solvated molecule. The extent that the models attempt to simulate the real systems can vary.

The continuum models treat the solvent as a continuous homogeneous medium whose only relevant property is its bulk, frequency-dependent dielectric correlation function $c(t)$, defined by the expression [32–35]:

$$c(t) = \frac{E(t) - E_\infty}{E_0 - E_\infty} \tag{1}$$

where $E(t)$ is the time-dependent solvation free energy. This energy can be obtained by evaluating the reaction field of the polar solvent at the site of the solute. In such models, the dynamics are simply related to the solvent dielectric properties and the detailed solute-solvent interactions are ignored. For point dipole solvation, the continuum model predicts that the time-correlation function for solvation is a single exponential with a time constant of [34]

$$\tau_L^d = \left(\frac{2\varepsilon_\infty + 1}{2\varepsilon_0 + 1} \right) \tau_D. \tag{2}$$

For the solvation of an ion, the continuum model predicts a slightly faster solvation time, which is equal to the longitudinal polarization relaxation time of the unperturbed solvent τ_L

$$\tau_L = \frac{\varepsilon_\infty}{\varepsilon_0} \tau_D \tag{3}$$

where ε_∞ and ε_0 are the dielectric constant of the solvent in an electric field of infinite frequency and in static field respectively. τ_D is the Debye relaxation time of the solvent. For typical values of these dielectric constants, the difference between τ_L^d (dipole) and τ_L (ion) is unimportant. In essence, the longitudinal relaxation time accounts for difference in the motion of an unperturbed fluid and the response of the fluid to a charge. The prediction of the continuum model that the solvation time in a fluid should be equal to τ_L of that fluid has served as an important benchmark against which to compare experimental data.

Recent experiments of the solvation dynamics occurring around dipoles have revealed that the correlation function $c(t)$ may be non-exponential in several solvents [13, 19, 36, 37]. These observations are not consistent with the predictions of the simple continuum theory. As mentioned earlier, the experimentally measured rate of decay of c(t) lies between the values of τ_L and τ_D but with the value close to τ_L.

There are many approximations that may lead to the failure of the simple continuum model to quantitatively predict the experimental data [34, 35]. The assumption of the Debye form of the frequency-dependent permittivity $\varepsilon(\omega)$ may oversimplify the solvent. In practice, one of the following three functional forms of $\varepsilon(\omega)$ is used in describe the experimental data [38].

$$\varepsilon(\omega) = \varepsilon_\infty + \sum_{j=1}^{n} \frac{(\varepsilon_{oj} - \varepsilon_{\infty j})}{1 + i\omega\tau_{Dj}} \qquad \text{Multiple Debye} \tag{4}$$

$$\varepsilon(\omega) = \varepsilon_\infty + \frac{\varepsilon_0 - \varepsilon_\infty}{(1 + i\omega\tau_0)^\beta} \qquad \text{Davidson–Cole} \tag{5}$$

$$\varepsilon(\omega) = \varepsilon_\infty + \frac{\varepsilon_0 - \varepsilon_\infty}{1 + (i\omega\tau_D)^{1-\alpha}} \qquad \text{Cole–Cole.} \tag{6}$$

The effect of the non-Debye form of $\varepsilon(\omega)$ was studied using the expression given in (5) and (6) [39]. It was found that the short time dynamics were significantly affected even when the deviations from Debye behavior were small. The continuum model has recently been generalized to an inhomogeneous dielectric medium [40–42]. In these studies, the dielectric constant depends both on the frequency and the distance from the solute site. The model assumed that the portion of the solvent that is close to solute charge responds differently than does the bulk solvent. This is physically reasonable both because of the strong electric field of the ion or dipole and because of the existence of specific solute-solvent interactions such as hydrogen bonding.

The second group of theories goes beyond a simple continuum representation by considering the molecular nature of the solvent. One class of such models was proposed by Wolynes [43] and further developed by Rips *et al.* [24] and Nichols and Calef [44]. These theories model the solvent medium as a collection of hard polarizable spheres, whose dynamic properties are handled by the mean spherical approximation (MSA). The MSA model demonstrates that solvation dynamics proceeds on multiple time scales. In most cases the solvation dynamics can roughly be described by biexponential relaxation. The short time scale corresponds to the bulk relaxation time τ_L, while the long time scale is associated with the rearrangement of the first several solvation layers, being close to τ_D. Physically, these different time scales arise because the solvent will respond at different rates, depending on the distance from the charge solute. Furthermore, the precise mix of observed times depends on probe attributes such as size and electronic coupling.

Qualitatively, one might expect that the dynamic MSA model should be an improvement over the continuum models in predicting solvation in polar solvents. However this is not always true in practice. Recently, Chapman and Maroncelli have analyzed the observed dynamics in amide solutions [19]. The measurements were made using the time-dependent Stokes shift of the fluorescence. Although the correlation function $c(t)$ decays nonexponentially, the average solvation times are in better agreement with the predictions of the continuum models. The MSA models yield very poor quantitative agreement with experiment.

A different approach to a microscopic theory of solvation dynamics was taken by Calef and Wolynes [45] and Bagchi and Chandra [46, 47]. They derived a generalized Smoluchowski equation to describe the solvent relaxation. In this description the dipolar solvent molecules undergo rotational and translational diffusion in a potential of mean force, while feeling the mean force of the other surrounding solvent dipoles. In this model, the time-dependent solvent polarization is related to the number density of the solvent by the following expression:

$$\boldsymbol{P}(r,t) = \int \boldsymbol{\mu}(\omega)\rho(r,\omega,t)\,\mathrm{d}\omega \tag{7}$$

where $\rho(r,\omega,t)$ is the position, orientation and time-dependent distribution of the solvent and $\mu(\omega)$ is the dipole moment vector. The deviation of $\rho(r,\omega,t)$ from its equilibrium value $\delta\rho(r,\omega,t)$ satisfies the generalized Smoluchowski equation

$$\begin{aligned}\frac{\partial}{\partial t}\delta\rho(r,\omega,t) &= D_R\nabla_\omega^2\delta\rho(r,\omega,t) + D_T\nabla^2\delta\rho(r,\omega,t)\\ &\quad -[D_R\nabla_\omega\cdot\rho(r,\omega,t)\nabla_\omega + D_T\nabla\cdot\rho(r,\omega,t)\nabla]\beta F.\end{aligned} \tag{8}$$

D_R and D_T are the rotational and translational diffusion coefficients of the solvent, respectively; ∇_ω and ∇ are the usual angular and spatial gradient operators. Equation (8) contains the rotational and translational diffusion terms, a potential of mean force term and a term βF, which contains the mean-field contribution from intermolecular interactions and an external field term and is defined in (8′)

$$\beta F = \int \mathrm{d}r'\,\mathrm{d}\omega' C(r,\omega,r',\omega')\delta\rho(r',\omega',t) + \beta U_{ext}(\rho,t). \tag{8'}$$

$C(r,\omega,r',\omega')$ is the two-particle direct correlation function of dipole solvent [48]. U_{ext} is the external field that comes from the solute ion or dipole whose solvation is being investigated. The time-dependent solvation energy can be given by the following expression:

$$E(t) = -\frac{1}{2}\int \mathbf{d}r\,\mathbf{D}(r)\cdot\boldsymbol{P}(r,t) \tag{9}$$

where $D(r)$ is the bare electric field of the polar solute whose solvation properties are being studied, and $P(r,t)$ is defined in Equation (7). Hence the solution to the Equation (8) provides complete dynamical information of the physical process.

The solution of Equation (8) predicts the following general feature for the solvation of ions and the solvation of dipoles [31, 49]. First, in the absence of a translational contribution, the solvation of dipoles is always slower than the solvation of ions. This also agrees with the predictions of both the inhomogeneous continuum model and the dynamic MSA model. Second, the translational motions of the solvent accelerate the relaxation rate. Third, the solvation-time-correlation function will generally be non-exponential; however as the ratio of solute to solvent increases, the model reaches the continuum limit with an exponential time constant τ_L.

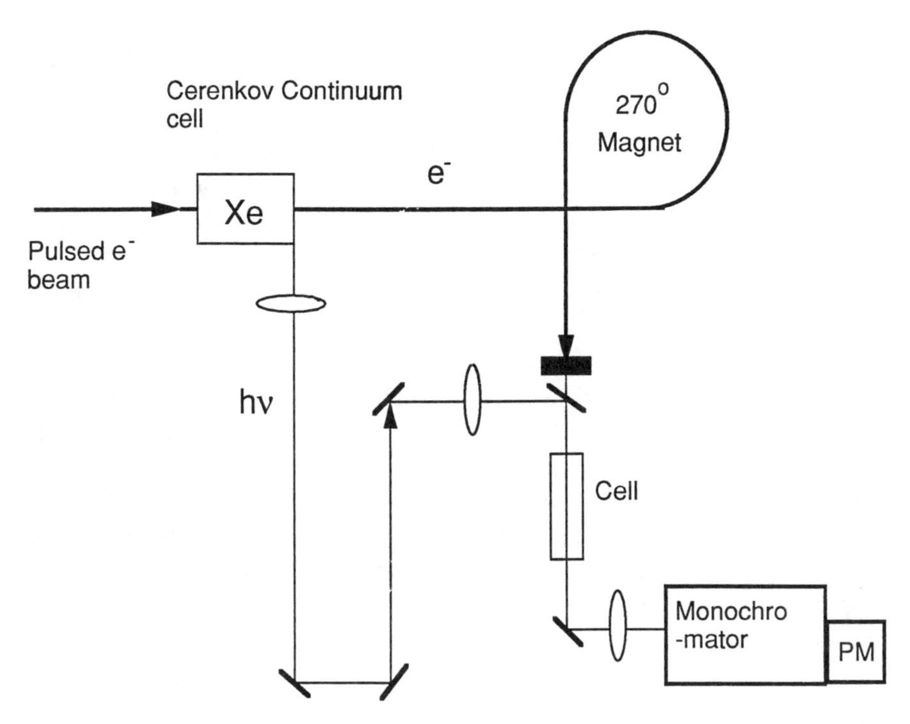

Fig. 1. Schematic of picosecond time-resolved spectroscopy apparatus employing single-pulse, picosecond electron beam and stroboscopic detection technique with Cerenkov continuum emission.

The generalized Smoluchowski equation is normally linearized around the equilibrium solvent density, which may not be valid if the solvent distribution is sufficiently distorted around the solute molecule. The recent molecular dynamics simulations in methanol [50] and acetonitrile [51] indicate that solvation dynamics cannot be completely described by a linear response model, especially if the relaxation is around a species where the charge distribution is created by the shift of a full electron charge.

4. Experimental Techniques

Experimental studies on solvation dynamics are carried out using the pulse radiolysis pump probe technique [52–54]. As shown in Figure 1, a 20-Mev electron LINAC delivers a short pulse of electrons. A portion of the electrons pass through a cell filled with 1 atm xenon to generate a picosecond Cerenkov continuum for a probe pulse. The remainder of the electrons enter the sample cell (through which the liquid is flowing to avoid thermal effects and chemical degradation) as the pump pulse. Spectra and kinetics of transients are followed by recording the absorption of the Cerenkov continuum as the delay time between the probe pulse and the pump pulse is varied using an optical delay line. The pulse width of this system is approximately 30 ps. The data acquisition system has a large dynamic range and permits very weak absorption signals to be reliably recorded from 250–800 nm.

If the solvated electron were present, its absorption spectrum would overlap with the absorption spectrum of the probe anion. However, only few solvated electrons are formed at 0.25 M benzophenone concentration. The contribution of the solvated electron to the anion absorption has been shown to be negligible [55].

The overall instrument response time was determined from the hydrated electron absorption at 600 nm. This electron absorption was monitored before and after each experiment to ensure no significant beam drift occurred during the experiment. Because the electron solvation process in water is much faster than our pump and probe pulses, the fast rise of the hydrated electron absorption can be used to determine a response function that can be used for subsequent data analysis. The initial hydrated electron concentration was approximately 20 μM. The time resolution of the measurements was always around 30–40 ps.

5. Ion Solvation and Solvent Structure

When a high energy electron beam is injected into a dense polar medium, it ionizes the solvent molecules and generates quasi-free electrons. The 'dry-electron' (defined as an electron before solvation) can react with an aromatic solute molecule such as benzophenone to form the corresponding anion. It has been shown that the 'dry electron' (or precursor of the solvated electron) doesn't react as a localized electron (the immediate precursor of the solvated electron – the species whose solvation time is measured) in the fluid, but instead the precursor of the localized electron would react with the benzophenone. Thus, if the formation of the solvated electron is quenched, one need not worry about the solvation time of the electron in the fluid [56]. As solvation of the charged anionic species progresses, the solvent structure undergoes a rearrangement, leading to a change in the energy of the probe ion that is reflected in the change of the emission or absorption spectrum. The benzophenone anion was selected as the microscopic probe for the following reasons: (1) The anion spectrum is separated from that of the triplet and excited state and is strongly shifted by a polar solvent. (2) The anion can be produced very quickly by a reaction with the 'dry electron', which can conveniently be produced by electron beam radiolysis and benzophenone is sufficiently soluble so that no solvated electrons are formed.

To illustrate what is involved in determining the solvation dynamics spectroscopically, typical transient absorption spectra of benzophenone anion is

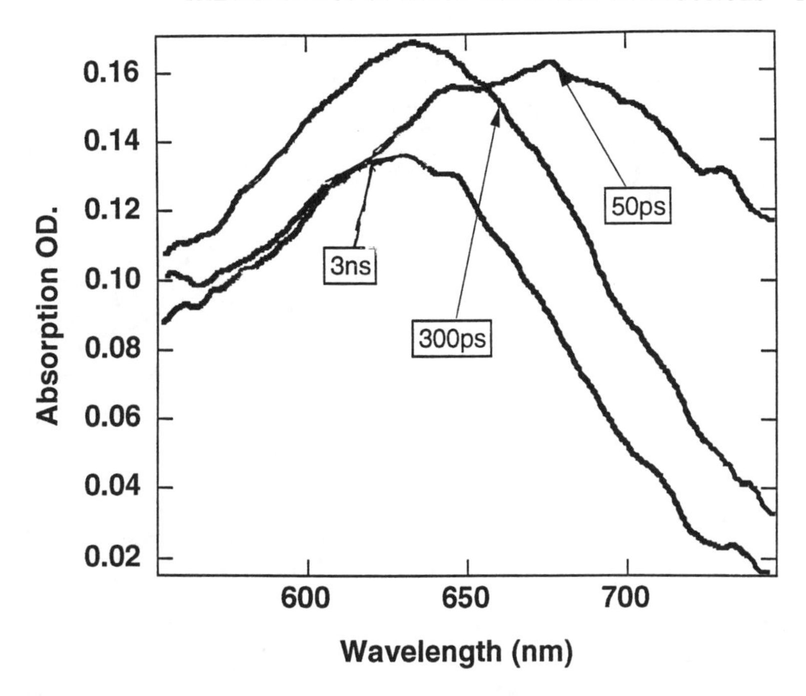

Fig. 2. Benzophenone anion absorption at 50 ps, 300 ps and 3 ns after the pulse excitation in n-octanol solution. The maximum of the absorption spectrum shifts from 675 nm to 635 nm and then to 625 nm for the three different times.

shown in Figure 2. The data recorded here are the benzophenone anion absorption at 50 ps, 300 ps and 3 ns after the electron beam pulse in n-octanol solvent. As shown in Figure 2, the spectrum blue shifts and narrows as a function of time. The final broad absorption band has been assigned to the relaxed benzophenone anion absorption in linear alcohol [57–59], while the spectra at early times are assigned to the anion before the rearrangement of the solvent [60, 61]. These data demonstrate that the amount of solvent reorganization strongly affects the electronic structure of the solute ions. The time-dependent behavior in other linear alcohols is similar, albeit faster for the smaller alcohols. As shown in Figure 3, the final absorption peak position and width are practically the same throughout the linear alcohol series, only the time scale for the blue shift depends on the chain length of the alcohol solvents. The similarities in the final spectrum of the linear alcohols indicate that the equilibrium local solvent environments are mainly determined by the structure and functional group of the solvent but not size of the solvent. Hence, the solvation energetics are largely determined by a relatively small number of molecules in the first solvation shell around the solute. This observation is in qualitative agreement with the Maroncelli's results for the computer simulation of a model dipolar system [62].

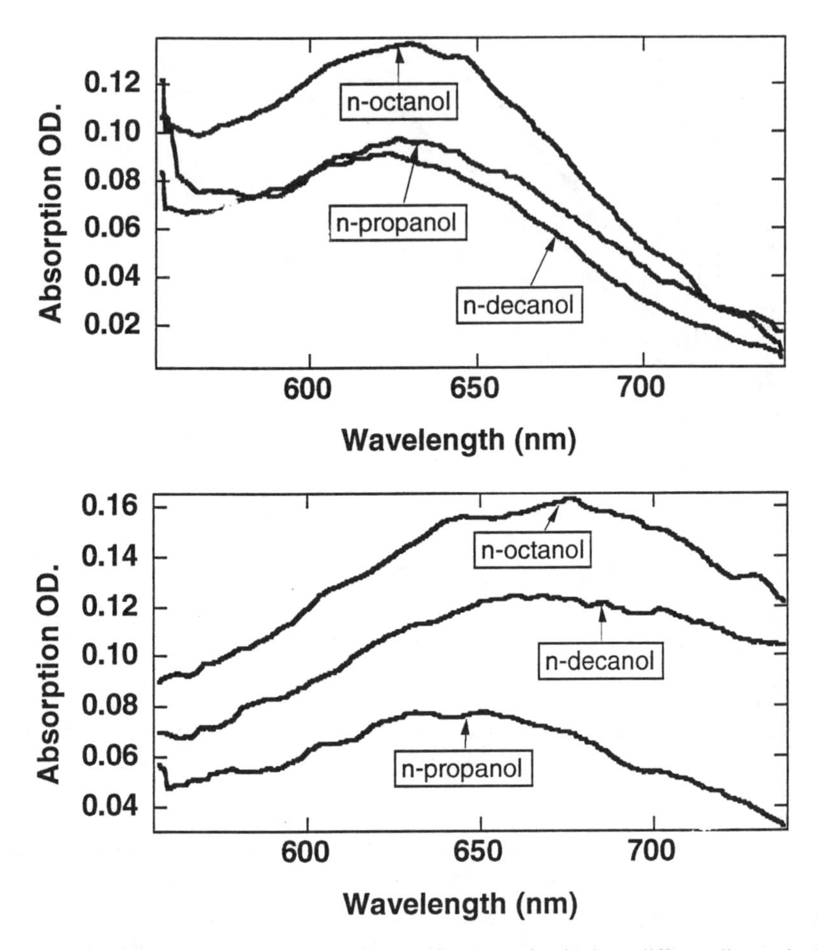

Fig. 3. The transient absorption spectra of benzophenone anion in three different linear alcohol solvents (n-propanol, n-octanol, and n-decanol) at 3 ns (upper) and 50 ps (bottom) after the pulse excitation. All three solvents show a very similar final absorption spectrum.

How does the structure of the solvent molecules determine the dynamics and energetics of the solvation? There are a few publications where the spectral changes in the solvation process caused by varying molecular structure of the solvent have been investigated [63–66]. In particular, a systematic exploration of the solvated electron spectrum in a variety of alcohols has been made by Hentz and Kenney–Wallace [65, 66]. They found that the peak of the final solvated electron absorption spectrum is sensitive to the number and size of branches and distance of the branch point from OH group. In a recent publication, Fujisaki *et al.* [63] also found that the absorption spectra of the solvated electron in cis and trans isomers of methylcyclohexanol can

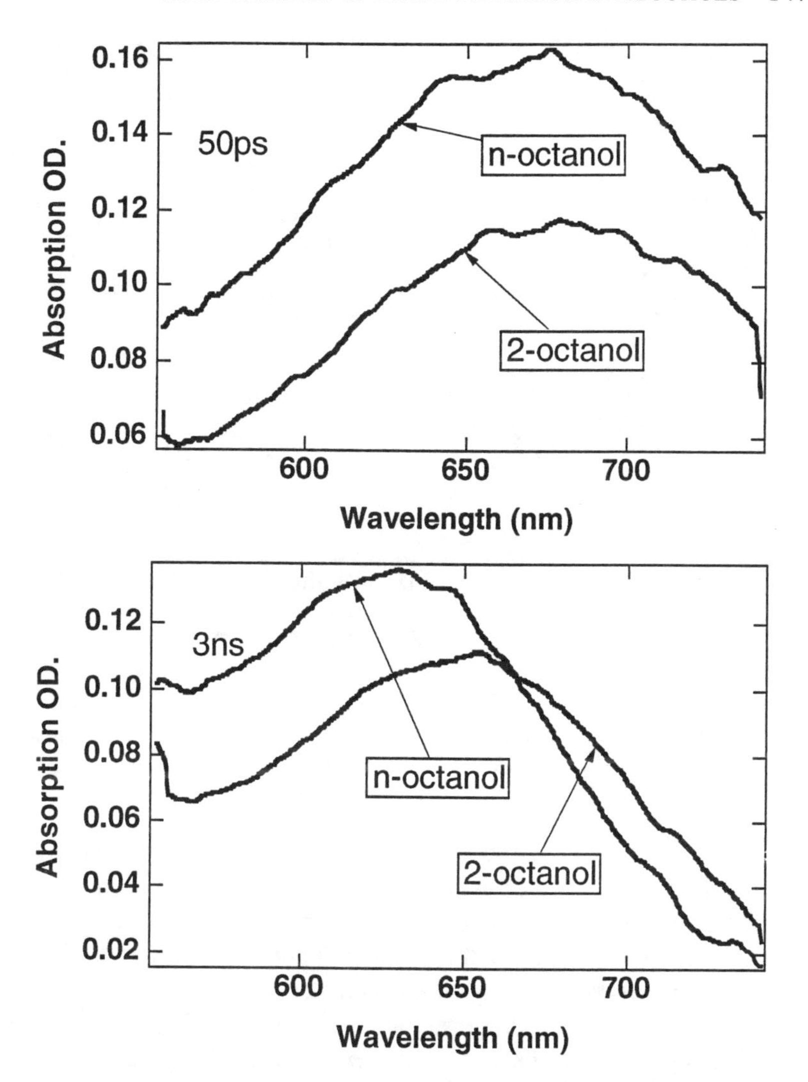

Fig. 4. The transient absorption spectra of benzophenone anion in n-octanol and 2-octanol solutions at 50 ps (upper) and 3 ns (bottom) after the pulse excitation.

be different as much as 100 nm, irrespective of the position where a hydrogen atom on the cyclohexanol ring is substituted for a methyl group.

Figure 4 plots the absorption of the benzophenone anion in n-octanol and 2-octanol solutions. As shown in Figure 4a, the absorption spectra are very similar for the primary and secondary alcohols at very early times, times prior to solvation. Both reflect newly created ion species in random (unrelaxed) solvent configurations. The absorption spectra are very different for the fully solvated species in a linear alcohol and in a branched alcohol. The shift of the

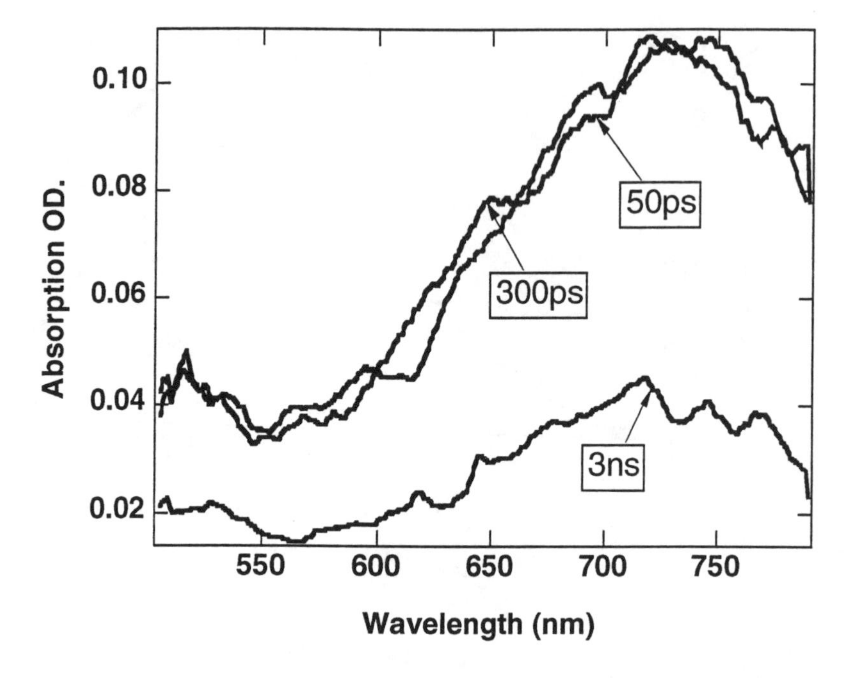

Fig. 5. The transient absorption spectra of benzophenone anion in acetonitrile solution at 50 ps, 300 ps and 3 ns after the pulse excitation. The dynamics of the solvation in this solvent are too fast to be observed. The center of the absorption peak is at 720 nm.

spectrum of the benzophenone anion in 2-octanol is smaller than in n-octanol and final spectral position is about 35 nm red-shift from what is observed in the normal alcohol. Similar solvation behavior was also observed in other branched alcohol system such as 2-butanol and 2-propanol.

In addition to the linear and branched alcohols, we have examined solvation in acetonitrile, dimethylformamide, diisopropyl amine. In this discussion we shall only include the acetonitrile experiments. The transient absorption spectra of the benzophenone anion in acetonitrile are shown in Figure 5. Because the absorption maxima measured at 50 ps, 300 ps and 3 ns are the same, the time-dependent solvation process in this system is too fast for us to observe. Using the time-dependent fluorescence Stokes shift, recently Fleming and coworkers measured the solvation time of LDS-750 in acetonitrile to be approximately 0.1 ps [67], which is faster than our instrument response. Thus the measurements in acetonitrile described are presumably solvated. The final position of the benzophenone anion absorption in acetonitrile shows that the equilibrium solvation structure and energetics in acetonitrile are significantly different from those in linear alcohols. The absorption spectrum at 3 ns is 90 nm red-shifted compared to those in linear alcohols and 60 nm red-shifted compared to the branched alcohols. The effect of structure is even greater on the solvated electron spectrum. The absorption maximum of the solvated electron is strongly shifted to the red in 2-octanol or 2-butanol compared to what is observed in the primary alcohols [8]. The solvated electron absorption in acetonitrile is very difficult to record since the peak of the absorption is beyond the dynamical range of our detection system. The difference in the amount of the spectral shift among the solvents and the way this difference changes between benzophenone anion and solvated electron reflect the strong effect in the different solvent-solute coupling mechanism.

The physical picture that we use to explain the data can be described as follows. Before the electron attaches to the benzophenone molecules, the solvent molecules around the benzophenone are arranged to solvate a neutral molecule with a dipole moment. After a benzophenone molecule attaches an electron, the solvent dipole configuration is inappropriate for an anion. As the ion solvates, the orientation and distance of the solvent molecules from the solute ion changes, and the coupling between the solvent and solute ion also changes. These changes will thus lower the total energy of the system with electron at its ground state, or equivalently, the electron will experience a deeper effective potential and the level spacing in the anion absorption spectrum will increase and the anion absorption will shift towards blue. As ion solvation proceeds, the distribution of the local solvent structure around the probing anion becomes more organized, the anion sees a better defined potential and thus the width of anion absorption becomes narrower. These factors lead to the observation of the blue shift and the narrowing of the anion absorption spectrum.

The differences between the amount of spectral shift in primary and secondary alcohols cannot be explained by the dipole densities, because the dipole density of n-octanol and 2-octanol are the same. The difference can however be explained from the steric factors that prevent a close packing of secondary alcohols around an anion. For a primary alcohol, there is only a single carbon chain attached to the carbon atom bonded to the OH dipole. The OH dipole can point towards the anion and the carbon chain extends in the opposite direction. Structurally there will thus be little interference between alcohol molecules. However, for the secondary alcohol, there are two carbon chains attached to the carbon atom bonded to the OH dipole. The second carbon chain (in the experiments described here, a methyl group) will interfere with a neighboring alcohol molecule, and thus the OH moieties will not be as near to the anion for a secondary alcohol as for a primary alcohol. Computer modeling studies in support of these ideas are given in the next section.

6. Monte Carlo Simulations of Solvation Energetics

As was discussed above, we ascribe the difference between primary and secondary alcohols to the differences in the packing of the solvent molecules around an anion. To confirm this hypothesis, a Monte Carlo simulation of the system was performed. In this simulation, we use a simple model that includes the linear and branched chain alcohols and acetonitrile within a simple framework. With this generic solvent model, we can map out the energetics of the system as a function of the relevant solvent properties and thus characterize a given class of solvents.

We have explored the equilibrium properties of the anion solvation process in a model dipolar cluster consisting of N model solvent molecules around a central charged entity (anion). The use of the finite cluster approximation rather than the quasi-infinite system often studied in liquid simulations avoids the complications introduced in dealing with the truncation of the long-range dipole-dipole interaction [68]. The characteristics of the solvent arrangement around the anion that we expect to determine the shift of the anion spectrum, such as the size of the first solvation shell and the alignment of the nearest molecules, do not change appreciably as a function of N ($N \leq 1330$).

The solvent molecules are modeled as three linearly connected hard spheres. Three different dipole distributions of the solvent molecules were used. In the first case, the dipole is at the end of the molecule with the dipole positive charge at the exposed end. This describes the linear alcohols. The second case corresponds to the situation in acetonitrile, where the dipole is at the end of the molecule, and the exposed end is negatively charged. The third case simulates the situation in the branched alcohols, where the dipole is located in the center of the molecule (center ball in our model) and is perpendicular to the axis of the molecule.

The degrees of freedom that are involved in the calculation of the energetics are the position and orientation of the solvent molecules and the position and orientation of the solvent dipole moments. All electrostatic interactions among the solvent and solute molecules are considered explicitly. Other molecular interactions are simply replaced by hard-sphere repulsive potentials that keep molecules from overlapping.

The process of 'relaxation to equilibrium' was monitored through the calculation of the solvation energy V_s, which is the electrical potential at the solute site. V_s was calculated as a function of the number of configurations for several different initial configurations. After the system has reached equilibrium, the values of solvation energy obtained from different initial configurations are within the statistical fluctuation. The influence of the initial configuration disappeared after 30,000 configurations. Such complete calculations have been done for a series of cluster sizes from 26 to 342.

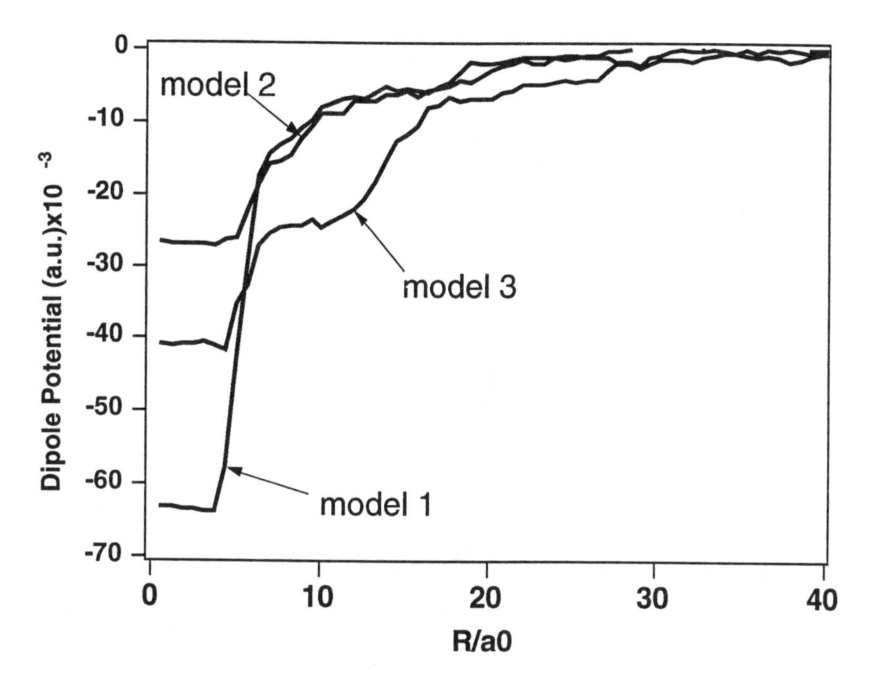

Fig. 6. Monte Carlo simulation of the distance dependence dipole potential energy for three different model fluids. (1) Solvent has end dipole group and positive charge terminus. (2) Solvent has end dipole and negative charge terminus. (3) Solvent has center dipole group. The parameters used in these calculations are q_0 =–1 (a.u.), $p_1 = 1.5\,D, T = 300$ K and $N = 124$.

With the Monte Carlo recipe for generating a series of configurations [68–71], the probability of the appearance of particular configuration C_k in the series is normalized and proportional to the Boltzmann factor. The ensemble average for any quantity X is then

$$\langle X \rangle = \lim_{k \to \infty} \left(\frac{1}{k} \right) \sum_{k=1}^{K} X(C_k). \tag{10}$$

Details on the method will be given in the references [71].

The potential well arising from the arrangement of the solvent molecules around the anion was calculated. The results of these calculations are show in Figure 6. The differences between the potentials are evident for the three different placements of the dipole group in the model solvent molecule. As can be seen in the Figure 6, curve 1 (primary alcohol) shows the deepest and narrowest potential well; hence, the absorption of the anion in this kind of solvent would occur at higher energy. This indeed is what we observed experimentally. Curve 2 (acetonitrile) displays a relatively shallow and wide potential well. Curve 3 (branched alcohols) shows a deeper potential than for acetonitrile but the potential is not as narrow nor as deep as curve 1. From

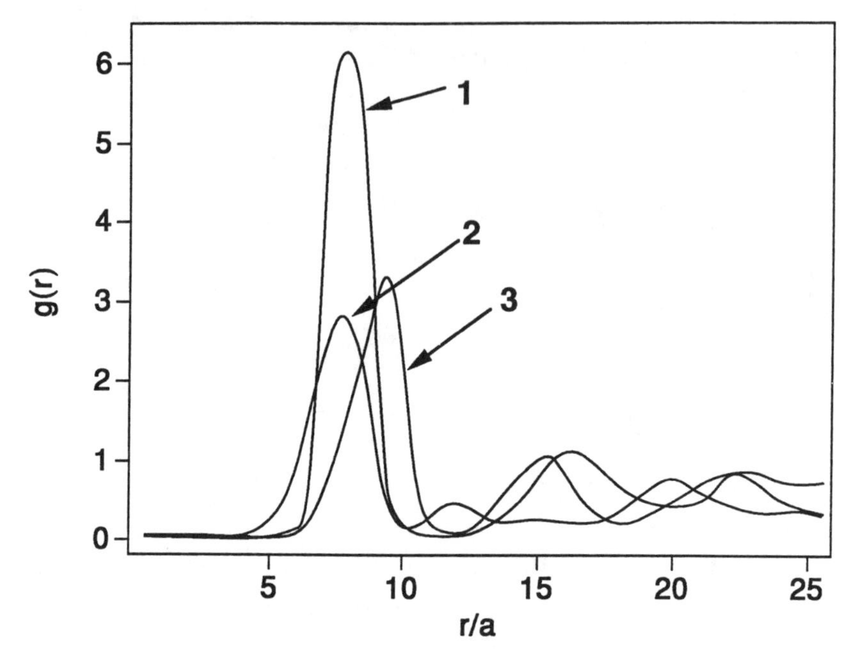

Fig. 7. Radial distribution function of the model solvent systems. (1) Solvent has end dipole group and positive charge terminus. (2) Solvent has end dipole and negative charge terminus. (3) Solvent has center dipole group. The parameters used in these calculations are q_0 =–1 (a.u), $p_1 = 1.5\,D$, $T = 300$ K and $N = 342$.

Figure 6 a red shift is expected to occur on going from linear alcohols to branched alcohols to acetonitrile. Our simulation thus qualitatively explains the observed spectral shift as a function of solvent.

In addition to the potential energy $V_s(r)$ arising from the arrangement of the solvent molecules around the anion, we also calculated the solute solvent radial distribution function (rdfs). Figure 7 plots the solute-solvent rdfs for the three model solvents mentioned above, which correspond to three arrangements of the dipole group in a molecule. These simulations show that the size of the first solvation shell varies with solute charges and the distribution of the solvent dipoles. The model 1, linear alcohol, has a smaller first solvation shell than the other two solvents. The figure also shows that the number of solvent molecules in the first solvation shell also varies considerably with the position and orientation of the solvent dipole.

7. Solvation Times and Solvation Dynamics

Figure 8 gives a typical example of the time-dependent behavior of the absorption of the benzophenone anion in an alcohol. These data are for

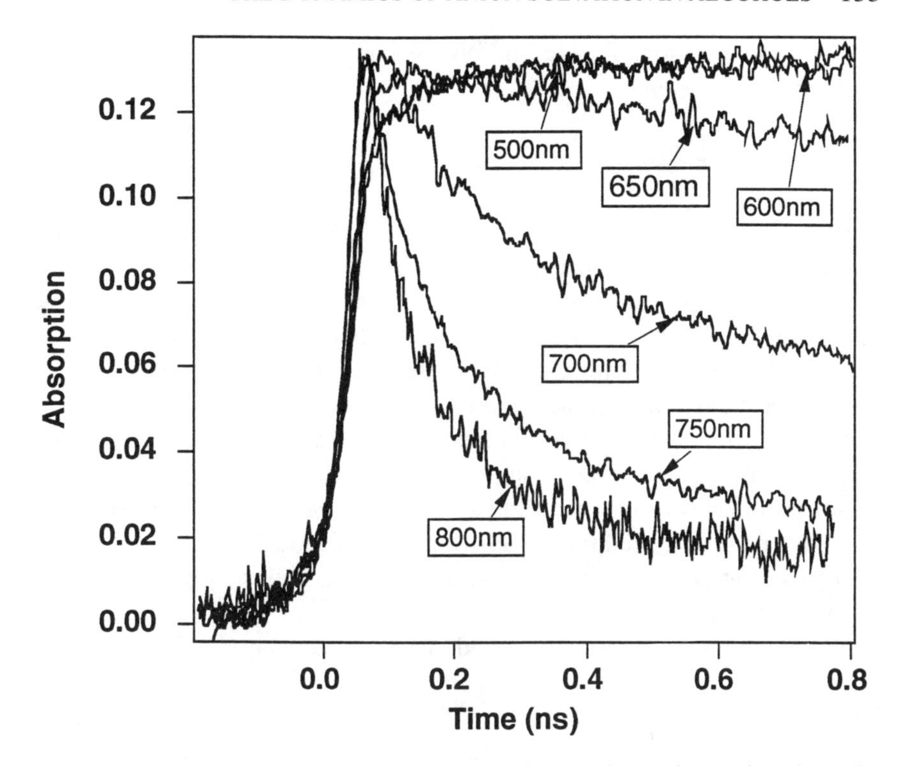

Fig. 8. Time-dependent transient absorption spectra of 0.25 M benzophenone in n-decanol at various wavelengths. The intensities are scaled to show the rate of the decay.

benzophenone in n-decanol. The lack of significant ion recombination on the time scale of these experiments is shown by the slow decay of the absorption at 600 nm. Different kinetics are seen at different wavelengths. The rate of decay is considerably faster at 800 nm than at 750 nm, which is faster than at 700 nm. Also there is a small growth followed by a decay at 650 nm. Similar kinetic behavior was also found in all the other primary alcohols [55].

These results clearly show that the time-dependent spectral behavior cannot be described by an interconversion of one species to another, which approximates the description of the electron solvation process in room temperature alcohols and in water [7–9, 72, 73]. Instead, the shift of the spectra, which shows different maxima at different times, suggests that the spectra are continuously evolving. The different kinetics at different wavelengths are also consistent with this picture.

As discussed above, the dynamics are described by the normalized response function, $c(t)$ (define in Equation (1)), which describes the temporal evolution of the solvation free energy (E) subsequent to the solute perturbation. The majority of experimental studies of solvation dynamics have focused on the shift of the spectral maximum, $\nu(t)$ as a function of time because under reasonable assumptions, as has been discussed elsewhere, the following expression,

$$c_\nu(t) = \frac{\nu(t) - \nu(\infty)}{\nu(0) - \nu(\infty)}$$

can be shown to be equivalent to $c(t)$ [28]. This can conveniently be done by fitting each fluorescence spectrum with an appropriate lineshape function and determining its maximum numerically. Alternatively, in a technique developed by Barbara and co-workers, the fluorescence decay at a single wavelength, together with the solvent-dependent spectral density function, can also be used to determine $c_\nu(t)$ [18, 36].

While these techniques are convenient for extracting information from fluorescence data, alternative approaches are more efficient in extracting data from the absorption experiments. Because there is no interference from excited state decay, the kinetics as a function of time are a good measure of the concentration as a function of time. For this reason, we choose to use these data to determine $c(t)$.

We assume that the anion ground state relaxes as $c(t)$. This means that the spectral maximum and width will relax with the same time-dependent function. During the solvent reorganization process, the center and width of the anion absorption can be described by Equations (11a) and (11b)

$$\lambda(t) = \lambda_\infty + (\lambda_0 - \lambda_\infty)e^{-\tau/\tau_s} \tag{11a}$$

$$\Delta(t) = \Delta_\infty + (\Delta_0 - \Delta_\infty)e^{-\tau/\tau_s} \tag{11b}$$

where λ_0 and λ_∞ are the absorption maxima for the initial unsolvated anion and the final solvated anion respectively while Δ_0 and Δ_∞ are the corresponding absorption band widths. Theoretically, the time response of the system can be approximated over a limited time regime by an exponential with a characteristic time response τ_s [45, 74]. Note that some of the nonexponential behavior found in the fluorescence experiments may be due to the change in lifetime as a function of wavelength [28]. The time- and wavelength-dependent anion absorption can be written as

$$I(\lambda, t) = \frac{1}{(\sqrt{2\pi})\Delta(t)} e^{-\frac{(\lambda - \lambda(t))^2}{2(\Delta(t))^2}} e^{-k_r t} \tag{12}$$

where k_r is the reaction rate of the solvated benzophenone anion. At short times, this expression must be convoluted with the spectral response function.

The convoluted expression can be fit to the experimental decay curves. Several of the spectral parameters can be determined from other experiments.

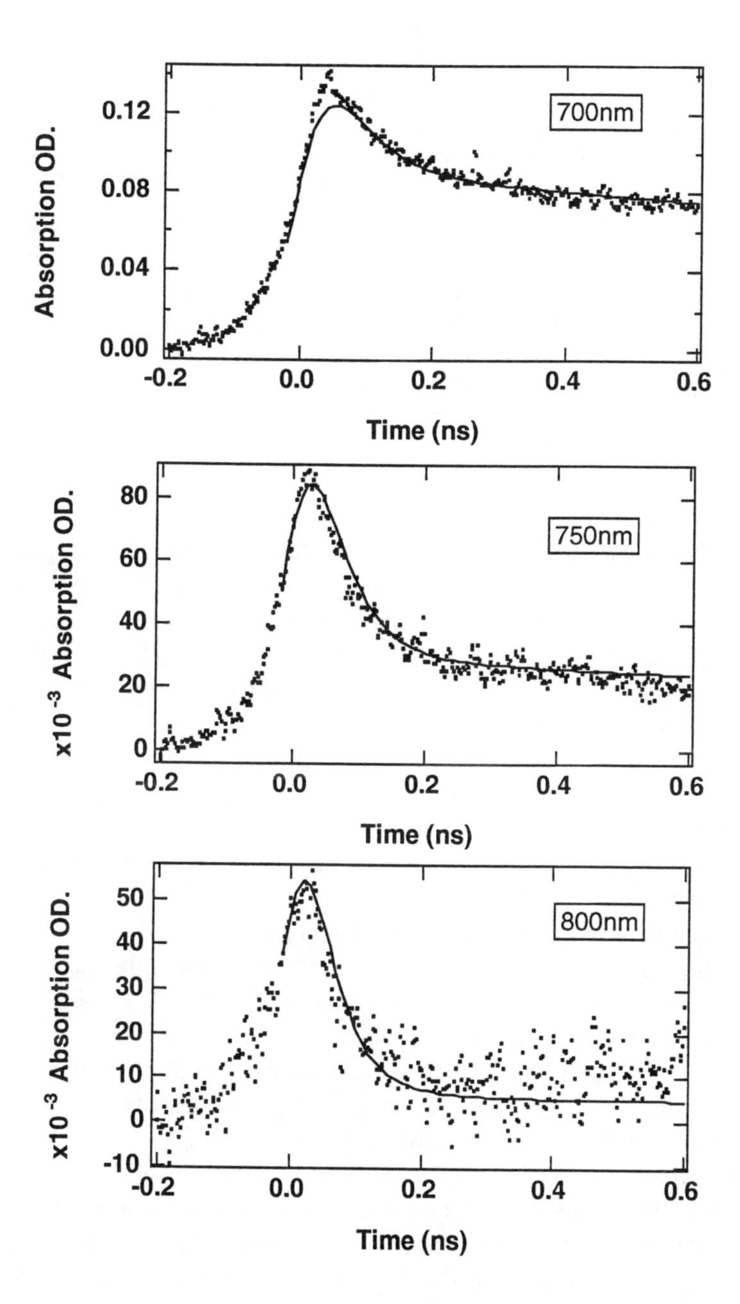

Fig. 9. Time-dependent absorption of the benzophenone anion at several different wavelengths in n-octanol. The lines correspond to the calculations from the Equation (13) and dots are the experimental data. In these calculations $\tau_s = 55 \pm 5$ ps, $\Delta_\infty = 68 \pm 2$ nm and $(1/k_r) = 2.8 \pm 0.3$ ns.

TABLE I

The characteristic benzophenone anion solvation time, τ_s, and final spectal line width, Δ_∞, for several alcohols. The rates for the subsequent chemical reaction are also given.

Alcohol	τ_s (ps)	Δ_∞ (nm)	k_r(1/ns)
n-butanol	35	66	0.5
n-octanol	55	70	0.36
n-decanol	90	70	0.33
2-octanol	57	70	0.45

For example Δ_0 and λ_0 are determined from the spectrum of benzophenone in cyclohexane, a nonpolar solvent while λ_∞ is determined from the spectrum at 3 ns. The value of Δ_∞ cannot be conveniently determined from long time spectrum because of the interference of the ketyl radical and the triplet benzophenone.[61] Figure 9 compares the fits using Equation (12) to the experimental data for benzophenone in n-octanol. The fitted parameters for τ_s, Δ_∞, and k_r are given in Table I for the different alcohols. As expected Δ_∞ is similar for the three linear alcohols, indicating that the energetics of the solvation is primarily determined by the first solvation shell. The value of $1/k_r$ is approximately 3 ns, which is two orders of magnitude longer than the solvation time in the solvent.

While this procedure provides a good reproduction of the experimental data, differences do exist at early times and for wavelengths shorter than 750 nm. There are several possible reasons for these differences. (1) The description of the spectral relaxation process as a single exponential process is almost certainly an oversimplification. The experimental determinations of $c(t)$ using fluorescence spectroscopy have shown multi-exponential behavior [19, 37]. While this behavior may partially be due to a change in lifetime as the spectrum evolves, the MSA-type theories also predict non single exponential behavior [45, 49]. Thus the relaxation time derived for the anion transient absorption data may be a single exponential average of a multi-exponential response function. (2) The assumption of a gaussian form for the anion absorption spectrum was made for mathematical convenience. We have found that the absorption of unsolvated species can be well represented by a gaussian while the absorption of the solvated species is better described by a form where the low energy side is a gaussian and the high energy side is a lorentzian. This form is similar to that used to describe the spectrum of the solvated electron in water and in alcohols [75, 76]. (3) Finally, the assumption that the initial unsolvated species has the same absorption spectrum as the anion in hydrocarbons is only an approximation. Because the neutral species can partially align the solvent molecules before the anion creation, one would expect the hydrocarbon spectrum to be red-shifted from the unsolvated anion in alcohols. This would most affect the simulation of the early time kinetics and will tend to make the calculated solvation time too short.

For the three reasons given above, the calculated solvation kinetics is likely to be at variance from the true dynamics during the first 10 ps. However, because the slow relaxation dominates the long-chain alcohols, we expect the solvation time determined from this model should be very close to the average relaxation time and thus should well characterize the process.

The other charged species whose solvation dynamics have been well studied is the solvated electron. In both water and alcohols, the experimental data suggest that the electron solvation process can be described by a two-state model [7–9, 72, 73]. Shortly after the formation of the solvated electron, an absorption band in the infrared is observed. This absorption band is assigned to an electron localized in a preexisting solvent trap. As time goes on, the infrared band decays and a band grows in the visible with very similar kinetics. The visible absorption band is characteristic of the solvated electron. The formation of the infrared band is very fast. The growth of the visible absorption band is slower and takes places over tens of picoseconds in alcohols.

For room temperature alcohols, there is no evidence for intermediate species; the infrared absorbing species is transferred to the visible band [8, 25]. However there does not appear to be a well-defined isobestic point. Whether this is due to chemical reactions of the electron or whether it indicates more complex solvation dynamics is still a matter of conjecture. At lower temperatures in alcohols there is clear evidence of intermediate species. The temperature where the added complexity occurs depends on the alcohol [2–6].

For our anion studies, the neutral solute molecule benzophenone traps the electron. The potential for attaching the electron is dominated by the valence electrons and the atomic cores of the benzophenone. The potential minimum will be better defined by the molecular structure of the benzophenone rather than the fluctuating structure of the solvent dipoles. Hence the presolvated anion spectrum will be narrower and better defined than that of an electron.

The arrangement of the solvent around the presolvated electron would be expected to be closer to the final solvation configuration than the solvent arrangement around the presolvated benzophenone anion. The electron selects a region where the solvent is appropriately arranged – if the solvent molecules were not appropriately arranged, the electron will not solvate there. However, the benzophenone molecule provides the major attraction for the electron and the solvent structure will not strongly affect the electron-molecule reaction.

TABLE II

Comparison of the solvation time of the benzophenone anion with the solvation time of the electron and the dielectric relaxation times of the solvent.

Solvent	τ_s (ion) (ps)	τ_s (elec) (ps)	Dielectric relaxation times τ_1	τ_2	τ_3	Longitudinal τ_{L1}	τ_{L2}
n-butanol	35	30	670	27	2.4	127	
n-octanol	55	45	1780	39	3.2	406	30
n-decanol	90	51	2020	48	3.3	565	42
2-octanol	57						

Thus, the presolvated benzophenone anion will be in a solvent region where any order will be defined by the initial benzophenone-solvent interaction and not by the anion-solvent interaction. We would then expect that the electron solvation should be faster because the initial solvent structure is more appropriate to a negative species.

The appropriate solvent motions and the rates of these motions can be estimated. Table II compares (1) the benzophenone anion solvation time that we have measured; (2) the electron solvation time; (3) the solvent dielectric relaxation times. As expected, the electron solvation time is faster than the anion solvation. We can compare the measured times to the dielectric dispersion times to get an idea of what solvent motions may be responsible for the solvation. The dielectric dispersion behavior in alcohols is complex. It has been found that the relaxation processes can be described by three times; τ_1, τ_2, and τ_3 [77, 78]. For decanol these times are 2 ns, 48 ps and 3.3 ps respectively [77]. Recent measurements of the dielectric constants of some common polar liquids by Barthel and coworkers show only slight differences [79, 80]. The time τ_1 is generally thought to be the time necessary for an alcohol molecule to rotate where hydrogen bonds must be broken while τ_2 is the time necessary to rotate a non-hydrogen-bonded molecule. The rotation of the terminal C—OH molecule is described by τ_3. As Table II clearly shows, the electron solvation time correlates best with τ_2 while the benzophenone-anion solvation time is slightly longer than τ_2.

The longitudinal relaxation times (τ_L) as described in Equation (3), are faster than the Debye relaxation time since they take into account the effect of a strong charge-solvent coupling in the relaxation process. Because of the complicated behavior of the dielectric dispersion in alcohols, a single longitudinal relaxation time does not exist [77]. The value for τ_{L1} is given in Table II for several alcohols [77, 79, 80]. The values for τ_{L2} and τ_{L3} are within

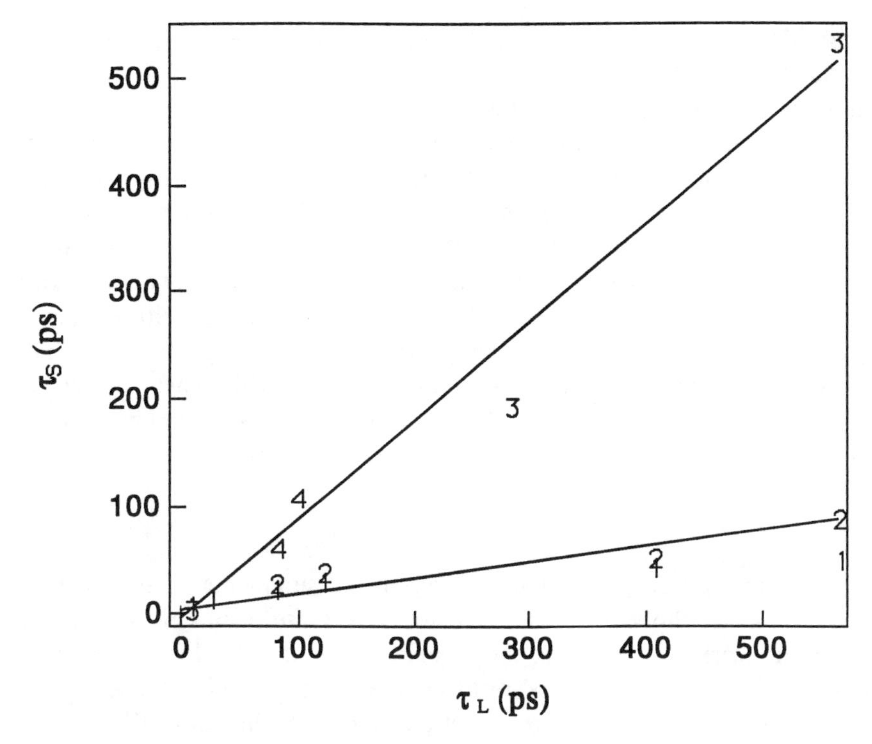

Fig. 10. The correlation between the experimental measured solvation time τ_s and the longitudinal relaxation time τ_L of the solvents. Data set 1 are the electron solvation times obtained from [8]. Data set 2 are the anion solvation times obtained from [12, 69]. Points 3 and 4 are the dipole solvation times obtained from [28, 26], respectively.

20% of τ_2 and τ_3 except for the smallest alcohols. The measured solvation times both for the electron and the benzophenone anion are considerably faster than τ_{L1}. As can be seen in Table II, the solvation time of the benzophenone anion is between τ_{L1} and τ_{L2}. These results again suggest that more solvent motion will be necessary to solvate the benzophenone than is necessary to solvate the electron.

One of the major advances of the present work is the measurement of the solvation time of anions in room temperature solvents. Most of the previous work has involved dipole solvation. Figure 10 compares the solvation time of anions, the electron and dipoles to the longitudinal relaxation time τ_{L1}. For most of the probes, the dipole solvation times are quite close to the longitudinal relaxation times [26, 28] and thus are considerably longer than the solvation times obtained from electron solvation and anion solvation experiments. The great similarity between the solvation time of the anion and the electron (despite the quantum nature of the latter) shows up clearly as do the differences between the solvation around an anion and a dipole (despite the similar molecular framework). The substantial difference shown in these measurements indicates that there are fundamental differences between the solvation of charged entities and dipoles.

Dipole solvation times do not depend only on the solvent; the probe molecule also will affect the solvation times. Su and Simon [81] compared the solvation of dimethylaminobenzonitrile (DMABN) and diethylaminobenzonitrile (DEABN) in propanol at several temperatures and found a small but consistent difference in their solvation times. They observed that the solvation of DMABN is $\cong$ 15% faster than DEABN. Recently Maroncelli has examined a wide range of probes in two reference solvents, n-propanol (253 K) and propylene carbonate (220 K), and found much larger differences in dynamics [28]. These results show that the solvation dynamics depend both on the solvent motions and the specific solvent-solute interactions.

8. Summary

In this review we have discussed our recent measurements of the solvation of anions in alcohols. We have tried to place these results in the context of present work by reviewing the recent studies of solvation dynamics in polar liquids. The experimental results can be summarized: (1) The amount of spectral shift on solvation does not depend on the length of the alcohol. All primary alcohols show a similar final solvated structure. Instead these differences depend on the structure of the alcohol. (2) The spectral solvent dynamics show an apparent continuous shift as a function of time. This suggests that solvation takes place through many intermediate states. These results are different from the results obtained for electron solvation. (3) The dynamics of solvent localization depend on the length of the alcohol and not on the branched structure of the alcohol. The time scale for solvation is quite similar to the solvation of the electron in the same solvent and considerably different from the solvation of a dipole in similar solvents.

From these results we have shown that (1) That a simple molecular simulation that mimics the structural changes can describe the role of solvent structure on the energetics. (2) Solvation times and spectral shifts between a continuum of intermediate states can predict the time-dependent behavior.

With measured solvation times available for anions in room-temperature fluids, we expect that both experimental and theoretical advances will be catalyzed. Such studies will provide insight into the energetics of reactions where charged entities are created.

References

1. Reichardt, C., *Solvents and Solvent Effects in Organic Chemistry,* 1988, VCH, Weinheim, FRG.
2. Baxendale, J. H. and Wardman, P., *J. Chem. Soc. Faraday Trans.*, **169**, 584 (1973).
3. Baxendale, J. H. and Wardman, P., *Can. J. Chem.*, **55**, 1996 (1977).
4. Gilles, L., Bono, M. R., and Schmidt, M., *Can. J. Chem.*, **55**, 2003 (1977).
5. Okazaki, K. and Freeman, G. R., *Can. J. Chem.*, **56**, 2305 (1978).
6. Kevan, L., *J. Phys. Chem.*, **84**, 1232 (1980); *Acct. Chem. Res.*, **14**, 138 (1981).
7. Chase, W. J. and Hunt, J. W., *J. Phys. Chem.*, **79**, 2835 (1975).
8. Kenney-Wallace, G. A. and Jonah, C. D., *Chem. Phys. Lett.*, **39**, 596 (1976); *J. Phys. Chem.*, 2572 (1982).
9. Migus, A., Gauduel, Y., Martin, J. L., and Antonetti, A., *Phys. Rev. Lett.*, **58**, 1559 (1987).
10. Wang, Y., McAuliffe, M., Novak, F., and Eisenthal, K. B., *J. Phys. Chem.*, **85**, 3736 (1981).
11. Huppert, D., Rand, S. D., Rentzepis, P. M., Barbara, P. F., Struve, W. S., and Grabowski, Z. R., *J. Chem. Phys.*, **75**, 5714 (1981).
12. Castner, Jr., E. W., Maroncelli, M., and Fleming, G. R., J. *Chem. Phys.*, **86**, 1090 (1987).
13. Maroncelli, M. and Fleming, G. R., *J. Chem. Phys.*, **86**, 6221 (1987).
14. Simon, J. D. and Su, S. G., *J. Chem. Phys.*, **87**, 7016 (1987).
15. Su, S. G. and Simon, J. D., *J. Phys. Chem.*, **91**, 2693 (1987).
16. Declémy, A., Rullière, C., and Kottis, Ph., *Chem. Phys. Lett.*, **133**, 448 (1987).
17. Declémy, A. and Rullière, C., *Chem. Phys. Lett.*, **146**, 1 (1988).
18. Nagarajan, V., Brearley, A. M., Kang, T.-J., and Barbara, P. F., *J. Chem. Phys.*, **86**, 3183 (1987).
19. Chapman, C. F., Fee, R. S., and Maroncelli, M., *J. Phys. Chem.*, **94**, 4929 (1990).
20. Jauquet, M. and Laszlo, P., *Influence of Solvents on Spectroscopy,* in Dack, M. R. J. (ed.) *Solutions and Solubilities.* Vol. VIII, Part I of Weissberger, A. (ed.) *Techniques of Chemistry*, Wiley-Interscience New York, 1975.
21. Amos, A. T. and Barrows, B. L., *Adv. Quantum Chem.*, **7**, 289 (1973).
22. Nicol, M. F., *Appl. Spectrosc. Rev.*, **8**, 183 (1974).
23. Davis, K. M. C., *Solvent Effects on Charge-Transfer Complexes* in Forster, R. (ed.) *Molecular Association,* Academic Press, London, New York (1975), VI.
24. Rips, I., Klafter, J., and Jortner, J., *J. Chem. Phys.*, **89**, 3246 (1988); *J. Chem. Phys.*, **89**, 4288 (1988).
25. Kenney-Wallace, G., *Adv. Chem. Phys.*, **47**, 535 (1981).
26. Simon, J. D., *Acc. Chem. Res.*, **21**, 128 (1988).
27. Fleming, G. R., *Chemical Applications of Ultrafast Spectroscopy*, (1986). New York, Oxford.
28. Maroncelli, M., *J. Mol. Liquids.*, (1992), in press.
29. Barbara, P. F. and Jarzeba, W., *Acc. Chem. Res.*, **21**, 195 (1988).
30. Kosower, E. M. and Huppert, D., *Annu. Rev. Phys. Chem.*, **37**, 127 (1986).
31. Bagchi, B., *Annu. Rev. Phys. Chem.*, **40**, 115 (1989).
32. Bakhshiev, N. G., *Opt. Spectrosc.* (USSR), **16**, 446 (1964).
33. Mazurenko, Yu. T., *Opt. Spectrosc.* (USSR), **36**, 283 (1974).
34. Bagchi, B., Oxtoby, D. W., and Fleming, G. R., *Chem. Phys.*, **86**, 257 (1984).
35. Van der Zwan and Hynes, J. T., *J. Phys. Chem.*, **89**, 4181 (1985).
36. Kahlow, M. A., Kang, T.-J., and Barbara, P. F., *J. Chem. Phys.*, **88**, 2372 (1988).
37. Jarzeba, W., Walker, G. C., Johnson, A. E., Kahlow, M. A., and Barbara, P. F., *J. Chem. Phys.*, **92**, 7039 (1988).
38. Bottchev, C. J. F. and Bordewijk, P., *Theory of Electric Polarization,* Vol. 2, (Elsevier, New York, 1978).
39. Castner, E. W. Jr., Fleming, G. R., and Bagchi, B., *Chem. Phys. Lett.*, **143**, 270 (1988).
40. Ehrenson, S., *J. Phys. Chem.*, **91**, 1868 (1987).

41. Bagchi, B., Castner, E. W. Jr., and Fleming, G. R., *J. Mol. Struc. Theor. Chem.*, **194**, 171 (1989).
42. Castner, E. W. Jr., Fleming, G. R., Bagchi, B., and Maroncelli, M., *J. Chem. Phys.*, **89**, 3519 (1988).
43. Wolynes, P. G., *J. Chem. Phys.*, **86**, 5133 (1987).
44. Nichols III, A. L. and Calef, D. F., *J. Chem. Phys.*, **89**, 3783 (1988).
45. Calef, D. F. and Wolynes, P. G., *J. Chem. Phys.*, **78**, 4145 (1983).
46. Chandra, A. and Bagchi, B., *Chem. Phys. Lett.*, **151**, 47 (1988).
47. Bagchi, B. and Chandra, A., *J. Chem. Phys.*, **90**, 7338 (1989).
48. Gray, C. G. and Gubbins, K. E., *Theory of Molecular Fluids,* Vol. 1. Oxford, Clarendon (1984).
49. Chandra, A. and Bagchi, B., *J. Phys. Chem.*, **93**, 6996 (1989).
50. Fonseca, T. and Ladanyi, B. J,. *Phys. Chem.*, **95**, 2116 (1991).
51. Maroncelli, M., *J. Chem. Phys.*, **94**, 2048 (1991).
52. Lin, Y. and Jonah, C. D., *Chem. Phys. Lett.*, **191**, 357 (1992).
53. Jonah, C. D., *Rev. Sci. Instrum.*, **46**, 62 (1975).
54. Jonah, C. D. and Kenney-Wallace, G. A., in *Proceedings of Sixth International Congress of Radiation Research,* Okada, S. (ed.); Tokyo (1979).
55. Lin, Y. and Jonah, C. D., *J. Phys. Chem.*, (1992).
56. Lewis, M., *Trans. Faraday Soc.*, **57**, 1686 (1961).
57. Beckett, A. and Porter, G., *Trans. Fraraday Soc*, **59**, 2038 (1963).
58. Adams, G. E., Baxendale, J. H., and Boag, J. W., *Proc. Roy. Soc.*, **A277**, 549 (1964).
59. Marignier, J. L. and Hickel, B., *Chem. Phys. Lett.*, **86**, 95 (1982).
60. Marignier, J. L. and Hickel, B., *J. Phys. Chem.*, **88**, 5375 (1984).
61. Papazyan, A. and Maroncelli, M., *J. Chem. Phys.*, **95**, 9219 (1991).
62. Fujisaki, N., Comte, P., Infelta, P. P., and Gaumann, T., *J. Phys. Chem.*, **95**, 6259 (1991).
63. Leu, A. D., Jha, K. N., and Freeman, G. R., *Can. J. Chem.*, **60**, 2342 (1982).
64. Hentz, R. R. and Kenney-Wallace, G. A., *J. Phys. Chem.*, **76**, 2931 (1972).
65. Hentz, R. R. and Kenney-Wallace, G. A., *J. Phys. Chem.*, **78**, 514 (1974).
66. Rosenthal, S. J., Xie, X., Du, M., and Fleming, G. R., *J. Chem. Phys*, **95**, 4715 (1991).
67. Levesque, D., Weis, J.-J., and Hansen, J.-P., *Simulation of Classical Fluids*, Binder, K. (ed.) *Monte Carlo Methods in Statistical Physics*, Springer-Verlag, Berlin, Heidelberg (1986).
68. Lin, Y. and Jonah, C. D., submitted to *J. Phys. Chem.*.
69. Jansoone, V. M., *Chem. Phys.*, **3**, 78 (1974).
70. Allen, M. P. and Tildesley, D. J., *Computer Simulation of Liquids*, Oxford, Clarendon, (1990).
71. Long, F. H., Lu, H., and Eisenthal, K. B., *J. Chem. Phys.*, **91**, 4413 (1989).
72. Long, F. H., Lu, H., and Eisenthal, K. B., *Phys. Rev. Lett.*, **64**, 1469 (1990).
73. Loring, R. F. and Mukamel, S., *J. Chem. Phys.*, **87**, 1272 (1987).
74. Dye, J. L., DeBacker, M. G., and Dorfman, L. M., *J. Chem. Phys.*, **52**, 6251 (1970).
75. Dorfman, L. M. and Jou, F. Y., in *Electrons in Fluids,* Jortner, J. and Kesner, N. R. (ed.) Springer-Verlag, New York, p447 (1973).
76. Garg, S. K. and Smyth, C. P., *J. Phys. Chem.*, **69**, 1294 (1965).
77. Davies, M., *Dielectric Properties and Molecular Behavior,* Hill, N. E.; Vaughan, W. E.; Price, A. H.; Davies, M., (ed.), Van Nostrand, London, (1969).
78. Barthel, J., Bachhuber, K., Buchner, R., Gill, J. B., and Kleebauer, M., *Chem. Phys. Lett.*, **167**, 62 (1990).
79. Barthel, J., Bachhuber, K., Buchner, R., and Hetzenauer, H., *Chem. Phys. Lett.*, **165**, 369 (1990).
80. Su, S.-G. and Simon, J. D., *J. Phys. Chem.*, **93**, 753 (1989).

5. Surface Electron Transfer Dynamics at Semiconductor Interfaces

J. LANZAFAME and R. J. DWAYNE MILLER*
Department of Chemistry and Institute of Optics, University of Rochester, Rochester, NY 14627, U.S.A.

1. Introduction

Interfacial electron transfer between a discrete molecular state and a conducting surface is the simplest of all surface reactions in that no bonds are broken; it involves only the exchange of an electron. For this reason, we have the greatest chance of coming to a detailed understanding of this surface reaction coordinate. In addition to its fundamental importance, research in this area is strongly motivated by the large number of practical applications involving interfacial charge transfer. For example, this process is the heart of almost all modern day imaging (photography and xerography), lithographic processes, the entire field of electrochemistry, and has been implicated in many surface catalytic processes [1–3]. Interfacial charge transfer also harbours great potential as an efficient mechanism for solar energy conversion [4]. This latter application is primarily due to the great facility with which conducting solid state materials separate charge relative to an all molecular approach. Despite the great technological importance of this problem, our understanding of surface electronic transitions is lagging well behind the analogous problem of homogeneous electron transfer. The simple reason for this difference is the greater inherent experimental difficulties in studying surfaces.

This review will focus on the use of novel time domain spectroscopies to directly probe the dynamics of interfacial charge transfer. Two different initial conditions will be explored. For one set of experimental conditions, the reactive carriers are optically prepared spatially within the solid state. These studies exploit the properties of semiconductor/liquid junctions as an optical switch to turn on the charge transfer processes with high quantum yield. In this case, the overall dynamics of charge carrier lifetimes at the surface are related to the electronic coupling across the interface and the relaxation dynamics in both the solid and liquid state sides of the surface. The key photophysical processes governing the electron transfer step are shown in Figure 1. Of these processes, the emphasis here will be on studies that probe the acceleration

* A. P. Sloan Fellow and Camille and Henry Dreyfus Teacher-Scholar

J.D. Simon (ed.), Ultrafast Dynamics of Chemical Systems, 163–204.

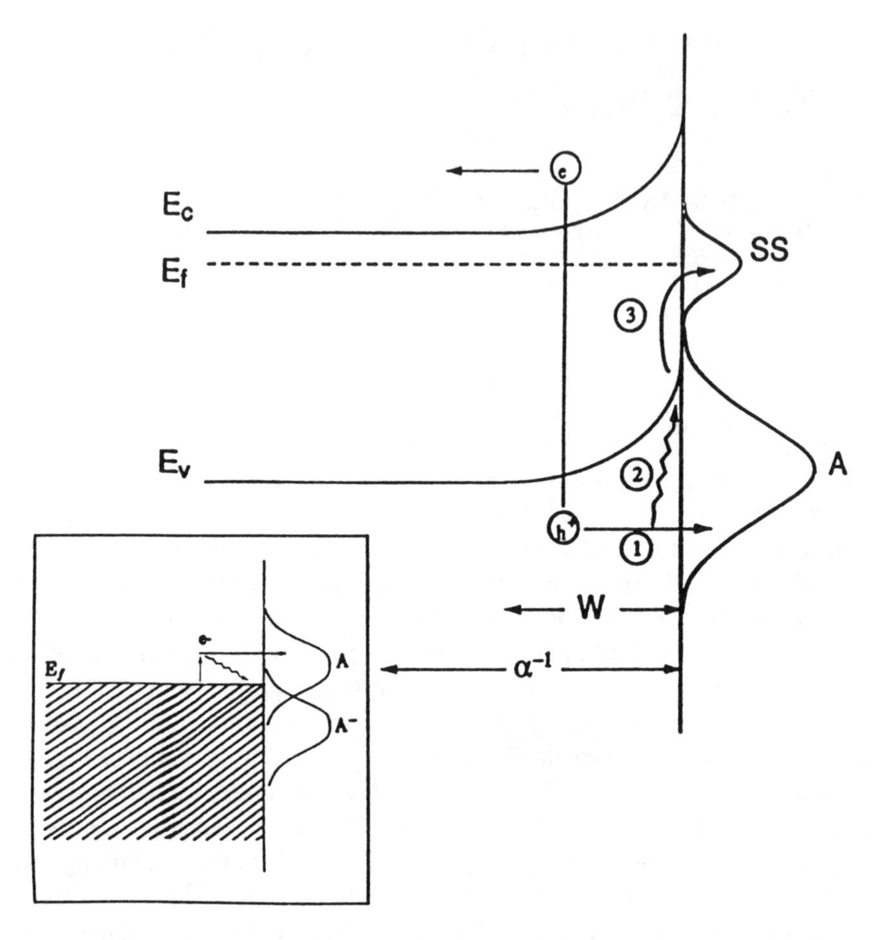

Fig. 1. The main figure outlines the possible photophysical processes at an n-type semiconductor interface. Pathway 1 represent ballistic transfer, pathway 2 represents thermalization, and pathway 3 denotes surface state trapping. W corresponds to the depletion layer, and α^{-1} is the optical penetration depth. SS denotes the surface states and A is representative of some hole acceptor on the surface. E_c, E_v, and E_f are the conduction band, valence band, and Fermi level respectively. Interfacial hole transfer can occur at any point in energy depending on the degree of wave function overlap with the acceptor state and competing dynamics. The inset shows charge transfer at a metal surface which always occurs near the Fermi level. Non-equilibrium studies at metals are limited by extremely rapid electron thermalization through electron-electron scattering.

of photo-generated carriers in the surface space charge region and the carrier depletion dynamics in the surface region. This information is important for an understanding of the surface photochemistry and provides a real time view of the electron trajectory as it crosses the interface. Specifically, the GaAs(100) surface will be discussed as a model interface for understanding the surface reaction dynamics in that it has well characterized electronic, optical, and electrochemical properties relative to other surfaces [5].

The other initial condition will involve the optical preparation of molecular excited states resonant with conduction band states at dye sensitized semiconductor surfaces. Electron injection under these conditions probes the complementary process of electron transfer into an electronic continuum. The general features are shown in Figure 2. This boundary condition provides a direct probe of the electronic coupling or wavefunction overlap between a discrete molecular state and the delocalized solid state electronic levels. The information gained from these studies provides the closest analogy to the time evolution of transition states at metal electrode surfaces. The main problem is to come up with a system which is both well defined and experimentally accessible. Because of the great importance of this process in the field of photography, numerous dye-sensitized systems have been explored. However, even silver halide surfaces suffer from the very photochemical processes that make them good photographic agents. Other more robust surfaces exhibit extremely low quantum yields for electron transfer, prohibiting the extension of the wavelength range of dye-sensitized semiconductors for optical devices [6]. The recent development and characterization of SnS_2 surfaces has enabled studies of electron transfer at nearly perfect surfaces. SnS_2 is representative of layered 2D semiconductors which are the only single crystal surface at the present time which exhibits high quantum efficiency for photoinduced electron transfer ($> 80\%$) [7]. The high quantum yield, chemically inert nature of the surface, and fortuitous optical properties make this a model surface for understanding the effects of an electronic continuum on the electron transfer reaction coordinate. A fundamental understanding of the electron transfer at this surface should provide the general principles for maximizing the efficiency of interfacial electron transfer.

Electron transfer at surfaces is unique among reaction mechanisms in that it is the only reaction type in which a dense manifold of electronic levels is intimately connected to the reaction coordinate. The different $t = 0$ boundary conditions being explored provide critical information concerning the role that both the electronic continuum and nuclear continuum play in localizing electrons (or hole carriers) on discrete molecular states at surfaces.

2. General Considerations

At a surface, the electronic coupling is occurring across an abrupt phase discontinuity. The very nature of the phase boundary dictates that the electronic coupling involves states derived from localized molecular potentials (for the adsorbate) and extended states derived from the periodic potential of the solid state. For metal and semiconductor surfaces, the extended states are highly delocalized and are best described by the crystal band structure as defined in terms of Bloch plane waves (**k** states). These electronic states are only localized by very weak electron-phonon scattering and as such can extend over 100 Å spatial dimensions. The highly delocalized nature of these states and the enormous number of states that can couple to the reactive coordinate introduce significant differences in the problem of electron transfer at surfaces relative to homogeneous electron transfer (discrete two state problem).

In addition to these considerations, the effect of the structure of the surface layer needs to be incorporated explicitly into the problem. Defects and impurities in the surface structure create surface states that are involved in the overall electron transfer mechanism. This point is particularly relevant to semiconductor surfaces. The energetics of the electron transfer step are also modified by the surface potential. These points and modified versions of them are the key issues underlying the mechanism of surface electron transfer.

MODELS FOR SURFACE ELECTRON TRANSFER

The emphasis in electron transfer theory has been in correlating the electron transfer rates to nuclear and electronic factors of the reaction coordinate. Early works by Marcus [8], Levich [9], and Gerischer [10] have led to relatively simple expressions amenable to experimental tests. The rate equation, as formulated for the problem of electron transfer at electrode surfaces, assumes harmonic displacements for the solvent modes involved in solvent reorganization along the reaction coordinate. These theories are in the weak electronic coupling limit where the electron transition probability is largely determined by the Boltzmann statistics or activation barrier in attaining resonance between the two coupled electronic states. For a semiconductor surface, the problem of charge transfer is simplified by assuming it involves thermalized charge carriers at the valence or conduction band edges [10]. With these assumptions, the electron transfer rate constant is (k) is given by,

$$k = \nu\kappa(r) \exp \frac{-(E - E_{\text{redox}} - \lambda)^2}{4kT\lambda} \qquad (1)$$

where λ is the medium reorganization energy, E_{redox} is the redox potential of the acceptor/donor redox couple in solution, ν is the effective frequency of the nuclear coordinate or reorganization, and $\kappa(r)$ is the distance dependent electronic coupling. This expression is essentially a Fermi Golden rule calculation where the transition probability is weighted by the Frank-Condon factors contained in the Boltzmann statistics of attaining resonance.

Perhaps, the most significant understanding gained from the solution phase studies of homogeneous electron transfer is that the approximation of linear response (harmonic approximation) for the solvent coordinate is basically correct. It is not evident a priori that the simplest approximation should capture the essential details of the nuclear coordinate. However, the dependence of electron transfer rate constants on the energetics has been found to scale quadratically for homogeneous electron transfer [11, 12] and more recently at surfaces by Miller and Gratzel [13], and Chidsey [14] using well defined, chemically modified electrodes. This observation reflects the large number of nuclear degrees of freedom in the repolarization volume that are part of the reaction coordinate. All the displacements are small and correspond to linear regions of the intermolecular potentials. The nuclear barrier and relaxation can then be approximated to an excellent extent by assuming a dielectric continuum to calculate λ and ν – as generally done for Equation (1).

The weak coupling limit given by Equation (1) will serve as a discussion point. For technical applications in solar energy conversion, it is desirable to have as small a barrier as possible to the electron transfer process. By proper choice of redox potentials this can be accomplished; the exponential term in Equation (1) can be made close to unity. The real issue is the degree of electronic coupling that occurs between molecular states and extended band states, i.e., the degree of adiabaticity. If the electronic coupling is large enough, the κ term in also becomes unity, and the electron transfer occurs adiabatically for a range of **k** state energy levels within the crystal band structure. In this event, the electron transfer process would occur on the same time scale as the fluctuations in nuclear coordinates that lead to stabilization of charge on the molecular acceptor (ν^{-1}). This time scale is approximately equal (but not identical) to the longitudinal relaxation time of the solution phase [15, 16] adjacent to the surface or the intramolecular vibrational relaxation time [17–19] for reactions with appreciable intramolecular reorganization. The time scale of this effective bath mode is on the order of 10^{-13} sec for many molecular systems and will serve as the lower limit for the electron transfer time. It will also serve as our yard stick to indicate the occurrence of electron transfer in the adiabatic limit.

The electronic coupling between the molecular potential and the periodic lattice potential is the fundamental issue. At surfaces, the coupling is complicated by any intervening solvent layer or contamination that would act as an insulating barrier to the electronic overlap. The electronic coupling has been estimated by using a triangular potential for the surface space charge region and a square well model for the molecular potential. This model illustrated that, even with an intervening barrier the thickness of the solvation shell, the electron tunneling time is on the 100 fsec time scale, demonstrating appreciable electronic coupling. This model was originally used by Nozik *et al.* to demonstrate the feasibility of hot carrier transitions at surfaces: for 100 fsec electron transfer times, the transfer step would be faster than the carrier thermalization dynamics [20]. This time scale is compatible with solvent relaxation processes needed to stabilize the charge separation. (Note: If this upper limit for electron transfer can be realized in practice, it would reduce energy loss in semiconductor liquid junction solar cells and could approximately double conversion efficiencies.)

In addition, Schmickler has recently reworked the problem for variable electronic coupling at the surface in which the associated Hamiltonian includes the bath phonons [21]. Using typical electronic couplings observed for homogeneous electron transfer, it is fairly straight forward to arrive at solutions in which the electronic coupling is so enhanced at the surface that it should fall in the strong coupling limit. In comparison to homogeneous electron transfer, a molecule at a surface has many more electronic levels which act as sources of electrons. In this case, the process is activationless and the energy distribution, rather than give the above gaussian distribution (Equation (1)), is strongly perturbed by the electronic coupling. Thus, there are well founded reasons to believe that electron transfer at surfaces can occur in the strong coupling limit on very fast time scales.

Since the early prediction of the possibility of unthermalized electron transfer, there have been a number of experiments which support the concept of extremely fast electron transfer at surfaces [22–24]. Without measuring the dynamics directly, however, these steady state experiments alone are always subject to speculation regarding other kinetic mechanisms that would give the same experimental result [25]. It is clearly desirable to observe the carrier dynamics directly, which is the major focus of this review. Furthermore, measurements of the various photophysical processes, as shown in Figure 1, will enable a kinetic rationalization of the photochemistry observed at semiconductor interfaces.

3. Theory: Incorporating the Electronic Continuum

The theory in this section will primarily treat the $t = 0$ boundary conditions for the time domain studies in which the electron is localized initially on a discrete molecular state at the surface. The conditions leading to electron transfer in the reverse direction are equivalent to the back electron transfer aspects of this problem. The effect of the electronic continuum on the reaction coordinate and the role of surface fields will become apparent in this discussion.

The classic starting point in electron transfer theory is the Marcus-Levich derivation. This theory begins by postulating that the electronic levels of the donor and acceptor can be modelled as single discrete levels in a potential well. The nuclear potentials of the reactants and products are assumed to be parabolic (see Figure 3) with a splitting due to coupling between sites. In this normal formulation of electron transfer [26–29], a nuclear continuum is necessary to stabilize the charge transfer. Dynamically, the electron

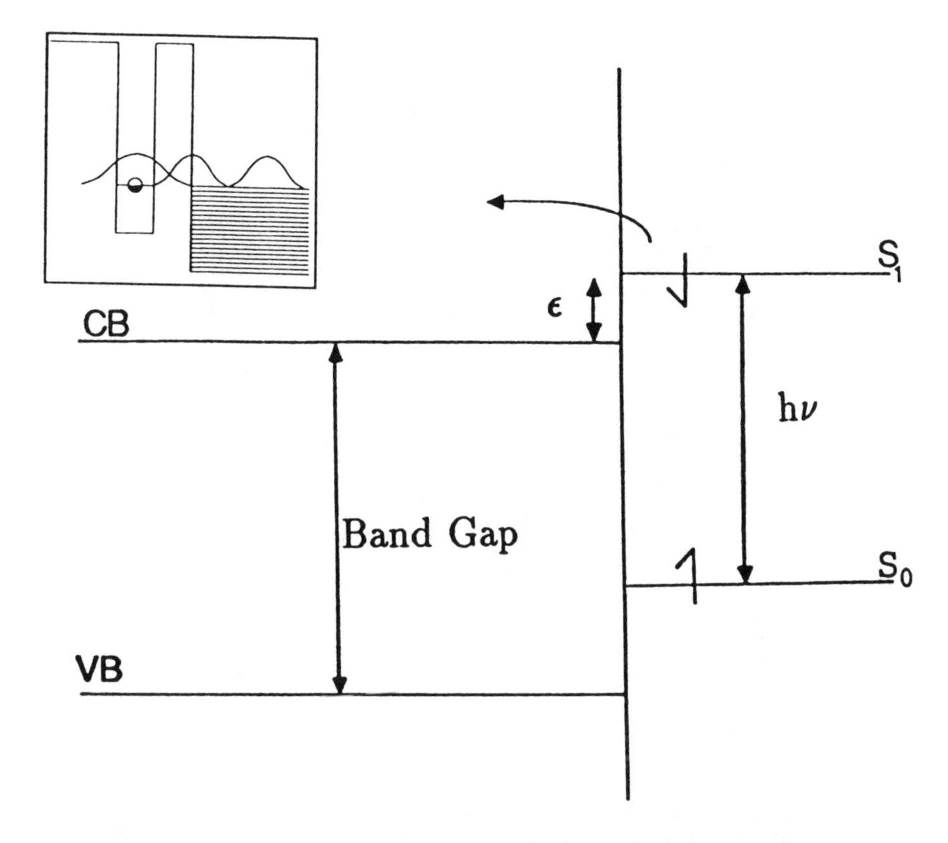

Fig. 2. A general photo-induced charge injection process. CB and VB are the conduction and valence bands respectively. S_0 and S_1 are the ground state and first excited singlet states of the chromophore, separated in energy by $h\nu$, with ε representing the difference in energy between the injecting level and the conduction band edge. The inset shows an electronic state representation of the process. The electronic wavefunction of the chromophore is a discrete state in a single well which is coupled to a huge number of quasi-unbound states in the semiconductor.

tunnels resonantly from the reactant channel to the product channel and the resonance is subsequently broken by relaxation in the nuclear continuum of either species. This nuclear reorganization process plays a crucial role in the Marcus-Levich theory, providing the upper limit for the electron transfer rate.

For electron transfer reactions between molecular species in solution, this treatment has been quite successful. Molecular species have a relatively simple, well defined electronic structure with a very large continuum of nuclear degrees of freedom supplied by the reacting species and the bath. The situation is significantly different at heterogeneous interfaces. A molecular species with a discrete electronic spectrum and a continuum of nuclear degrees of freedom exists at the surface of the semiconductor. However, the semiconductor

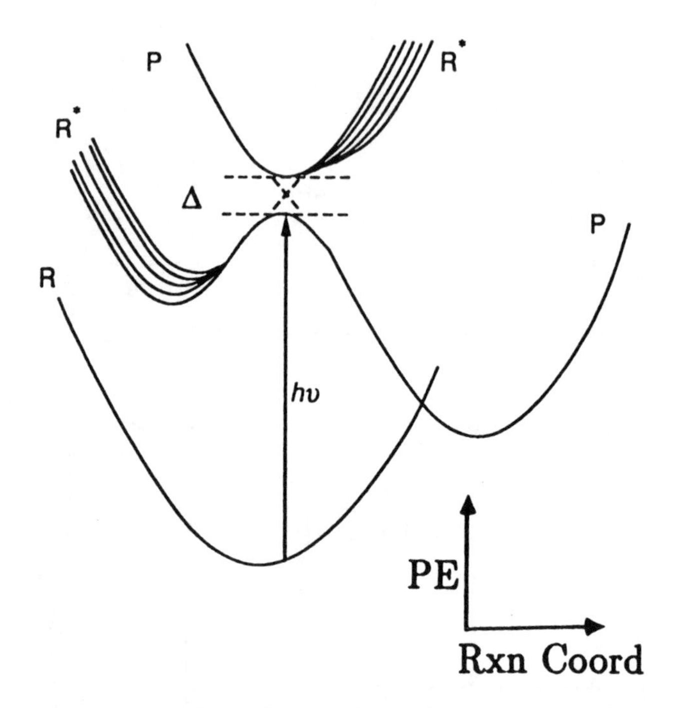

Fig. 3. The normal Marcus theory diagram of potential energy vs. reaction coordinate. R and P represent the reactants and products; Δ indicates the degree of mixing between R and P, namely the electronic coupling. The absorption of a photon of energy $h\nu$ promotes the system to the adiabatic crossing point between the reactant and product channels. The multitude of levels in the reactant coordinate is meant to schematically represent the electronic level mismatch across the interface.

has a quasi-continuum of electronic states, with considerable corresponding delocalization of the wavefunction over many electronic states.

The semantics of the discussion become critical at this juncture, a charge placed at one site in the conduction band coherently samples the entire band on a timescale given by the root mean square coupling of that state to all other states. This places the charge carrier in a mixed state of all the electronic states of the band on an incredibly short timescale. This delocalization is analogous, but not equivalent, to electronic dephasing in the band. True electronic dephasing is normally taken to be a population renormalization due to scattering processes. In the delocalization discussed here, there is no actual population renormalization, the carrier is delocalized due solely to the volume of phase space the mixed state occupies. This can be viewed as "scattering free" dephasing in the sense that coherent recurrence on the initial site is stretched so much in time by the delocalization that the scattering processes occur much faster than the recurrences and, hence, the carrier essentially undergoes "scattering free" dephasing on the timescale of the delocalization – not the scattering. However, the charge transfer event is not deemed to have taken place as long as there is any coherent coupling to the donor, even if the recurrence time for the charge carrier is years long. Defining the charge transfer to include the population relaxation dynamics removes the ambiguity present in the system between a carrier in a mixed state and a carrier actually changing states.

For a carrier transferring across the interface, the scattering free dephasing will be much, much faster than the interfacial transfer time. (If it were not, then the coupling between the donor and the conduction band would be so large that the donor could no longer be thought of as a distinct species.) The shear volume of phase space available in the band causes scattering free dephasing, significantly lengthening the time between coherent recurrences on the donor. When the scattering processes are much faster, the electronic population renormalizes, by scattering in the solid phase or nuclear relaxation in the donor half-space, before it has time to reoccur on its initial site, effectively localizing the carrier in the semiconductor half-space. In the absence of any scattering processes, the wavefunction propagation would remain coherent and the wavefunction eventually recurs at the initial site. Hence, the phase space delocalization does not obviate the need for some kind of relaxation process to localize the carrier in the substrate half space. It simply lengthens the time between recurrences so that the nuclear relaxation of the molecular species is no longer the rate limiting step. Therefore, even though this model suggests that the fundamental limit on electron transfer rates need not be the nuclear relaxation timescales as the Marcus equation implies, relaxation or scattering processes cannot be eliminated as an integral part of the charge transfer event.

The earliest theoretical work on charge injection at interfaces centered around a diffusional model [30, 31] based on the Onsager model [32, 33]. In these models the carrier motion is taken to be diffusive in the band, once separation from the parent ion has occurred, and is influenced by the applied field and the induced image potential at the surface. These models are essentially classical in nature and attempt to determine the current generated in materials after charge injection. These models do not attempt to answer any questions about the actual carrier dynamics in these systems.

The system's dynamics can be discussed in terms of the density matrix and the complete Liouville equation. With a few simple approximations, the rate equation can be shown to be essentially Fermi's Golden Rule in nature. The nature of these approximations and their effect on the calculated rate is discussed in depth and justification given for the final expression.

In attempting to formulate an analytical model, the complete molecular Liouville equation was employed within the Mori- Zwanzig projection operator formalism [34, 35] to derive the rate equation for the electron transfer. The

Liouville equation describes the evolution of the density matrix in time, which is directly correlated to the populations of various electronic states within the model described here. This approach maps the time evolution of the system, but requires no detailed knowledge of the eigenvalues and eigenvectors.

The system is modelled as a single electronic level in the molecular half space [the states denoted by k in space β] and a dense quasi-continuum of states in the semiconductor half space [denoted by j in space α]. We can then write the combined Hamiltonian (within the rotating wave approximation) in the time independent form:

$$H = \sum_{j\alpha} |j\alpha\rangle E_{j\alpha}\langle j\alpha| + |k\beta\rangle E_{k\beta}\langle k\beta| + \sum_{j\alpha} |j\alpha\rangle V_{jk}^{\alpha\beta}\langle k\beta| \tag{2}$$

where $E_{j\alpha}$ and $E_{k\beta}$ are the energies of the dye states and the conduction band states and $V_{jk}^{\alpha\beta}$ is the coupling between the discrete molecular state $k\beta$ and the conduction band electronic states $j\alpha$.

The state of the system at any time is given by the complete molecular density matrix, $\rho(t)$, satisfying the Liouville equation:

$$\frac{d\rho}{dt} = -i[H, \rho] = -iL\rho \tag{3}$$

where L ($L = L_0 + L'$) is the Liouville operator corresponding to commutation of H with the operand. The density matrix contains the complete information regarding the state of the system at any time, much of which is redundant or superfluous. Only the populations are pertinent to this discussion, so the reduced equations of motion for these populations are constructed by projecting the density matrix onto the populations using the following set of orthonormal molecular operators:

$$A_{\phi\phi} = \frac{1}{\sqrt{d_\phi}} \sum_i |i\phi\rangle\langle i\phi| \tag{4}$$

$$\langle A_{\alpha\alpha}, A_{\beta\beta}\rangle = \mathrm{Tr}(A_{\alpha\alpha}^\dagger A_{\beta\beta}) = \delta_{\alpha,\beta} \tag{5}$$

where d_ϕ is the number of states in the space ϕ, and the summation over i goes from 1 to N, N being the total number of electronic levels considered in the spaces α and β (N for α, M for β). The density matrix can now be expanded as:

$$\rho(t) = \sum_\phi \sigma_\phi(t) A_{\phi\phi} + \rho', \quad \phi = \alpha, \beta, \tag{6}$$

where $\sigma_\phi(t)$ are complex numbers corresponding to the $\mathrm{Tr}[A_{\phi\phi}^\dagger \rho(t)]$ and where ρ' is constructed orthogonal to $A_{\phi\phi}$. The density matrix is now effectively partitioned into diagonal ($\sum_\phi \sigma_\phi(t) A_{\phi\phi}$) and off diagonal ($\rho'$) elements.

The total information has now been reduced to the populations of the various levels, P_ϕ, where:

$$P_\phi = \sqrt{d_\phi}\ \mathrm{Tr} A^\dagger_{\phi\phi}\rho(t) = \sqrt{d_\phi}\sigma_\phi(t). \tag{7}$$

The Mori-Zwanzig projection operator technique is employed to derive the reduced equations of motion for the populations. The vector σ whose components are σ_α, σ_β is now introduced along with the Mori projection operator, P, whose action on operator $\rho(t)$ is given by:

$$P\rho(t) = \sum_\phi \sigma_\phi(t) A_{\phi\phi} \tag{8}$$

and its complement, Q, given by $Q = 1 - P$. The Mori-Zwanzig formalism gives *exactly* the closed reduced equation of motion for our population vector σ (for any choice of P):

$$\frac{d\sigma}{dt} = -i\langle L\rangle\sigma - \int_0^t d\tau \langle R(t-\tau)\rangle \sigma(\tau) \tag{9}$$

The operator $R(t-\tau)$ is a tetradic operator (as is L) given by:

$$R(t-\tau) = L\ \exp[-iQL(t-\tau)]QL \tag{10}$$

and the term $\langle X\rangle$ denotes the Tr($A^\dagger \times A$). For our particular projection P, $PLP = 0$, and $R(t-\tau)$ can be expanded in power series in QL' [see reference 36] with only the even powers in QL' nonzero. This allows us to write the reduced equation of motion for σ :

$$\frac{d\sigma}{dt} = -\int_0^t d\tau \langle R(t-\tau)\rangle \sigma(\tau) \tag{11}$$

$$\langle R(t-\tau)\rangle = \langle R^{(2)}(t-\tau)\rangle + \langle R^{(4)}(t-\tau)\rangle + \ ... \tag{12}$$

For illustrative purposes only, we choose a very simple model for the coupling in the system. The coupling frequency between the discrete molecular states and all relevant semiconductor states is assumed to be constant, V_0. This is a reasonable assumption for a narrow energy width of semiconductor band states. Setting the molecular donor state as the origin, we consider the semiconductor states between $-\Delta$ and Δ, and we consider the density of states to be isotropic over this range. This coupling scheme is illustrated in Figure 4 for the case in which the S_1 level of an adsorbed dye molecule is the origin. If this range is taken as the full width at half maximum of the Gerischer curve of the adsorbed molecule and lies well above the conduction band edge of the semiconductor, these assumptions are not unrealistic and will give a qualitative understanding of the real physics of the system.

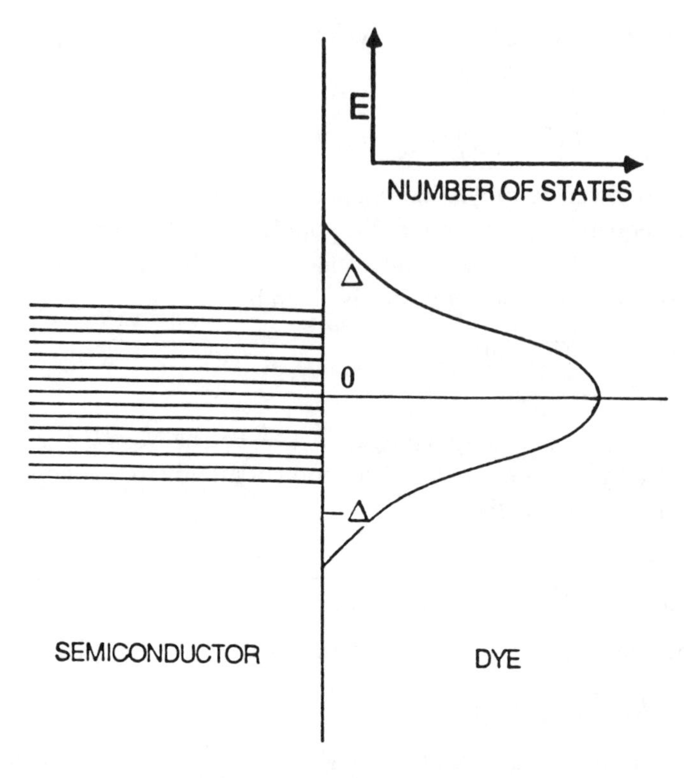

Fig. 4. The coupling scheme for the Mori-Zwanzig formalism. A quasi-continuum of semiconductor states is coupled to a Gerischer distribution of dye molecules. The peak of the Gerischer curve is arbitrarily chosen as the origin.

Constructing the second order terms of $\langle R(t-\tau)\rangle$ gives:

$$\langle R^{(2)}_{\alpha\alpha,\beta\beta}(t-\tau)\rangle = -\frac{1}{\sqrt{d_\alpha d_\beta}}\sum_{j,k}[L'^{jj,jk}_{\beta\beta,\alpha\beta}(t-\tau)L'^{jk,kk}_{\alpha\beta,\beta\beta}(0)+\text{c.c.}] \quad (13)$$

where c.c. refers to the complex conjugate, and:

$$L'^{\alpha\beta,\gamma\delta}_{ij,kl} \equiv \langle\langle i\alpha, j\beta|\, L'\, |k\gamma, l\delta\rangle\rangle \quad (14)$$

$$|i\alpha, j\beta\rangle\rangle \equiv |i\alpha\rangle\,\langle j\beta| \quad (15)$$

This can also be written:

$$\langle R^{(2)}_{\alpha\alpha,\beta\beta}(t-\tau)\rangle = -\frac{1}{\sqrt{d_\alpha d_\beta}}\sum_{j,k} V_0^2\, \exp[i\omega_{kj}(t-\tau)] + \text{c.c.} \quad (16)$$

Assuming the semiconductor quasi-continuum is dense enough to be considered a real continuum (a fair approximation at room temperature), the summation over levels may be replaced with an integration with respect to energy. The integration is trivial and gives a rate expression, exact to second order in the coupling:

$$\frac{d\sigma_\alpha}{dt} = \frac{V_0^2 2\pi\hbar\rho(E_j)}{\sqrt{d_\alpha d_\beta}} \int_0^t \delta(t-\tau)\sigma_\beta(\tau)d\tau \qquad (17)$$

The populations are related to the vector components of σ by the relation $P_n = \sqrt{d_n}\sigma_n$, so:

$$\frac{dP_\alpha}{dt} = \frac{V_0^2 4\pi\hbar\rho(E_j)}{d_\beta} P_\beta(t) \qquad (18)$$

This can be considered as a simple rate equation. Consider that in this model, $\rho(E_k) = d_\beta/2\Delta$, and one can write the rate of going from space β to space α to second order in the coupling, $k^{(2)}_{\alpha\beta}$, given by:

$$k^{(2)}_{\alpha\beta} = 4\pi\hbar V_0^2 \rho(E_j) \qquad (19)$$

The higher order terms ($R^{(4)}$, $R^{(6)}$, ...) can also be constructed, but within the random phase approximation [37, 38] they are zero and the above expression is exact to second order. This is essentially Fermi's Golden Rule which is anticipated once the constant coupling assumptions is made. The key step in reaching the final equation becomes the $\delta(t-\tau)$ in Equation (17). This δ-function implies that the system equilibrates faster than any net change in populations – the system is nonadiabatic. It is this nonadiabaticity that gives rise to a single net rate for the charge transfer (Equation (19)). If the system is adiabatic, then the charges are moving faster than any equilibration is taking place and the transfer rate cannot be written so simply.

The electron resonantly tunnels across the interface into the very large electronic phase space of the semiconductor and samples the whole quasi-continuum of the conduction band on the time scale of the electronic coupling between band states (10–50 fsec for semiconductors). At this point the electron exists in a mixed state of all of the relevant coupled band states and the very large number of these states prevents any recurrence on the donor molecule before scattering and nuclear relaxation processes break the resonance with the donor.

The transfer rate depends only on the electronic density of states and the electronic coupling. These conditions are sufficient for the propagation of the electron into the semiconductor and separation from the initial molecular site. Further, it is not necessary to invoke the nuclear degrees of freedom of the system to stabilize the charge transfer event since the transfer results in delocalization of the electronic wavefunction over the conduction band on the sub-50 fsec timescale of the electronic dephasing. In a normal Marcus approach, the relaxation of the nuclear degrees of freedom, both intramolecular and intermolecular, is the rate limiting step in an electron transfer event – on the order of 100 fsec for most molecular species [15–19, 39, 40]. The "scattering free" dephasing described above is significantly faster than typical nuclear relaxation processes and the limit for the electron transfer is now the electron dephasing time of the semiconductor (as fast as 10 fsec).

Only four assumptions have been made in the derivation of the rate equations: rotating wave approximation, random phase approximation, constant coupling assumption, and the approximation of the system to a continuum. The rotating wave approximation is made in writing the initial Hamiltonian for the system. The Hamiltonian used is a common one and there is nothing about this system that suggests it is anything but perfectly reasonable.

The random phase approximation is a natural conclusion of the continuum assumption. Clearly, the semiconductor electronic states are virtually a continuum at room temperature – this is barely an assumption at all. The random phase approximation should prove equally mild stating that for a high density of states system, the Fourier components of the density operator:

$$\rho_k = \sum_j e^{ik \cdot r_j} \tag{20}$$

are so dense that in the equation of motion of ρ the different phases of k cancel in the cross term and can be neglected. The density of states here is so high that this is an excellent approximation.

The final assumption is the constant coupling model for the coupling between molecular and semiconductor electronic states. This is probably the least physically reasonable assumption, especially if one were to include the transient fields at the surface or different molecules in an ensemble of donors. For such an ensemble, near the peak of the energy distribution for the molecular states, this distribution is relatively flat. The inhomogeneous broadening of adsorbed species, suggests a variety of orientations and environments of specific dye molecules. The coupling between different dye molecules and the semiconductor electronic states would reflect these differences in environment and would not be "constant". While such an "inhomogeneously broadened" coupling would give rise to a distribution of transfer times for the ensemble, it would not affect the fundamental proposition. The transient fields would be more problematic and represent a very important dynamical variable; however, they have been excluded to simplify the mathematics and allow the investigation of general trends. The coulombic effects can be treated separately.

The back transfer rate can be written as well, but keep in mind that the number of semiconductor states is huge relative to the number of molecular states. In fact, the molecular state is really only one state within a large linewidth (2Δ), this is formally equivalent, for an ensemble of molecules, to a large number of narrow electronic states. In any case, the forward rate will be dominated by the number of participating states in the semiconductor and should be significantly greater than the back transfer rate.

In addition, the back transfer will likely require nuclear relaxation to stabilize the charge transfer event. This is obvious if one examines the most likely mechanisms for the back electron transfer – thermalized band edge transfer or electron transfer from an impurity or dopant band. (It would not be the case if the back transfer were occurring by hole injection into the valence band, but that pathway should not be energetically favorable.) The density of states of the molecular species is too small to effectively stabilize the charge transfer by delocalization alone. The electronic resonance between semiconductor band state and dye molecular state needs some nuclear relaxation phenomenon to break the resonance and localize the charge transfer – as in normal Marcus–Levich type charge transfer. In the absence of this relaxation, the electronic states remain resonantly coupled and the carrier would simply transfer back to the semiconductor. This has important implications for the potential use of a system such as this to probe electron transfer as both an activationless process (forward electron injection) and as an activated process (back electron transfer).

To make these concepts as transparent as possible, a simulation of the electron wavefunction propagation was conducted. The model used to simulate the conduction band is a rigid cubic lattice of N identical atoms. This system can be treated quantum mechanically within the Dirac interaction picture and the time dependence of the system calculated using the time-dependent Schroedinger equation:

$$i\hbar \frac{\partial \left|\Psi(t)\right\rangle}{\partial t} = H \left|\Psi(t)\right\rangle \tag{21}$$

where $\hbar = h/2\pi$ (h being Planck's constant), $|\Psi(t)\rangle$ is the time-dependent state function of the system, and H is the Hamiltonian, written:

$$H = \sum_{n} H_{0,n} + \sum_{n,m} V_{nm} \tag{22}$$

The unperturbed Hamiltonian is $H_{0,n}$, with eigenvalues E_n and eigenvector ϕ_n, and V_{nm} represents the coupling of state m to state n. The summation is performed over all of the states in the cluster and the eigenvectors are taken as a basis set for the entire space. Within the Dirac Interaction picture, the time evolution of the state vector can be shown to satisfy the following matrix equation:

$$i\hbar\frac{\delta}{\delta t}\begin{pmatrix} c_1 \\ c_2 \\ \vdots \\ c_N \end{pmatrix} = \begin{pmatrix} V_{11} & V_{12}e^{i\omega_{12}t} & \dots & V_{1N}e^{i\omega_{1N}t} \\ V_{21}e^{i\omega_{21}t} & V_{22} & \dots & V_{2N}e^{i\omega_{2N}t} \\ \vdots & \vdots & \vdots & \vdots \end{pmatrix}\begin{pmatrix} c_1 \\ c_2 \\ \vdots \\ c_N \end{pmatrix} \tag{23}$$

where:

$$\omega_{nm} = \frac{E_n - E_m}{\hbar} \tag{24}$$

$$c_n(t) = \langle \phi_n | \Phi(t) \rangle \tag{25}$$

The solution of a general equation of the form:

$$\frac{\partial A}{\partial t} = VA \tag{26}$$

where A is a vector of dimension N and V is a matrix of dimension $N \times N$ which can be diagonalized, is given by:

$$A(t) = Xe^{\Lambda t}X^{-1}A_0 \tag{27}$$

where X is the eigenvector matrix whose N columns correspond to the N eigenvectors of the matrix V and $X^{-1}$is its inverse. A_0 is the initial condition, the vector A evaluated at time $t = 0$, and $e^{\Lambda t}$ is a diagonal matrix whose diagonal elements, k_{ii}, are given by:

$$k_{ii} = e^{\lambda_i t} \tag{28}$$

with the i-th value λ corresponds to the i-th eigenvalue of the original matrix V.

Computationally, the problem reduces to the determination of the eigenvalues and eigenvectors of the complex coupling matrix, V^c_{nm}, of Equation (24). This is easily solvable using standard computer library (IMSL Library) routines for solving complex linear equations.

Initially, the cluster is taken to consist of neutral elements only, so there are no coulombic fields to consider. The electron is placed at one site, i, so that the initial condition is:

$$c_n(0) = \begin{cases} 1 & n = i \\ 0 & n \neq i \end{cases} \tag{29}$$

In order to demonstrate the concept of "scatter free" dephasing, Figure 5A shows the probability for being on the initial site as a function of time for

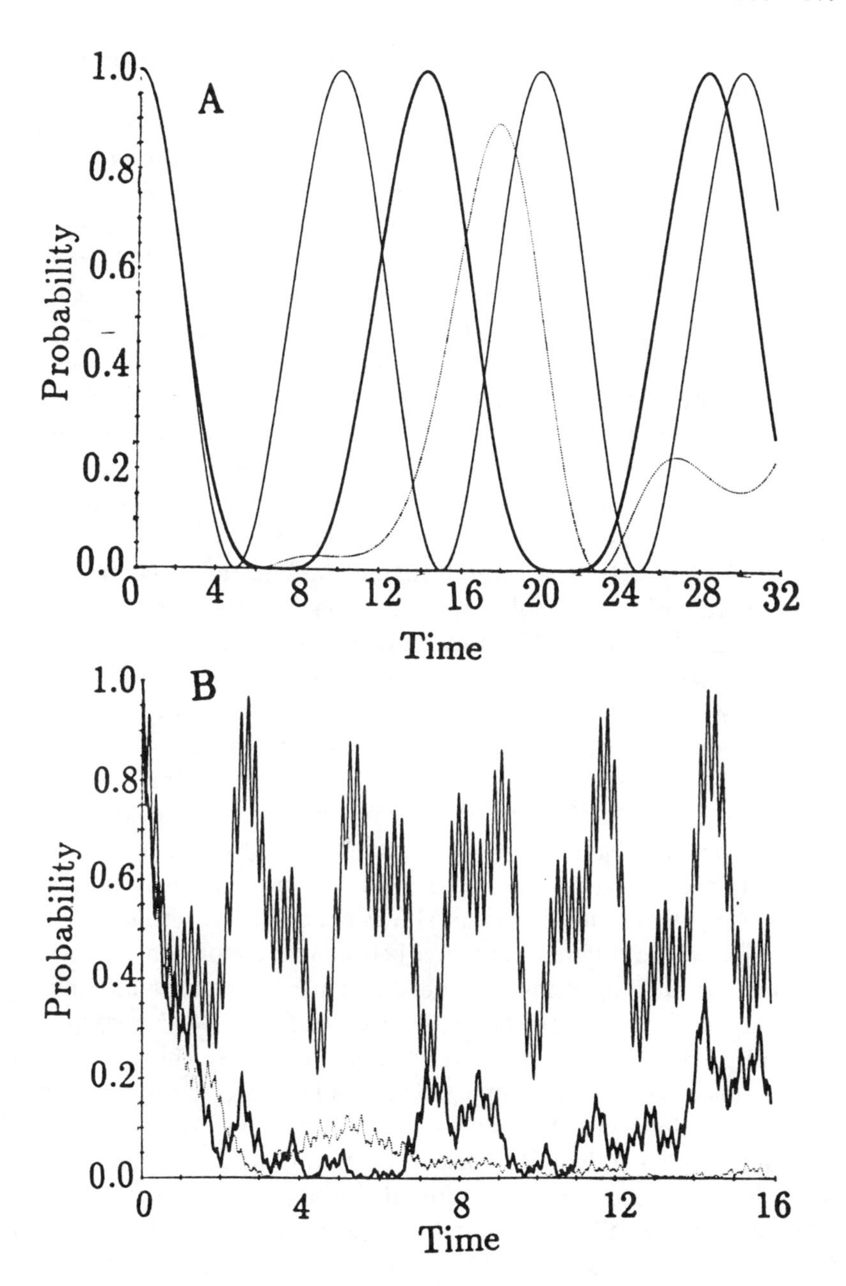

Fig. 5. Phase space volume delocalization. Graph A shows the probability of being on the initial site for systems of 2 (thin line), 3 (thick line), and 4 (dotted line) sites coupled in a linear fashion. Graph B shows the probability of being on the initial site for cubic lattices of N atoms. The thin line represents 27 atoms ($3 \times 3 \times 3$ lattice), the thick line represents 343 atoms, and the dotted line represents a 512 atom cluster. The time scale corresponds to periods of the lattice eigenvalues.

2, 3, and 4 atoms arranged in a line. The coupling between adjacent sites is considered to be equal and constant. The results for the 2 atom problem clearly demonstrate Rabi oscillations, as would be expected. It must be remembered that this calculation is completely coherent, there are no dephasing processes considered. Notice that as the atoms are added to the line, from 2 to 3 to 4, the time it takes for the electron to recur on its initial state gets progressively longer. Clearly, if an infinite number of states were coupled, the recurrence time itself would become infinite. More practically, Figure 5B shows the results for propagation of the wavefunction in a cubic lattice of 27, 343, and 512 atoms (3^3, 7^3, and 8^3). The coupling between sites is taken to decrease exponentially with distance with a 1/e point at the nearest neighbors. The edge atoms are fitted with periodic boundary conditions to soften the edge effects. It is clear that for the 27 atom clusters recurrences are large and frequent, but for the larger clusters the recurrences have become far less frequent and much smaller. This clearly displays what was termed "scattering free" dephasing earlier; the recurrence times become infinitely long due to the phase space volume available for propagation. Eventually, for large systems, they become longer than all of the normal dephasing processes: scattering, nuclear relaxation, etc. In all of these simulations, the time axis is delineated by periods corresponding to the eigenvalues of the system.

The simulation is now performed for a charge transfer process. The semiconductor is modelled as a cubic lattice, as above. A single atom is now placed on one face of the cube, this site represents the absorbed species. The coupling between lattice sites remains as it was, the external site is coupled weakly to one site on the face of the cube, the wavefunction is initially localized on the external site. Figure 6A shows representative results for a 512 + 1 site simulation for different couplings between the donor atom and the lattice. The transfer rate is approximately quadratic in coupling, as one would expect from the Golden Rule calculation. Notice that for the weaker couplings the wavefunction never completely vacates the initial site. This can be thought of in terms of coupled harmonic oscillators with two spring constants, one between the donor and the lattice and some effective lattice constant. The harmonic motion of the donor pushes the "spring" that represents the lattice. If the harmonic motion between the donor and lattice is sufficiently slow relative to the effective lattice oscillation, the lattice begins pushing back on the donor before the donor completes its harmonic motion.

To investigate the complementary process of charge transfer from the lattice to an acceptor on the surface, the same simulation is run, only the wavefunction is initially located in the lattice. The probability of being on the acceptor is monitored in time and the initial lattice site is taken to be two lattice sites removed from the acceptor. The lattice atoms are coupled as described above and the adsorbate is coupled to one site (adjacent to the

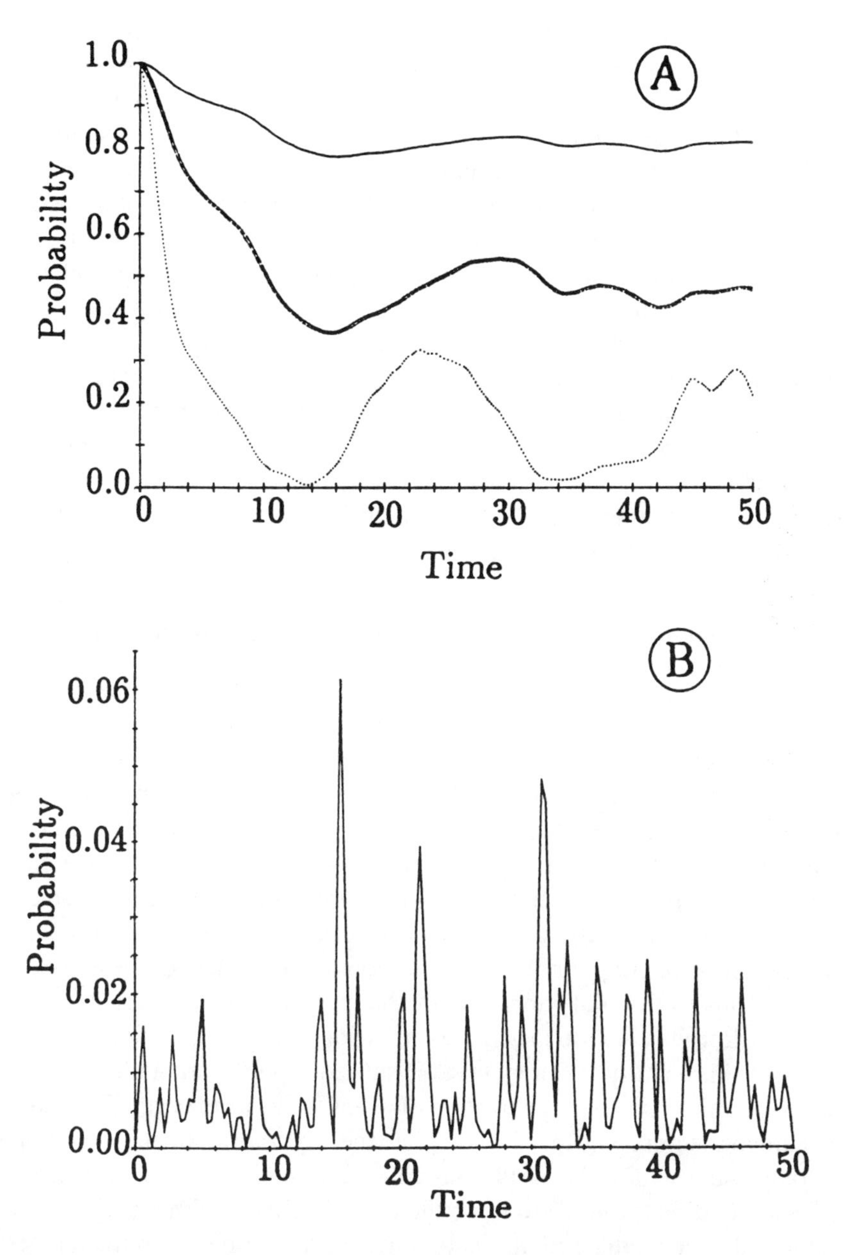

Fig. 6. Electron transfer across the interface. Plot A shows the probability of being on the initial site as a function of time for a system consisting of one site coupled to a 512 atom lattice. The single line corresponds to coupling of the single site to one site on the lattice with coupling = 1/56 of the lattice atom's root mean square coupling. The double line corresponds to coupling = 1/28 the lattice's rms coupling, and the dotted line corresponds to coupling = 1/14 the rms coupling. Plot B shows the complementary process. The electron is initially located in the lattice and the graph monitors the probability of being on the single external site which is coupled to one atom with a coupling = 1/3 the rms coupling.

initial site) with a coupling equal to one-third that of the root mean square (rms) coupling of the lattice atoms to each other. It should be remarked that this is a very strong coupling condition for the adsorbate, much stronger than for the charge injection simulation discussed above, the results are shown in Figure 6B. There exist frequent occurrences of the wavefunction on the adsorbate. They remain, however, on the order of 5 percent at a maximum and are transient in nature. Since there is no localization process and the phase space volume of the lattice is over 500 times that of the adsorbate, the lattice is far too inviting to the wavefunction for it to stay on the adsorbate for long. This clearly illustrates the need for a Marcus-type deactivation for the back transfer as discussed above, i.e. some kind of nuclear relaxation in the adsorbate half-space is required to localize the carrier on the adsorbate and stabilize the back transfer. Since the occurrences on the adsorbate are frequent in nature but require nuclear relaxation to stabilize the transfer, the time scale for the back transfer can be no faster than the nuclear relaxation phenomena as Marcus theory suggests.

4. Experimental Approaches to a Real Time View of Electron Transfer

From an experimental point of view, one would like to determine the spatial propagation of an electron across an interface under zero barrier conditions, i.e., at the adiabatic crossing point of the reaction surface. At this critical point, the electron transfer time is not determined by the activation barrier which is well described by Boltzmann statistics. Under zero barrier conditions, the upper limits to electron transfer can be determined as well as the degree of electronic coupling between states. This information is related to the so called promotor modes for the nuclear coordinate (ν^{-1} in Equation (1)) and the wavefunction overlap for the electronic coordinate. The barrierless condition corresponds to the transition state of the reaction. Optical preparation of a non-equilibrium electron that is resonant with both band and molecular states accesses this condition. For metals, the subsequent relaxation is generally too fast to permit a large enough fraction of reactive events to occur. The bandgap of semiconductors prevents the complete thermalization and the large electric fields present within the surface space charge region (absent in metals) act to spatially direct the electron into the reaction channel. The work discussed below takes advantage of the photophysical properties of semiconductors to optically prepare the system at the adiabatic crossing point for electron transfer.

This model problem must be studied under low electron densities to avoid excessive electron-electron scattering and coulombic effects between carriers which have nothing to do with the reaction coordinate but would complicate the dynamics. This condition will be referred to as the *one electron limit* and necessitates the detection of a reactive flux involving less than 10^{-3} to 10^{-4} of a monolayer of acceptor/donor states. The high sensitivity and time resolution needed to study these transition state process at surfaces has only recently become available with the development of proper laser sources and new optical spectroscopies developed specifically to address this problem. In what follows is a purely experimental approach to map out the electron trajectory across an interface.

4.1 Solid State Donor Condition

4.1.1 Surface Restricted Grating Studies of Carrier Dynamics

The problem of sensitivity and time resolution has been overcome by adopting the transient grating approach to follow the electron dynamics at the surface [41, 42]. This technique owes its high sensitivity to the optical interferometric nature of the spectroscopy. The grating experimental setup and pulse sequence are shown in Figure 7. Two above band gap excitation pulses are used to write a sinusoidal intensity pattern on the surface which becomes encoded in the form of electron-hole pairs. There is a very large change in the index of refraction with carrier formation. In essence, the optically induced carrier distribution represents a "wire" diffraction grating. The carrier dynamics are monitored by diffracting a below band gap probe well away from the band edge to probe the intraband carrier transitions and the carrier populations. Both the electron and hole carrier contribute to the signal through their independent contributions to the index of refraction at the probe wavelength.

The key element in this experiment is the space charge field which focuses the minority carriers, photogenerated within the space charge region, to the surface on 100 fsec time scales (determined by electrooptic sampling, as discussed below). For the experimental excitation wavelengths (absorptivity $\sim 10^5$ cm^{-1}) and space charge widths ($\sim 10^{-5}$ cm) this gives more than a 50% surface contribution to the signal. The minority carrier becomes confined to within 10 Å of the surface reaction plane by the surface field. Thus, the carriers are driven to the surface to within tunneling distance of the molecular acceptors on the right time scale to observe even unthermalized electron transfer processes.

The main innovative features of this spectroscopy are: the use of the surface space charge field to eliminate transport limitations on the determined carrier dynamics; and the surface grating diffraction geometry for signal detection. With this approach, the diffracted signal is spatially isolated from the other beams so the signal is detected against essentially zero background. A small amount of probe scatter limits the detection at 10^{-5} to 10^{-4} of a monolayer of reactive carrier flux. By direct comparison, the signal sensitivity is found to

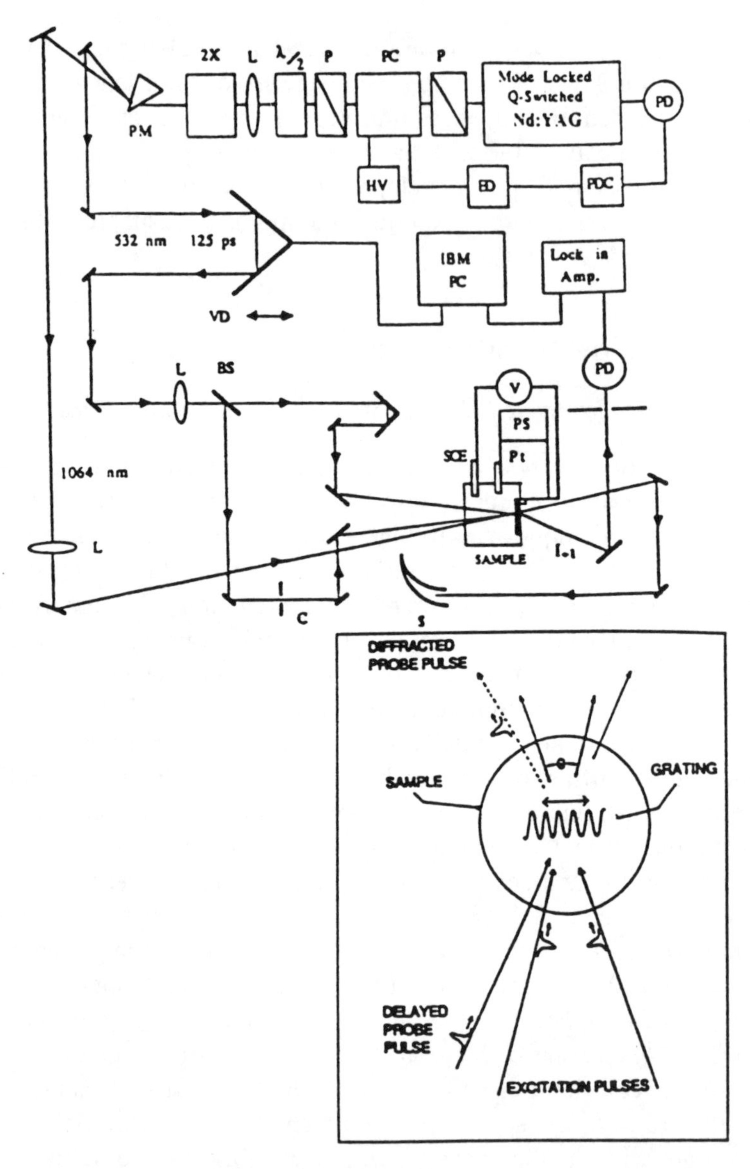

Fig. 7. The Surface Restricted Transient Grating setup. P = polarizer, L = lens, λ/2 = half-wave plate, 2X = doubling crystal, PD = photodiode, PDC = pulse discriminator circuit, ED = electronic delay, VD = variable optical delay, PC = Pockel's cell, PM = prism, BS = 50/50 beamsplitter, SCE = saturated calomel reference electrode, PT = platinum counterelectrode, and C = mechanical chopper. The insert shows the grating pulse sequence, where θ denotes the angle between excitation beams.

be two orders of magnitude better than strictly pump/probe analogues. This sensitivity is important in minimizing carrier-carrier coulombic interactions

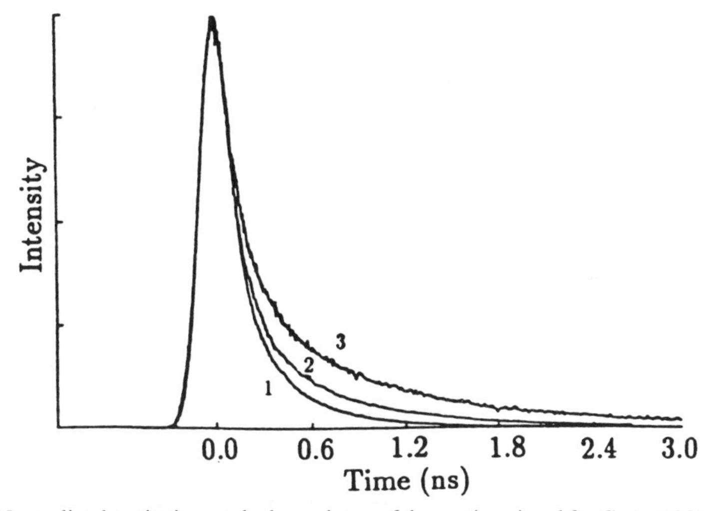

Fig. 8. Normalized excitation angle dependence of the grating signal for GaAs(100) surfaces. The grating excitation was 532 nm (2×10^{14} photons/cm^2) and the probe was 1.064 μm. The grating fringe spacing was varied: curve 3 = 28 μm, curve 2 = 10 μm, and curve 1 = 6 μm. The observed fringe spacing dependence gives an ambipolar diffusion constant of 7 ± 2 cm^2/sec. This result demonstrates the grating decay is determined by the carrier population dynamics.

and space charge screening. Grating imaging also allows a clean distinction between mobile and trapped states.

The different applications of grating spectroscopy to understanding the surface dynamics [42] are shown in Figures 8, 9, and 10. A grating fringe spacing dependence at zero surface field shows that the grating is sensitive to carrier populations (Figure 8). The signal decays faster at narrower fringe spacings due to spatial transport of the photocarriers parallel to the surface that washes out the grating pattern. The observed fringe spacing dependence gives a direct measure of the electron-hole diffusion constant which is exactly that expected for GaAs. In the study of surface state trapping, studies of n-GaAs(100)/oxide interfaces found a hundred fold enhancement in the grating signal (Figure 9). The enhancement was attributed to surface state transitions near resonance to the probe wavelength that increase the dispersive contribution (Δn) of the surface states over the free carriers [42]. The surface state nature of the signal was confirmed by the effect of surface treatments on the signal, absence of a fringe spacing dependence (i.e., a trapped state), and signal saturation at surface state number densities. The main distinction of this signal component from carrier depletion is that it *shows a rise rather than a decay*. The rise time of the surface state related signal was faster than 30 picoseconds. Similar findings have been made using time resolved photoemission to selectively study surface state dynamics at Si surfaces [43]. This fast surface state trapping time is expected based on the barrierless nature

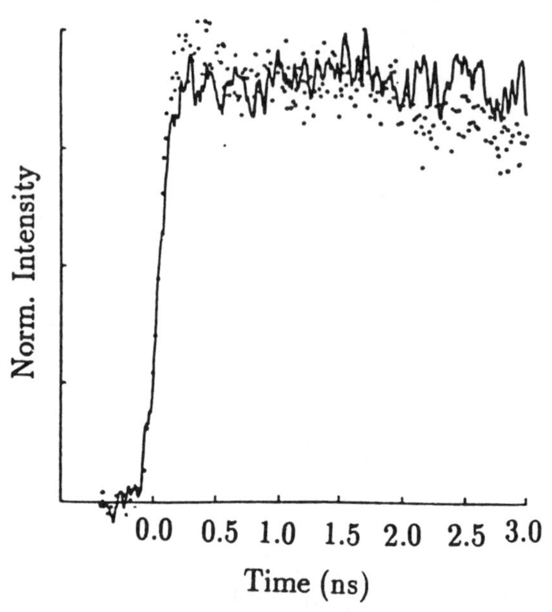

Fig. 9. Surface state trapping dynamics. The curve represents the surface grating for GaAs(100) in air at low photocarrier injection (1×10^{11} photons/cm^2). At this injection level, there is substantial surface field present due to oxide induced surface states that drives the hole carrier to the surface. The surface state trapping of the hole carrier is observable as a rise in the grating signal. There is a two orders of magnitude enhancement of the grating signal with the surface state trapping. The solid line corresponds to an excitation angle of 3.1 degrees, the dotted line corresponds to an angle of 1.3 degrees. The lack of a fringe spacing dependence shows this to be a trapped state.

of carrier trapping at oppositely charged trapping centers [42]. This trapping step is only limited by the carrier thermalization.

In-situ studies of charge transfer have been demonstrated for the n-GaAs(100)/[Se^{-2}/Se^{-1}] interfaces. This surface is known to be stabilized against photooxidation and degradation in the presence of high concentrations of Se^{-2} which acts as an efficient hole acceptor to yield an 80% photocurrent quantum yield [4]. The grating signal, with and without the Se^{-2} under identical conditions, shows a factor of 2 reduction in the signal amplitude at $t = 0$, i.e., a decay component exists which is beyond the experimental resolution of the particular laser used in this study. The fast initial grating decay is assigned to interfacial hole transfer based on the dependence of the signal on Se^{-2} concentration (Figure 10A) and on the field dependence (Figure 10B).

The most important finding is that the hole transfer to Se^{-2} is occurring faster than 30 picoseconds. The interfacial charge transfer is occurring in competition with surface state trapping and appears to be a direct process as the enhanced surface component is not observed. The exact branching ratio between unthermalized versus thermalized and surface state mediated can be

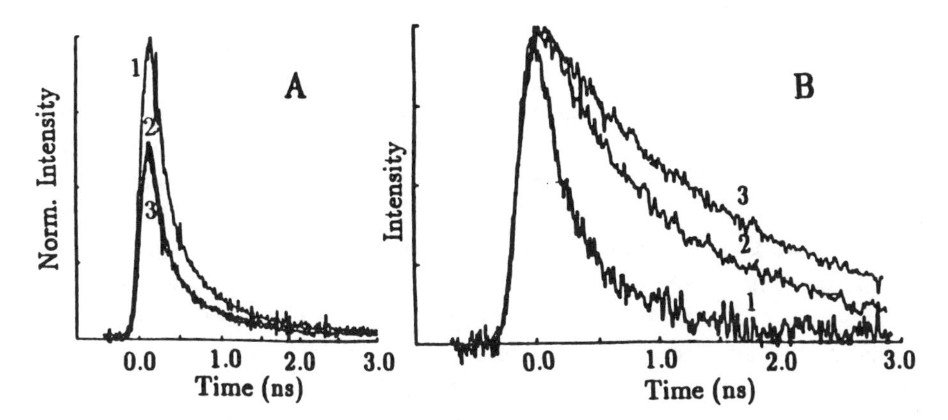

Fig. 10. Graph A shows the dependence of the grating signal on the concentration of Se^{2-} under low excitation conditions and an approximate space charge field of 10^5 V/cm. Curve 1 $\equiv$ 0.21 M, curve 2 $\equiv$ 0.56 M, and curve 3 $\equiv$ 1 M. The concentration dependence demonstrates carrier depletion involves the selenium acceptors. Graph B show the dependence of the signal on applied voltage. Curve 1 $\equiv$ −1.00 V vs. SCE (same as 10A), Curve 2 $\equiv$ −1.5 V, and Curve 3 $\equiv$ −1.7 V. The flat band potential corresponds to −1.95 V vs. SCE. The field dependence demonstrates it is the hole carrier which is depleted in the presence of Se^{-2} and that field focusing of hole carriers is achieved.

resolved with higher time resolution. Nevertheless, the observed time scale is approaching that of solvent relaxation. This result indicates that electronic coupling at surfaces can approach the strong coupling or adiabatic limit. In addition, space charge acceleration imparts approximately 1 eV of excess energy to the hole carriers which should take a picosecond to thermalize with the lattice. The sub-30 picosecond hole depletion time supports the concept that at least a small fraction of hole carriers are transmitted across the interface prior to thermalizing.

The above findings should be compared to other work in this general area. Cowin et al. using CH_3Br at Ni(111) surfaces as a probe molecule under UHV conditions found evidence for an electron attachment reaction that competed efficiently with direct photodissociation [24]. The involvement of an electron transfer step from the metal surface was implicated in a series of studies using multilayers of CH_3Br. Using the approximate dissociation time of the competing process of photodissociation as an internal clock, this measurement gave evidence for electron transfer occurring on the 10 fsec time scale. Work from a number of other groups exploring different adsorbates and surfaces have found similar evidence for electron induced desorption [44–48]. The photoproducts are found to come off vibrationally hot and involve 100 fsec time scales for the electron transfer induced desorption [48]. The quantum yields are generally low such that the observed photoproducts are in line with a small fraction of electron transfer processes occurring within the non-adiabatic limit (weak coupling) in competition with extremely fast electron thermalization.

Recently a very efficient electron transfer induced desorption process has been observed by Prybyla et al. for CO on Cu(111) with quantum yields up to 15% [49]. An elegant experiment using time resolved surface second harmonic generation to monitor the fraction of surface bound CO found that the CO dissociation was complete in less than 325 fsec. Based on the non-linear dependence of the quantum yield, this electron induced desorption is believed to occur through multiple electron transfer processes involving the non-equilibrium electrons in the metal and the $2\pi^*$ CO electronic level. The electron transfers to the CO and back transfers to the metal after partial nuclear relaxation. Multiple electron transfers are required to deposit enough energy in the CO adbond to dissociate it. This experiment gives supportive evidence for electron transfer processes occurring in the adiabatic strong coupling limit.

The other work which should be mentioned is that of Rosetti and Brus [50], and related work by Kamat *et al.* [51] using colloidal semiconductors to follow electron transfer by time resolved Raman and transient absorption respectively to monitor the product formation. The colloidal systems have interesting photophysics that distinguish them from single crystal surfaces. These systems have also been observed to exhibit picosecond interfacial electron transfer times.

The in-situ surface grating studies, taken together with recent findings from the UHV community, and studies of interfacial charge transfer dynamics at semiconductor colloids illustrate that electron transfer processes at surfaces can occur on very fast time scales.

4.1.2 Space Charge ElectroOptic Sampling of Carrier Transport

As schematically shown in Figure 1, the overall surface photochemistry involves the field assisted transport of the minority carrier to the surface. From the above grating studies, it is apparent that the charge transfer dynamics are approaching time scales associated with the carrier transport to the surface. The surface field also has a dramatic effect on the surface populations. This aspect of the electron "trajectory" needs to be characterized. The time resolution of electrooptic sampling has been improved by an order of magnitude and made surface selective by studying field changes directly in the surface space charge region [52, 53]. The separation of photogenerated electron and hole carriers in this region leads to a decrease in surface field; the dynamics of which is directly related to the spatial transport of the electron and hole carriers via Poisson's equation for the field changes.

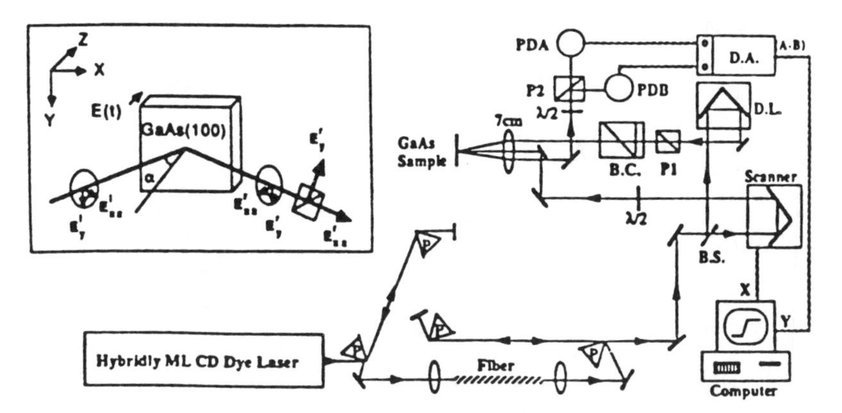

Fig. 11. The Surface Field Electro-optic Sampling Experimental setup. B.C. = Babinet compensator; P1, P2 =polarizers; $\lambda/2$ = half-wave plates; B.S. = 70/30 beam splitter; PDA, PDB = 1 cm^2 silicon photodiodes; D.A. = differential amplifier. The scanner is a galvanometer delay line and D.L. is a stepper motor delay line with an optical encoder for determining position.

By taking advantage of the intrinsic electrooptic effect in GaAs(100), the field changes are monitored optically through changes in polarization of either reflected or transmitted probes. The experimental setup is shown in Figure 11. In this manner, the field changes are monitored exactly where they occur such that there is no broadening in the time response from transmission lines. This experiment has attained the highest possible resolution through the electrooptic effect. The time resolution is dictated by the highest frequency polar phonon that is displaced by the field changes. For GaAs, this phonon frequency is 8.55 THz which corresponds to approximately 30 fsec time resolution. This development is equivalent to having a 10 THz sampling oscilloscope.

The high time resolution of this new spectroscopy, along with the ultrashort distances the electron and hole carriers propagate (100 Å length scales – comparable to the electron mean free path) make this technique ideally suited for the study of ballistic motion of the electrons. In addition, the space charge transit time is needed to determine the relative energies of the carriers as they arrive at the surface. Representative results are shown in Figure 12 using fiber optic compressed pulses of 120 femtoseconds. The time evolution of the field should be biphasic: the higher mobility electrons giving rise to the fastest component and the lower mobility holes giving rise to slower components. Only one effective time constant of 200 fsec is observed in the present experiment which is assigned to the hole carrier motion based on the factor of 10 lower mobility of the holes relative to the electrons [53, 54]. This result demonstrates that the hole carriers are driven to the surface on a

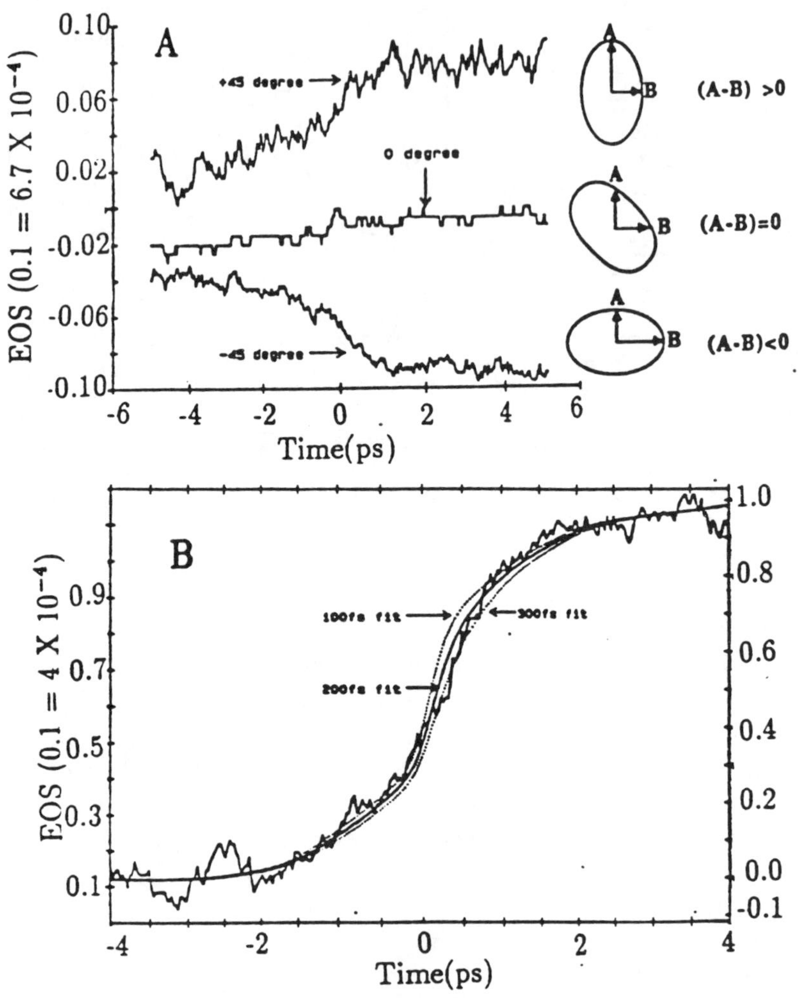

Fig. 12. Reflective Electro-optic Sampling (REOS) data for n-GaAs(100)/oxide interface. The crystal was doped to 7.8×10^{17} cm^{-3} and the photocarrier injection level was 4×10^{17} cm^{-3}. Graph A shows the dependence of the signal on the crystal orientation relative to the analyzer polarizer axis. This result demonstrates the signal arises from an electrooptic effect. The maximum magnitude of the polarization rotation at higher excitation corresponds to .8 eV of band bending. Graph B shows the REOS signal with fittings to three different rise times: 100, 200, and 300 fs.

hundred fsec time scale. This time scale is faster than carrier thermalization such that the initial distribution does in fact arrive at the surface with large amounts of excess energy through field acceleration.

This space charge electrooptic sampling method also provides a contactless method of determining the amplitude of the surface fields under in-situ conditions. The electrooptic signal can be calibrated with the known electrooptic coefficients of the crystal to directly determine the amplitude of the surface field and the transient field changes. This is difficult to do by any other means and may prove extremely valuable for many interfaces which are not amenable to conventional surface field probes. Most important, the combination of Surface EO sampling and Surface Restricted Grating spectroscopy used in conjunction provide a real time view of the spatial propagation of the electron (or hole carrier) across the interface.

4.1.3 Monte Carlo Modelling of Surface Reaction Dynamics

As can be appreciated from Figure 1, the photophysics and photochemistry that occurs at semiconductor surfaces are more complex than in molecular systems. The two biggest differences between molecular and semiconductor photophysics are the spatial transport and coulombic effects on the excited state populations. In the case of semiconductors, the problem is concerned with highly mobile photocarriers whose spatial distribution with the surface reaction plane is strongly influenced by the transport and transient field effects. The transient field conditions involve many particle interactions which can not be treated simply.

To gain quantitative details concerning the carrier spatial and energy distributions, an Ensemble Monte Carlo approach to the problem was undertaken which explicitly takes into account the band structure of GaAs and includes the $t = 0$ boundary conditions imposed by the short pulse excitation of photocarriers [55]. Since the carrier distributions are strongly influenced by carrier-carrier scattering and coulombic interactions between the oppositely charged electron and hole carriers, the Monte Carlo calculations must be done with a large number of particles simultaneously. Up to 10,000 electron and hole carriers were used for the ensemble size. The code keeps track of the carriers in both **k** and **r** space which is essential to the problem. The different scattering mechanisms that lead to changes in the carrier energy (**k**) distribution are included as well as a Poisson equation solver to update the surface field in 1 fsec intervals. In these initial studies, the electron was primarily treated, as the hole carrier is essentially stationary relative to the much higher mobility electron. The parameters were refined and the calculations compare favorably to experimental studies of carrier thermalization and other more standard checks.

Calculations of this type provide tremendous insight to the surface dynamics. The different types of information that can be obtained are shown in Figures 13 and 14. The magnitude and dynamics of the surface field changes can be correlated to the electron and hole distributions. In addition, the energetics of the carriers as well as velocity distributions can be obtained. A comparison of the calculated field dynamics to the electrooptic sampling studies was the

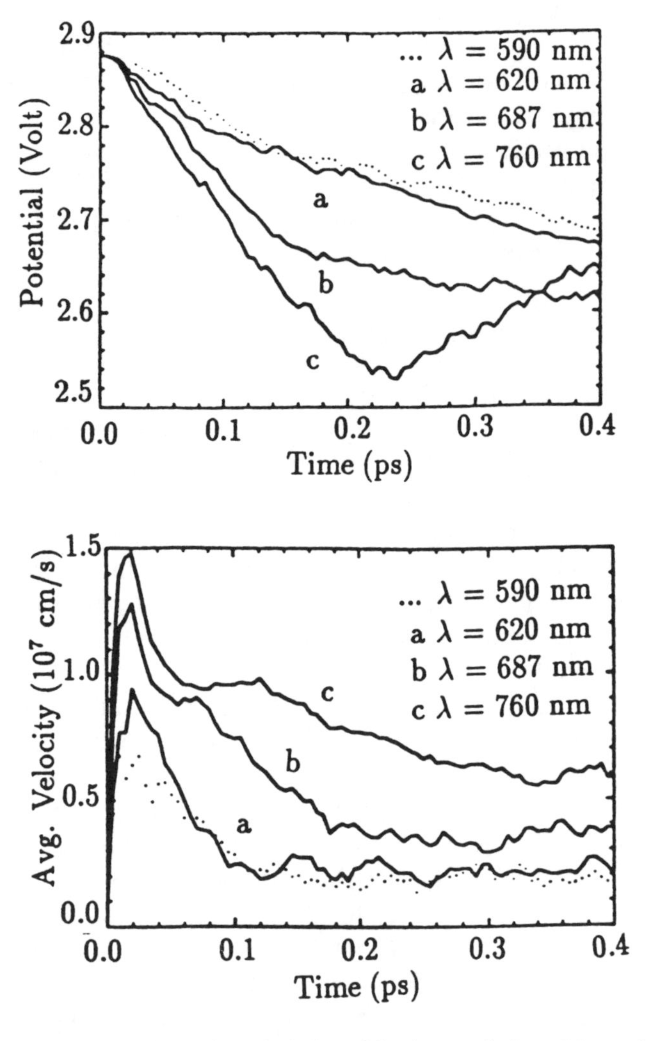

Fig. 13. The upper graph shows the calculation of the time evolution of the surface potential at different excitation wavelengths. The lower graph shows the velocity profiles in time for the same wavelengths.

original motivation for this work. As discussed above, the field changes can be directly related to the spatial transport of the carriers to the surface. A comparison of the experimental results to Monte Carlo simulation is found in Figure 13A ($\lambda_{ex} = 590$nm). Both the calculated dynamics and the magnitude of the field changes for the particular excitation conditions are in good agreement with observations. The excellent extent of agreement between the carrier thermalization dynamics and surface field transport with experiment indicates

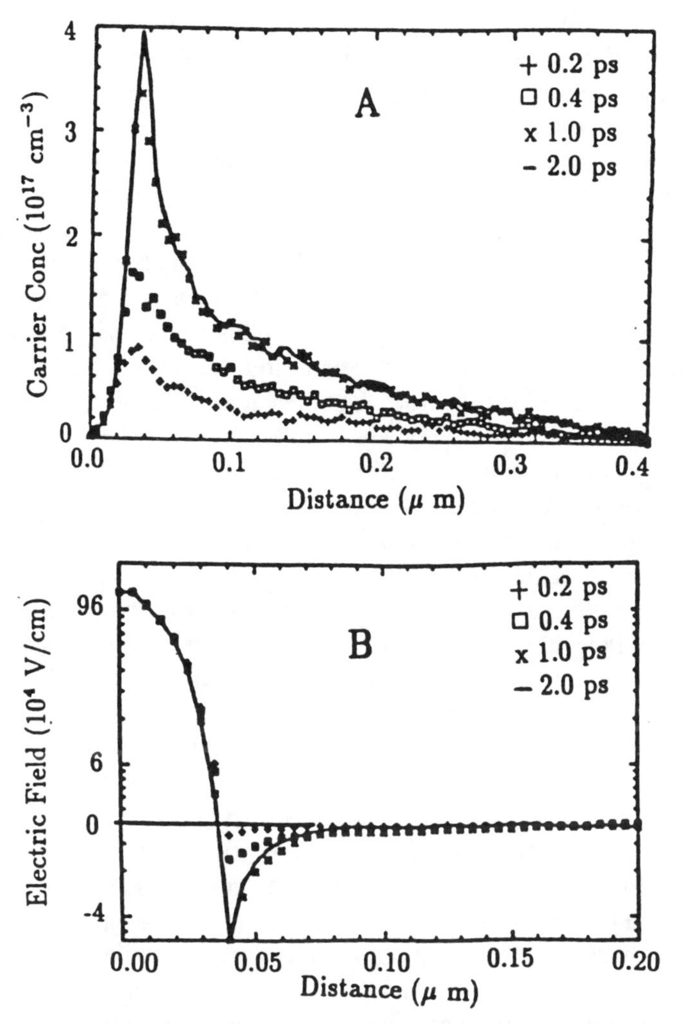

Fig. 14. Plot A shows the time dependent electron spatial distribution in the surface region. Plot B shows the calculated field distributions. Note that a back retarding field develops due to the exponential tail in the spatial distribution of hole carriers that extends past the space charge region. This field is responsible for the velocity overshoot in Figure 13, not the normal intervalley scattering into low mobility satellite valleys.

the parameters used in the calculation are providing good approximations to the actual carrier dynamics.

A detailed analysis of the carrier velocity distribution as a function of time finds the so called "velocity overshoot". There is an initial rapid increase in the carrier velocity through field acceleration which is related to the ballistic (scatter free) phase of the motion. This phase of the transport should be observable with higher time resolution in the electrooptic sampling studies.[1] Following the acceleration by the field, the electron velocity undergoes a dramatic decrease (the overshoot). Normally, this phenomenon occurs in GaAs when the electrons reach a critical energy after field acceleration that places them isoenergetic with the lower mobility L and Γ satellite valleys. Intervalley scattering puts them into these lower mobility **k** states and hence lowers the velocity. The factor of ten higher density of states in these satellite valleys keeps them statistically trapped in this low mobility condition. This aspect of the carrier transport is seen in the calculated wavelength dependence. However, what is unique in the surface problem is that the normal velocity overshoot is dominated, not by intervalley scattering, but by coulombic effects. The separation of the electrons from the hole carriers creates a transient field that retards the electron motion away from the surface. The occurrence of the transient field line and the relation to the carrier spatial distribution can be seen by comparing Figures 14A and 14B.

Most important, the transient field that develops creates an approximately 0.1 eV barrier at the edge of the space charge boundary to the diffusion of hole carriers to the surface. This is an important finding as it was not at all clear why the surface grating and electrooptic experimental studies seem to measure only a fast component to the surface dynamics without an equally important contribution from carrier diffusion from outside the space charge region. The only distributed kinetics that could be related to diffusion occur on a ten nanosecond time scale. It is clear from the Monte Carlo calculations that the transient electric field barrier is responsible for this observation. It acts to greatly retard the diffusion of minority carriers into the space charge region and thus access to the surface. The coulombic barrier that develops away from the surface is fortuitous in that it enables a clean distinction of the reaction dynamics of the minority photocarriers generated within the space charge region from those generated outside this region. Only the minority photocarrier within the space charge region undergoes fast enough transport to the surface to remove transport limitations on the dynamics. This fortunate occurrence permits a direct correlation of the initially rapid carrier population depletion to the fundamental processes related to the electronic transition across the interface.

An important future direction for Monte Carlo studies of carrier dynamics at surfaces will be to include molecular acceptors on the surface with an **r** dependent coupling to the carriers. The energetic dependence for the electronic coupling can be handled through a **k** dependence on the coupling and an assumed energy distribution for the molecular acceptors. Given the complex

---

[1] It is this phase of the electron motion which is needed to understand recent observations of electron transfer induced desorption at metal surfaces. references [39–44]

dependence of the overall surface kinetics on surface fields and energetics in the time scale under investigation, these calculations are needed to gain insight into the problem. Calculations of this type should play an important role in improving our understanding of ultrafast surface reaction dynamics.

4.2 Excited State Donor Condition

Interfacial charge transfer is a two way street. The complementary process to charge transfer out from a semiconductor (or metal), as discussed above, is charge injection from a surface species into an appropriate substrate. Dye sensitization provides a molecular species that can be efficiently probed optically, allowing for picosecond and even sub-picosecond probes of charge injecting systems. The earliest studies of time-resolved fluorescence decay curves as probes of interfacial charge injection were performed on dye-sensitized silver halide by Muenter [56] in 1976. These nanosecond experiments showed a marked decrease in both fluorescence quantum yield and lifetime when the dye was adsorbed on AgBr and AgCl relative to the dye alone in gelatin. Ascribing this decrease to electron injection dynamics, indicated a rate constant for electron injection of 10^9–10^{10} s^{-1} in this system.

Only four years later, in 1980, the first picosecond experiments were performed by Nakashima, Yoshihara and Willig [57] on Rhodamine B sensitized organic crystals (anthracene, phenanthrene, and naphthalene). They noted two different types of electron transfer dynamics in the different systems. In the case of the anthracene crystal, they assign a 35 ± 7 ps fluorescence decay to electron transfer kinetics alone. The phenanthrene and naphthalene crystals showed much longer electron injection kinetics since the dynamics are endoenergetic, but they did show significant 50 ps fluorescence quenching due to energy transfer to dimers at the surface. This work demonstrates the importance of taking into account the effect of competing dynamical pathways on the overall observed kinetics. The branching ratios between different decay channels is much more easily observed in picosecond and sub-picosecond optical experiments than with the more traditional voltammetric techniques which have relatively long RC time constants and, therefore, do not discriminate between competing ultrafast kinetics.

Since that time, a large number of researchers have investigated a wide variety of different photosensitized systems [58–71]. These systems employed a wide range of sensitizers and an even wider range of substrates, from insulators [58, 59] to semiconductors [60–71]. All of these photosensitized systems showed fluorescence decays, due to electron injection from the S_1 state of the sensitizing dye to the substrate, of 100 ps to 2 ns.

One of the most important of these studies was performed by Kemnitz, Nakashima, Yoshihara, and Matsunami [61] using Rhodamine B to sensitize

SiC and GaP single crystals. This study demonstrates that the fluorescence decay rates for the Rhodamine B are independent of temperature, all the way down to 4 K. This indicates conclusively that the electron injection is a not an activated process and, therefore, not subject to the kind of dynamical picture that would emerge from the Marcus–Levich equation.

While the observed electron injection rates in the studies mentioned above (50–2000 ps) are significant, they remain relatively slow relative to the nuclear relaxation rate which can be as fast as 100 fs. This is likely due to poor coupling between the sensitizing agent and the charge accepting species. The nature of the sensitizer and the acceptor may not be compatible with strong coupling or the coupling may be disrupted by the poor surface quality of the materials which are sensitized. While metals in ultra high vacuum have very clean surface constructions, most materials, especially oxide semiconductors, have a very large surface state density and very poor surface quality.

These considerations are not a factor for a somewhat different type of semiconductor substrate, SnS_2. SnS_2 is a layered semiconductor with a two dimensional lattice. The 2-D nature of the material means that all of the bonds are terminated in each layer. This leaves the surface remarkably inert to oxidation, preventing formation of oxide surface states, and the perfect surface construction leaves the semiconductor free from lattice defect surface states. Typically, quantum yields for photocurrent in dye-sensitized single crystal oxide semiconductors are on the order of 3 percent per absorbed photon. The low quantum yields have been attributed to weak electronic coupling or rapid back electron transfer processes which limit the injection efficiency. In contrast, the quantum yield for electron transfer and collected current is greater than 80 percent at SnS_2 surfaces [7]. The perfect surface quality suggests that the observed high quantum efficiency for electron transfer arises from strong electronic coupling between the dye and the surface. In order to fully understand the reasons for the high quantum yield, one would like to measure both the forward and backward electron transfer rate constants.

Oxazine 1 sensitized SnS_2 was studied as a model photosensitized system. Oxazine 1 had been shown by Kaiser *et al.* to have a very low aggregation propensity [72] which should allow for strong dye/semiconductor coupling without the complication of dimer formation at the surface. In addition, the first excited single state of the oxazine dye lies approximately 0.35 eV above the conduction band edge so that the electron injection should be barrierless (see Figure 2 with $\epsilon = 0.35$ eV). This system was studied using both time-resolved fluorescence decay curves and the fluorescence quenching of the oxazine on SnS_2 versus oxazine on an inert substrate. The fluorescence quenching is calculated using the Stern-Vollmer formula [73] by comparing the fluorescence yield for the Oxazine 1/semiconductor system to an Oxazine 1/insulator system.

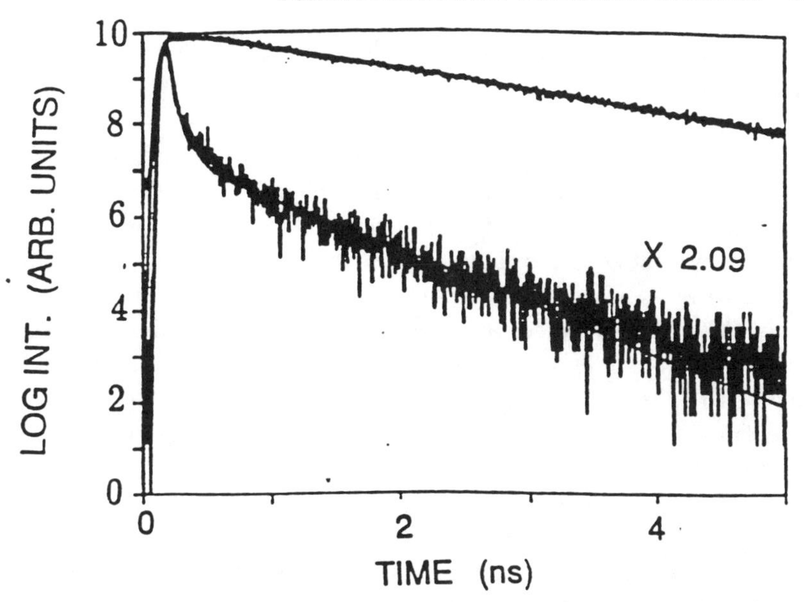

Fig. 15. The fluorescence quenching data from the time-correlated single photon counting setup. The upper decay curve corresponds to oxazine on an inert reference substrate, the lower decay curve (normalized at the peak) corresponds to oxazine on SnS_2. Integration shows a fluorescence quenching ratio of 9.0×10^4.

Both the fluorescence decays and the quenching rate [71] were determined using time-correlated single photon counting to take advantage of its high sensitivity and low background noise. The instrument response time of the photon counting electronics was 40 ps. The SnS_2 samples were unintentionally doped samples with a low donor concentration (less than 10^{15} cm^{-3}). Doped samples are prone to the formation of intense surface space charge fields. While these surface fields assist electron separation from the parent cation, enhancing the quantum yield, it also creates a barrier to the ground state recovery of the dye, effectively bleaching its ground state absorption. Consequently, knowledge of the timescale of the ground state recovery is important for verifying that the fluorescence quenching results are not distorted by bleaching of the ground state. Further, the ground state recovery reflects the back electron transfer which is important for determining the overall photocurrent yield.

A typical fluorescence decay curve is shown in Figure 15. The decays of the oxazine on an inert substrate were fit to a single exponential decay with a 2.6 ± 0.1 nsec lifetime. The decay curves for the dye sensitized semiconductor are more complex and were fit to two exponentials: an instrument response limited decay of 40 ps, and a longer decay of 900 ± 100 ps.

The longer decay shows a concentration dependent intensity and virtually disappears at low coverages, it is only 1% of the integrated intensity in Figure 15. This component corresponds to dye molecules either weakly adsorbed or not strongly coupled electronically to the surface. These dye molecules are most likely isolated from the surface either by intervening dye molecules due to the formation of aggregates or microcrystals of dye, uncoupled surface sites, or are simply poorly coupled due to orientational effects.

The instrument response limited decay of the SPC fluorescence decay curves is attributed to the fluorescence from adsorbed molecules strongly coupled to the surface, indicating that the electron injection rate is significantly faster than 40 psec. In order to better determine the exact electron injection rate, the integrated intensity of the fluorescence was studied. When compared to the oxazine on inert substrate samples, the fluorescence was found to be quenched by a factor of $9 \times 10^4 \pm 5 \times 10^4$. Assuming electron injection is the only new nonradiative channel in the semiconductor samples, this corresponds to an electron injection rate of 3×10^{13} (40 ± 20 fs). In addition, the absorption spectrum of the dye on the semiconductor was observed using both photocurrent detection and with an integrating spectrometer under the experimental conditions employed. The conserved spectral properties of the dye at the surface (save slight broadening roughly corresponding to the estimated excited state lifetime) demonstrate that the S_1 excited state is localized in the dye and the initially excited state is not a continuum analogue of a charge transfer band or some other hybrid state [71].

A one color pump-probe experiment was conducted on the oxazine/semiconductor system to monitor the important back electron transfer rate. The experimental setup was similar to that shown in Figure 10. The important element was the scanning galvanometer which enabled high speed data acquisition at frequencies above the dominant laser noise. Ten minutes of signal averaging permitted the detection of changes in probe intensity of 10^{-7}. This level of sensitivity was essential for following the ground state recovery.

The ground state recovery reflects the back electron transfer which is important for determining the overall photocurrent yield. It must also be noted that while the theory demonstrates that nuclear reorganization is not the rate limiting step for the charge transfer, this is not true for the back transfer. If the back transfer is viewed as occurring from the condition that the relaxation processes lower the energy of the dye state such that a potential barrier is created to re-injection of the carrier. Hence, the back transfer step can be used as a probe of nuclear relaxation processes and their role in interfacial charge transfer.

The ground state recovery was studied in reflection (see Figure 16A) in order to prevent bulk dynamics from interfering with the detection of surface dynamics. The pump excites the ground to first excited singlet state transition, partially bleaching this absorption. The probe pulse then monitors this same absorption. The signal has two features of note. The dominant feature of the

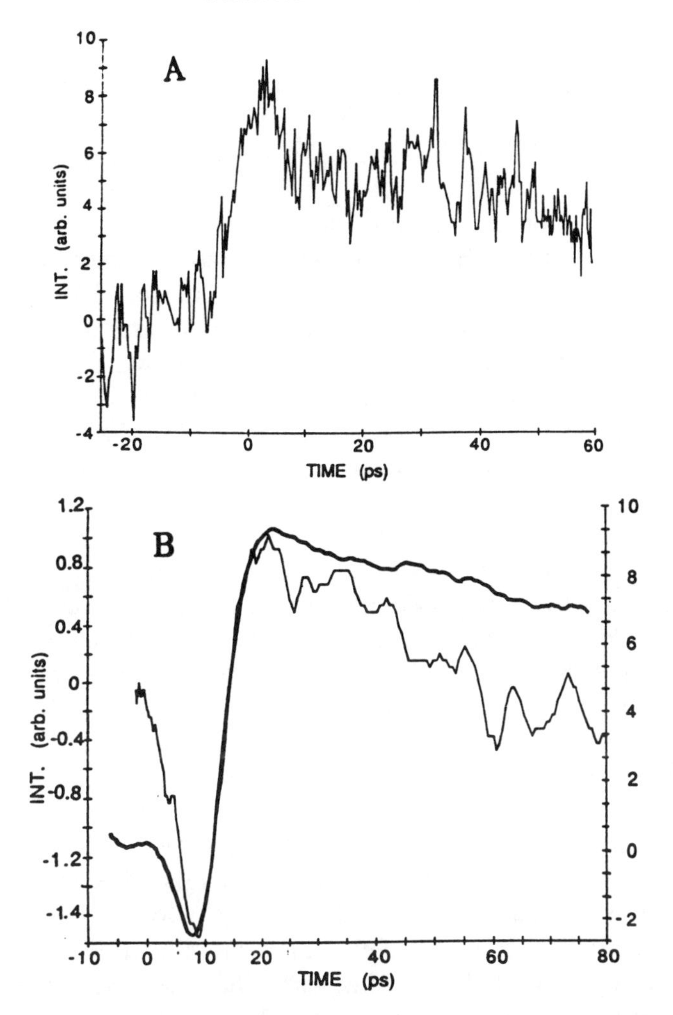

Fig. 16. Graph A shows the ground state recovery of the oxazine dye on SnS_2 probed by reflective one color pump-probe spectroscopy. The decay is nonexponential with a fast 10 ps decay and a longer, position sensitive, decay of several hundred picoseconds. Graph B shows transmissive one color pump-probe spectroscopy of SnS_2 single crystals. The single line corresponds to the intrinsic semiconductor (left axis), the thicker line corresponds to a low As doped sample (right axis). Both signals appear to be the convolution of two different signals of opposite phase. The As doped sample shows an order of magnitude greater signal. The fast initial decay is attributed to rapid carrier trapping as confirmed by grating studies of free carrier lifetimes.

signal is a very fast, approximately 10 ps, decay which appears to be the dominant recovery time. However, there is also a longer decay of several hundred picoseconds whose magnitude relative to the fast decay is generally small and varies, disappearing completely in many cases, as different positions on the surface are studied. The most logical explanation for this observation is that the fast decay is due to the ground state recovery of the strongly coupled dye molecules and that the longer component is due to the ground state recovery of less strongly coupled dye molecules, an observation consistent with both the emission spectra and SPC fluorescence decay curves discussed above.

The other important feature of this experiment is that the ground state recovery demonstrates different dynamics relative to those of the excited state. From the time resolved fluorescence, the excited state life time is definitely less than 10 ps and estimated to be 40 fsec based on the fluorescence quenching. The fact that the ground state shows slower recovery dynamics than the excited state rules out energy transfer to surface state traps as an alternative explanation for the enhanced nonradiative channel at the surface. An intermediate state is implicated which involves electron transfer rather than energy transfer.

If the unidirectional charge transfer is so strongly favored by the density of states mismatch as discussed in the theory section above, how does the electron tunnel from the conduction band back to the parent cation of the dye molecule on the surface? The most likely answer lies in the rapid relocalization of the electronic wavefunction after the initial forward transfer and delocalization. Essentially, the fast ground state recovery dynamics imply that the electron does not propagate far from the surface before it is localized on a trap state (defect, impurity, bound exciton) or dynamically localized in a polaron state. This relaxation step would remove the electron from the electronic continuum and change the problem to an electron exchange between discrete states. This possibility was investigated by studying carrier lifetimes in SnS_2. The free carriers were generated both by optically exciting impurities and using above band gap excitation. From both pump/probe studies and transient grating studies, the free carrier lifetime was found to be on the picosecond to subpicosecond time scale. Representative data are shown in Figure 16B.

In the above experiments, both the reactant and product populations have been monitored, as well as the relaxation dynamics. We now have a fairly complete picture of the electron dynamics at SnS_2 interfaces. There is a fast forward electron transfer step from the discrete state defined by the adsorbed dye. The dynamics of this electron transfer implicate a purely electronic process. The reaction coordinate at the adiabatic crossing point is displaced towards product formation by the electronic continuum of the conduction band. The products consist of a free electron generated in the solid state and the ionized parent molecule at the surface. For SnS_2 crystals, the free electron would become trapped near the surface on a subpicosecond to picosecond time scale. The subsequent back transfer process from this localized state regenerates the ground state through a distribution of time scales with a dominant component of 10 ps. These recovery dynamics are consistent with an electron transfer coordinate directed by nuclear relaxation and repolarization.

Since SnS_2 and other layered semiconductors are the only well defined single crystal surface that demonstrate a high quantum yield for anodic photocurrent, it is important to understand the above dynamics with respect to the role of surface fields in separating the electron from the coulombic trap of the parent cation. The fast trapping time observed for the electron in probably important in removing the electron from the conducting state so it does not spatially sample the large coulombic cross section of the parent cation and become trapped at the surface. An intense surface space charge field would impart a travelling wave component to the free electron product state which would propagate the electron further from the surface than in the zero field conditions of the above experiments. As long as the electron makes it outside the Onsager escape distance (10 Å typically), the field will prevent back transfer and the electron current is produced by low mobility electron hopping processes in an applied field. To understand these surface dynamics in detail, the missing information is the energetics of the trapped electron and the relative positions of the parent cation electronic levels with respect to the surface bands. However, the observed dynamics illustrate the general guidelines for optimizing surface charge transfer.

5. Concluding Remarks

Two different $t = 0$ boundary conditions for electron transfer at surfaces have been experimentally explored at conditions approximating the time evolution of the adiabatic crossing point. The SnS_2 dye sensitization studies correspond to $t = 0$ conditions with the electron localized at a discrete state at the interface under barrierless conditions. The dynamics of electron transfer are consistent with the involvement of an electronic continuum in promoting unidirectional passage across the reaction critical point. The electronic continuum in this case is defined by the dense manifold of extended electronic states of the solid state. For sufficient electronic coupling between the molecular donor and band acceptor states, the transition can become completely electronic. The traditional guidelines for the upper limit on the rate constants from transition state theory for discrete states no longer apply. The rate constant can be completely determined by the electronic dephasing and not by motion along the nuclear relaxation coordinate. This concept was supported by a reevaluation of electron transfer in the presence of a manifold of closely spaced electronic levels and has been treated in a different context concerning the lifetimes of hydrogenic states at metal surfaces [75]. The involvement of an electronic continuum in the reaction coordinate is the feature that distinguishes electron transfer at surfaces from homogeneous electron transfer.

The opposite $t = 0$ boundary condition corresponds to the initial localization of the charge in the solid state. From the principle of microscopic reversibility, the transmission probability across the interface to a discrete state should be strongly attenuated by the electronic density of states mismatch across the interface. This is seen for zero field conditions. However, systems exhibiting efficient electron transfer have surface space charge fields which alter the reaction conditions. Based on the SnS_2 studies and theoretical analysis, the space charge field must act to break up the electronic continuum and localize the electron (or hole carrier) at the surface. For typical surface fields, the electron probability distribution is restricted to 10 Å of the surface. The wavefunction now approximates something closer to a molecular state and the rate of passage across the critical point becomes controlled by the nuclear coordinate, i. e. approximates more closely homogeneous electron transfer. The issue then reduces to the degree of electronic coupling across the interface for the molecular acceptor and the contributing states of the solid state. The coupling will decay exponentially with the separation of the acceptor from the surface and a wide range of electronic factors and time scales are expected. Based on the in-situ surface grating studies of hole transfer at n-GaAS(100) interfaces, colloidal semiconductor studies, recent UHV observations, and the equivalent back electron transfer observed for oxazine/SnS_2, this electronic coupling can approach the strong coupling limit with charge transfer dynamics occurring on picosecond and potentially even subpicosecond time scales. There is no fundamental barrier to attaining ultrafast time scales for electron transfer from surfaces. In fact, since the optical approach necessitates relatively low carrier densities, this work hints that the actual electron dynamics at metal electrode surfaces, with high electron donor densities, should exhibit adiabatic processes more as a general feature than the analogous homogeneous problem.

The current experimental approach has given a real time view of the electron trajectory across an interface. Both the spatial distributions and surface populations have been studied. These studies are starting to reveal details of the electron transfer coordinate that are inaccessible with conventional methods used in electrochemical approaches. In situ studies capable of giving structure and energetics are now needed to rationalize the observed dynamics. Even without this information, it is clear that a reexamination of the various tenets of electron transfer theory as it pertains to surfaces is in order.

Acknowledgements

The research on GaAs was funded by DOE (grant number DE-FG02-91ER14185) and the SnS_2 studies by the NSF Science and Technology Center for Photoinduced Charge Transfer (grant number CHE-8810024). RJDM would like to acknowledge additional support from a John Simon Guggenheim fellowship during the writing of this chapter. The authors would like to thank A. A. Muenter, S. Mukamel, and B. A. Parkinson for numerous enlightening discussions that contributed to this work.

References

1. Miller, R. J. D., McLendon, G., Schmickler, W., and Willig, F., *Surface Electron Transfer Processes*, (VCH, New York) in press.
2. Morrison, *Electrochemistry at Semiconductor Surfaces and Oxidized Metal Electrodes* (Plenum Press, New York 1980).
3. Finklea, H. O., in *Semiconductor Electrodes*, H. O. Finklea, ed. (Elsevier, Amsterdam, 1988), and related articles.
4. Heller, A., *Acc. Chem. Res.* **14**, 154 (1981).
5. Blakemore, J. S., *J. Appl. Phys.* **53**, R123, (1982).
6. Parkinson, B. A. and Spitler, M. T., *Electrochimica Acta* **37**, 943 (1992).
7. Parkinson, B. A., *Langmuir* **4**, 967 (1988).
8. Marcus, R. A., *Ann. Rev. Phys. Chem.* **15**, 155 (1964).
9. Levich V. G., in *Physical Chemistry: an Advanced Treatise*, eds. H. Eyering, D. Henderson, and W. Jost (Academic, New York 1970), **9B**, 985.
10. Gerisher H., in *Physical Chemistry: an Advanced Treatise*, eds. H. Eyering, D. Henderson, and W. Jost (Academic, New York 1970), **9A**, 463.
11. Closs, G. L., Johnson, M. D., Miller, J. R., and Pitrowiak, P., *J. Am. Chem. Soc.* **111**, 3751 (1989).
12. Bader, J. S., and Chandler, D., *Chem. Phys. Lett.* **157**, 501 (1989).
13. Miller, C. and Gratzel, M., *J. Phys. Chem.* **95**, 5225 (1991).
14. Chidsey, C. E. D., *Science* **251**, 919 (1991).
15. Barbara, P. F. and Jarzeba, W., *Adv. Photochem.* **15**, 1 (1990).
16. Rosenthal, S. J., Xie, X., Du, M., and Fleming, G. R., *J. Chem. Phys.* **95**, 4715 (1991).
17. Simon, J. D. and Su, S.-G., *J. Chem. Phys.* **87**, 7016 (1987).
18. Sumi H. and Marcus, R. A., *J. Chem. Phys.* **84**, 4894 (1986).
19. Jortner, J., Bixon, M., Heitele, H., and Michel-Beyerle, M. E., *Chem. Phys. Lett.* **197**, 131 (1992).
20. Boudreaux, D. S., Williams, F., and Nozik, A. J., *J. Appl. Phys.* **51**, 2158 (1980).
21. Schmickler, W., *J. Electroanal. Chem.* **204**, 31 (1986).
22. Cooper, G., Turner, J. A., Parkinson, B. A., and Nozik, A. J., *J. Appl. Phys.* **54**, 6463 (1983).
23. Koval C. A. and Segar, P. R., *J. Phys. Chem.* **94**, 2033 (1990).
24. Marsh, E. P., Tabares, F. L., Sneider, M. R., Gilton, T. L., Meir, W., and Cowin, J. P., *J. Chem. Phys.* **92**, 2004 (1990).
25. Lewis, N. S., *Ann. Rev. Phys. Chem.* **23**, 176 (1990).
26. Hush, N. S., *J. Chem. Phys.* **28**, 962 (1958).
27. Marcus, R., *Faraday Disc. Chem. Soc.* **74**, 7 (1982).
28. Marcus R. and Sutin, N., *Biochim. Biophys. Acta* **811**, 265 (1985).
29. Marcus, R., *J. Phys. Chem.* **94**, 1050 (1990).
30. Spitler, M., *J. Electroanal. Chem.* **228**, 69 (1987).
31. Charle, K. P. and Willig, F., *Chem. Phys. Lett.* **57**, 253 (1978).
32. Onsager, L., *Physical Review* **54**, 554 (1938).
33. Blossey, D. F., *Phys. Rev. B* **9**, 5183 (1974).
34. Tokuyama, M. and Mori, H., *Prog. Theor. Phys.* **55**, 411 (1976).
35. Zwanzig, R., *Physica (Utrecht)* **30**, 1109 (1964).
36. Mukamel, S., *J. Chem. Phys.* **71**, 2012 (1979).

37. Mayer, J. E. and Mayer, M. G., *Statistical Mechanics* (Wiley, New York 1977), Chapter 13.
38. Feynman, R. P., *Statistical Mechanics* (W. A. Benjamin; Reading, Massachusetts 1972), Chapter 9.
39. Huppert, D., and Ittah, V., *Jerusalem Symp. Quant. Chem. Biochem.* **22**, 301 (1990).
40. Weiner, A. M., De Silvestri, S., and Ippen, E. P., *J. Opt. Soc. Am. B* **2**, 654 (1985).
41. Kasinski, J. J., Gomez-Jahn, L. A., Faran, K. J., Gracewski, S. M., and Miller, R. J. D., *J. Chem. Phys.* **90**, 1253 (1989).
42. Gomez-Jahn, L. A. and Miller, R. J. D., *J. Chem. Phys.* **96**, 3982 (1992).
43. Halas, N. J. and Bokor, J., *Phys. Rev. Lett.* **62**, 1679 (1989).
44. Zhou, X.-L., Zhu, X.-Y., and White, J. M., *Surface Sci. Rep.* **13**, 76 (1991).
45. Dixon-Warren, St. J., Jensen, E. T., Polanyi, J. C., Xu, G.-Q., Yang, S. H., and Zeng, H. C., *Faraday Discuss. Chem. Soc.* **1991**, 91.
46. Ying, Z. C. and Ho, W., *Phys. Rev. Lett.* **65**, 741 (1990)
47. Buntin, S. A., Richter, L. J., Cavanaugh, R. R., and King, D. S., *Phys. Rev. Lett.* **61**, 1321 (1988).
48. Budde, F., Heinz, T., Loy, M., Misewich, J., Rougemont, F., and Zacharias, H. *Phys. Rev. Lett.* **66**, 3024 (1991).
49. Prybyla, J. A., Tom, H. W. K., and Aumiller, G. D., Phys. *Rev. Lett.* **68**, 503 (1992).
50. Rosetti, R. and Brus, L. J., *J. Phys. Chem.* **90**, 558 (1986)
51. Kamat, P. V., Ebbesen, T. W., Dimitrijevic, N. M., and Nozik, A. J., *Chem. Phys. Lett.* **157**, 384 (1989).
52. Min, L. and Miller, R. J. D., *Chem. Phys. Lett.* **163**, 55 (1989).
53. Min, L. and Miller, R. J. D., *Appl. Phys. Lett.* **56**, 524 (1990).
54. Min, L., Ph. D. thesis, University of Rochester, 1991.
55. Zhou, X., Hsiang, T. Y., and Miller, R. J. D., *J. Appl. Phys.* **66**, 3066 (1989).
56. Muenter, A. A., *J. Phys. Chem.* **80**, 2178 (1976).
57. Nakashima, N., Yoshihara, K., and Willig, F., *J. Chem. Phys.* **73**, 3553 (1980).
58. Liang, Y., Ponte Goncalves, A. M., and Negus, D. K., *J. Phys. Chem.* **87**, 1 (1983).
59. Liang, Y. and Ponte Goncalves, A. M., *J. Phys. Chem.* **89**, 3290 (1985).
60. Hashimoto, K., Hiramoto, M., Kajiwara, T., and Sakata, T., *J. Phys. Chem.* **92**, 4636 (1988).
61. Kemnitz, K., Nakahima, N., Yoshihara, K., and Matsunami, H., *J. Phys. Chem.* **93**, 6704 (1989).
62. Hashimoto, K., Hiramoto, M., and Sakata, T., *J. Phys. Chem.* **92**, 4272 (1988).
63. Hahimoto, K., Hiramoto, M., Lever, A. B. P., and Sakata, T., *J. Phys. Chem.* **92**, 1016 (1988).
64. Itoh, K., Chiyokawa, Y., Nakao, M., and Honda, K., *J. Am. Chem. Soc.* **106**, 3553 (1980).
65. Crackel R. L. and Struve, W. S., *Chem. Phys. Letters* **120**, 473 (1985).
66. Kamat, P. V., Chauvet J. P., and Fessenden, R. W., *J. Phys. Chem.* **90**, 1389 (1986).
67. Spitler, M. and Parkinson, B. A., *Langmuir* **2**, 549 (1986).
68. Eichberger, R. and Willig, F., *Chem. Phys.* **141**, 159 (1990).
69. Willig, F., Eichberger, R., Sundrasen, N. S., and Parkinson, B. A., *J. Am. Chem. Soc.* **112**, 2702 (1990).
70. Lanzafame, J. M., Min, L., Miller, R. J. D., Muenter, A. A., and Parkinson, B. A., *Mol. Cryst. Liq. Cryst.* **194**, 287 (1991).
71. Lanzafame, J. M., Miller, R. J. D., Muenter, A. A., and Parkinson, B. A., *J. Phys. Chem.* **96**, 2820 (1992).
72. Seilmeier, A., Scherer, P. O., and Kaiser, W., *Chem. Phys. Lett.* **105**, 140 (1984).
73. Kirkbright, G. F. and Sargent, M., *Atomic Absorption and Fluorescence Spectroscopy* (Academic Press, New York 1974), Chapter 2.
74. George, J., Valsala Kumari, C. K., and Joseph, K. S., *J. Appl. Phys.* **54**, 5347 (1983).
75. Norlander, P. and Tully, J. C., *Phys. Rev. Lett.* **61**, 990 (1988).

6. Ultrafast Transient Raman Investigations of Condensed Phase Dynamics

JOHN B. HOPKINS and JUNBO CHEN
Department of Chemistry, Louisiana State University, Baton Rouge, LA 70803, U.S.A.

During the last several years major advances in ultrafast laser technology have taken place. In the earlier days of this field femtosecond lasers had to be painstakingly built from scratch in the laboratory. Today, much better lasers are available commercially. State-of-the-art titanium sapphire lasers routinely generate 100 fs pulses. Amplification to millijoule pulses at repetition rates of 1 kHz (and microjuole pulses at 250 kHz) is now routine using titanium sapphire regenerative amplifiers. These advances in laser technology have greatly increased the sensitivity of femtosecond spectroscopy. As a result, powerful new ultrafast spectroscopic methods have been developed.

One of the methods developed by the first ultrafast pioneers was transient absorption spectroscopy where transients are probed by measurements of the electronic absorption spectrum. This technique has proven to be extremely powerful due to its excellent sensitivity. However, absorption spectroscopy is not ideal since no direct information is obtained about the chemical nature of the transient. Many questions are difficult to difficult to answer from absorption data alone. What chemical bonds are being broken and what new bonds are forming? What conformational changes are taking place? What affect is the solvent having on the structure and charge distribution in the transient?

There is another aspect of ultrafast chemistry which becomes increasing important as shorter and shorter time scales are investigated. Much of ultrafast chemistry is concerned with photoproducts which are born with internal energies in excess of the average thermal energy. It is therefore important to characterize the internal energy of the transient as well as which vibrational modes contain the energy. Finally, it is also important to understand how the non-thermal internal energy affects the subsequent chemistry of the transient. To some extent these questions can be addressed through transient absorption spectroscopy. However this method is severely limited since it provides no direct information about the vibrational degrees of freedom.

Ultrafast transient infrared and Raman spectroscopy are relatively new methods which nicely complement some of the limitations of transient absorption spectroscopy. Infrared and Raman techniques provide a direct probe of the vibrational degrees of freedom. In addition, the partitioning of vibra-

J.D. Simon (ed.), Ultrafast Dynamics of Chemical Systems, 205–222.

tional energy among the internal modes can be directly measured. Infrared and Raman spectroscopy are obviously complementary in that the vibrational selection rules often select different allowed vibrational modes. The advantage of Raman spectroscopy is that very low frequency vibrational modes can be observed. Whereas, it is currently very difficult to produce short pulses of infrared light below 2000 cm^{-1}. The advantage of infrared spectroscopy is that it is generally applicable to all transients that have an allowed infrared spectrum. Raman spectroscopy however has the added requirement that only molecules with strong resonant Raman enhancements can be observed. This has the practical limitation that the resonant electronic transition must be accessible within the range of tunability of the laser. Secondly, the upper state of the resonant transition cannot strongly fluoresce. Even with these limitations, ultrafast Raman spectroscopy has proven to be a powerful technique. Several examples of the application of both infrared and Raman techniques can be found in this volume.

This article describes the application of ultrafast Raman spectroscopy to probe photodissociation reactions in the liquid phase. Measurements of the liquid dynamics on dissociation are made. The internal energy of the photofragments are characterized. In addition, the effect of the internal energy on the chemistry of the photoproduct is determined.

Photodissociation and Geminate Recombination in Iodine

The photodissociation and geminate recombination reaction of iodine in solution is an excellent probe of condensed phase reaction dynamics. In particular, ultrafast laser techniques have been extensively used to investigate this reaction [1–7]. Recent elegant experiments by Harris and co-workers [5, 6, 7] have shown that geminate recombination occurs in ≤ 5 ps. By comparing transient absorption data to calculated transition strengths, these same experiments have inferred that vibrational cooling is very strongly dependent on the vibrational energy gap. Following recombination at the top of the X-state potential it was found that complete relaxation into the lowest energy vibrational levels occurs on a roughly 100 ps time scale.

Transient Raman experiments in our group have been performed to investigate this same reaction. The advantage of the Raman technique is that direct information is obtained about the vibrational coordinates. At the dissociation limit of the X state potential the energy spacing between vibrational levels is approximately zero. As vibrational cooling takes place it is expected that time dependent Raman frequencies will be observed from zero frequency all the way out to the fundamental I_2 frequency [8] of 212 cm^{-1}. For a vibrationally hot I_2 molecule to relax, it must transfer a quanta of vibrational energy to the solvent. The beauty of studying vibrational cooling by transient Raman spectroscopy lies in the fact that the Raman frequencies correspond *directly* to the quanta of energy that must be lost to depopulate the level. The time dependence of the Raman frequencies can then be used to probe vibrational energy decay as a function of energy gap. Using the vibrationally resolved spectrum we have been able to separate several different transients and investigate vibrational relaxation of the excited electronic A/A' and ground X state. These experiments indicate that geminate recombination on the X state and A/A' state occurs in less than 5 ps. Vibrational population was detected in energy levels from 210 cm^{-1} to 9300 cm^{-1} above the vibrationless level. This indicates that vibrational relaxation is already 42% of the way down the X state potential in the first 5 ps. This result compares well with theoretical predictions [11]. These results reveal great detail about vibrational relaxation mechanisms. In addition, we have investigated the mechanism of the electronic decay of the excited A/A' state by measuring the solvent dependence of electronic relaxation upon isotopic substitution of the solvent.

Figure 1 shows the pure transient difference spectrum (with background Raman peaks subtracted out) of iodine probed at various times following photodissociation as indicated in the figure. The band at 107 cm^{-1} in the 100 ps spectrum has been assigned to the transient A/A' electronic state which is also formed as a result of geminate recombination. The frequency of this band is very close to the gas phase value [13, 14] of 90/105 cm^{-1} for the A/A' electronic state. In addition, this band has the same time dependence found in earlier transient absorption experiments [4–7] for the species assigned to these same excited electronic states.

The broad band from 130 to 210 cm^{-1} in the 0 ps spectrum is assigned to hot vibrational states in the ground electronic state. Based on gas phase spectroscopic constants [8] these states correspond to vibrational energies of 740 to 9300 cm^{-1}. This band appears featureless as a result of Raman scattering from a large distribution of vibrational states with slightly different frequencies. In fact, the difference in the Raman frequencies expected for the hot vibrational states can be calculated from the known gas phase spectroscopic constants [8]. For the vibrational states observed in Figure 1 the frequencies for Raman scattering from any two adjacent vibrational levels $v \rightarrow v+1$ and $v+1 \rightarrow v+2$ differ by a maximum of 2.3 cm^{-1}. Since the spacing between Raman bands from adjacent vibrational levels is so much smaller than the observed line width of an individual Raman band (23 cm^{-1}), only a distribution of states is observed.

The arrows in the figure illustrate the dynamics of the vibrational decay. From 0 ps to 100 ps the peak of the distribution shifts towards the higher frequency vibrational levels located in the lower energy regions of the X state potential. It might seem curious that Figure 1 does not show the eventual repopulation of the lowest energy level with a 212 cm^{-1} energy gap. However,

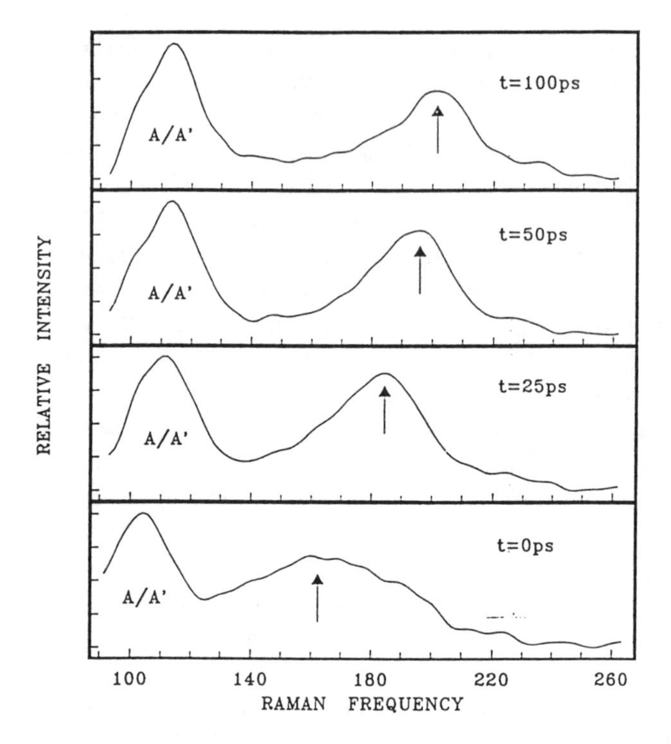

Fig. 1. Pure transient spectrum of I_2 in cyclohexane obtained by subtracting the one color probe only background spectrum from the raw two color spectrum. In order to subtract, spectra have been normalized for transient absorption at 354.7 nm using solvent bands as an internal reference. Probe is 354.7 nm, 20 μJ/pulse, 0.5 mm beam waist. Pump is resonant with the B–X transition at 53 nm; 50 μJ/pulse 0.5 mm beam waist. Time delay between pump and probe lasers is given in the figure. The band labeled A/A' in each frame corresponds to the excited electronic states denoted as such. Frequencies are in units of cm^{-1}.

it must be remembered that Figure 1 is a subtracted pure transient spectrum showing only features which are different from the normal relaxed Raman spectrum. As soon as the distribution cools into the lowest energy levels, the spectrum becomes identical to the relaxed Raman spectrum and is removed by the subtraction procedure.

These results prove that the analysis of the transient absorption spectrum by Harris and co-workers [5–7] is remarkably accurate considering the difficulty of interpreting the transient absorption data. That is, the chemical dynamics on the 100 ps time scale is dominated by vibrational cooling on the X state potential.

The quantum number dependence of vibrational relaxation was investigated by comparing relaxation rates for the X state and A/A' state for vibrational levels with identical vibrational frequencies. In the X state the vibrational frequency near $\nu = 64$ has an energy gap of 106 cm^{-1} which is similar to that in the A/A' state near $\nu = 2$. Relaxation of the X state $\nu = 64$ levels is about 60 times faster than that of the $\nu = 2$ level of the A/A' state. This result strongly agrees with theoretical models [11, 12] of vibrational relaxation which predict that the relaxation rate is strongly dependent on vibrational quantum number due to the larger amplitude motion for the higher quantum number vibration.

The solvent dependence of vibrational relaxation was found to be accounted for by the mass dependence of the solvent with the lighter, faster moving solvents being the more efficient quenchers. In complicated hydrocarbon solvents (pentane, hexane, cyclohexane, heptane) vibrational relaxation rates become difficult to predict. Relaxation to vibrational modes of the solvent strongly depends on specific mode couplings. For example, cyclohexane has no vibrational modes low enough in frequency to be nearly resonant with the 212 cm^{-1} X state vibration. However, the deuterated analog does have vibrations which are resonant. The I_2 relaxation rates are nearly identical in both solvents suggesting that not all solvent modes have the necessary coupling strengths to be efficient quenchers. This same result was found for the solvents hexane and neohexane where neohexane has a much larger density of resonant vibrational modes for I_2 X state vibrations with frequencies near 190 cm^{-1}. However, n-hexane is a far more efficient quencher. Evidently, the low frequency torsional motions of neohexane are not strongly coupled to the I_2 vibration and therefore the solvent modes do not enhance vibrational relaxation.

The A/A' electronic state relaxation rate is very strongly solvent dependent. The fundamental mechanism for the electronic relaxation was investigated by several experiments. Raman spectroscopic measurements of vibrational frequencies and anharmonicities provide estimates of the shape of the potential in various solvents. These experiments indicate that the potential well is not significantly perturbed and the solvent dependence of the lifetime of the A/A' state cannot be accounted for by simple thermal dissociation. Franck-Condon calculations were performed to determine if solvent shifts of the X state and A/A' state could enhance the nonradiative intersystem crossing rate by creating more favorable vibrational overlap. These calculations indicated only negligible changes in the respective Franck-Condon factors. The most crucial experiment was the observation that the A/A' state lifetime was significantly enhanced when the solvent was deuterated. This result suggests that electronic decay takes place through the formation of solvent exciplexes with the more efficient decay pathways involving higher frequency oscillators in the solvent. This is the same mechanism responsible for electronic relaxation in atoms [13, 14].

One way to picture the $E - V$ relaxation pathway is to consider the A' state to be lowered in energy (with respect to X) by one quanta of the solvent accepting mode. It is reasonable to presume that curve crossing from the A' to

X state takes place at the outer turning point. According to this picture, in the I_2 solvent complex the barrier height between the bottom of the A' state and the crossing point between the A' and X states is approximately 425 cm^{-1} for the C—H stretch in n-hexane. The barrier height suggested from the temperature dependence is 430 cm^{-1} which is in reasonable agreement with the (425 cm^{-1}) barrier expected for the $E - V$ relaxation. It is important to realize that this result is only suggestive since the temperature dependence may be a combination of an activation factor and a probability for transfer in the activated region. In addition, we have considered only one quanta transitions involving solvent vibrations. This simplification would cause the calculated barrier for E-V relaxation to be higher than actual. For CCl_4 the barrier height for A' state decay in the solvent complex is 1404 cm^{-1} and the experimental value from the temperature dependence 680 cm^{-1}. Here again, the temperature dependence is consistent with the proposed E-V relaxation mechanism if multiple transitions in the solvent vibration are considered.

Photodissociation and Geminate Recombination of Hemoglobin

The photochemistry of hemoglobin has been investigated for many years as a means to first dissociate the oxygen ligand and subsequently observe the back reaction of ligand rebinding. In this way it is possible to investigate the effects of protein dynamics on heme chemistry [15–29].

The most studied topics include the photodeligation of the heme, subsequent ligand (CO, O_2 and NO) rebinding with the heme, and the structural changes in the coupled heme/protein system. Knowledge of the time dependence of the dissociation and recombination processes provides insight into the potential surface along the heme-ligand reaction coordinate. However the absorption of a photon by a protein-bound chromophore can, in general, trigger a very complicated series of events involving electronic, vibrational, conformational, and photochemical dynamics. All of these types of relaxation phenomena will cause transients in the visible and ultraviolet absorption spectra of a complex biomolecule such as hemoglobin. Therefore, it is often difficult to confidently link an ultrafast absorption transient with a specific relaxation mechanism. This difficulty obscures the available information about the reaction coordinate for ligand rebinding.

However, the mode-specific dynamics provided by picosecond pump-probe Raman spectroscopy allow us to sort out the complex dynamics in a large biomolecule, specifically hemoglobin. Raman spectroscopy is an ideal tool for the study of photoinitiated dynamics in a large molecule, since the vibrational spectrum is very sensitive to small changes in the structure of the molecule. In addition, we have recently shown [30, 31] that transient Raman spectroscopy can quantitatively characterize excess vibrational energy in a photoexcited molecule. The spectral signature of vibrational energy relaxation (VER) in the Stokes and anti-Stokes transient Raman spectra is quite distinct from the spectral features which result from chemical or conformational changes in the molecule. It will also be shown that ligand binding has specific effects on the vibrational spectrum which allow us to further separate ligand dynamics from other structural changes (such as those in the protein).

Several attempts have been made to characterize the ultrafast dynamics of vibrational energy dissipation in photoexcited heme proteins [32, 33]. Previous Raman experiments were unable to probe the transient anti-Stokes spectrum and were interpreted using indirect analyses. An accurate measurement of the vibrational energy relaxation (VER) which follows photoexcitation is necessary to correctly interpret ultrafast electronic absorption data. Our experiment is unique in that it *directly* and *unambiguously* detects hot vibrations in both low- and high-frequency porphyrin modes. We are also able to estimate the vibrational temperature from anti-Stokes transient data alone. Confusion also exists in the literature over the time scale for ligand recombination following photoexcitation. We have applied our mode-specific picosecond Raman technique to oxyhemoglobin in order to separate some of the complex vibrational and conformational dynamics.

Vibrational Relaxation In Deoxyhemoglobin

We have studied deoxyHb by exciting at 532 nm and probing at 355 nm. A pure transient Raman spectrum is produced by subtracting the background probe-only components from the spectrum obtained with the pump-probe sequence. Figure 2A shows the ground state, Stokes resonance Raman spectrum of deoxyHb. Figures 2B and 2C show transient Stokes and transient anti-Stokes spectra at time zero, which only show features due to excited molecules. Negative transient Stokes bands indicate removal of population (by the 532 nm pump photons) from those levels thermally populated at room temperature. Positive transient anti-Stokes bands indicate that there is more population in the low-lying excited vibrational modes of the photoexcited heme than in unexcited heme at room temperature. Notice that the frequency-resolved transient data show opposite and complementary behavior for the Stokes and anti-Stokes cases.

We have observed the VER dynamics of photoexcited deoxyHb by monitoring the *Stokes* and *anti-Stokes* transient signals for specific deoxyHb vibrational modes as the time delay between pump and probe pulses is varied [31]. Deconvolution of the 8 ps laser pulse width yields a VER exponential time constant of 2–5 ps for heme vibrations at 671, 1424, and 1563 cm^{-1}. The complementary behavior of the Stokes and anti-Stokes transients is illustrated by the dynamics observed for the heme ν_2 band (not shown here). That

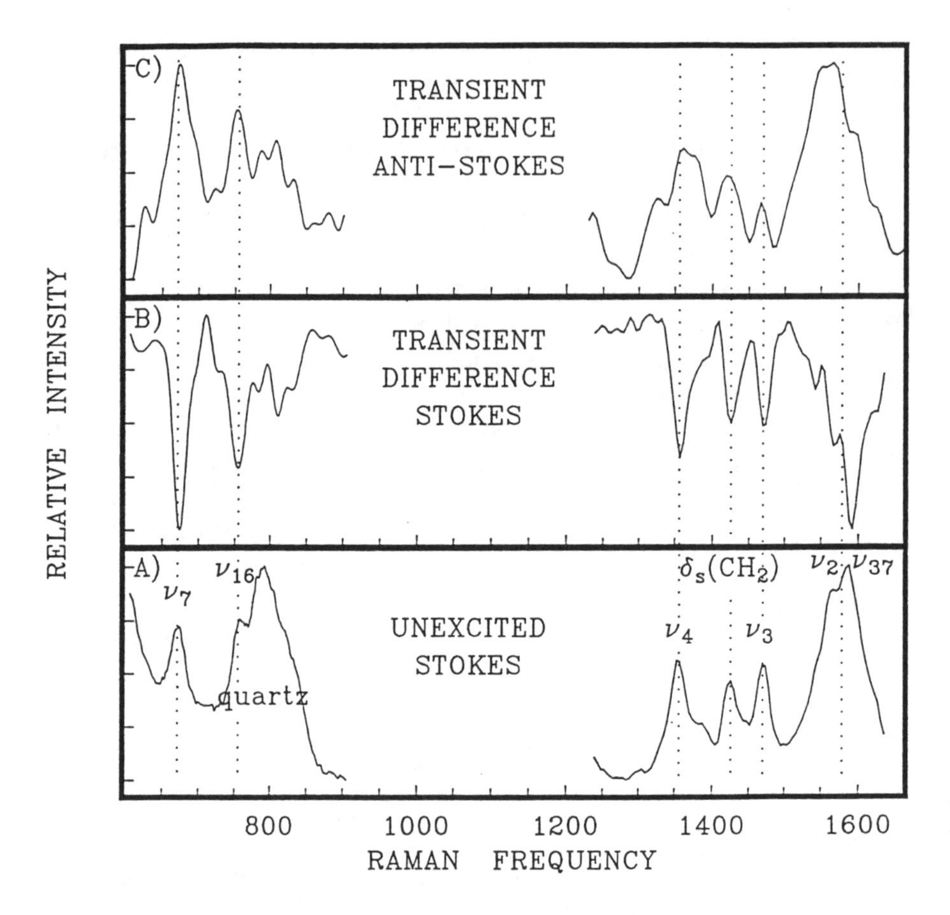

Fig. 2. Time zero Raman spectra of deoxyHb in 100 mM phosphate buffer at pH 7.5. This figure demonstrates the spectral signature of a vibrationally hot deoxyHb transient. Figure 2A shows the ground state Stokes probe-only Raman spectrum. Figures 2B and C show pure transient picosecond Raman spectra obtained by subtracting the 355 nm probe-only spectrum from the two-color spectrum. Figure 2B shows depleted Stokes Raman scattering. Positive anti-Stokes transient bands at the position of ν_7, ν_{16}, ν_4, the vinyl mode, ν_3, ν_2, and ν_{37} argue for Boltzmann distribution of heme internal modes, since all observed modes appear to be hot. Each of the six individual spectra are scaled independently to fill the height of the plot. The marked bands are assigned as ν_7 at 672 cm^{-1}, ν_{16} at 755 cm^{-1}(shoulder on quartz band), broad quartz window bands centered at 790 cm^{-1}(clearly labeled), weak ν_{32} on top of quartz bands at 790 cm^{-1}, ν_4 at 1359 cm^{-1}, a vinyl mode at 1430 cm^{-1}, ν_3 at 1475 cm^{-1}, and a broad group of overlapping bands including ν_2 plus ν_{37} plus others centered at 1580 cm^{-1}.

is, the anti-Stokes transient signal rises promptly and decays with 2–5 ps time constant, while the Stokes transient signal drops negative promptly and recovers to the baseline with a 2–5 ps time constant [31]. Identical cooling dynamics for vibrational modes widely spaced in frequency implies that the heme internal modes may be thermal equilibrium with each other. With this assumption, we can estimate the heme vibrational temperature by comparing the magnitudes of the anti-Stokes transient signals at the ν_7 and ν_4 band positions [30]. Our temperature jump estimate for the photoexcited deoxy heme, averaged over the convolution of our 8 ps pump and probe pulses, is $\approx$36 K above room temperature. A 532 nm photon can produce a heme internal temperature jump in excess of 400 K. Therefore the 2–5 ps VER time constant represents the loss of most of the heme internal energy.

HEME-LIGAND DYNAMICS IN OXYHEMOGLOBIN

The dynamics of photoexcited oxyHb are more complicated due to presence of the O_2 ligand. Following photoexcitation, heme-oxygen deligation and geminate recombination may occur in addition to vibrational relaxation. We can separate the heme-ligand dynamics from the vibrational relaxation dynamics by comparing *Stokes* and *anti-Stokes* Raman dynamics. VER with $\tau \approx 5$ ps is observed for both hot oxyHb (which could result from photoexcited oxyHb which did not lose its ligand or from geminate recombination that occurs faster than VER) and hot deoxyHb (resulting from photodeligated oxyHb). Our results indicate that O_2 recombination occurs either faster than 2 ps or slower than 1 ns.

Photodeligation of oxyHb yields a deoxy form of hemoglobin which has a vibrational spectrum very similar to that of oxyHb. The differences in the resonance Raman spectra for the ligated and unligated forms consist of band frequency shifts and/or band intensity changes. These differences in the resonance Raman of oxyHb and deoxyHb will dramatically affect the dynamics observed for the photodeligation reaction of oxyHb. For example, the oxyHb ν_4 band appears at 1376 cm^{-1}while the deoxyHb band is shifted to 1359 cm^{-1}. The oxyHb and deoxyHb ν_{16} bands are at roughly the same Raman shift. For Stokes Raman scattering, the ν_4 band is almost a factor of three more intense for oxyHb than for deoxyHb, while ν_{16} is about a factor of three stronger for deoxyHb than for oxyHb. In the anti-Stokes spectrum, both ν_4 and ν_{16} are more intense for deoxyHb than for oxyHb.

Figure 3 shows both Stokes and anti-Stokes dynamics of the ν_4 vibrational mode at 1376 cm^{-1}. Following a pulse-width limited bleaching of the Stokes signal in Figure 3A, we clearly observe two dynamics as the transient returns toward the baseline (established by the signal at negative time). The anti-Stokes dynamics of Figure 3B rises promptly with the laser pulse width, and then decays rapidly to a level *below* the baseline. Taken together, these dynamics indicate both vibrational cooling and slow recombination of the O_2 ligand. Photodeligation will remove a certain number of oxyHb molecules from the sample volume, leaving deoxy-like Hb in their place. Because the deoxyHb signal is weak at 1376 cm^{-1}(especially on the Stokes side), photo-

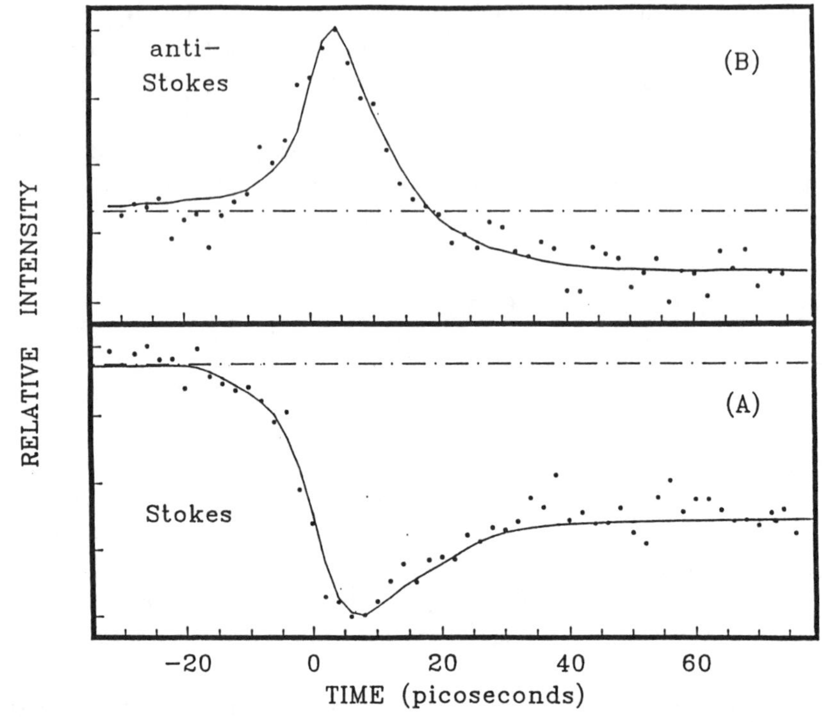

Fig. 3. Dynamics of VER and geminate recombination for ν_4 at 1377 cm^{-1} of oxyHb. The Stokes and anti-Stokes transient Raman intensity at ν_4 (1377 cm^{-1}) is plotted as a function of delay time between pump and probe pulses. At this band oxyHb dominates the observed dynamics because the resonance Raman enhancement is stronger for oxyHb than for deoxyHb and because the oxyHb ν_4 band is shifted 22 cm^{-1} higher than the corresponding deoxyHb band. A) The Stokes dynamics of the ν_4 band. The fits show that the deconvolved time constant of VER is about 2–5 ps, while the slow geminate recombination occurs with $\approx$ 1000 ps $\pm$500 ps time constant. No fast geminate recombination is observed on 2–100 ps time scale. The baselines are shown by the dashed lines. The circles are the experimental data points. The solid line is the fit to the dynamics. The spectra of frames A and B are independently scaled.

deligation will result in negative transient signals for both the Stokes and anti-Stokes bands. This is clearly observed in the long time dynamics, which represents heme-ligand recombination on the nanosecond time scale. Because the deoxyHb ν_4 band is shifted 20 cm^{-1} off of the position we are observing, *the oxyHb signal dominates the observed dynamics.*

The early time dynamics of Figure 3 is assigned to VER. The complementary negative/positive behavior in the Stokes/anti-Stokes early time dynamics is exactly the same as for the earlier study of deoxyHb. Because the oxyHb resonance Raman signal dominates the spectrum of Figure 3A, the early-time fast dynamics is believed to result from VER in hot *oxyHb* which either never lost its ligand or else underwent geminate recombination faster than 2 ps.

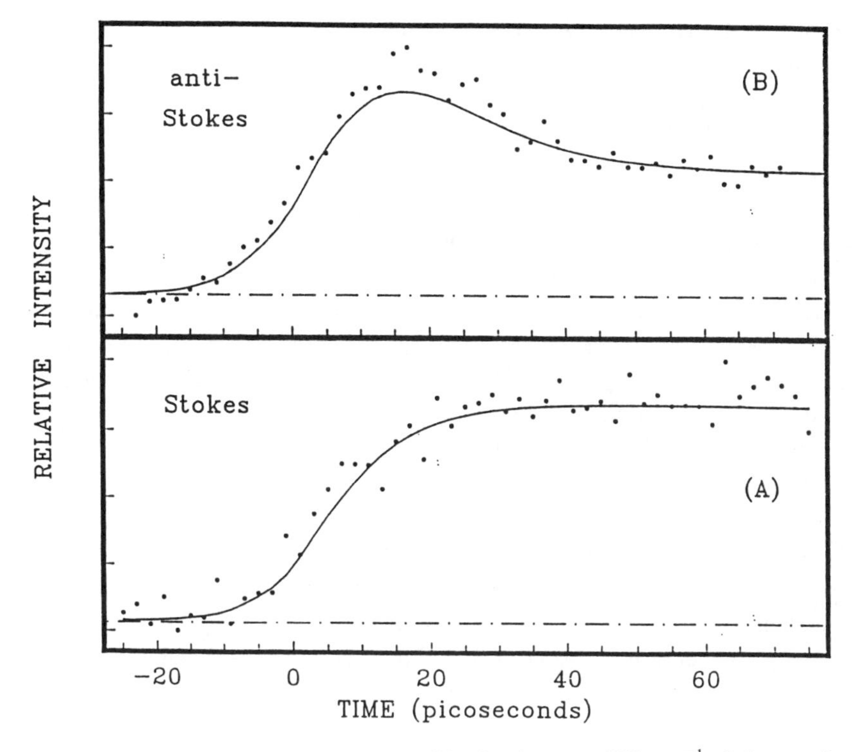

Fig. 4. Dynamics of VER and geminate recombination for ν_{16} at 755 cm^{-1} of photoproduct deoxy-like hemoglobin (deoxyHb′). The Stokes and anti-Stokes transient Raman intensity at ν_{16} (755 cm^{-1}) is plotted as a function of delay time between pump and probe pulses. Because the resonance Raman enhancementis much larger for deoxyHb than for oxyHb at ν_{16} in both the Stokes and anti-Stokes regions, the observed dynamics are dominated by the photoproduct deoxyHb′. A) The Stokes dynamics of the ν_{16} band. The fit shows that the deconvolved time constant of VER is about 2–5 ps, while the geminate recombination time constant is $\approx$ 1000 ps $\pm$500 ps. No fast geminate recombination is observed on 2–100 ps time scale. The spectra of frames A and B are independently scaled. The baselines are shown by the dashed lines.

Deconvolution of both Stokes and anti-Stokes dynamics gives the same VER rate of $\tau \approx 5$ ps. On the other hand, fast recombination would appear as a transient moving in the positive direction in both the Stokes and anti-Stokes dynamics. This is not observed.

Both the Stokes and anti-Stokes dynamics of the oxyHb ν_{16} band in Figures 4A and 4B show *positive* transient signal at long times, in contrast to the ν_4 dynamics in Figure 3. The transient is positive at long times because the resonance Raman enhancement for the ν_{16} Stokes and anti-Stokes bands is larger for deoxyHb than for oxyHb, so *the observed dynamics of ν_{16} are dominated by the deoxy photoproduct.* These deligated hemes will be vibrationally hot, as shown in the early time dynamics. The anti-Stokes transient signal, shown in Figure 4B, rises as the deoxyHb low-lying excited levels become populated; it then decays to the long time offset as the deoxyHb population cools to thermal equilibrium at room temperature. The Stokes transient for this band, shown in Figure 4A, serves as a window to observe the cold vibrationless level of deoxyHb; thus the dynamics rises as the deoxyHb vibrationless level is populated. We see no evidence for fast geminate recombination in the Stokes dynamics, which would appear as a negative-going feature in Figure 4A.

Geminate recombination for photolyzed oxyHb has been observed on the nanosecond time scale using transient absorption [26]. Petrich *et al.* [27] observed a 2.5 ps dynamics which they propose to be a deligated but highly reactive electronic state that relaxes back to the ligated form. However, we see no evidence of ligand recombination on a 2.5 ps time scale. A 2.5 ps geminate recombination is on roughly the same time scale as the VER we observe in deoxyHb, so it should appear in the dynamics of Figures 3 and 4 if it exists.

The vibrational dynamics indicate that no fast geminate recombination of O_2 or CO with the original heme occurs on a 2 to 100 ps time scale. The slow ligand rebinding occurring on a nanosecond time scale is observed for both oxyHb and carbonylHb.

Geminate recombination is affected by many factors. In an oversimplified picture geminate recombination is largely affected by energetic and diffusional factors. The lack of fast ($\approx$ 2–100 ps) geminate recombination suggests that the barrier for ligand rebinding forms promptly in oxyHb and carbonylHb. This barrier can be divided into a structural, electronic, and heating barrier.

The structural change of the heme has been discussed by Friedman and coworkers [21, 25, 34]. Upon photodeligation, the low spin Fe quickly changes to high spin, thus the central Fe becomes too large to be in the original position [18]. The iron rapidly moves out of the heme plane within several hundreds of femtoseconds [22] in order to reduce the repulsive potential between the imidazole and pyrrole nitrogen. This doming effect dramatically affects the dynamics of geminate recombination. For geminate recombination to occur, the iron in the heme has to move back onto the heme plane. The repulsive energy between the histidine and the heme must increase, and this raises the barrier height for ligand rebinding. This fast structural change of the heme will obviously influence the dynamics of fast geminate recombination.

A hot oxyHb molecule was observed despite the lack of geminate recombination. This was interpreted as evidence for the importance of non-radiative processes which quench the electronically excited state. This observation supports the previous hypothesis [17] that non-radiative processes rather than chemical barrier heights control the quantum yield of photodeligation.

Photodissociation of $Cr(CO)_6$

The photochemistry of $Cr(CO)_6$ has long been studied as the prototype of metal-carbonyl compounds. In the gas phase, $Cr(CO)_6$ has been known to lose up to 6 carbonyl groups depending on the excitation energy. In contrast, only one CO group dissociates [35] under ultraviolet irradiation in solution or matrix. With the development of new techniques such as pico and femtosecond laser spectroscopy, the ultrafast dynamics of $Cr(CO)_6$ photodissociation have recently begun to be unraveled. However, there exists a controversy over recent studies of $Cr(CO)_5$ solvation following the photodissociation reaction in solution.

Simon *et al.* [36] first reported that photoexcitation of $Cr(CO)_6$ results in the formation of solvated $Cr(CO)_5$ which appears in a time of ≤ 0.8 ps in hydrocarbon solvents. Their picosecond transient absorption results have been supported by similar femtosecond experiments by Nelson *et al.* [37] However, the appearance time of solvated $Cr(CO)_5$ has been questioned by other workers using both UV-visible and IR transient absorption spectroscopy [38–39]. Spears *et al.* [39] claim a much longer solvation time of 100 ps. Lee and Harris [38] suggest that vibrational relaxation controls solvation dynamics on a 17 ps time scale. We have used transient picosecond Raman spectroscopy to remove the present controversy by directly probing the vibrational coordinates of the photoproducts. There are two processes of interest in the Raman scattering. If vibrational relaxation occurs, the intensity of the Stokes signal is expected to grow while that of the corresponding anti-Stokes signal diminishes in time. Thus the complementary observations of a growing Stokes band and a decaying anti-Stokes band give unambiguous evidence of cooling in that particular vibrational mode. This technique is therefore very sensitive to vibrational relaxation [40].

Figure 5 shows the picosecond transient Raman spectra obtained at various optical delays for 266 nm excitation of $Cr(CO)_6$ in cyclohexane. Two 5 ps pulses at 266 nm are used as the pump and probe pulses in this experiment. Both pulses are produced in the same laser by injecting two seed pulses into the regenerative laser. The delay between the pulses depends on the cavity length of the regenerative amplifier and is measured by a simple autocorrelation technique. The first pulse photodissociates $Cr(CO)_6$ and the second probes the resonance Raman spectrum of $Cr(CO)_5$. The reason for using this wavelength is that $Cr(CO)_6$ has a strong absorption band corresponding to ligand field excitation around 280 nm, while $Cr(CO)_5$ has a strong ultraviolet absorption band for the metal to ligand charge transfer absorption at 240 nm($\varepsilon = 3 \times 10^4$) [41]. The pentacarbonyl absorption is known to be relatively insensitive to the solvent at this wavelength. As a result, the magnitude of the transient Raman signal should not depend on solvation. The pump and probe pulses

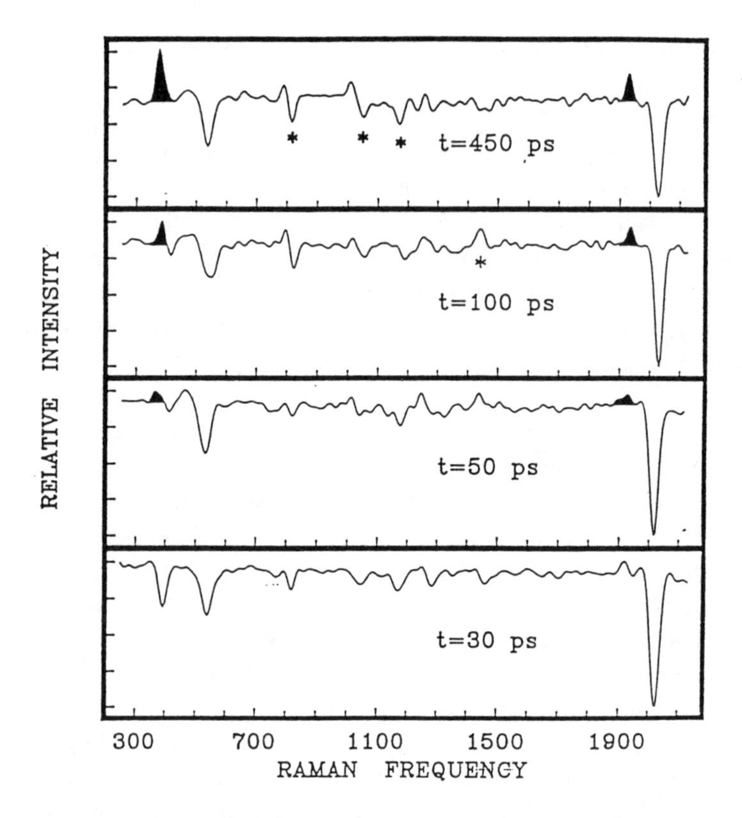

Fig. 5. Pure transient picosecond Stokes Raman spectrum in cyclohexane obtained by two pulse pump and probe at 266 nm as described in the text. The Raman intensity represents a temporal average over the pulse width of the laser which is specified in each frame. The colored-in bands are those assigned to solvent coordinated $Cr(CO)_5$. Asterisks are used to denote noise due to Raman bands of the solvent molecules which have been subtracted out of each spectrum. Frequency is in units of cm^{-1}.

are separated by a variable optical delay which is quoted separately in each frame of Figure 5. The spectra contain only pure transient Raman bands after removal of solvent and ground state $Cr(CO)_6$ bands using a spectrum differencing technique. This was achieved by alternately exciting the sample with a single interrogation laser pulse or the double pump-probe laser pulse sequence mentioned above. The pure transient spectrum shown in Figure 5 is obtained by subtracting the one pulse background spectrum from that obtained in the double pulse sequence. Other pertinent experimental details are given in the figure caption.

Comparing frames A-D in Figure 5, it is readily apparent that a single transient appears with a time scale of roughly 100 ps. The metal-CO stretch at 381 cm^{-1} and the CO vibration at 1935 cm^{-1} indicate that the transient is a metal carbonyl complex which is assigned to $Cr(CO)_5$. The bands marked

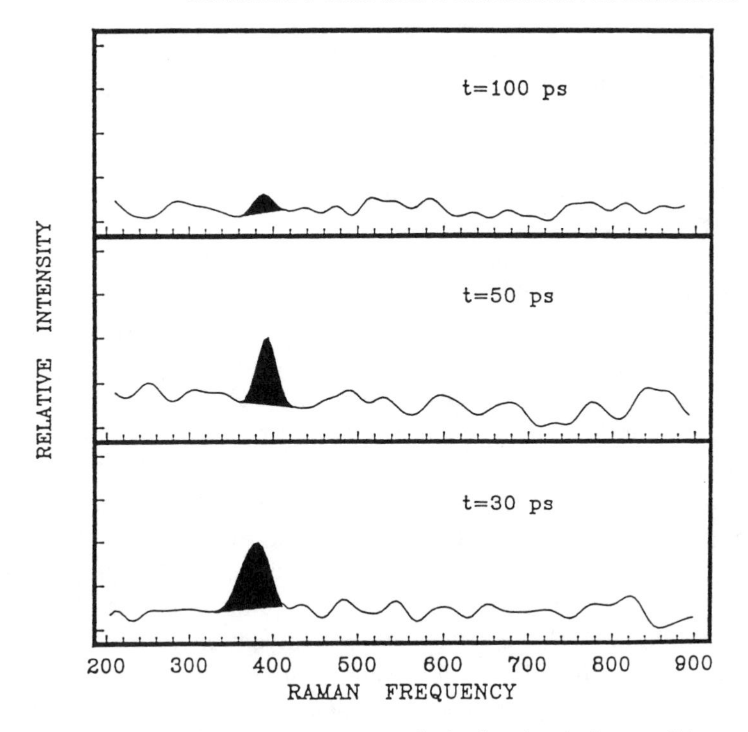

Fig. 6. Transient anti-Stokes Raman spectrum obtained under similar conditions to those given in Figure 1. Time delay between pump and probe pulses is given separately in each frame. Ground state bands have been subtracted out of the spectrum as described in the text. Spectra are normalized to the intensity of the ground state chromium band at 532 cm^{-1} in the un-subtracted spectrum. Frequency is in units of cm^{-1}.

with asterisks are the result of noise generated by the spectrum differencing technique at the frequencies of the cyclohexane solvent bands. The negative going peaks are ground state $Cr(CO)_6$ bands which appear in the transient spectrum as a result of population bleaching. The ground state 383 cm^{-1} metal-CO stretch of $Cr(CO)_6$ appears as a bleach in the 30 ps spectrum and gradually fills in at later times due to the growth of the $Cr(CO)_5$ transient band at 381 cm^{-1}. The dynamics of vibrational cooling can be investigated by comparing the Stokes and anti-Stokes band intensities. Figure 6 illustrates the transient anti-Stokes spectrum in the region of the 381 cm^{-1} band assigned to $Cr(CO)_5$. The results indicate that the anti-Stokes signal originates from the hot vibrational state which decays in 100 ps. The observation that the Stokes and anti-Stokes spectrum have complementary dynamics is consistent with vibrational relaxation. It may be appear that vibrational relaxation continues until 450 ps in the Stokes region. However, it should be noted that as the band grows from the bleach at 30 ps to the maximum at 450 ps, more than 70% of the transient intensity has developed in 100 ps. The appearance time of ground state $Cr(CO)_5$ is therefore believed to represent the time required for the photoproduct to approach thermal equilibrium with the solvent. This conclusion clearly removes the present controversy and also demonstrates the importance of non-equilibrium vibrational energy with relatively long relaxation times in the condensed phase.

Large internal energies were found in the photoproduct. Assuming statistical distribution of the internal energy (which is not a good assumption), an estimate of the nascent vibrational temperature of $Cr(CO)_5$ gives 1000 K in cyclohexane and 500 K in THF. Vibrational frequency shifts were observed and attributed to transitions from high quantum excitation in low frequency modes which are anharmonically coupled to the allowed Raman modes.

These results demonstrate that $Cr(CO)_5$—X remains very hot for relatively long times (10's of picoseconds). This conclusion comes entirely from the complementary behavior of the *intensities* of the Stokes and anti-Stokes band. (A decaying anti-Stokes band and a growing Stokes band was observed. The observed band shifts are consistent with vibrational cooling but it is the dynamics of the intensities that establishes vibrational cooling.) It is therefore quite possible that a large fraction of the solvent coordination and solvent reorientation (coordination with solvent H atom rearranging to coordination with solvent O atom) occurs at a rate similar to the vibrational cooling rate. In fact, solvent coordination may be responsible for *part* of the frequency shift observed for the Cr—CO vibration. If so, the time scale for the frequency shift of 20–50 ps is consistent with solvent coordination occurring at a similar rate to vibrational relaxation.

The appearance of the Raman spectrum for ground state $Cr(CO)_5$—X can be used to determine the electronic relaxation time. The results suggest that electronic relaxation occurs within the risetime of the apparatus with an uncertainty of 5 ps.

Mechanism Of Vibrational Relaxation

There is considerable interest in understanding the mechanism of vibrational relaxation in $Cr(CO)_5$. If the long relaxation times of 80 ps is relaxation of the high frequency CO stretching vibrations, this represents an incredibly long time for localization of vibrational energy for molecule in the liquid phase. This behavior is completely unexpected. However, $Cr(CO)_5$—X has a heavy atom at the center and it is precisely this kind of molecule which might be expected to exhibit weak coupling between the internal vibrational modes [42, 43, 44]. Is it reasonable for a vibrational mode such as the CO stretch with a moderate density of bath states to survive for 100 ps in the liquid phase where the collision frequency is on the order of $10^{12}s^{-1}$. Our results cannot completely answer this question.

In a 266 nm pump with 213 nm probe experiment we have measured the Stokes and anti-Stokes dynamics of low frequency metal-CO stretching vibrations in $Cr(CO)_5$—X. We have determined that the anti-Stokes levels of 478 and 318 cm^{-1} appear promptly and then decay with 50 ps and 80 ps time constants, respectively. The resonance enhancement for the Stokes bands favors signal originating from the vibrationless level which grows in with a time constant of 100 ps. These results prove that low frequency vibrations could be responsible for the 100 ps risetime observed for the vibrationless level. One possible mechanism of vibrational relaxation that fits these results is that high frequency modes quickly relax into low frequency modes where the decay rate slows down and approaches 100 ps decay rates for the lowest frequency vibrations.

Alternatively, there is no evidence in the anti-Stokes dynamics of the 478 and 318 cm^{-1} metal-CO modes stretching modes which suggests that these modes are being populated from high frequency CO stretching modes where the latter decay with 100 ps time constants into the bath of intramolecular $Cr(CO)_5$—X modes. This indicates that if the high frequency CO stretching modes decay with a 100 ps time constant, these modes must decay by pathways which avoid populating metal-CO modes.

It is interesting to consider that our time scale is approximately the same as that observed [45] for the CO stretching vibration in $Cr(CO)_6$, where a relaxation time of 145 ± 25 ps was observed in n-hexane. Similar rates would be expected in these two experiments if the latter dynamics represents the time required for energy randomization followed by vibrational relaxation through the entire manifold of vibrational levels. There is no direct way to compare our results to the faster dynamics attributed to vibrational decay in the transient absorption experiment [36–39].

References

1. Harris, A. L., Brown, J. K., and Harris, C. B., *Ann. Rev. Phys. Chem.* **39**, 341 (1988).
2. Chuang, T. J., Hoffman, G. W., and Eisenthal, K. B., *Chem. Phys. Lett.* **25**, 201 (1974).
3. Bado P. and Wilson, K. R., *J. Phys. Chem.* **88**, 655 (1984).
4. Kelly, D. F., Abul-Haj N. A., and Jang, D. J., *J. Chem. Phys.* **80**, 4105 (1984).
5. Harris, A. L., Berg M., and Harris, C. B., *J. Chem. Phys.* **84**, 788 (1986).
6. Paige, M. E., Russell D. J., and Harris, C. B., *J. Chem. Phys.* **85**, 3699 (1986).
7. Smith, P. E. and Harris, C. B., *J. Chem. Phys.* **87**, 2709 (1987).
8. Coxon, J.A., *J. Quant. Radiat. Transfer* **11**, 443 (1971).
9. Tellinghuisen, J., *J. Mol. Spect.* **94**, 231 (1982).
10. Viswanathan, K. S., Sur, A., and Tellinghuisen, J., *J. Mol. Spect.* **86**, 393 (1981).
11. Nesbitt D. J. and Hynes, J. T., *J. Chem. Phys.* **76**, 6002 (1982).
12. Nesbitt D. J. and Hynes, J. T., *J. Chem. Phys.* **77**, 2130 (1982).
13. Houston, Paul J., *Adv. Chem. Phys.* **47**(2), 381 (1981).
14. Tully, J. C., 'Dynamics of Molecular Collisions', in *Modern Theoretical Chemistry*, Vol. 2, W. H. Miller ed., Plenum, New York, 1976.
15. Shank, C. V., Ippen, E. P., and Bersohn, R., *Science* **193**, 50 (1976).

16. B.I. Greene, B. I., Hochstrasser, R. M., Weisman, R. B., and Eaton, W. A., *Proc. Natl. Acad. Sci. USA* **75**, 5255 (1978).
17. Chernoff, D. A., Hochstrasser, R. M., and Steele, A. W., *Proc. Natl. Acad. Sci. USA* **77**, 5606 (1980).
18. Terner, J., Stong, J. D., Spiro, T. G., Nagumo, M. Nicol, M., and El-Sayed, M. A., *Proc. Natl. Acad. Sci. USA* **78**, 1313 (1981).
19. Terner, J., Spiro, T. G., Voss, D. F., Paddock, C., and Miles, R. B., *J. Phys. Chem.* **86**, 859 (1982).
20. Cornelius P. A., and Hochstrasser, R. M., In *Picosecond Phenomena III*; K.B. Eisenthal, R.M. Hochstrasser, W. Kaiser, and A. Laubereau, Eds.; Springer-Verlag: New York, 1982, p. 288.
21. Friedman, J. M., Stepnoski, R. A., and Noble, R. W., *FEBS* **146**, 278 (1982).
22. Martin, J. L., Migus, A., Poyart, C., Lecarpentier, Y., Astier, A., and Antonetti, A., *Proc. Natl. Acad. Sci. USA* **80**, 173 (1983).
23. Cornelius, P.A., Hochstrasser, R. M., and Steele, A. W., *J. Mol. Biol* **163**, 119 (1983).
24. Hutchinson J. A. and Noe, L. J., *IEEE, J. Quant. Electr.* **QE-20**, 1353 (1984).
25. Scott T. W. and Friedman, J. M., *J. Am. Chem. Soc.* **106**, 5677 (1984).
26. Findsen, E. W., Friedman, J. M., Ondrias, M. R., and Simon, S. R., *Science* **16**, 661 (1985).
27. Petrich, J. W., Poyart, C., and Martin, J. L., *Biochemistry* **27**, 4049 (1988).
28. Janes, S. M., Dalickas, G. A., Eaton, W. A., and Hochstrasser, R. M., *Biophys. J.* **54**, 545 (1988).
29. Miller, R. J. D., *Ann. Rev. Phys. Chem.* **42**, 581 (1991).
30. Lingle, Jr., Robert, Xu, Xiaboing, Zhu, Huiping, Yu, Soo-Chang, and Hopkins, J. B., *J. Am. Chem. Soc.* **113**, 3992 (1991).
31. Lingle, Jr., Robert, Xu, Xiaobing, Zhu, Huiping, Yu, Soo-Chang, and Hopkins, J. B., *J. Phys. Chem.* **95**, 9320 (1991).
32. Petrich J. W. and Martin, J. L., *Chemical Physics* **131**, 31 (1989).
33. Alden, R. G., *J. Am. Chem. Soc.* **112**, 3241 (1990).
34. Friedman, J. M., Scott, T. W., Stepnoski, R. A., Ikeda-Saito, M., and Yonetani, T., *J. Biol. Chem.* **258**, 10564 (1983).
35. J.A. Welch, J. A., Peters, K. S., and Vaida, V., *J. Phys. Chem.* **86**, 1941 (1982).
36. Simon J. D. and Xie, X., *J. Phys. Chem.* **90**, 6751 (1986).
37. Joly A. G. and Nelson, K. A., *J. Phys. Chem.* **93**, 2876 (1989).
38. Lee M. and Harris, C. B., *J. Am. Chem. Soc.* **111**, 8963 (1989).
39. Wang, L., Zhu X., and Spears, K. G., *J. Am. Chem. Soc.* **110**, 8695 (1988).
40. Yu, Soo-Chang, Xu, Xiaobing, Lingle, Jr., Robert, and Hopkins, J. B., *J. Am. Chem. Soc.* **112**, 3668 (1990).
41. Graham, M. A., Poliakoff, M., and Turner, J. J., *J. Chem. Soc.(A),Inorg. Phys. Theor.* **29**, 39 (1971).
42. Lederman, S. M., Lopez, V. F., Voth, G. A., and Marcus, R. A., *Chem. Phys.* **139**, 171 (1989).
43. Lederman S. M. and Marcus, R. A., *J. Chem. Phys.* **88**, 6312 (1988).
44. Uzer T. and Hynes, J. T., *J. Chem. Phys.* **90**, 3524 (1986).
45. Heilweil, E. J., Cavanagh, R. R., and Stephenson, J. C., *Chem. Phys. Lett.* **134**, 181 (1987).

7. Diffusional Quenching of *trans*-Stilbene by Fumaronitrile: Role of Contact Radical Ion Pairs and Solvent Separated Radical Ion Pairs

KEVIN S. PETERS
Department of Chemistry and Biochemistry, University of Colorado, Boulder, CO 80309, U.S.A.

Introduction

A fundamental problem in bimolecular electron transfer reactions in solution between a photoactive acceptor molecule, A^*, and an electron donor molecule, D, is ascertaining under what conditions the electron is transferred at molecular contact to give contact radical ion pairs (CRIP) or at long range to give solvent separated radical ion pairs (SSRIP).

$$\begin{array}{ccc} A^* + D & \underset{k_{diss}}{\overset{k_{ass}}{\rightleftharpoons}} & A^*D \\ k_{let}\downarrow & & \downarrow k_{cet} \\ \text{SSRIP} & & \text{CRIP} \end{array}$$

The competition between these two processes is dependent upon the relative magnitudes of the rate of contact electron transfer, k_{cet}, and the rate of long range electron transfer, k_{let}, as well as the rate of association, k_{ass}, and dissociation, k_{diss}, of the reacting species. The rates for electron transfer, k_{let} and k_{cet}, are a function of the electronic coupling between the reacting species, the intramolecular vibrational reorganization energy, the solvent reorganization energy and the free energy change for the electron transfer reaction [1]. Although there are theoretical formulations [1, 2] for the rate of electron transfer that incorporate these various parameters, for a given molecular system it is not possible to accurately determine each of these quantities so that it is not feasible to always predict under what circumstances a CRIP or a SSRIP will be formed.

J.D. Simon (ed.), Ultrafast Dynamics of Chemical Systems, 223–233.

During the past twenty five years there have been numerous experimental studies that have sought to address these issues. The original and pioneering work of Weller [3] found that fluorescence quenching of aromatic hydrocarbons by diethylanaline in non polar solvents leads to the formation of a new light emitting species, the exciplex, which has alternatively been described as a CRIP. The property of light emission from an exciplex reflects strong electronic coupling which can only occur at molecular contact between the radical ion pair, a distance that is estimated at 3.5 Å. In polar solvents such as acetonitrile, where the free energy change for the electron transfer reaction is ΔG =–0.6 eV, no new light emitting species were observed upon fluorescence quenching suggesting that the intermediate formed upon electron transfer has a radical ion pair structure different from the exciplex or CRIP. By examining the quantum yield for fluorescence quenching as a function of donor concentration, the distance for reaction at which electron transfer occurred was estimated to be of the order of 7 Å which is consistent with the formation of a SSRIP. These studies led to the general view that for electron transfer reactions in polar solvents that if there is no exciplex emission, then SSRIP are produced.

Support for this model can be found in the studies of Mataga and coworkers [4]. For example, in acetonitrile, pyrene-pryomelletic dianhydride form a ground state charge transfer complex which upon 532 nm excitation produces directly CRIP which decays by charge recombination, $k_{\mathrm{CR}} = 10^{11}\ \mathrm{s}^{-1}$, as determined by picosecond absorption spectroscopy. However when the first excited singlet state of pyrene is diffusionally quenched by pyromelletic dianhydride to produce radical ions, the radical ion pair decays with substantially different kinetic behavior through a pathway of charge recombination, $k_{\mathrm{CR}} = 2.5 \times 10^9\ \mathrm{s}^{-1}$, and radical ion pair separation, $k_d = 2.6 \times 10^9\ \mathrm{s}^{-1}$. Given the difference in the kinetic behavior of the CRIP produced through irradiation of the ground state charge transfer complex and the radical ion pair formed upon diffusional quenching by electron transfer, Mataga concluded that the SSRIP is produced in this latter process.

The generality of the proposal that electron transfer quenching reactions leads to the formation of SSRIP in polar solvents has recently been questioned by Farid and coworkers [5]. In their study of the fluorescence quenching of tetra-cyanoanthracene (TCA) by a variety of alkyl substituted benzenes in acetonitrile, exciplex emission was observed. The efficiency of exciplex (CRIP) formation is dependent upon the driving force for the reaction which varies from unity to zero for a change in reaction free energy of –0.3 eV to –0.65 eV. Thus it appears that the efficiency for the CRIP formation depends not only on the solvent polarity but also upon the free energy change for the electron transfer reaction.

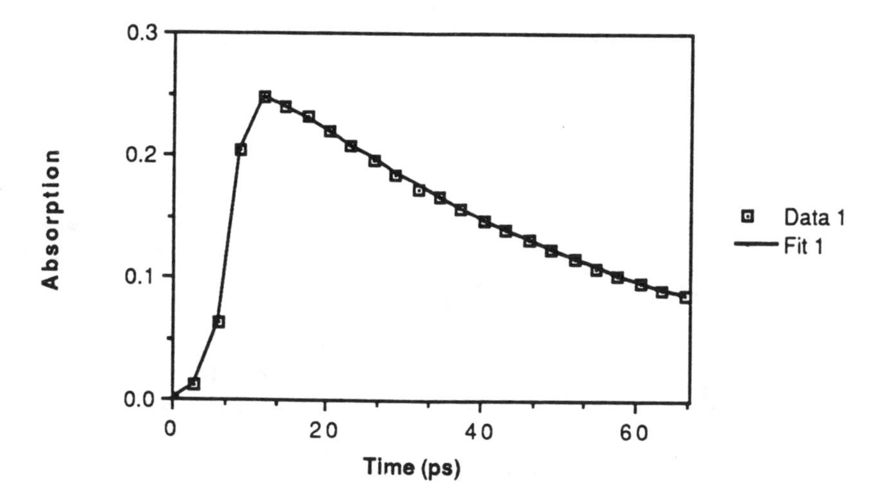

Fig. 1. Transient absorbance at 570 nm following the 300 nm irradiation of *trans*-stilbene in acetonitrile, 1 ps time resolution. The fit is a convolution of a single-exponential decay with the instrument response function, Equation (1) The lifetime is 42.5 ps. The squares are data, the solid line is the calculated fit.

In this article we will examine the question of whether CRIP or SSRIP are formed upon the quenching of the first excited singlet state of *trans*-stilbene (S_1) by fumaronitrile (FN) in acetonitrile; this work has been discussed in a series of three publications [6–8]. The free energy change for this reaction is estimated to be –0.76 eV. In the first section the kinetics for the quenching of S_1 by FN will be examined by picosecond absorption spectroscopy. The intrinsic rate of electron transfer is separated from the diffusion rate by applying the Collins and Kimball formalism for time-dependent rate constants [9]. From this analysis a distance parameter for the electron transfer reaction is obtained which is found to be consistent with the formation of a SSRIP. In the next section the dynamics of the CRIP and SSRIP are examined. The CRIP is formed by irradiation of the trans-stilbene (TS)–FN ground state charge transfer complex. The CRIP decays by back electron transfer and diffusional separation to produce SSRIP whose dynamics in turn are measured. Finally in the last section, the dynamics of the radical ion pair formed upon the diffusional quenching of S_1 by FN are studied and are found to be characteristic of a CRIP.

Excited State Dynamics of *Trans*-Stilbene/Fumaronitrile

The dynamics of the decay of the first excited singlet state of *trans*-stilbene (S_1) in acetonitrile following 300 nm excitation are monitored at 570 nm near the $S_n \leftarrow S_1$ absorption maximum with a time resolution of 1 ps, Figure 1.

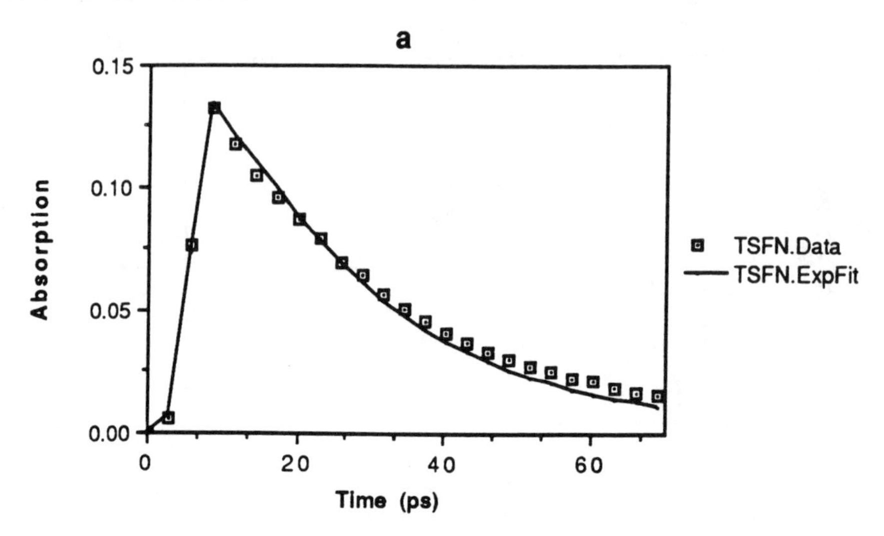

Fig. 2. The transient absorbance at 570 nm following the 300 nm irradiation of *trans*-stilbene in the presence of 0.33 M fumaronitrile in acetonitrile, 1 ps time resolution. Squares – experimental data; solid line – calculated fit using Equation (2).

Assuming the isomerization decay of S_1 is characterized by an exponential relaxation, the lifetime for S_1 is 42.5 ± 2.5 ps where the best fit is shown in Figure 1. The calculated curve $A(t)$ results from the convolution of the instrument response function $I(t)$ with the chemical kinetics signal $F(t)$.

$$A(t) = \int_0^t I(t-x)F(x)\,\mathrm{d}x\,. \tag{1}$$

The decay of S_1 in the presence of 0.33 M fumaronitrile (FN) in acetonitrile is shown in Figure 2. The simplest analytical expression for the rate of decay of S_1 in the presence of FN is

$$k_{\text{obs}} = k_{\text{iso}} + k_{\text{et}}[\text{FN}]_0\,, \tag{2}$$

where k_{iso} is the unimolecular decay constant of the excited singlet state, determined above, and k_{et} is the rate of electron transfer which is assumed to be time independent. The best fit of the experimental data to the kinetic expression given by Equation (2) is shown in Figure 2. It is evident the dynamics for the decay of S_1 in the presence of FN are not well characterized by the convolution of a single time-independent exponential decay with the instrument response function. As the experimental decay is faster at early times and slower at later times, the rate constant for the decay of S_1 in the presence of FN is time dependent.

The origin of the time-dependent rate constant for the quenching of S_1 by FN is a result of the nature of the original distribution of FN about S_1. Prior to the excitation of TS, there is an equilibrium distribution of FN about TS so that at high concentrations some of the TS will have a molecule of FN within the first solvent shell. Upon excitation, those S_1 with a molecule of FN in the vicinity will be rapidly quenched and the rate of the reaction will be limited by k_{et}. However, for those S_1 which do not have a molecule of FN in the vicinity, the quenching of S_1 will depend upon the diffusion of FN and will be limited by k_d. A theory for solution phase kinetics when $k_{et} \geq k_d$ was developed by Collins and Kimball [9]. Solving Fick's second law of diffusion with the boundary condition $c(R) = (D/\kappa)(\delta c/\delta r)_R$ yields

$$k_{CK}(t) = [k_{et}^{-1} + k_d^{-1}]^{-1} \, [1 + (k_{et}/k_d) \exp(y^2) \operatorname{erfc}(y)] \tag{3}$$

where

$$\operatorname{erfc}(y) = (2\pi)^{-1/2} \int_y^{\infty} \exp(-u^2)\, du \tag{4}$$

and

$$y = \{(Dt)^{1/2}/R\}\{1 + (k_{et}/k_d)\} \tag{5}$$

The parameter $D = D_A + D_D$ is the sum of the diffusion coefficients for the acceptor and donor molecules which can be determined by electrochemical measurements [6]. The value R is the distance at which the electron transfer reaction occurs. Since k_d can be described as

$$k_d = 4\pi R N_A (D_D + D_A) \tag{6}$$

where N_A is Avogadro's number, the only unknown parameters in the above equations are k_{et} and R since D_D and D_A are experimentally determined [6].

The best fit of the kinetic model given in Equation (3) to the quenching of S_1 by 0.31 M FN, 0.46 M FN and 0.58 M FN is shown in Figure 3. The fit to the experimental data is given as a normalized absorption (NA) as a function of time where NA is defined as NA $= 1 + (A_{exp} - A_{cal})/A_{exp}$ and A_{exp} is the experimental absorbance and A_{cal} is the calculated absorbance at a given time. The rate for electron transfer is $k_{et} = (3.0 \pm 1.0) \times 10^{11}$ s^{-1} and the distance for the electron transfer is 8.7 Å [6].

The question raised in the Introduction is whether the quenching of S_1 by FN produces CRIP or SSRIP. Given the distance at which the electron transfer reaction occurs, 8.7 Å, and the suggestion by Weller [3] that a distance of this magnitude corresponds to a SSRIP, it would appear that SSRIP are formed upon the quenching of S_1 by FN in acetonitrile.

A Dynamic Description of the CRIP and SSRIP

The methodology that is employed in ascertaining whether a CRIP or a SSRIP is formed upon diffusional quenching of S_1 by FN is to compare the resulting

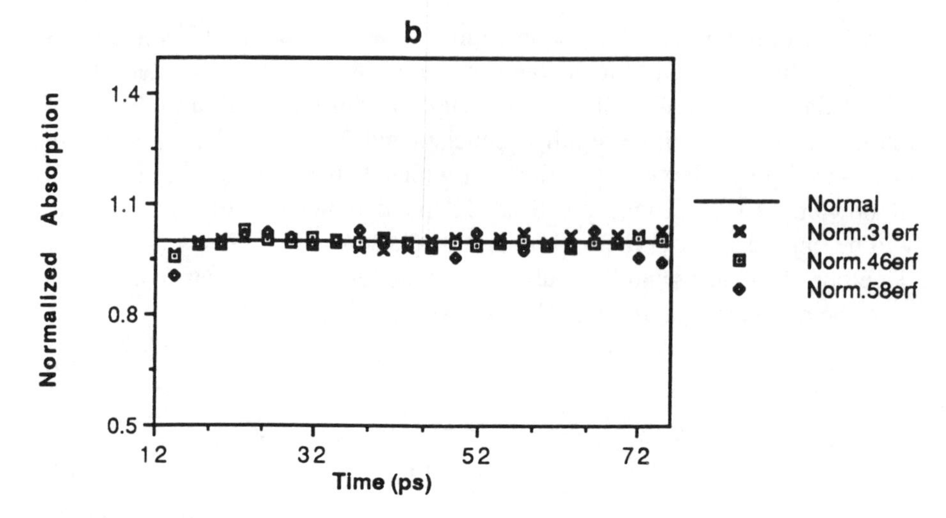

Fig. 3. Normalize fit to the experimental data for the quenching of S_1 in acetonitrile by 0.31 M FN (×), 0.46 M FN (□), 0.58 FN (♦) using Equation (3). Normalized absorbance is defined in text.

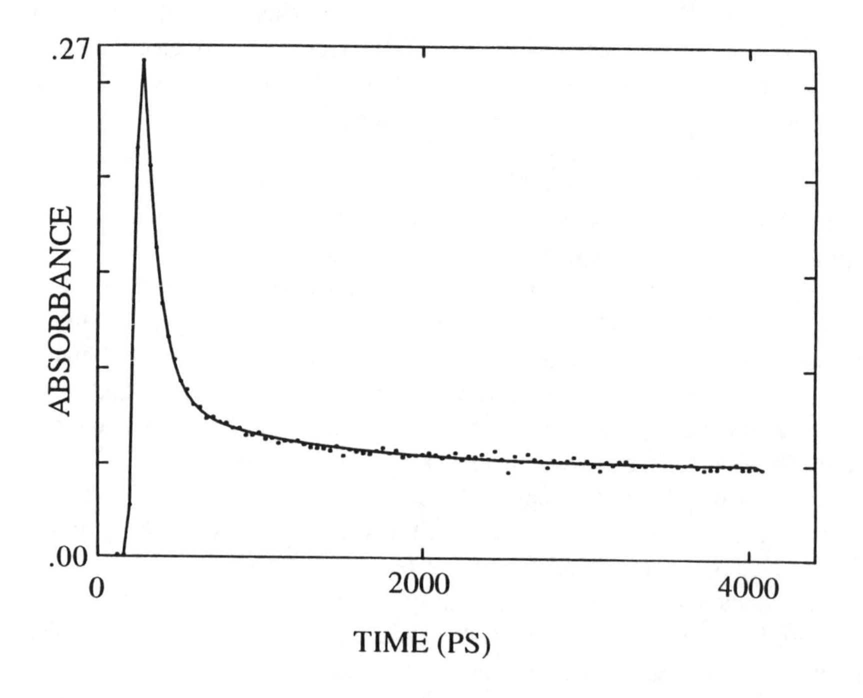

Fig. 4. Dynamics of the *trans*-stilbene radical cation, monitoring at 480 nm, following the 355 nm irradiation of a solution of 0.125 FN, 0.025 M TS in acetonitrile. Points – experimental data which are the average of five experiments, 40 ps time increments. Solid curve: calculated kinetics based upon Scheme 1 and the associated rate constants given for TS/FN in Table 1, Set 2.

radical ion pairs dynamics to the dynamics of a CRIP and a SSRIP [7]. The kinetic behavior for the CRIP is obtained by directly irradiating the TS/FN charge transfer complex at 355 nm which produces directly the CRIP. The dynamics of the CRIP are measured by monitoring the time evolution of the *trans*-stilbene radical cation (TS^+) absorbing at 480 nm. As shown in Figure 4 the kinetics are characterized by an initial rapid decrease in the 480 nm absorbance during the first 300 ps followed by a slower decrease in the absorbance during the subsequent 3 ns.

Two processes may contribute to the 480 nm decay. The first process is back electron transfer within the CRIP to form ground state reactants depleting the TS^+ population and thus decreasing its absorbance. A second possible pathway is collapse of the CRIP to form a diradicaloid species which leads to the cyclobutane adduct found in $2+2$ photocycloaddition reaction. However, the quantum yield for the photocycloaddition of TS with FN approaches zero in acetonitrile [7]; thus it is assumed that the second pathway does not contribute to the overall decay of the CRIP. A third decay pathway is diffusional separation to SSRIP which will not effect the 480 nm absorbance as the TS^+ concentration is not changed by this process. The SSRIP, in turn, may evolve into free radical ions (FI), undergo long range back electron transfer to yield ground state reactants or collapse to the CRIP; again, only long range electron transfer will lead to a decrease in the TS^+ absorbance.

The following kinetic scheme was used to analyze the data in Figure 4.

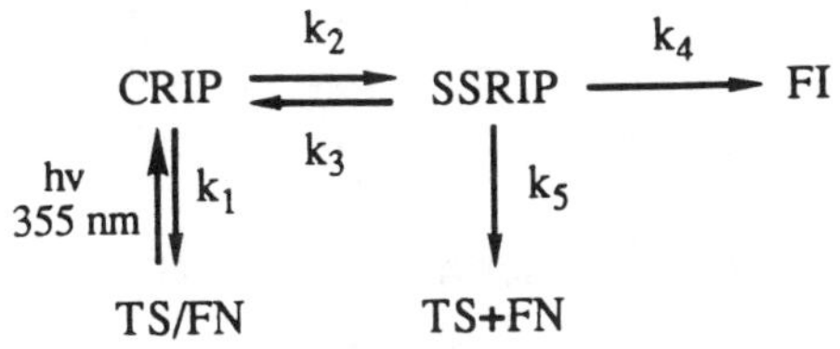

$$\begin{array}{ccccc} & & k_2 & & & k_4 & \\ & \text{CRIP} & \rightleftarrows & \text{SSRIP} & \longrightarrow & \text{FI} \\ h\nu\ 355\ \text{nm} & \uparrow\downarrow k_1 & k_3 & \downarrow k_5 & & \\ & \text{TS/FN} & & \text{TS+FN} & & \end{array}$$

Scheme 1.

Irradiation of the TS/FN charge transfer complex produces the CRIP which decays by two pathways: charge recombination, k_1, and radical ion pair separation to the SSRIP, k_2. The SSRIP decays by radical ion pair recombination to CRIP, k_3, or separation to FI, k_4, and charge recombination to uncomplexed TS and FN, k_5.

The fitting procedure examined two limiting sets of rate constants. The first set, Set 1, assumed that $k_3 \gg k_5$ so that k_5 was set to zero. The second set, Set 2, assumed that $k_5 \gg k_3$ so that k_3 was set to zero. Both Set 1 and Set 2 gave the same quality of fit to the experimental data and thus it was not possible to determine the relative contributions of k_3 and k_5 to the decay

TABLE I

Set 1. Parameters for the best fit of the kinetic model depicted in Scheme 1 holding $k_5 = 0$.

	$k_1 \times 10^9$ s^{-1}	$k_2 \times 10^9$ s^{-1}	$k_3 \times 10^9$ s^{-1}	$k_4 \times 10^9$ s^{-1}
ST/DF	13.5 ± 1.43	1.84 ± 0.28	0.843 ± 0.242	0.905 ± 0.095
TCS/DF	9.14 ± 0.33	2.00 ± 0.18	0.521 ± 0.070	0.753 ± 0.122
ST/FN	7.59 ± 0.30	1.98 ± 0.18	0.508 ± 0.069	0.686 ± 0.115
TCS/FN	4.52 ± 0.24	1.85 ± 0.17	0.361 ± 0.084	0.541 ± 0.116

Set 2. Parameters for the best fit of the kinetic model depicted in Scheme 1 holding $k_3 = 0$.

	$k_1 \times 10^9$ s^{-1}	$k_2 \times 10^9$ s^{-1}	$k_5 \times 10^9$ s^{-1}	$k_4 \times 10^9$ s^{-1}
ST/DF	13.8 ± 1.42	2.01 ± 0.26	0.829 ± 0.222	0.859 ± 0.097
TCS/DF	9.29 ± 0.31	2.12 ± 0.16	0.469 ± 0.075	0.736 ± 0.127
ST/FN	7.67 ± 0.33	2.11 ± 0.15	0.439 ± 0.077	0.659 ± 0.110
TCS/FN	4.57 ± 0.24	2.00 ± 0.19	0.291 ± 0.087	0.516 ± 0.120

of the SSRIP. The best fit for Set 1 and Set 2 are given in Table 1, and the calculated kinetics based on Set 2 are shown in Figure 4. With either $k_3 = 0$ or $k_5 = 0$, the lifetime of the CRIP ranges from 105 ps to 120 ps. The CRIP decays by either charge recombination ($k_1 = 7.6 \times 10^9$ s^{-1}) or radical ion pair separation ($k_2 = 2.0 \times 10^9$ s^{-1}). Similarly, the lifetime of the SSRIP ranges from 0.68–1.73 ns.

Since it was not possible to determined the relative contributions of k_3 and k_5 to the decay of the SSRIP based upon the quality of the fit to the kinetic model for the experimental data, we have examined how the decay of the SSRIP as well as the CRIP depends upon molecular structure. Thus we have also measured [8] the radical ion pair dynamics following irradiation of the charge transfer complex between *trans*-4-chloro-stilbene/fumaronitrile (TCS/FN), *trans*-stilbene/dimethylfumarate (TS/DF) and *trans*-4-chloro-stilbene/dimethylfumarate (TCS/DF). The experimental data was analyzed with the model depicted in Scheme 1 again with the limiting conditions of $k_5 = 0$ in Set 1 and $k_3 = 0$ in Set 2. The results of the analysis are given in Table 1.

The goal of this study is determined whether the dominate decay pathway for the SSRIP is collapse to the CRIP, k_3, or long range back electron transfer, k_5, in addition to separation to FI, k_4. If it is assumed that $k_3 \gg k_5$ then the appropriate set of rate constants associated with Scheme 1 are given in Table 1, Set 1. Under this assumption the ratio of k_2 to k_3 represents the free

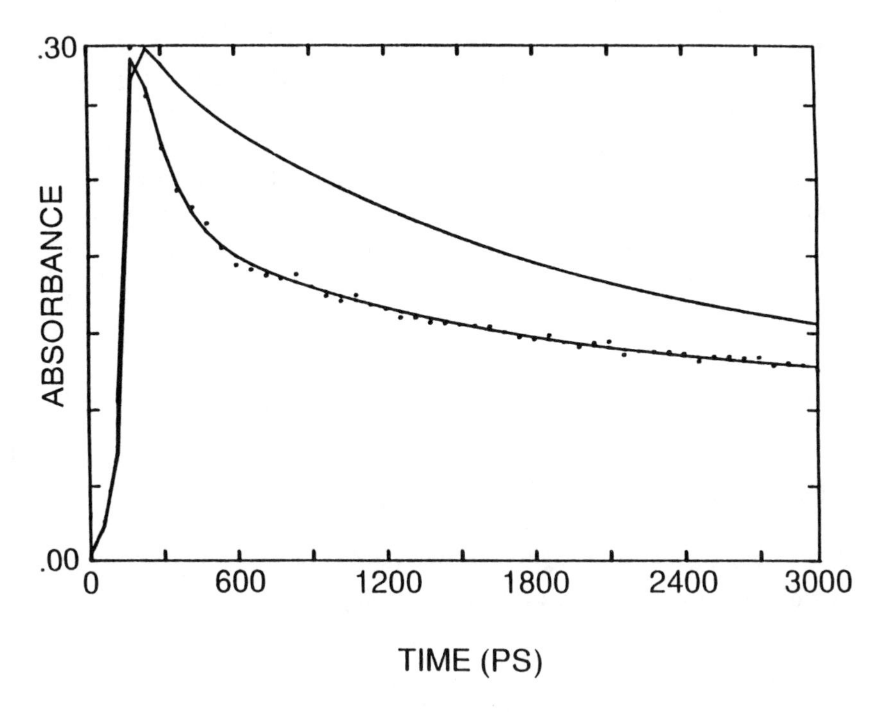

Fig. 5. The dynamics of the *trans*-stilbene radical cation, monitoring at 480 nm, following the 266 nm excitation of 0.6 mM TS and 0.2 M FN in acetonitrile. Points – experimental data, an average of three experiments. Top solid curve – Calculated decay based upon Scheme 2 with $k_6 = 0.0$ and $k_7 = 2.63 \times 10^{10}\ s^{-1}$ and the rate coefficients listed for TS/FN in Table 1, Set 2. Lower solid curve – Calculated decay based upon Scheme 3 with $k_6 = 2.63 \times 10^{10}\ s^{-1}$ and $k_7 = 0.0$ and the rate coefficients for TS/FN listed in Table 1, Set 2.

energy change for CRIP separation to the SSRIP which is –0.46 kcal/mole for ST/DF, –0.79 kcal/mole for TCS/DF, –0.79 kcal/mole for ST/FN and –0.95 kcal/mole for TCS/FN. Although there is virtually nothing known about the free energy change associated with the interconversion of a CRIP into a SSRIP in acetonitrile from experiment, the free energy change can be estimated through theoretical considerations. Weller [10] has derived an expression for the free energy change for radical ion pair interconversion based upon the Kirkwood–Onsager model for the solvation of a dipole and the Born model for solvation of a ion where the solvent is modeled as a dielectric continuum. The free energy change, ΔG in eV, is

$$\Delta G = (\mu^2/\varrho^3)\,((\varepsilon - 1/2\varepsilon + 1) - 0.19) + 2.6\ \text{eV}/\varepsilon - 0.51\ \text{eV} \qquad (7)$$

where μ is the dipole moment of the CRIP, ϱ the radius of the solvation cavity of the CRIP and ε is the static dielectric constant of the solvent. The estimated free energy change for the conversion of a CRIP into a SSRIP in acetonitrile $(\varepsilon = 35)$ is –5 kcal/mole if it is assumed that the cavity radius is 6.5 Å and the dipole moment is 15 Debye, a value determined from exciplex emission from complexes between an aromatic compounds and olefins [10]. The value of –5 kcal/mole is significantly larger than the values determined from k_2/k_3 assuming $k_3 \gg k_5$. Therefore, given discrepancy between the magnitude in the free energy change predicted by theory and Scheme 1 with $k_3 \gg k_5$, it would appear that k_5 is larger than k_3. Thus the dominate pathway for the decay of a SSRIP is through separation to free ions and long range back electron transfer.

Dynamics of the Radical Ion Pair Formed upon Quenching of S_1 by FN

Having characterized the kinetic behavior of the TS/FN CRIP and SSRIP in acetonitrile, the kinetics of the radical ion pair formed upon the quenching of S_1 by FN are probed by monitoring at 480 nm following 266 nm irradiation of TS in the presence of 0.2 M FN, Figure 5. The following kinetic scheme is employed in the analysis of the data displayed in Figure 5.

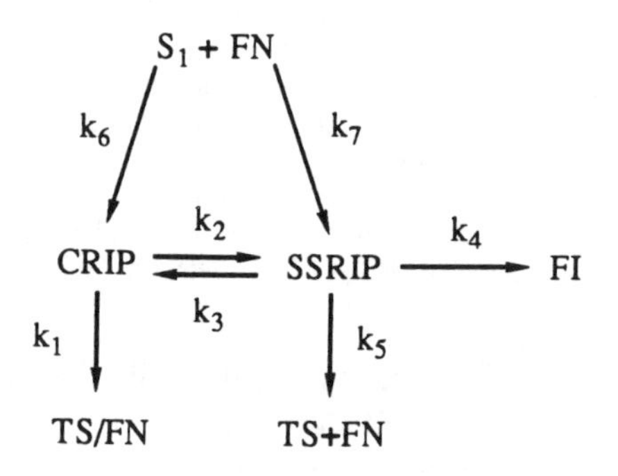

Scheme 2.

The diffusional quenching of S_1 by FN may proceed by two pathways: k_6 to give CRIP and k_7 to give SSRIP. From previous measurements [7], in the presence of 0.2 M FN, S_1 is quenched by electron transfer with a rate constant of $2.63 \pm 0.23 \times 10^{10}\ s^{-1}$. Therefore in the kinetic analysis employing Scheme 2, the sum of k_6 and k_7 was held constant at $2.63 \times 10^{10}\ s^{-1}$. Also during the fitting procedure the remaining rate constants, k_1–k_5, were held constant at the values given in Table 1, Set 2. The only parameters varied were k_6 and k_7. The best fit to the experimental data was for $k_6 = 2.63 \times 10^{10}\ s^{-1}$ and $k_7 = 0$, Figure 5. Based upon analysis of the error in the experiment it is estimated that the diffusional quenching of S_1 by FN leads to greater than 90% CRIP formation.

Conclusions

These experiments reveal that for diffusional quenching of an excited singlet state by electron transfer where the driving force for the reaction is –0.76 eV in acetonitrile, the reaction occurs upon molecular contact producing CRIP. This observation differs from the predominate view of practitioners in the field of photochemical electron transfer reactions [4]. Also, the present study brings into question the validity of deducing radical ion pair structures from the distance parameter contained within the time dependent rate equations, Equation (3), used to analyze reaction dynamics in solution. The reaction distance for the diffusional quenching of S_1 by FN is 8.6 Å which is consistent with the formation of a SSRIP. However, from the kinetic properties of the radical ion pair formed by the diffusional quenching of S_1 by FN, the predominate radical ion pair produced is the CRIP. Clearly great care must be exercised in deducing microscopic properties from theories that treat phenomena in terms of macroscopic parameters such as bulk diffusion coefficients.

Acknowledgement

This work is supported by a grant from the National Science Foundation., CHE 9120355.

References

1. Marcus, R.A., *J. Chem. Phys.* **81**, 167 (1984).
2. Rips, I. and Jortner, J., *J. Chem. Phys.* **87**, 2090 (1987).
3. Knibbe, H., Rehm, D., and Weller, A., *Ber. Bunsenges. Physik Chem.* **72**, 257 (1968).
4. Mataga, N., Shioyama, J., and Kanda, Y., *J. Phys. Chem.* **91**, 314 (1987).
5. Gould, I.R., Young, R.H., Moody, R.E., and Farid, S., *J. Phys. Chem.* **95**, 2068 (1991).
6. Angel, S.A. and Peters, K.S., *J. Phys. Chem.* **93**, 713 (1989).
7. Peters, K.S. and Lee, J., *J. Phys. Chem.* **96**, 8941 (1992).
8. Peters, K.S. and Lee, J., *J. Am. Chem. Soc.* **115**, 3643 (1993).
9. Collins, F.C. and Kimball, G., *J. Colloid Sci.* **4**, 425 (1949).
10. Weller, A., *Zeitschrift für Physik. Chemie Neue Folge* **133**, 93 (1982).

8. The Molecular Basis of Solvent Caging

BENJAMIN J. SCHWARTZ, JASON C. KING and
CHARLES B. HARRIS
Department of Chemistry, University of California at Berkeley, Berkeley, CA 94720, U.S.A.

Introduction

One of the central themes of modern physical chemistry is to elucidate the elementary steps associated with a chemical reaction. While understanding of gas phase reactions has reached a high level of sophistication, many open questions remain concerning reactivity in the liquid phase. Reactant species suffer 10^{13} collisions per second in a room temperature solvent. These collisions can cause energy exchange between the solvent and the reactant, perturb or mix the reactant potential energy surfaces and spatially localize or separate reacting species, all on picosecond or subpicosecond time scales. Thus, ultrafast laser technology provides an outstanding tool for the study of condensed phase reaction dynamics. With ultrafast laser techniques, it will be possible to unveil the fundamental principles of condensed phase reactivity, including a detailed picture of reactive energy flow and the microscopic basis for dissociation quantum yields.

Perhaps the most striking difference between the gas and solution environments lies in the ability of a photodissociating species to become trapped in a "cage" of solvent molecules. Molecules, which dissociate with unit efficiency in the gas phase, can collide with this solvent cage and back react to form the parent species when in solution (illustrated schematically in Figure 1), an effect first proposed by Franck and Rabinowitch in the 1930's [1–3]. Subsequent work by Noyes and others in the 1950's quantified important parameters [4–8] and emphasized three qualitatively different types of dissociation/recombination events [9]. First, the recoiling photofragments could lose enough energy upon their initial collisions with the surrounding solvent molecules to recombine on their ground state potential energy surface without ever leaving the solvent cage. This process is referred to as *primary geminate recombination*, where the word "geminate" denotes the recombination of the initially excited photofragment pair. Second, some of the photodissociating fragment pairs could escape from the solvent cage until they are separated by one or two solvent molecules. These fragments can then diffuse through the solution until they meet again and recombine in a newly formed cage. This process is called *secondary geminate recombination*, or diffusive

J.D. Simon (ed.), Ultrafast Dynamics of Chemical Systems, 235–248.

Gas Phase: Photodissociation

Solution Phase: Photodissociation and Recombination

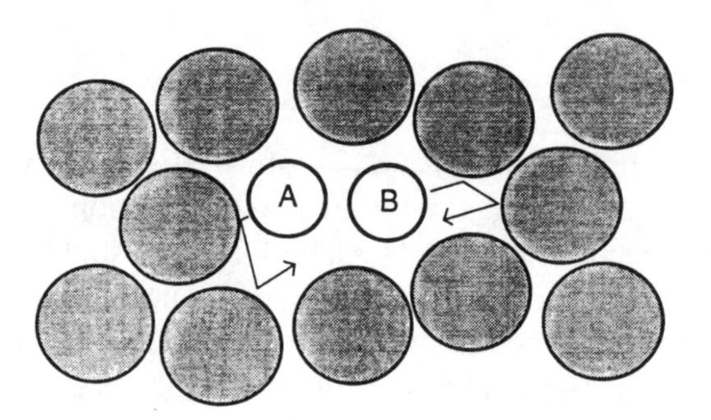

Fig. 1. Schematic illustration of solvent caging showing the difference between gas and solution phase photodissociation reactions.

geminate recombination. Finally, some of the photofragments which escape the initial solvent cage may never recombine geminately (with their original partners). These fragments will diffuse through solution until they recombine with photofragments from neighboring photolyses, a process known as *non-geminate recombination*. As will be discussed in detail below, these three recombination processes are distinguished by the time scales on which they take place. Non-geminate recombination proceeds at a rate controlled by long-range translational diffusion, and takes place on the nanosecond to microsecond time scale in normal, room temperature solvents. Secondary geminate recombination, the diffusive recombination of original partners that have left the solvent cage, takes place in tens to hundreds of picoseconds [10]. Primary geminate recombination is known [11–15] to take place on a picosecond or subpicosecond time scale, and as will be argued below, most likely occurs in just a few hundred femtoseconds, enough time for the fragments to collide only once or twice with the surrounding solvent cage [16].

Recombination Dynamics of Molecular Iodine

The prototypical system for the study of geminate recombination has been the photodissociation of I_2 [1–3, 9–15, 17–20]. Density dependant studies of the I_2 [21–32] (and to a lesser extent Br_2 [32, 33]) photodissociation reaction in compressed rare gases have led to great progress in understanding the microscopic physics underlying non-geminate recombination. Radical recombination rate constants can be predicted for solvent densities ranging from the gas to solution phases [12, 21, 32, 34–37]. Under gas phase conditions, non-geminate recombination is controlled by single molecule behavior and the transfer of energy during isolated binary collisions between the bath molecules and the reacting fragment: the reaction rate is proportional to the concentration of "solvent" molecules [21, 23, 32]. At intermediate densities, the formation of van der Waals clusters dominates the recombination dynamics [38–42, 43, 44]. Diffusion becomes important in the control of recombination kinetics as liquid densities are approached, where recombination rates become roughly inversely proportional to the solvent viscosity [9, 45–47]. At these high densities, however, the role of non-geminate recombination decreases in importance as solvent caging dynamics control much of the microscopic physics of recombination [9, 16, 19, 21, 23, 32].

The importance of solvent caging dynamics in the geminate recombination of I_2 has been the subject of controversy. Early picosecond bleach experiments [48] indicated a $\sim$ 100 ps recovery time, suggesting that geminate recombination in this system is diffusion controlled, i.e., that the recombination is predominantly secondary. These results were modelled theoretically using diffusion [49], modified diffusion [14, 50], Langevin [14] and other descriptions [51]. These conclusions were at odds, however, with the results of early molecular dynamics simulations which indicated that "cage recombination appears to be primarily a direct mechanical effect rather than a diffusive one", and noted "the relative unimportance of truly diffusive effects, involving solvent intervention, under all conditions" [11]. More sophisticated simulations [12–14] also found that geminate recombination was rapid and assigned the longer ground state bleach recovery times to slow vibrational cooling of the newly recombined iodine molecule. This assignment has since been confirmed in transient absorption [19, 52–55] and Raman [56–58] experiments which have shown that most of the geminate recombination is essentially complete in $\leq$ 2ps and therefore is primary. It is likely that some amount of secondary geminate recombination also occurs on a slightly longer time scale in this system [10]. This secondary recombination is difficult to observe with transient spectroscopy, however, due to the slow vibrational relaxation rate and to interfering signals from the vibrationally hot ground state molecular iodine produced via primary recombination [19, 20].

To the best of our knowledge, there have been no direct measurements of the primary geminate recombination time for the molecular iodine system. Recent work by Scherer *et al.* [59] has shown that the predissociation lifetime of I_2 in hexane solution is 200 fs. Since we believe the time scale for primary geminate recombination is also on this order [16], this makes I_2 a poor model system for studies of primary recombination due to the difficulty of distinguishing predissociation dynamics from primary geminate recombination. Molecular dynamics simulations of the I_2 photodissociation in liquid Xe performed by Brown *et al.* [60, 61], however, have modelled I_2 as a directly dissociative system. In these simulations, the recoiling I atoms usually lost enough energy on their initial collisions with the solvent cage to recombine on a very short time scale: most of the geminate recombination occurred within a few hundred femtoseconds after photoexcitation. Those atoms that escaped the solvent cage could eventually undergo recombination with their original partners on a much longer time scale where the rate becomes controlled essentially by diffusion. For iodine, the emerging picture is that most of the geminate recombination is primary and occurs after just a few collisions with the solvent cage, followed by some degree of secondary recombination characterized by diffusive behavior.

Recombination Dynamics of Polyatomics

For photodissociation reactions producing polyatomic fragments, recombination dynamics are expected to be much more complex. Internal vibrational energy redistribution and vibrational cooling of polyatomic radicals takes place on the picosecond time scale [62, 63]; this rapid removal of excess photofragment energy could significantly enhance the probability of geminate recombination at early times. Recombination yields could also be enhanced by the low mobility of large radical fragments which cannot escape the solvent cage as easily as atomic iodine. On the other hand, steric considerations may significantly decrease the probability of pair recombination: simple confinement within the solvent cage may not be sufficient to ensure recombination if a specific orientation is required for the photofragment back reaction. If the recombination reaction is orientation-specific, then a small amount of rotation of one of the photofragments may completely separate reactive sites on fragment species which are still spatially localized, lowering the probability of recombination. Unfortunately, the experimental study of polyatomic geminate recombination reactions is complicated by the ultrafast time scales involved as well as by the problem of overlapping absorptions of the parent and fragment species. Thus, despite the wealth of information these reactions could provide about fundamental processes in condensed phase reactivity, there have been relatively few detailed studies performed.

One of the first direct measurements of polyatomic singlet radical geminate recombination has been in the photolysis of diphenyl disulfide (Ph–S:S–Ph). Upon absorption of a near UV photon, the sulfur-sulfur bond of diphenyl disulfide is cleaved, forming a phenylthiyl (Ph–S·) free radical pair [64, 65]. Since the phenylthiyl radical has an absorption which is not present in the parent molecule, the presence of the phenylthiyl radical can be monitored as a function of time with transient absorption spectroscopy. Experiments with 35 ps time resolution have followed the decay of the Ph–S· absorption after photolysis of Ph–S:S–Ph in a variety of hydrocarbon liquids at different temperatures [66, 67]. The data show a disappearance of the phenylthiyl absorption with time which fits well to a diffusive model of geminate recombination with parameters input from the macroscopic properties (temperature and viscosity) of the solvents. Thus, the roles of fragment vibrational temperature and orientation do not appear to be important in the secondary geminate recombination of diphenyl disulfide: simple translational diffusion is adequate to describe the observed recombination behavior. These experiments did not have sufficient time resolution, however, to examine the role of primary geminate recombination in this system.

Additional picosecond studies of secondary geminate recombination have been performed on azocumene (Ph–CH_2N–NCH_2–Ph) and 3,8-diphenyl-1,2-diaza-1-cyclooctene, which has the same structure as azocumene but with the two benzylic carbon atoms connected by a 4-carbon alkyl chain, forming a "tethered" azocumene [68]. Photoexcitation of both of these molecules cleaves one of the C—N bonds [69] which is followed by rapid loss of N_2 leaving a carbon-centered cumene radical pair [70]. This makes these two compounds unsuitable for study of primary geminate recombination since the multiple steps involved in radical formation may take longer than the primary recombination time. The two cumene (or tethered cumene) radicals produced by photolysis, however, are formed in sufficient time ($\leq$ 10 ps) to provide an excellent system for the study of secondary geminate recombination. As with the phenylthiyl radicals, the recombination kinetics of cumene radicals follow diffusive behavior very nicely. Since the tethered cumene radicals can never separate by diffusion and must eventually recombine, the tethered system shows a different behavior with a pair survival probability which decays exponentially. For both cases, the recombination kinetics depend only on the probability of an encounter between the two radicals; additional orientational and energetic considerations do not appear to be important.

The situation has not been interpreted quite as simply, however, in the photodissociation of tetraphenylhydrazine (Ph_2N–NPh_2). Early experiments failed to find any evidence for diffusive recombination of the diphenylaminyl (DPA) radicals (Ph_2–N·) produced upon photolysis of tetraphenylhydrazine [71]. Subsequent picosecond transient absorption experiments have found evidence for a viscosity dependent recombination of DPA radicals [72]. The pair survival probability for the DPA radicals fit well to the same diffusion model used to describe the recombination of phenylthiyl and cumene radicals discussed above. The question of DPA radical recombination was considered in another way: if solvent dynamics temporarily prevent further separation of the newly formed radicals, they are effectively trapped in shallow potential wells and the recombination reaction can be viewed as a barrier crossing problem. With some standard assumptions, the viscosity dependence of the DPA recombination fits reasonably well to Kramers' theory [73, 74]. Although both the Kramers' and diffusion models describe the data, the effects of radical orientation, not considered by these models, could still be important. Rotational diffusion could separate the N atoms in DPA much more rapidly than translational diffusion which should be slow for the large DPA radical fragments. The problem of geminate recombination of spherical molecules caused by rotational diffusion has recently been addressed [75], and the results fit the tetraphenylhydrazine photodissociation data very well.

All of these studies suggest that for polyatomic photodissociation reactions, diffusion of the photofragment species is the rate-limiting step for secondary geminate recombination. The microscopic processes which lead to recombination after a diffusive encounter of the fragments must take place rapidly enough for the reaction to occur with near certainty before the fragments have time to diffuse apart again. This implies that secondary geminate recombination is controlled primarily by the nature of the solvent, and that the molecular details of specific photodissociation reactions are relatively unimportant. The success of diffusion theories based on continuum hydrodynamics shows that even the molecular nature of the solvent plays little role in the dynamics of secondary geminate recombination.

Microscopic Structure and Recombination

Although the molecular details of the photodissociating molecule and the solvent appear to be unimportant after the photofragments leave the solvent cage and begin diffusional motion, the recombination dynamics immediately following dissociation could be strongly influenced by both the molecular properties of the photofragments and the molecular nature of the surrounding solvent cage. The microscopic details of the nascent dynamics leading to geminate recombination can be better understood with molecular beam cluster techniques, where it is possible to build a solvent shell around a dissociating species one molecule at a time. In this spirit, the photodissociation and recombination dynamics of I_2 in argon clusters [20] and I_2^- in $I_2^- - (CO_2)_n$ clusters, $0 \leq n \leq 22$ [43, 44, 76], have been studied in great detail. Upon absorption of a 720 nm photon, I_2^- photodissociates into $I + I^-$ with unit quantum yield. The addition of up to 5 CO_2 molecules has no effect on the dissociation yield of I_2^-, indicating either the presence of too few molecules for effective caging, or that the "boiling" off of 5 CO_2 molecules does not remove enough energy to allow for recombination on the diiodide ground state potential energy surface. The addition of a sixth CO_2 molecule, however, changes the quantum yield: some parent diiodide anion is reformed after the initial photolysis. The fraction of I_2^- that recombines after photodissociation steadily increases with increasing number of CO_2 molecules in the cluster above 6, and becomes unity (i.e. the dissociation yield becomes 0) for $n \geq 16$. The photofragment mass spectrum shows that most of the recombined diiodide from $I_2^-(CO_2)_n$ clusters appears in $I_2^-(CO_2)_{n-7}$ fragments, indicating that the energetics of CO_2 evaporation are responsible for the lack of observed recombination for $n \leq 5$. The steady increase in recombination yield from $n = 6$ to 16 must then reflect the effectiveness of solvent caging, suggesting that the important factor in determining the recombination yield is simply the probability that the recoiling photofragments strike one of the solvent molecules. The onset of complete caging at $n = 16$, in combination with the unusual stability for formation of this cluster ion, supports the notion that the first solvation shell closes at $I_2^-(CO_2)_{16}$ and that the dissociating photofragments must collide with the surrounding solvent cage.

The photodissociation and recombination dynamics for the $I_2^-(CO_2)_n$ clusters were also studied with picosecond pump-probe spectroscopy [43, 44]. The bleach of the 720 nm I_2^- absorption was found to recover in $\sim$ 10 to 30 ps, depending on cluster size. As with I_2, this "long" bleach recovery time is most likely due to vibrational relaxation, in agreement with the observation of faster recovery times in larger clusters. The most recent set of experiments [44] places an upper limit of 2 ps on the recombination time: the observation of a "bump" in the transient bleach signal has been assigned to coherent motion of the recombined I_2^-, implying that recombination (possibly onto an excited electronic surface) is already complete at a delay of 2 ps. The disappearance of this recurrence in the bleach in smaller clusters may have implications about the structure of the recombined diiodide ion and the solvent shell, including the possibility of formation of an I–I^- solvent-separated pair.

Although pictures of cage structures developed from cluster studies are tantalizing, large differences between the molecular beam and liquid environments demand caution when extrapolating the results of cluster work to reactions which take place in solution. Solution phase studies of the I_2^- photodissociation [77] find that bleach recovery takes place in < 1.3 ps in water and in 5 ps in alcohols, indicative of ultrafast recombination followed by rapid vibrational relaxation. Contrary to the cluster work, there is no evidence for a recurrence in the bleach signal in the solution studies. In the solution case, the recombination/cooling dynamics may be strongly influenced by the nature of the electrostatic interactions between the charged photofragments and the polar solvent. In both the cluster and solution environments, however, the available evidence suggests that the recombination event takes place exceedingly fast.

Primary Geminate Recombination

One of the most important aspects missing from all of the above studies is information concerning the molecular physics underlying primary geminate recombination. The indirect evidence from the I_2 and I_2^- work discussed above suggests that primary recombination takes place in less than a picosecond. Molecular dynamics simulations have indicated that primary recombination occurs very rapidly after dissociation and that diffusion is unimportant [11–13]. Since the nature of the initial collision(s) of the photofragments with the solvent cage is likely to determine the resulting primary recombination dynamics, femtosecond time resolution is essential to the study of primary geminate recombination.

The first direct observation of primary geminate recombination has recently been reported in the photodissociation of methylene iodide (CH_2I_2) in room temperature solutions [16]. Typical transient absorption results for the photodissociation of CH_2I_2 in CCl_4 are displayed in Figure 2a. The data show an instrument limited absorption rise followed by a 350 fs decay with a subsequent, slower rise of 10 ps. Since this absorption measures the presence of the CH_2I radical fragment, the data of Figure 2a can be interpreted in the following manner. The rapid absorption rise indicates that CH_2I is produced within 120 fs, in agreement with resonance Raman data which have led to a calculated 80 fs dissociation time for CH_2I_2 [78]. The slow, 10 ps rise time is typical of cooling times for small molecules in molecular liquids and can be assigned to vibrational relaxation [62, 63]. The 350 fs decay after the initial absorption rise must then correspond to a disappearance of the hot CH_2I radicals due to primary geminate recombination of CH_2I and I reforming the parent methylene iodide molecule. This result implies that most of the geminate recombination is controlled non-stochastically: there is not enough time for equilibrium diffusion to explain the observed dynamics.

This extremely short measured recombination time places severe constraints on the microscopic physics underlying primary geminate recombination. Direct photodissociation of small molecules in the gas phase takes place in 50–200 fs. Resonance Raman data on alkyl iodides show that the early dissociation dynamics do not change significantly between the gas and solution phases [79]. The time scale for a single collision with a solvent molecule in a room temperature liquid is 100–150 fs. If the photofragments need an additional 100-150 fs to recombine on their ground electronic state, the observed recombination time of 350 fs allows for only a single collision of the photofragments with the surrounding solvent. Even if the translationally hot fragments have time to collide more than once with the solvent cage, primary geminate recombination is determined to be an essentially mechan-

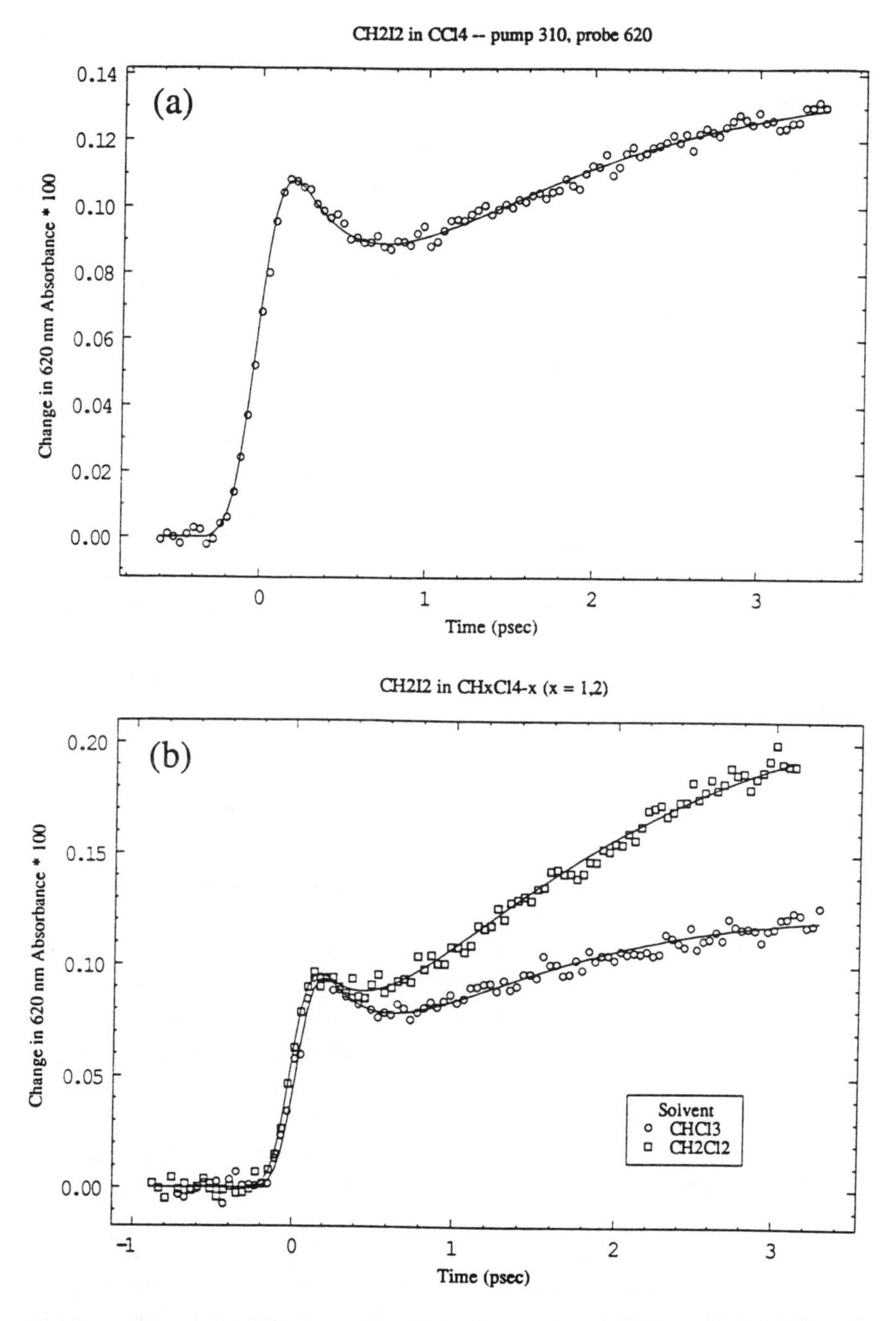

Fig. 2. Femotsecond 620 nm transient absorption spectra of CH_2I_2 excited at 310 nm in different solutions. Symbols are data points, solid lines are fits to the data. All three scans show a fast 350 fs decay indicating dissappearance of CH_2I, the 620 nm absorbing species, and a slower (5–10 ps) rise due to vibrational relaxation. (a) For CCl_4 solutions, the fast 350 fs decay has 0.38 the amplitude of the initial rise. (b) For $CHCl_3$ (circles) and CH_2Cl_2 (squares) solutions, the amplitude ratios of the 350 fs decay to the initial rise are 0.33 and 0.24, respectively. These scans have been scaled to have the same absorbance after the initial rise for better comparison.

ical effect: the first collision of the photofragments with the surrounding molecules controls the resulting recombination dynamics.

If this single collision interpretation of primary geminate recombination holds for the CH_2I_2 dissociation, it should be possible to vary the amount of recombination by changing the structure of the solvent cage without significantly altering the recombination rate. This idea was tested by studying the dissociation dynamics of methylene iodide in $CHCl_3$ and CH_2Cl_2, shown in Figure 2b. If the lighter chloroform and methylene chloride solvent molecules are more easily brushed aside by the recoiling photofragments, a lower primary recombination yield is expected since more fragments should end up escaping the solvent cage. Those atoms that do recombine, however, should do so on the same single cage collision time scale of 350 fs. The results shown in Figure 2b are in agreement with these predictions; the time of the 350 fs decay due to recombination is unchanged between the three solvents, while the amplitude diminishes as the solvent is hydrogenated from CCl_4 to CH_2Cl_2 indicating a decreased primary recombination yield. Secondary recombination may also be occurring in this system, but its presence is likely to be masked under the strong absorption changes due to vibrational cooling.

Nothing in the above argument that primary recombination is dominated by the dynamics of a single collision with the solvent cage depends strongly on the nature of either the photodissociating molecule or the solvent. Thus, the same type of results should be obtained for the solution photolyses of many different molecules. Thus it is interesting to compare the femtosecond dynamics of photodissociating CH_2I_2 (Figure 2) with those of photodissociating $Cr(CO)_6$ (see especially Figure 2 of [80]). The two traces (Figure 2, [80] and Figure 2, above) are remarkably similar. The transient absorption data for chromium hexacarbonyl, however, has been interpreted as follows: the fast initial absorption rise is due to electronically excited $Cr(CO)_6$; this absorption then decays in 350 fs reflecting the predissociation time for $[Cr(CO)_6]^*$ to fragment into $Cr(CO)_5 + CO$ (this interpretation of predissociation cannot be used to explain the CH_2I_2 data since CH_2I_2 is known to directly dissociate within 100 fs [78]). The subsequent absorption changes on the picosecond time scale were assigned to solvation and vibrational cooling of the pentacarbonyl fragment: geminate recombination was not considered. Many studies [81–85] have indicated that the quantum yield for $Cr(CO)_6$ dissociation is ~ 0.7 in simple alkane and alcohol solvents; the 30% of the dissociated molecules which recombine should have a marked effect on the femtosecond transient absorption dynamics. Additional femtosecond studies [86] of the $M(CO)_6$ family (M = Cr, Mo, W) show similar fast absorption decays for all 3 species, indicating that either curve crossing is very rapid, or more likely that the dissociation is direct since pre-dissociation times are expected to be different for the different metal atom centers. This makes the interpretation of the fast decay in the $M(CO)_6$ data as single collision dominated recombination plausible, and supports the idea that the control of primary geminate recombination by the initial collision dynamics after photodissociation is "universal".

Additional photodissociation studies have also provided data which are consistent with this picture of extremely rapid primary recombination. Femtosecond transient absorption studies [16, 87] of photodissociating $M_2(CO)_{10}$ (M = Mn or Re) in solutions show a fast 300–500 fs component, consistent with the interpretation of single collision caging and recombination. Work on photochromic spiropyrans [88], in which the photoexcitation cleaves a ring system leaving the reactive fragments tethered together (similar to the tethered azocumene system [66] discussed above) shows a 200 fs decay which has been interpreted as primary geminate recombination. For this example, the photofragments are confined by both tethering and solvent caging, leading to the slightly higher primary recombination rate. Recent experiments on the photodissociation of I_3^- to I_2^- + I in ethanol solution [89, 90] have found a 400 fs partial recovery of the I_3^- bleach with a corresponding 400 fs decay at the wavelength of I_2^- absorption, indicating possible disappearance of I_2^- and reformation of I_3^- on this time scale. Molecular dynamics simulations of this system support the assignment of this 400 fs decay as primary geminate recombination [91]. In all these system, the decay of the photoproducts on the 200–500 fs time scale supports the picture of primary geminate recombination controlled by the initial collisions of the fragments with the surrounding solvent cage.

Conclusions

One of the fundamental conclusions that emerges from all these studies is that the geminate recombination dynamics following photodissociation reactions are controlled by the properties of the solvent. This is in sharp contrast to the gas phase, where the details of molecular energetics and structure determine the resulting dynamics upon photoexcitation. In very low density (gas phase) "solutions", cooling and recombination dynamics are determined by single collisions of the photofragments with the solvent molecules; the concentration (collision rate) of solvent molecules controls the dynamics. At intermediate densities, the formation of van der Waals' clusters [20, 21, 44] determines the caging and cooling dynamics of photodissociating species. At liquid densities, primary recombination occurs as a mechanical effect: the separating photofragments collide almost immediately with neighboring solvent molecules and rebound back together. Since solvent molecules cannot diffuse appreciably in 300 fs, the static structure of the liquid determines the effective solvent cage and controls the dynamics of primary recombination.

Solvent motions on the tens of picoseconds time scale, on the other hand, are well described by continuum hydrodynamics in which the molecular nature of the solvent is relatively unimportant. This leads to a diffusive description of secondary recombination for those fragments that escape the solvent cage; the dynamics are determined by macroscopic bulk liquid properties such as the temperature and viscosity. Thus, the dynamics of geminate recombination at all time scales are determined primarily by simple properties of the solvent and transcend the molecular details of a given photodissociation reaction.

Acknowledgements

This work was supported by the National Science Foundation. B. J. S. wishes to acknowledge the National Science Foundation and the W. R. Grace and Co. Foundation for their generous support with graduate fellowships.

References

1. Franck J. and Rabinowitch, E., *Trans. Faraday Soc.* **30**, 120 (1934).
2. Rabinowitch E. and Wood, W. C., *Trans. Faraday Soc.* **32**, 547 (1936).
3. Rabinowitch E. and Wood, W. C., *Trans. Faraday Soc.* **32**, 1381 (1936).
4. Zimmerman J. and Noyes, R. M., *J. Chem. Phys.* **18**, 658 (1950).
5. Marshall R. and Davidson, N., *J. Chem. Phys.* **21**, 2086 (1953).
6. Lampe F. W. and Noyes, R. M., *J. Am. Chem. Soc.* **76**, 2140 (1954).
7. Booth D. and Noyes, R. M., *J. Am. Chem. Soc.* **82**, 1868 (1960).
8. Meadows L. F. and Noyes, R. M., *J. Am. Chem. Soc.* **82**, 1872 (1960).
9. Noyes, R. M., *Prog. React. Kinet.* **1**, 129 (1961).
10. Nesbitt D. J. and Hynes, J. T., *J. Chem. Phys.* **77**, 2130 (1982).
11. Bunker D. L. and Jacobson, B. S., *J. Am. Chem. Soc.* **94**, 1843 (1972).
12. Murrell, J. N., Stace, A. J., and Dammel, R., *J. Chem. Soc. Faraday Trans. 2* **74**, 1532 (1978).
13. Bado, P., Berens, P. H., and Wilson, K. R., *Proc. Soc. Photo-Optic. Instrum. Eng.* **322**, 230 (1980).
14. Hynes, J. T., Kapral, R., and Torrie, G. M., *J. Chem. Phys.* **72**, 177 (1980).
15. Smith D. E. and Harris, C. B., *J. Chem. Phys.* **87**, 2709 (1987).
16. Schwartz, B. J., King, J. C., Zhang, J. Z., and Harris, C. B., *Chem Phys. Lett.* **203**, 503 (1993).
17. Bado, P., Berens, P. H., Bergsma, J. P., Coladonato, M. H., Dupuy, C. G., Edelsten, P. M., Kahn, J. D., and Wilson, K. R., *Photochemistry and Photobiology*, A. Zewail, Ed., (Harwood Academic, New York, 1983), p. 615.
18. Chuang, T. J., Hoffman, G. W., and Eisenthal, K. B., *Chem. Phys. Lett.* **25**, 201 (1974).
19. Harris, A. L., Brown, J. K., and Harris, C. B., *Ann. Rev. Phys. Chem.* **39**, 341 (1988).
20. Potter, E. D., Liu, Q., and Zewail, A. H., *Chem. Phys. Lett.* **200**, 605 (1993).
21. Schroeder J. and Troe, J., *Ann. Rev. Phys. Chem.* **38**, 163 (1987).
22. Otto, B., Schroeder, J., and Troe, J., *J. Chem. Phys.* **81**, 202 (1984).
23. Troe, J., *Ann. Rev. Phys. Chem.* **29**, 233 (1978).
24. Hippler, H., Luther, K., and Troe, J., *Ber. Bunsenges. Phys. Chem.* **77**, 1104 (1974).
25. Hippler, H., Luther, K., and Troe, J., *Chem. Phys. Lett.* **16**, 174 (1972).
26. Van den Bergh H. and J. Troe, J., *J. Chem. Phys.* **64**, 736 (1976).
27. Dutoit, J. C., Zellweger, J. M., and Van den Bergh, H., *J. Chem. Phys.* **78**, 1825 (1983).
28. Luther, K., Schroeder, J., Troe, J., and Unterberg, J., *J. of Phys. Chem.* **84**, 3072 (1980).

29. Luther K. and Troe, J., *Chem. Phys. Lett.* **24**, 85 (1974).
30. Zellweger J. M., and Van den Bergh, H., *J. Chem. Phys.* **72**, 5405 (1980).
31. Dupuy C. and Van den Bergh, H., *Chem. Phys. Lett.* **57**, 348 (1978).
32. Hippler, H., Otto, B., Schroeder, H., Schubert, V., and Troe, J., *Ber. Bunsenges. Phys. Chem.* **89**, 240 (1985).
33. Hippler, H., Schubert, V., and Troe, J., *J. Chem. Phys.* **81**, 3931 (1984).
34. Hynes, J. T., *Ann. Rev. Phys. Chem.* **36**, 573 (1985).
35. Zawadski A. G. and Hynes, J. T., *J. Phys. Chem.* **93**, 7031 (1989).
36. Valentini J. J. and Cross, J. B., *J. Chem. Phys.* **77**, 572 (1982).
37. Amar F. G. and Berne, B. J., *J. Phys. Chem.* **88**, 6720 (1984).
38. Bunker D. L. and Davidson, N., *J. Am. Chem. Soc.* **80**, 5090 (1958).
39. Bunker, D. L., *J. Chem. Phys.* **32**, 1001 (1960).
40. Stace A. J. and Murrell, J. N., *Molec. Phys.* **33**, 1 (1977).
41. Stace, A. J., *J. Chem. Soc. Faraday Trans. 2* **75**, 1657 (1979).
42. Amar F. G., and Berne, B. J., *J. Phys. Chem.* **88**, 6720 (1984).
43. Papanikolas, J. M., Gourd, J. R., Levinger, N. E., Ray, D., Vorsa V., and Lineberger, W. C., *J. Phys. Chem.* **95**, 8028 (1991).
44. Papanikolas, J. M., Vorsa, V., Nadal, M. E., Campagnola, P. J., Gord, J. R., and Lineberger, W. C., *J. Chem. Phys.* **97**, 7002 (1992).
45. Hynes, J. T., *The Theory of Chemical Reactions*, M. Baer, Ed., (CRC Press, Boca Raton, 1985).
46. Kapral, R., *Adv. Chem. Phys.* **48**, 71 (1981).
47. Lorand, J. P., *Prog. Inorg. Chem.* **17**, 207 (1972).
48. Chaung, T. J., Hoffman, G. L., and Eisenthal, K. B., *Chem. Phys. Lett.* **25**, 201 (1974).
49. Langhoff, C. A., Moore, B., and DeMeuse, M., *J. Chem. Phys.* **78**, 1191 (1983).
50. Evans G. T. and Fixman, M., *J. Phys. Chem.* **80**, 1544 (1976).
51. Schell M. and Kapral, R., *Chem. Phys. Lett.* **81**, 83 (1981).
52. Bado, P., Dupuy, C., Magde, D., Wilson, K. R., and Malley, M. M., *J. Chem. Phys.* **80**, 5531 (1984).
53. Smith D. E. and Harris, C. B., *J. Chem. Phys.* **87**, 2709 (1987).
54. Berg, M., Harris, A. L., Brown, J. K., and Harris, C. B., *Ultrafast Phenomena IV*, D. Auston and K. Eisenthal, Eds., (Springer-Verlag, New York, 1984) p. 300.
55. Berg, M., Harris, A. L., and Harris, C. B., *Phys. Rev. Lett.* **54**, 951 (1985).
56. Xu, X. B., Lingle Jr., R., Yu, S.-C., Chang, Y. J., and Hopkins, J. B., *J. Chem. Phys.* **92**, 2706 (1990).
57. Lingle Jr., R., Xu, X. B., Yu, S.-C., Chang, Y. J., and Hopkins, J. B., *J. Chem. Phys.* **92**, 4628 (1990).
58. Xu, X. B., Yu, S.-C., Lingle Jr., R., Zhu, H., and Hopkins, J. B., *J. Chem. Phys.* **95**, 2445 (1991).
59. Scherer, N. F., Ziegler, L. D., and Fleming, G. R., *J. Chem. Phys.* **96**, 5544 (1992).
60. Brown, J. K., Harris, C. B., and Tully, J. C., *J. Chem. Phys.* **89**, 6687 (1988).
61. Brown J. K. and Harris, C. B., unpublished results, 1988.
62. Elsaesser T. and Kaiser, W., *Ann. Rev. Phys. Chem.* **42**, 82 (1991).
63. Seilmeier A. and Kaiser, W., *Ultrashort Laser Pulses and Applications*, W. Kaiser Ed. (Springer-Verlag, Berlin, 1988) p. 279.
64. Burkey, T. J., Majewski, M., and Griller, D., *J. Am. Chem. Soc.* **108**, 2218 (1986).
65. Schaafsma, Y., Bickel, A., and Kooyman, E. C., *Tetrahedron* **10**, 76 (1960).
66. Scott T. W. and Liu, S. N., *Ultrafast Phenomena V*, G. R. Fleming and A. E. Siegman, Eds., (Springer-Verlag, Berlin, 1986) p. 338.
67. Scott T. W. and Liu, S. N., *J. Phys. Chem.* **93**, 1393 (1989).
68. Scott T. W. and Doubleday Jr., C., *Chem. Phys. Lett.* **178**, (1991).
69. Nelson S. F. and Bartlett, P. D., *J. Am. Chem. Soc.* **88**, 143 (1966).
70. Skinner, K. J., Hochater, H. S., and McBride, J. M., *J. Am. Chem. Soc.* **96**, 4301 (1974).
71. Anderson R. W. and Hochstrasser, R. M., *J. Phys. Chem.* **80**, 2155 (1974).

72. Hyde M. G. and Beddard, G. S., *Chem. Phys.* **151**, 239 (1991).
73. Kramers, H. A., *Physica* **7**, 284 (1940).
74. Grote R. F. and Hynes, J. T., *J. Chem. Phys.* **73**, 2715 (1980).
75. Beddard G. S. and Masters, A. J., *Chem. Phys. Lett.* **188**, 513 (1992).
76. Ray, D., Levinger, N. E., Papanikolas, J. M., and Lineberger, W. C., *J. Chem. Phys.* **91**, 6533 (1989).
77. Johnson, A. E., Levinger, N. E., and Barbara, P. F., *J. Phys. Chem.* **96**, 7841 (1992).
78. Zhang, J., Heller, E. J., Huber, D., and Imre, D. G., *J. Chem. Phys.* **89**, 3602 (1988).
79. Markel F. and Myers, A. B., *Chem. Phys. Lett.* **167**, 175 (1990).
80. Joly A. G. and Nelson, K. A., *J. Phys. Chem.* **93**, 2876 (1989).
81. Nasielsk J. and Colas, A., *J. Organomet. Chem.* **101**, 215, (1975).
82. Nasielsk J. and Colas, A., *Inorg. Chem.* **17**, 237 (1978).
83. Burdett, J. K., Grzybowski, J. M., Perutz, R. N., Poliakoff, M., Turner, J. J., and Turner, R. F., *Inorg. Chem.* **17**, 147 (1978).
84. Wieland S. and Eldik, R. V., *J. Phys. Chem.* **94**, 5865 (1990).
85. Nayak S. K. and Burkey, T. J., *Organomet.* **10**, 3747 (1991).
86. Joly A. G. and Nelson, K. A., *Chem. Phys.* **152**, 69 (1991).
87. Schwartz, B. J., King, J. C., Zhang, J. Z., and Harris, C. B., unpublished data, 1992.
88. Zhang, J. Z., Schwartz, B. J., King, J. C., and Harris, C. B., *J. Am. Chem. Soc.* **114**, 10921 (1992).
89. Banin, U., Waldman, A., and Ruhman, S., *J. Chem. Phys.* **96**, 2416 (1992).
90. Banin, U., Waldman, A., and Ruhman, S., *Proc. Ultrafast Phenomena. VIII*, Mourou G. A. and Zewail, A. H., Eds., (Springer-Verlag, Berlin, 1992).
91. Benjamin, Ilan, private communication, 1993.

9. Dielectric Continuum Models of Solute/Solvent Interactions

DAVID S. ALAVI
Department of Chemistry and Chemical Physics Institute, University of Oregon, Eugene, OR 97403, U.S.A.

and

DAVID H. WALDECK
Department of Chemistry, University of Pittsburgh, Pittsburgh, PA 15260, U.S.A.

Introduction

Electrostatic interactions between a solute and its environment are quite significant and can significantly modify the solute molecule's dynamics. In the case of chemical reactions, polarity effects of the medium can significantly alter reaction rates and the relative yields of different reaction products. However these interactions are not quantitatively understood for molecular systems. This failing is clearly indicated by the prevalence of semiempirical measures of these important interactions [1]. A focus of recent work from our group has been to better quantitate these important interactions. The methodology has been to first experimentally measure the rotational diffusion of rigid solute compounds in order to empirically determine the friction experienced when a molecule moves on molecular length and time scales. Secondly electrostatic models of the solute/solvent coupling have been explored and compared to the experimentally determined friction coefficients. The result of this effort has been to show the necessity of including the extended nature of the solute molecule's charge distribution when modeling the friction. It has been found that point source models may underestimate the magnitude of the friction by a hundred times for medium sized molecules (a displaced volume of a few hundred cubic angstroms). Furthermore the experimental results show that structural aspects of the solvent medium can be quite important to the solute molecule's dynamics. It is clear however that over a wide regime that continuum models of the solvent work fairly well in describing what is characteristically molecular behaviour. The following discussion focusses on these results in more detail.

J.D. Simon (ed.), Ultrafast Dynamics of Chemical Systems, 249–265.

Onsager Cavity Model

The basic physical model common to most continuum based treatments of dielectric interactions in liquids is the Onsager cavity model [2]. For a real molecular system, the solute molecule is surrounded by a continuously fluctuating bath of solvent molecules. In a cavity model the solute molecule is replaced by a charge distribution inside a cavity whose boundary the solvent is not allowed to penetrate. In this model the bath is replaced by a dielectric continuum which is completely characterized by its frequency and wavevector dependent dielectric constant $\varepsilon(\mathbf{k}, \omega)$. The dynamics of the bath fluctuations are contained in the frequency dependence of $\varepsilon(\mathbf{k}, \omega)$ and the variation in the dielectric response over different length scales is contained in the wavevector dependence of $\varepsilon(\mathbf{k}, \omega)$ [Note that $|k| = 2\pi/\lambda$ where λ is a wavelength]. The properties of interest for the system, such as solvation energy, solvation relaxation times, and rotational dielectric friction coefficients are calculated from the electrostatic potential inside the cavity. The procedure involves a standard boundary value calculation [3].

First, let the electrostatic potential for the solute charge distribution in vacuum be Φ_c. When this distribution is placed inside the cavity in the dielectric medium, its electric field polarizes the surrounding dielectric. This polarization gives rise to a polarization charge density at the cavity boundary, which acts as an additional source of electrostatic potential. This additional potential within the cavity is called the reaction potential Φ_{rxn}. The interaction between the solute charge distribution and the reaction potential accounts for the solute/solvent properties alluded to above.

The boundary value calculation is carried out as follows. The potential inside the cavity is $\Phi_{\text{in}} = \Phi_c + \Phi_{\text{rxn}}$. The potential outside is Φ_{out}. The appropriate boundary conditions are

$$\varepsilon_{\text{in}} \frac{\partial \Phi_{\text{in}}}{\partial n} = \varepsilon_{\text{out}} \frac{\partial \Phi_{\text{out}}}{\partial n}; \qquad \frac{\partial \Phi_{\text{in}}}{\partial t} = \frac{\partial \Phi_{\text{out}}}{\partial t} \tag{1}$$

where derivatives with respect to n and t are normal and tangential to the boundary surface, respectively, and are evaluated at the cavity boundary [3]. It should be pointed out that the dielectric constants used in (1) are the static (i.e., zero frequency) dielectric constants, since the charge distribution is stationary. Inclusion of the wavevector dependence of the medium adds an extra boundary condition, in which the nonlocal part of the polarization must vanish at the boundary [4]. The potential can be expressed as an eigenfunction expansion over an appropriate set of surface harmonics, determined by the symmetry of the cavity chosen to represent the molecule. For a given charge distribution $\varrho(\mathbf{x})$, which determines $\Phi_c(\mathbf{x})$, these boundary conditions are sufficient to determine $\Phi_{\text{rxn}}(\mathbf{x})$.

Once this static reaction potential has been obtained, it can be used to calculate various solute/solvent interactions. The equilibrium solvation energy of the solute in the solvent relative to the solute in vacuum is [3]

$$\frac{1}{2}\int d^3x\rho(\mathbf{x})\Phi_{\mathrm{rxn}}(\mathbf{x}) \tag{2}$$

where the integral is taken over the interior of the cavity. To calculate relaxation times for nonequilibrium solvation, one must consider the frequency dependence of ε in (2), which yields a frequency dependent solvation energy. A Fourier-Laplace transform gives a time dependent response function whose decay time is the solvation relaxation time τ_S.

Calculation of a rotational dielectric friction coefficient is somewhat more complicated. If the charges within the cavity move, this produces a time varying electric field in the surrounding dielectric. Because the dielectric response is not instantaneous, energy can be dissipated in the medium and the rotating solute experiences friction. This retarding frictional torque can be calculated from the reaction potential and the frequency dependence of the dielectric medium as follows. With the solute rotating at angular frequency ω, the various components of the reaction potential vary at ω or its harmonics. The amplitude and temporal lag of the polarization field in the dielectric can be determined from $\varepsilon(\mathbf{k},\omega)$. The spatial angular displacement between the charge distribution and its equilibrium position, relative to its reaction potential, is related to the temporal lag and hence may be used to calculate the retarding torque.

These various calculations are common to most continuum models of solute/solvent dielectric interactions. For any particular model, two things must be determined. Firstly, the form of the frequency and wavevector dependent dielectric constant, $\varepsilon(\mathbf{k},\omega)$, must be chosen for the solvent. Secondly, a cavity geometry and charge distribution must be chosen to model the solute.

The frequency dependent behavior of ε has been addressed by many studies, both experimental and theoretical, and is fairly well characterized [5]. The most commonly chosen form for the temporal dielectric response function is a single exponential decay (Debye relaxation). The resulting expression for $\varepsilon(\omega)$ is

$$\varepsilon(\omega) = \varepsilon_\infty + \frac{\varepsilon_s - \varepsilon_\infty}{1 + i\omega\tau_D} \tag{3}$$

where ε_s and ε_∞ are the static and high frequency dielectric constants, respectively, and τ_D is the dielectric relaxation time.

The wavevector dependence of ε is less well understood, although recent activity is focussing on the nature of this dependence [6]. Clearly, however, $\varepsilon(|k| = 0)$ must be the static dielectric constant ε_s, and the magnitude of the dielectric constant decreases as k becomes large. The transition from low wavevector to high wavevector behavior occurs where k corresponds to some characteristic length scale of microscopic structure in the solvent. For nonassociated liquids, such as a polar aprotic solvent, this length scale might correspond to the size of a single solute molecule, since the orientational polarization could not respond to field variations on smaller scales. For associated liquids, such as hydrogen bonded solvents, this characteristic length scale might be the average size of complexes of multiple solvent molecules. At present, the correct form for $\varepsilon(\mathbf{k})$ is not known. A simplistic model which has been used with success at a qualitative level is a step function [7], with $\varepsilon(\mathbf{k}) = \varepsilon_s$ for $|k| \leq |k_{\text{max}}|$ and $\varepsilon(\mathbf{k}) = \varepsilon_\infty$ for $|k| > |k_{\text{max}}|$, where $|k_{\text{max}}|$ corresponds to the characteristic solvent structure length scale. An additional complication arises when one considers the interaction between temporal and spatial aspects of the solvent dielectric response. It is commonly assumed that the two are independent, so that the temporal response is the same regardless of the length scale. This assumption may not be justified, particularly if translational motion of solvent molecules contributes significantly to the solvent dynamics.

The cavity geometry and charge distribution are the other main features of any model of solute-solvent dielectric interactions. A single point source at the center of a spherical cavity has commonly been used [8], but recently extended distributions of point charges have been developed [9]. It has been shown that models employing extended sources are generally superior to those using a single point source. Expressions for spheroidal and ellipsoidal cavities have also been exhibited.

In the following sections, the explicit forms for solvation energies and rotational dielectric friction coefficients are given for both point sources and extended sources. In all of the formulae ε_{in}, the dielectric constant inside the cavity, and ε_∞, the high frequency dielectric constant, are taken to be unity. Furthermore a spherical cavity is used so that the potential may be expressed as an expansion over spherical harmonics. Lastly, a step function form is chosen for $\varepsilon(\mathbf{k})$. In terms of the spherical harmonic expansion, the k dependence of ε translates into a dependence on L, the spherical harmonic index. This dependence will also be a step function, with $\varepsilon(L) = \varepsilon_s$ for $L \leq L_{\text{max}}$ and $\varepsilon(L) = 1$ for $L > L_{\text{max}}$.

Point Sources

The simplest dielectric continuum models replace the solute with a single point source at the center of a spherical cavity [8]. A single point charge can be used to represent an ionic solute, while neutral solutes are represented by a point dipole. In some cases asymmetric ionic solutes are also modeled by point dipoles. However, this is somewhat ambiguous, since the dipole moment of a charged particle depends on the choice of origin, and is therefore not well defined.

For a single point charge q located at the center of a cavity of radius a, the potentials within the cavity are

$$\Phi_c(r, \theta, \phi) = \frac{q}{r}; \quad \Phi_{\text{rxn}}(r, \theta, \phi) = \frac{-q}{a}\left(\frac{\varepsilon_s - 1}{\varepsilon_s}\right). \tag{4}$$

Using these expressions, along with (2), one obtains an expression for the solvation energy of the ion

$$E_{\text{solv}} = \frac{-q^2}{2a}\left(\frac{\varepsilon_s - 1}{\varepsilon_s}\right). \tag{5}$$

Since the reaction potential and charge distribution for the ion are spherically symmetric, no rotational dielectric friction can result.

For a point dipole μ directed in the $+z$ direction and located at the center of a cavity of radius a, the potentials within the cavity are

$$\Phi_c(r, \theta, \phi) = \frac{\mu \cos\theta}{r^2}; \quad \Phi_{\text{rxn}}(r, \theta, \phi) = \frac{-2\mu}{a^3}\left(\frac{\varepsilon_s - 1}{2\varepsilon_s + 1}\right) r \cos\theta. \tag{6}$$

Once again, these expressions along with (2) yield the solvation energy for the dipole

$$E_{\text{solv}} = \frac{-\mu^2}{a^3}\left(\frac{\varepsilon_s - 1}{2\varepsilon_s + 1}\right). \tag{7}$$

The reaction potential given in (6) is not uniform, and therefore the dipole will experience dielectric friction as it rotates within the cavity about a perpendicular axis. The negative gradient of the reaction potential is the reaction field **R**, given by

$$\mathbf{R} = \frac{2\mu}{a^3}\left(\frac{\varepsilon_s - 1}{2\varepsilon_s + 1}\right). \tag{8}$$

As the dipole rotates at angular frequency ω, the dielectric medium experiences a sinusoidally varying potential. The reaction field also oscillates at ω, but because of the frequency dependence of the solvent dielectric response, it lags behind the dipole by some temporal phase angle $\phi(\omega)$. This phase angle is given by

$$\phi(\omega) = \left|\text{phase}\left(\frac{\varepsilon(\omega) - 1}{2\varepsilon(\omega) + 1}\right)\right| \tag{9}$$

and will equal the geometric angle between the dipole and the reaction field. The magnitude of the reaction field also changes, which is reflected by the following substitution into (8):

$$\left(\frac{\varepsilon_s - 1}{2\varepsilon_s + 1}\right) \rightarrow \left|\frac{\varepsilon(\omega) - 1}{2\varepsilon(\omega) + 1}\right|. \tag{10}$$

This modification takes into account the change in amplitude of the dielectric response of the medium as a function of frequency. Since the reaction field and the dipole are no longer parallel, a torque develops and opposes the rotation of the dipole. This torque is the physical coupling which leads to energy dissipation in the medium, and hence dielectric friction.

The torque on the rotating dipole is given by the product of the dipole magnitude, reaction field magnitude, and the sine of the lag angle $\phi(\omega)$:

$$\begin{aligned} |\,\mathbf{N}(\omega)\,| &= \mu\,|\,R(\omega)\,|\ \sin\phi(\omega) \\ &= \frac{2\mu^2}{a^3}\left|\frac{\varepsilon(\omega) - 1}{2\varepsilon(\omega) + 1}\right| \sin\phi(\omega) \\ &= \frac{2\mu^2}{a^3}\left|\mathrm{Im}\left(\frac{\varepsilon(\omega) - 1}{2\varepsilon(\omega) + 1}\right)\right|. \end{aligned} \tag{11}$$

The frequency dependent friction coefficient $\xi(\omega)$ is defined as the torque divided by the angular frequency ω:

$$\begin{aligned} \xi(\omega) &\equiv \left|\frac{\mathbf{N}(\omega)}{\omega}\right| \\ &= \frac{2\mu^2}{a^3}\left|\frac{1}{\omega}\mathrm{Im}\left(\frac{\varepsilon(\omega) - 1}{2\,\varepsilon(\omega) + 1}\right)\right|. \end{aligned} \tag{12}$$

This expression is the general form for the frequency dependent dielectric friction coefficient on a rotating point dipole. If one assumes the Debye form for $\varepsilon(\omega)$ and takes the limit of zero frequency,

$$\xi = \lim_{\omega \to 0} \xi(\omega) = \frac{6\mu^2}{a^3}\frac{\varepsilon_s - 1}{(2\varepsilon_s + 1)^2}\tau_D \tag{13}$$

is obtained for the rotational dielectric friction coefficient [8, 9]. This expression is valid for a point dipole when the dielectric response of the medium is much faster than the characteristic rotational relaxation time of the solute, and yields rotational correlation functions which are single exponential decays. When the solvent dielectric relaxation is slow, the full frequency dependent friction coefficient must be used, and the rotational dynamics may not be single exponential.

Extended Sources

A significant improvement over models employing single point sources are those using extended sources. The simplest extended source is a rigid collection of point charges, although the equivalent expressions for a continuous charge distribution could be easily produced.

For a collection of N point charges q_i at position vectors $\mathbf{r}_i(= r_i, \theta_i, \phi_i)$ within a spherical cavity of radius a, the potentials inside the cavity are

$$\begin{aligned}\Phi_c(r, \theta\phi) &= \sum_{i=1}^{N} \frac{q_i}{\mid \mathbf{r} - \mathbf{r}_i \mid} \\ \Phi_{rxn}(r, \theta, \phi) &= \sum_{i=1}^{N} \sum_{L=0}^{L_{\max}} \sum_{M=-L}^{L} \frac{-4\pi}{2L+1} \frac{q_i}{a} \left(\frac{\varepsilon_s - 1}{\varepsilon_s + \frac{L}{L+1}} \right) \\ &\times \left(\frac{r_i}{a}\right)^L \left(\frac{r}{a}\right)^L Y^*_{LM}(\theta_i, \phi_i) Y_{LM}(\theta, \phi). \end{aligned} \tag{14}$$

Inserting the expression for Φ_{rxn} into (2) yields the solvation energy for the charge distribution

$$\begin{aligned}E_{solv} &= \sum_{i=1}^{N} \sum_{j=1}^{N} \sum_{L=0}^{L_{\max}} \sum_{M=-L}^{L} \frac{-2\pi}{2L+1} \frac{q_i q_j}{a} \left(\frac{\varepsilon_s - 1}{\varepsilon_s + \frac{L}{L+1}} \right) \\ &\times \left(\frac{r_i}{a}\right)^L \left(\frac{r_j}{a}\right)^L Y^*_{LM}(\theta_i, \phi_i) Y_{LM}(\theta_j, \phi_j). \end{aligned} \tag{15}$$

The $L = 0$ term in (15) is the same as (5), with the single charge being replaced by the total charge of the distribution. The $L = 1$ term is the same as (7), with the point dipole replaced by the net dipole moment of the charge distribution.

Since the reaction potential for the charge distribution is not uniform, the solute molecule will experience a frictional drag as it rotates. When the charge distribution rotates about the z axis at angular frequency ω, the LM^{th} component of the electric field applied to the dielectric also oscillates at $M\omega$, and the magnitude and phase lag of the reaction potential depend not only on the frequency but also on the indices L and M. This dependence results in a reaction potential that not only lags behind the charge distribution but is also distorted in shape. Once again the the temporal phase angle can be related to a geometric angle. If the charges are rotating so that the angular velocity vector points in the $+z$ direction, then the torque exerted on q_j by the LM^{th} reaction field component must be computed with the field evaluated at $\phi_j + \phi_L(M\omega)$, where $\phi_L(M\omega)$ is a positive angle given by

$$\phi_L(M\omega) = \left| \frac{1}{M} \text{ phase} \left(\frac{\varepsilon(M\omega) - 1}{\varepsilon(M\omega) + \frac{L}{L+1}} \right) \right| \tag{16}$$

and with the field magnitude scaled by the substitution

$$\left(\frac{\varepsilon_s - 1}{\varepsilon_s + \frac{L}{L+1}}\right) \rightarrow \left|\frac{\varepsilon(M\omega) - 1}{\varepsilon(M\omega) + \frac{L}{L+1}}\right|. \tag{17}$$

The friction from the LM^{th} component of the reaction potential as the charge distribution rotates about the z axis is

$$\begin{aligned}
\xi_{LM}(\omega) &= \frac{-[N_Z(\omega)]_{LM}}{\omega_Z} \\
&= \sum_{j=1}^{N} \frac{-q_j r_j \sin\theta_j}{\omega} \left[E_\phi(r_j, \theta_j, \phi_j + \phi_L(M\omega))\right]_{LM} \\
&= \sum_{J=1}^{N} \frac{q_j}{\omega} \left[\frac{\partial}{\partial\phi} \Phi_{\text{rxn}}(r_j, \theta_j, \phi_j + \phi_L(M\omega))\right]_{LM} \\
&= \sum_{i=1}^{N}\sum_{j=1}^{N} \frac{4\pi}{2L+1} \left|\frac{\varepsilon(M\omega) - 1}{\varepsilon(M\omega) + \frac{L}{L+1}}\right| \frac{-q_i q_j}{a\omega} \left(\frac{r_i}{a}\right)^L \left(\frac{r_j}{a}\right)^L \\
&\quad \times iMY^*_{LM}(\theta_i, \phi_i) Y_{LM}(\theta_j, \phi_j + \phi_L(M\omega))
\end{aligned} \tag{18}$$

where E_ϕ is the component of the reaction field in the ϕ direction, and the substitution from (17) has been used. Summation over L and M followed by some algebraic simplification of the spherical harmonics yields a general expression for the frequency dependent dielectric friction coefficient for rotation of the charge distribution about the z axis [9]:

$$\begin{aligned}
\xi(\omega) &= \sum_{j=1}^{N}\sum_{i=1}^{N}\sum_{L=1}^{L_{\max}}\sum_{M=1}^{L} \frac{2q_i q_j}{a\omega} \left|\text{Im}\left(\frac{\varepsilon(M\omega) - 1}{\varepsilon(M\omega) + \frac{L}{L+1}}\right)\right| \frac{(L-M)!}{(L+M)!} M \\
&\quad \times \left(\frac{r_i}{a}\right)^L \left(\frac{r_j}{a}\right)^L P_L^M(\cos\theta_i) P_L^M(\cos\theta_j) \cos M\phi_{ji}.
\end{aligned} \tag{19}$$

where $\phi_{ji} \equiv \phi_j - \phi_i$.

Assuming Debye relaxation for $\varepsilon(\omega)$ and taking the limit of zero frequency, one finds

$$\frac{1}{\omega}\left|\text{Im}\left(\frac{\varepsilon(M\omega) - 1}{\varepsilon(M\omega) + \frac{L}{L+1}}\right)\right| \rightarrow \left(\frac{2L+1}{L+1}\right) \frac{\varepsilon_s - 1}{\left(\varepsilon_s + \frac{L}{L+1}\right)^2} M\tau_D. \tag{20}$$

One final approximation is required to obtain an expression similar to the point dipole expression given in (13). The expression $[\varepsilon_S + (L/(L+1))]$ in the denominator of (19) has values ranging from $[\varepsilon_S + 1/2]$ for $L = 1$ up to $[\varepsilon_S + 1]$ for large L. For polar solvents with fairly large dielectric constants ($\varepsilon_S \geq 10$),

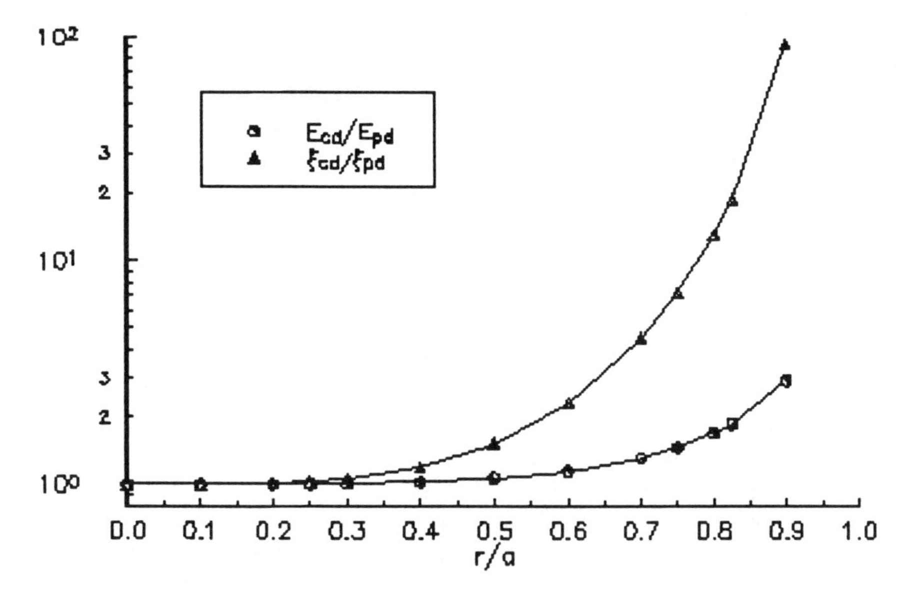

Fig. 1. This figure plots the solvation energy and dielectric friction ratios for an extended dipole versus the distance of the charges from the cavity boundary (r/a). The values of the energy and the friction are normalized to that of a point dipole.

the change is $\leq 5\%$. It seems reasonable to replace $[\varepsilon_S + (L/(L+1))]$ with $[\varepsilon_S+1/2]$. In this limit, the final form for the dielectric friction coefficient for rotation about the z axis is [9]:

$$\xi = \frac{(\varepsilon_s - 1)}{(2\varepsilon_s + 1)^2}\tau_D \sum_{j=1}^{N}\sum_{i=1}^{N}\sum_{L=1}^{L_{\max}}\sum_{M=1}^{L}\left(\frac{2L+1}{L+1}\right)\frac{(L-M)!}{(L+M)!}M^2 \times\frac{8q_iq_j}{a}\left(\frac{r_i}{a}\right)^L\left(\frac{r_j}{a}\right)^L P_L^M(\cos\theta_i)P_L^M(\cos\theta_j)\cos M\phi_{ji}. \quad (21)$$

This expression for the friction has the same dependence on the dielectric properties of the solvent as the expression for a point dipole, but a very different dependence on the electrical properties of the solute. In fact, the $L = 1$ term in (21) is nearly the same as the point dipole result (with μ^2 replaced by $\mu_x^2 + \mu_y^2$, assuming rotation about the z axis).

Some sample calculations were performed on two simple model systems to illustrate the quantitative differences between the calculated solvation energies and friction coefficients for point dipoles and extended charge distributions. The first test system had two charges $\pm q$ placed symmetrically on the x axis within a 5 Å cavity, simulating a neutral dipole rotating about the z axis. The charge and distance were varied so the net dipole moment remained constant. The other test system studied was a single point charge $q = 1e$ at

TABLE I

Solvation energy and dielectric friction on a neutral dipole (μ = 2qr = 4.8 D, a = 5 Å).

r (Å)	q (e)	$\frac{-E_{cd}(\epsilon+1/2)}{(\epsilon-1)}$ (10^{-14} erg)	E_{cd}/E_{pd}	$\frac{\xi_{cd}(2\epsilon+1)^2}{(\epsilon-1)\tau_D}$ (10^{-12} erg)	ξ_{cd}/ξ_{pd}
0.0	∞	9.216	1	1.105	1
0.5	1.0	9.216	1.000	1.106	1.001
1.0	0.5	9.230	1.002	1.118	1.012
1.25	0.4	9.252	1.004	1.136	1.028
1.5	0.3333	9.291	1.008	1.170	1.059
2.0	0.25	9.458	1.026	1.318	1.193
2.5	0.2	9.830	1.067	1.679	1.519
3.0	0.1666	10.58	1.148	2.555	2.312
3.5	0.1429	12.12	1.315	4.989	4.515
3.75	0.1333	13.48	1.463	7.959	7.203
4.0	0.125	15.60	1.693	14.47	13.10
4.125	0.1212	17.16	1.862	20.91	18.92
4.5	0.1111	26.79	2.907	102.4	92.67

$x = r$, with the friction calculated for rotation about z within a 5 Å cavity. For the point dipole model we assume $\mu = qr$. This sort of charge distribution has frequently been used to model polyatomic ions with a charge localized on one atomic center. The numerical results are listed in Tables 1 and 2 and plotted in Figure 1. It is apparent in both systems that the ratio r/a is very important. The point dipole expression is adequate for $r/a \leq 0.1$. But as r/a increases, the solvation energies from (15) exceed that of the equivalent point dipole by factors of 3 to 5, and the friction calculated by (21) exceeds that of a point dipole by up to two orders of magnitude for $r/a = 0.9$.

The physical basis for this sensitivity to r/a can be readily understood. The electric field of a point charge varies as $1/r^2$, so the strength of the electric field polarizing the dielectric increases dramatically as the charge is brought closer to the boundary of the cavity. In turn the degree of coupling between the dielectric and the charge is increasing, which results in a larger solvation energy and a larger friction coefficient. The friction coefficient increases more than the solvation energy because it also depends on the characteristic dielectric relaxation time. For the extended charge distribution the amount of phase lag depends strongly on the multipole moment (in particular the value of M). This additional effect causes the friction to increase more rapidly than does the solvation energy. It is clear that when calculating these quantities

TABLE II

Solvation energy and dielectric friction on an ion (q =1e, a = 5Å). $E_q(\epsilon + 1/2)/(\epsilon - 1) = -230.4 \times 10^{-14}$ erg is the $L = 0$ term, and has been subtracted from E_{cd} to allow a more direct comparison.

r (Å)	μ (D)	$\frac{-E_{pd}(\epsilon+1/2)}{\epsilon-1}$ (10^{-14} erg)	$\frac{-(E_{cd}-E_q)(\epsilon+1/2)}{(\epsilon-1)}$ (10^{-14} erg)	E_{cd}/E_{pd}
0.5	2.4	2.304	2.327	1.010
1.0	4.8	9.216	9.600	1.042
1.25	5.0	14.40	15.36	1.067
1.5	7.2	20.74	22.78	1.098
2.0	9.6	36.86	43.88	1.190
2.5	12.0	57.60	76.80	1.333
3.0	14.4	82.94	129.6	1.563
3.5	16.8	112.9	221.3	1.960
3.75	18.0	129.6	296.2	2.285
4.0	19.2	147.5	409.6	2.777
4.125	19.8	156.8	491.0	3.131
4.5	21.6	186.6	982.2	5.264
r (Å)	μ (D)	$\frac{\xi_{pd}(2\epsilon+1)^2}{(\epsilon-1)\tau_D}$ (10^{-12} erg)	$\frac{\xi_{cd}(2\epsilon+1)^2}{(\epsilon-1)\tau_D}$ (10^{-12} erg)	ξ_{cd}/ξ_{pd}
0.5	2.4	.2765	.2858	1.034
1.0	4.8	1.106	1.266	1.0145
1.25	6.0	1.728	2.140	1.238
1.5	7.2	2.488	3.401	1.367
2.0	9.6	4.424	7.861	1.777
2.5	12.0	6.912	17.74	2.567
3.0	14.4	9.953	42.52	4.272
3.5	16.8	13.55	118.8	8.768
3.75	18.0	15.55	220.5	14.18
4.0	19.2	17.69	460.1	26.01
4.125	19.8	18.82	708.7	37.66
4.5	21.6	22.39	4145	185.1

for a molecule in liquid solution, replacing the actual charge distribution by a point charge or point dipole located at the center of the molecule can lead to significant underestimations. It is the amount of charge close to the boundary of the cavity which is of primary importance in determining these dielectric solute/solvent interactions.

Selected Experimental Results

As stated in the Introduction, our work has involved measurement of rotational diffusion times for a variety of solute/polar solvent combinations as a function of temperature. In the experiments, the orientational correlation function $r(t)$, which has a correlation time τ_{or}, is directly measured. This relaxation time is related to the rotational friction through the Einstein relation, $\tau_{\mathrm{or}} = \xi/6\,kT$. The total friction will typically have multiple sources, which must be separated in order to evaluate them individually. Besides dielectric friction, the other main source of friction is of "mechanical" origin, which is modeled using hydrodynamics with a slip boundary condition, so that $\tau_{\mathrm{hyd}} \equiv \xi_{\mathrm{hyd}}/6\,kT$ [10]. The remainder of the friction is attributed to dielectric friction. Hence one can write [7]

$$\tau_{\mathrm{or}} - \tau_{\mathrm{slip}} = B\frac{\varepsilon_s - 1}{(2\varepsilon + 1)^2}\frac{\tau_D}{T}. \tag{22}$$

By plotting the data as $\tau_{\mathrm{or}} - \tau_{\mathrm{hyd}}$ vs. $(\varepsilon_s - 1)\tau_D/[(2\varepsilon_s + 1)2T]$, the slope can be modelled using (13) or (21). For the point dipole theory, an effective dipole moment can be calculated which would account for the observed friction. For the charge distribution theory, a model charge distribution is obtained from a quantum mechanical calculation of the electronic structure, and the cavity radius and L_{max} are adjusted to match the measured friction.

The group of solutes studied were three phenoxazine dyes (Figure 2). These were mechanically similar, but electronically distinct. Resorufin is an anion, oxazine a cation, and resorufamine is a polar neutral molecule. Reorientation times were measured for all three solutes as a function of temperature in dimethyl sulfoxide (DMSO, polar aprotic) [9, 11] and isopropanol (iPrOH, hydrogen bonding) [7]. Oxazine was also studied in two sets of binary solvent mixtures, n-propanol/water (nPrOH/water) and DMSO/water [12]. The results of the data analysis are shown in Table 3.

From the data, it is clear that point source models fail even at a qualitative level. For a point charge the magnitude of the friction should not depend on the sign of the charge, in a continuum model. However, the experiments show that the cation rotates more slowly than the anion (approximately three times). Furthermore, a point source model predicts no change in the magnitude of the dielectric friction with change of solvent, other than that given by changes in ε and τ_D. In contrast, the order of the friction for the solute molecules changes with a change in the solvent. In DMSO, oxazine experiences the most friction and resorufin the least friction, whereas in n-propanol, oxazine experiences the most friction but resorufamine experiences the least friction. In contrast to the point source models, the extended charge distribution model predicts these changes in an accurate manner.

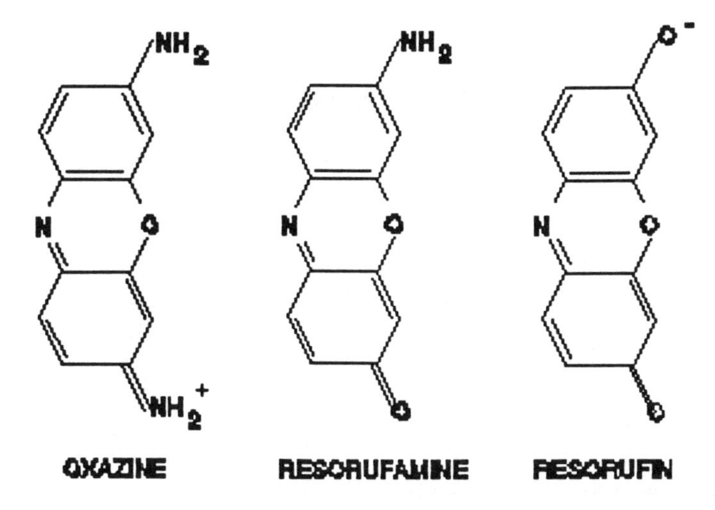

Fig. 2. Chemical structures and physical parameters are shown for the phenoxazine dyes used in the rotational diffusion studies.

It is apparent that the dipole moments required to produce the observed dielectric friction are much too large in all the solvents, and therefore the point dipole expression underestimates the friction by as much as two orders of magnitude. The charge distribution does a much better job accounting for the observed friction in DMSO, the polar aprotic solvent. $L_{\max} = 25$ was chosen based on the relative sizes of the solute and single solvent molecules, but in fact the series has nearly converged by that point. The cavity radii for all three solutes are very near the length of the major axis of the ellipsoid which best fits the molecules.

In isopropanol the charge distribution must be cutoff between $L_{\max} = 2$ and $L_{\max} = 4$, which indicates that the characteristic length scale for the solvent must be larger than that in DMSO. The dielectric relaxation behavior of DMSO is known to arise from reorientations of single polar molecules. The dielectric relaxation of water and the alcohols, on the other hand, is dominated by the intermolecular dynamics of large hydrogen bonded complexes, which is consistent with the observed dielectric friction.

The $L_{\max}$ required to reproduce the experimental slopes for the binary solvent mixtures varied with solvent composition in a manner consistent with what is known of their dielectric relaxation behaviour. For the

TABLE III

Results of experimental data analysis.

Solute	Solvent	meas'd slope (10^4K)	μ_{eff} (D)	calc'd slope (10^4K)	L	a (Å)
resorufin (S_0)	DMSO	10.9	26	10.9	25	6.69
resorufamine (S_0)	DMSO	31.1	44	31.1	25	6.30
oxazine (S_0)	DMSO	67.6	65	68.3	25	6.32
resorufin (S_0)	iPrOH	1.99	11	1.17	2	6.69
				2.77	3	
resorufamine (S_0)	iPrOH	1.54	10	.342	2	6.30
				1.73	3	
oxazine (S_0)	iPrOH	3.22	14	1.82	3	6.32
				3.77	4	
oxazine (S_1)	nPrOH	2.43	13	1.82	3	6.32
				3.77	4	
oxazine (S_1)	.5/.5 nPrOH/water	10.6	26	10.0	9	6.32
				14.2	10	
oxazine (S_1)	.2/.8 nPrOH/water	26.9	42	22.2	13	6.32
				28.7	14	
oxazine (S_1)	water	49.5	56	45.8	18	6.32
				51.4	19	
oxazine (S_1)	.32/.68 DMSO/water	30.6	44	28.7	14	6.32
				32.2	15	
oxazine (S_1)	.6/.4 DMSO/water	38.8	50	37.4	16	6.32
				42.4	17	
oxazine (S_1)	DMSO	76.5	70	68.3	25	6.32

n-propanol/water mixtures, L_{max} varied monotonically, from 3–4 in pure n-propanol to 18–19 for pure water. As the composition of the binary solvent changes from pure propanol to pure water, the basic character of the solvent does not change, but the average size of the transient hydrogen bonded complexes in the fluid decreases. This behaviour is evident in Figure 3A which shows the dielectric properties changing monotonically as the solvent composition is varied from propanol to water. The domains are the largest in propanol (L_{max} =3–4), decrease in size monotonically as the water concentration increases, and are the smallest for pure water (L_{max} = 18–19). Starting from pure water (L_{max} = 18–19), L_{max} decreases to 14–17 for the water/DMSO mixtures, and then increases to 25 for pure DMSO. One can

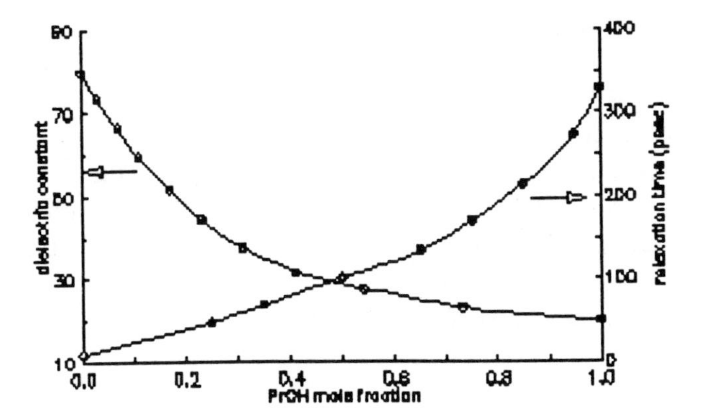

Fig. 3A. Dielectric properties of propanol/water binary mixtures are plotted here versus the mole fraction of propanol.

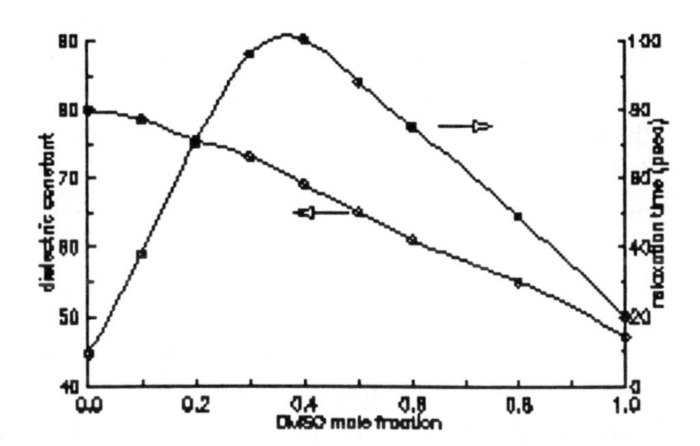

Fig. 3B. Dielectric properties of DMSO/water binary mixtures are plotted here versus the mole fraction of DMSO.

see in Figure 3B that while the variation in solvent dielectric constant is monotonic as the composition changes from pure DMSO to pure water, the dielectric relaxation time goes through a maximum at intermediate DMSO concentrations. What is the significance of this variation? It is well known that water and DMSO interact very strongly, primarily through hydrogen bonding interactions between the DMSO oxygen atom and a water molecule [13]. Beginning with pure water, with its multimolecular domains of a given size (which yields L_{max} = 18–19), addition of DMSO can result in larger complexes as the larger DMSO molecules are incorporated into them. This mechanism can explain both the larger relaxation times for water/DMSO mixtures and also the smaller L_{max} (14–15) required to reproduce the experimental data. As the DMSO concentration is increased further, the amount of DMSO present will eventually exceed the amount which can be accommodated into the complexes, and the dielectric relaxation begins to reflect the behavior of single DMSO molecules. The domain size will begin to decrease, and so the L_{max} begins to increase (L_{max} = 16–17). Finally, as the limit of pure DMSO is approached, the fast relaxation and large L_{max} ($L_{max} = 25$) characteristic of individual polar molecules is reflected in the dielectric relaxation and rotational diffusion data. By modeling both the solute electrical properties and the solvent dielectric properties in a sufficiently rigorous way, a consistent explanation for the experimental data in both the binary solvent mixtures and the pure solvents is obtained.

Conclusions

This work demonstrates clearly the utility of continuum models in describing the dynamics of molecular systems. For the case of diffusive motion as probed here many of the molecular aspects of the medium need not be treated explicitly. Consequently a rather simple model works well. When however the dynamics of the solute occur on time or length scales similar to the solvent, a more detailed treatment of the medium is required. In the case shown here it was found that structural aspects of the medium became important in modifying the strength of the solute/solvent interaction. Treating the charge distribution of the solute molecule in a realistic manner but maintaining a simple model for the solvent has provided enough rigor to describe the experimental data discussed herein. How well such a model performs on shorter length and timescales remains to be seen. It is clear however that the extended nature of the solute molecules charge distribution will be more important for larger molecules like those discussed here than for smaller molecules, such as diatomics. This change reflects the dependence of the electrostatic coupling on r/a in this model.

Although the model described here provides quantitative agreement with experiment, more emphasis should be placed on the qualitative aspects of the agreement between experiment and theory. Important limitations of the model are the need to know the charge distribution of the solute and the models strong dependence on the cavity radius (see Figure 1). Also, the model does not contain a polarizability for the solute, however, this may be mitigated by including a cavity with $\varepsilon_{in} \neq 1$ [14]. This work is being extended in two directions. Firstly the cavity used here is spherical which may not always describe the shape of the solute molecule very well. Similar expressions to those given here have been obtained for symmetric ellipsoids [15]. Secondly, the model used for the wavevector dependence of the dielectric response is crude. A more sophisticated model is presently being developed and contains the characteristics discussed here [16].

Acknowledgment

This work was supported by NSF grant CHE-8613468. We thank our colleague R. S. Hartman, who has been involved in this research.

References

1. Reichardt, C., *Solvents and Solvent Effects in Organic Chemistry* (VCH, NY, 1988) and references therein.
2. Onsager, L., *J. Am. Chem. Soc.* **58**, 1486 (1936).
3. Jackson, J. D., *Classical Electrodynamics*, second edition (Wiley, New York, 1975).
4. Urbakh M. and Klafter, J., *J. Phys. Chem.* **96**, 3480 (1992).
5. Böttcher, C. F. J., *Theory of Electric Polarization*, (Elsevier, New York, 1973).
6. Bagchi B. and Chandra, A., *Advances in Chemical Physics*, Vol LXXX, I. Prigogine and S. A. Rice, editors (Wiley, New York, 1991).
7. Alavi, D. S., Hartman, R. S., and Waldeck, D. H., *J. Chem. Phys.* **95**, 6770 (1991) and references therein.
8. a) Nee T. W. and Zwanzig, R., *J. Chem. Phys.* **52**, 6353 (1970); b) Hubbard J. B. and Wolynes, P. G., *J. Chem. Phys.* **69**, 998 (1978).
9. Alavi D. S. and Waldeck, D. H., *J. Chem. Phys.* **94**, 6196 (1991).
10. Youngren G. K. and Acrivos, A., *J. Chem. Phys.* **63**, 3846 (1975).
11. Alavi, D. S., Hartman, R. S., and Waldeck, D. H., *J. Chem. Phys.* **94**, 4509 (1991).
12. Alavi D. S. and Waldeck, D. H., *J. Phys. Chem.* **95**, 4848 (1991).
13. a) Franks, F., *Water: A Comprehensive Treatise* (Plenum, New York, 1973); b) Brink G. and Falf, M., *J. Mol. Struct.* **5**, 27 (1970); c) Walrafen, G. E., *J. Chem. Phys.* **52**, 4176 (1970); d) Rallo, F., Rodante, F., and Silvestroni, P., *Thermochim. Acta* **1**, 311 (1970); and e) Stokes R. H. and Robinson, R. A., *J. Phys. Chem.* **70**, 2126 (1966).
14. Hartman, R. S., Alavi, D. S., and Waldeck, D. H., *J. Phys. Chem.* **95**, 7872 (1991).
15. Alavi, D. S., PhD Thesis, University of Pittsburgh (1991).
16. Alavi, D. S., Waldeck D. H., and Urbakh, M., work in progress.

10. Analysis of Condensed Phase Photochemical Reaction Mechanisms with Resonance Raman Spectroscopy*

MARY K. LAWLESS, PHILIP J. REID and RICHARD A. MATHIES**
Department of Chemistry, University of California, Berkeley, CA 94720, U.S.A.

Introduction

Pericyclic photochemical rearrangements have enjoyed attention over the years due to their importance in a wide variety of photochemical and photobiological processes [1, 2]. The stereospecific nature of these reactions catalyzed the development of theories based on orbital symmetry conservation which successfully predict the structural outcome of both photochemical and thermal rearrangements [3–5]. Although many aspects of pericyclic reactions are understood, their reaction dynamics and product formation kinetics have remained largely unexplored. Specifically, the excited states which participate in these reactions, the nuclear dynamics composing the reaction coordinate, and the timescales for completion of pericyclic reactions are unknown. To address these questions, we have used resonance Raman (RR) intensity analysis and picosecond time-resolved UV resonance Raman spectroscopy to study the photoinduced reaction dynamics of ring-opening and sigmatropic shift reactions [6–13]. Measurement and analysis of RR intensities determine the change in equilibrium geometry upon excitation, thereby revealing the direction of the initial nuclear motion out of the Franck-Condon region [14–29]. Time-resolved RR spectroscopy provides information about the kinetics of photoproduct formation as well as the structures of intermediates and products that are produced [30]. The combination of these spectroscopies allows us to detail the structural dynamics of pericyclic photochemistry, from the production of the initial excited state to the final appearance of ground state products.

* This research was supported by a grant from the NSF (CHE 91–20254).
** Author to whom correspondence should be addressed.

J.D. Simon (ed.), Ultrafast Dynamics of Chemical Systems, 267–287.

Resonance Raman Intensity Analysis

The dependence of RR intensities on the properties of the excited-state potential surface can be used to study the femtosecond dynamics of photochemical reactions without relying on ultrafast lasers. By carefully measuring the RR intensity of each vibrational mode, we can determine how the excited-state equilibrium geometry differs from that of the ground state along that coordinate. These displacements (Δ's), the zero-zero energy, the transition moment, and the spectral broadening parameters are adjusted to permit an accurate calculation of both the RR intensities and the absorption spectrum. In certain instances, these data can also be used to study more subtle properties of the excited state, including frequency changes, anharmonicity, and Duschinsky rotation [17, 31, 32]. In most cases, this analysis is best performed using the time-dependent equations which describe these spectroscopies [28, 33]

$$\sigma_R(E_L) = \frac{8\pi e^4 E_S^3 E_L M_{eg}^4}{9\hbar^6 c^4} \left| \int_0^\infty \langle f \mid i(t) \rangle \, e^{i(E_L+\varepsilon_i)t/\hbar} D(t) \, \mathrm{d}t \right|^2 , \quad (1)$$

$$\sigma_A(E_L) = \frac{4\pi e^2 E_L M_{eg}^2}{6\hbar^2 cn} \int_{-\infty}^{\infty} \langle i \mid i(t) \rangle \, e^{i(E_L+\varepsilon_i)t/\hbar} D(t) \, \mathrm{d}t. \quad (2)$$

In these expressions for the Raman σ_R (E_L) and absorption σ_A (E_L) cross sections, E_L is the incident photon energy, ε_i is the energy of the initial vibrational level, n is the solvent refractive index, and $D(t)$ is a damping function which includes the homogeneous linewidth Γ.

The heart of these equations which relates RR intensities to Δ's is the time-dependent overlap $\langle f \mid i(t) \rangle$ found in the Raman cross section expression (Equation (1)). The value of this overlap is determined by evolution of the ground state vibrational wavepacket $|i\rangle$ on the excited-state potential surface depicted schematically in Figure 1. Immediately after virtual excitation, the overlap of $|i(0)\rangle$ with the final state $| f \rangle$ in the Raman transition is zero due to the orthogonality of the wavefunctions. After evolution of the wavepacket away from the Franck-Condon region, the $\langle f \mid i(t) \rangle$ overlap increases in value. Generation of overlap in this manner results in fundamental RR intensity. Thus, a difference between the excited- and ground-state equilibrium geometries resulting in the temporal evolution of the wavepacket is the primary mechanism for Raman intensity. Also, since symmetry dictates that only totally symmetric modes may be displaced, it is only these modes which can exhibit fundamental RR scattering in the simple harmonic excited-state surface model [34].

An example of the relationship between Δ's and RR intensity is illustrated in Figure 2 where the absorption spectrum and Raman excitation profile are calculated for a four-mode model of a linear polyene. The three sets of

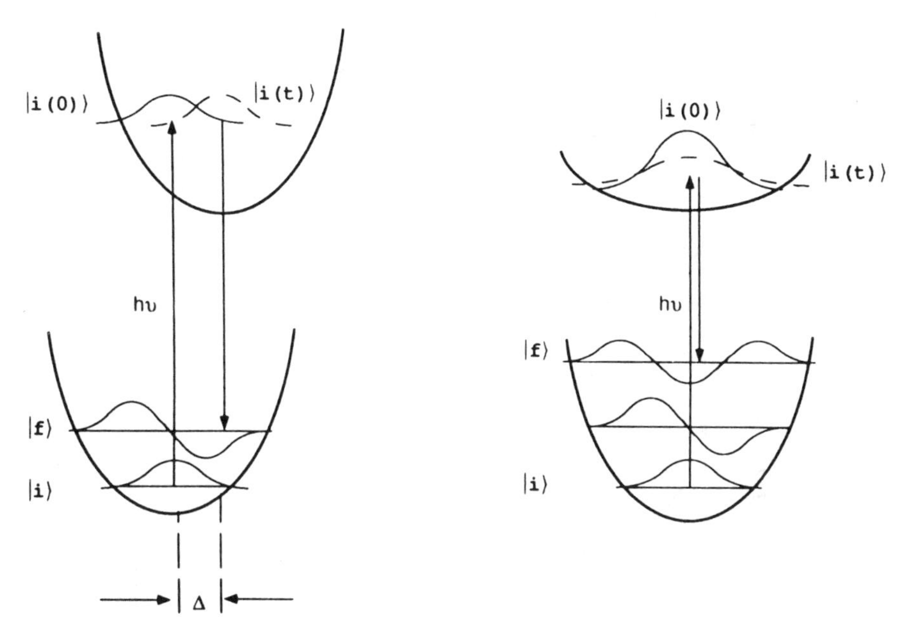

Fig. 1. Schematic ground and excited electronic state potential surfaces along totally symmetric and nontotally symmetric coordinates. The totally symmetric mode exhibits a substantial horizontal displacement (Δ) along the coordinate, permitting the wavepacket to move away from the Franck–Condon region, generating overlap with the wavefunction $|f\rangle$, corresponding to a fundamental transition. The nontotally symmetric mode cannot be displaced, but the frequency change in the excited state allows the wavepacket to distort and generate overlap with final wavefunctions corresponding to even-overtone transitions.

calculations differ only in the value of the Δ for the 900 cm^{-1} mode. The RR intensity in this mode increases quadratically as the Δ of this mode is increased from 0.1 to 0.4. However, only a slight change in the absorption spectrum is observed as Δ is increased, since the absorption lineshape in this example is dominated by displacements along the other three modes. This makes it clear that the RR spectrum will be dominated by those modes with the largest Δ's and that one can use RR intensities to determine these Δ's.

Evolution of the wavepacket along non-totally symmetric modes is evidenced by overtone intensity, demonstrating a change in excited state frequency along that coordinate (see Figure 1), [35]. If the excited-state frequency is lower (higher) than that of the ground state, $|i(t)\rangle$ will respond by spreading (compressing) along this coordinate, resulting in overlap between $|i(t)\rangle$ and the final vibrational wavefunctions corresponding to even overtone RR scattering ($|f\rangle = |2\rangle, |4\rangle$, etc.). Figure 3 illustrates the increase in intensity of

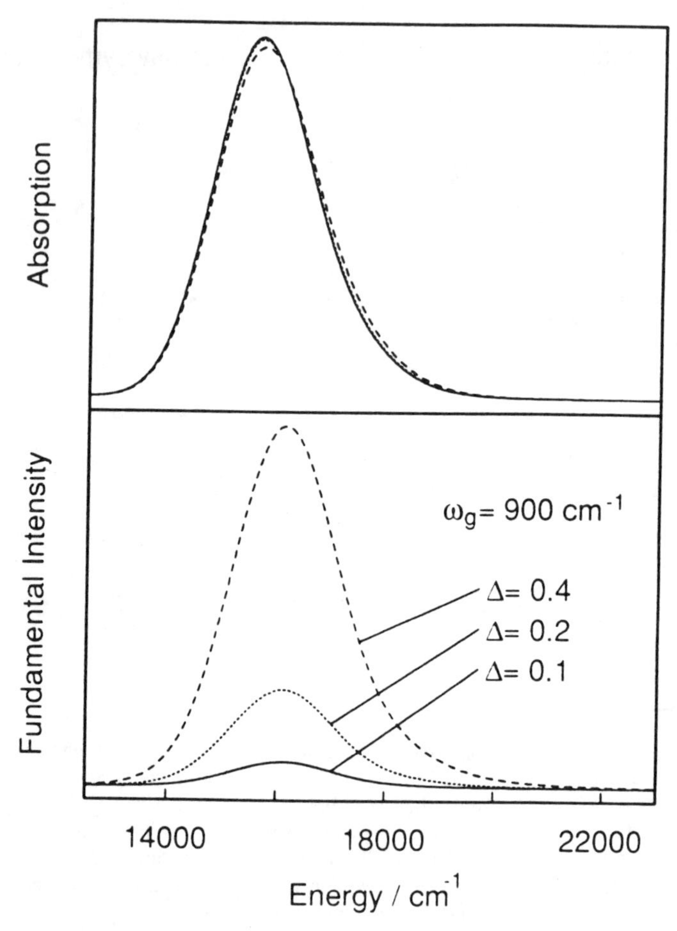

Fig. 2. Calculated absorption (*top panel*) and Raman (*bottom panel*) cross sections for a molecule with four totally symmetric modes: 500 cm^{-1} $\{\Delta = 0.15\}$, 900 cm^{-1} $\{\Delta = 0.10, 0.20, 0.40\}$, 1200 cm^{-1} $\{\Delta = 0.20\}$, and 1600 cm^{-1} $\{\Delta = 0.50\}$. Broadening parameters include $\Gamma = 600$ cm^{-1} and $\theta = 0$ cm^{-1}.

the first overtone for a 1000 cm^{-1} ground-state mode when the excited-state frequency is lowered from 1000 cm^{-1} (where the intensity = 0) to 750 and 500 cm^{-1}.

Central to RR intensity analysis is the dependence of both the absorption spectrum and RR intensities on the same parameters, allowing for the modeling of the experimental spectra to produce a single set of parameters describing the excited-state potential surface. From a chemical standpoint, a photochemical reaction occurs because the nuclear framework of a molecule moves under the influence of the excited-state potential. Therefore, mode-specific knowledge of the excited-state geometry changes provides a map of

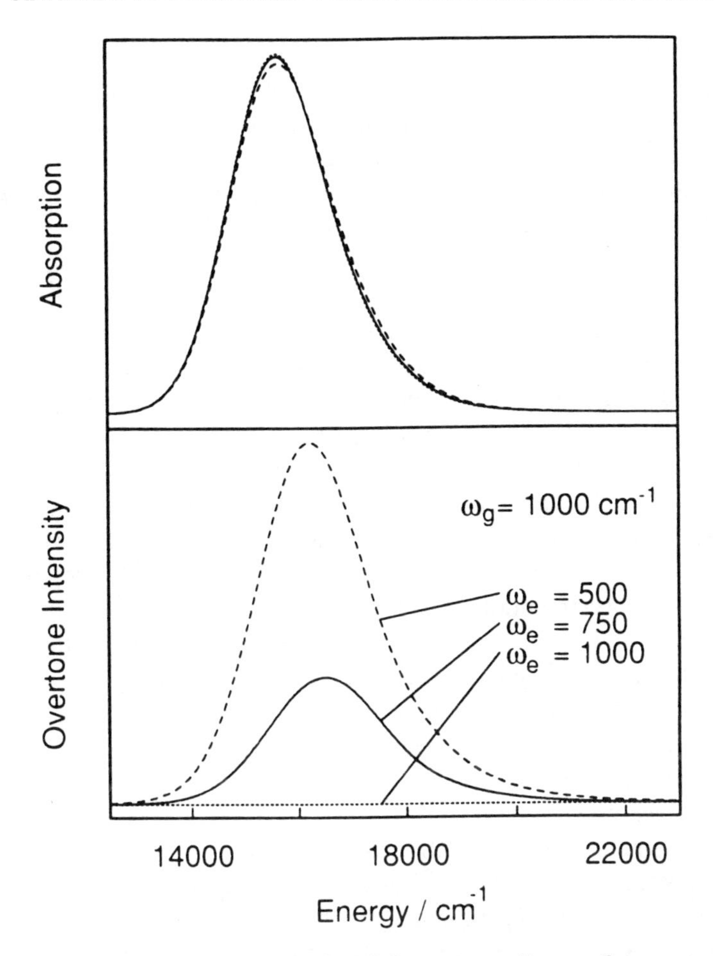

Fig. 3. Calculated absorption (*top panel*) and first overtone Raman (*bottom panel*) cross sections for a model with four totally symmetric modes (as defined in Figure 2) with one nontotally symmetric mode ($\Delta = 0$), ground-state frequency = 1000 cm^{-1}, and excited state frequency = 1000 cm^{-1} (dotted line), 750 cm^{-1} (solid line), and 500 cm^{-1} (dashed line).

how the atoms (and thus, bond lengths and angles) change as the molecule begins its journey from reactant to photoproduct. It is this intimate connection between the Δ's and the RR intensities that makes RR intensity analysis such a powerful tool for understanding the first stages of photochemical reactions.

The conrotatory ring-opening of CHD. Our initial investigations of pericyclic photochemical rearrangements focused on the photochemical ring-opening of 1,3-cyclohexadiene (CHD), [6, 7]. The Woodward–Hoffmann rules predict that this reaction proceeds with a conrotatory motion (C_2-symmetry axis preserved) of the methylene groups. The RR spectrum of CHD excited at 257 nm is presented in Figure 4. Significant RR intensity is observed for several modes demonstrating that the excited state of CHD is displaced relative to the ground state along these normal coordinates. The Raman intensities allow us to describe the structural evolution from the Franck-Condon geometry along the reaction coordinate. The largest peak in the spectrum at 1578 cm^{-1}, corresponding to the symmetric double-bond stretch, indicates a significant reduction in bond order. The second largest peak is the 1321 cm^{-1} mode assigned to the symmetric methylene twist. This mode has a large projection on the conrotatory motion of the methylene groups along the reaction coordinate. This intensity shows that a large Δ exists along this coordinate, demonstrating that the stereospecificity of this reaction is established as the molecule leaves the Franck–Condon region. The 948 cm^{-1} mode also carries large RR intensity. This mode is assigned as the methylene C—C single bond stretch. A large Δ along this coordinate shows that the stretching and eventual breaking of this bond begins immediately after excitation. The intensity of the 507 cm^{-1} mode, assigned to the double-bond torsion, indicates that planarization of the CHD ring occurs during this reaction. To summarize, the RR spectrum has shown that in concert with changes in the olefinic bonding and torsion angles, the methylene single bond stretches as it begins to break while the methylene groups twist in the Woodward–Hoffmann-predicted conrotatory fashion. This evolution occurs immediately after excitation, demonstrating that *the stereochemistry of this photochemical reaction is established in the first stages of the apparently concerted reaction.*

Measurement of absolute RR intensities allows for an exact determination of the molecular geometry changes that accompany this reaction as well as the optical dephasing time T_2 [7]. Analysis of these data showed that this dephasing time for CHD is only 10 fs. This, in conjunction with the 10^{-6} fluorescence quantum yield, indicates rapid depopulation of the initially prepared excited state to a lower-lying electronic surface of A symmetry. This excited-state ordering is similar to that found in other olefinic systems [36–39].

The disrotatory ring-opening of COT. 1,3,5-cyclooctatriene (COT) also undergoes a photochemical ring-opening reaction. Here, the ring-opening is believed to occur by a disrotatory motion of the CH_2 groups [5]. The RR spectrum of COT excited at 257 nm is shown in Figure 5 [8]. The fundamental region shows many modes with significant intensity, indicating that the reaction coordinate is multi-dimensional. While large intensity is again observed in the ethylenic modes at 1610 and 1640 cm^{-1}, several low-frequency modes at 140, 339, and 404 cm^{-1}, corresponding to ring planarizations and double-bond torsions, are particularly intense. This shows that in the early stages of this photochemical reaction, the ring system undergoes substantial planarization. Significant intensity is also seen in the 959 cm^{-1} mode, which is

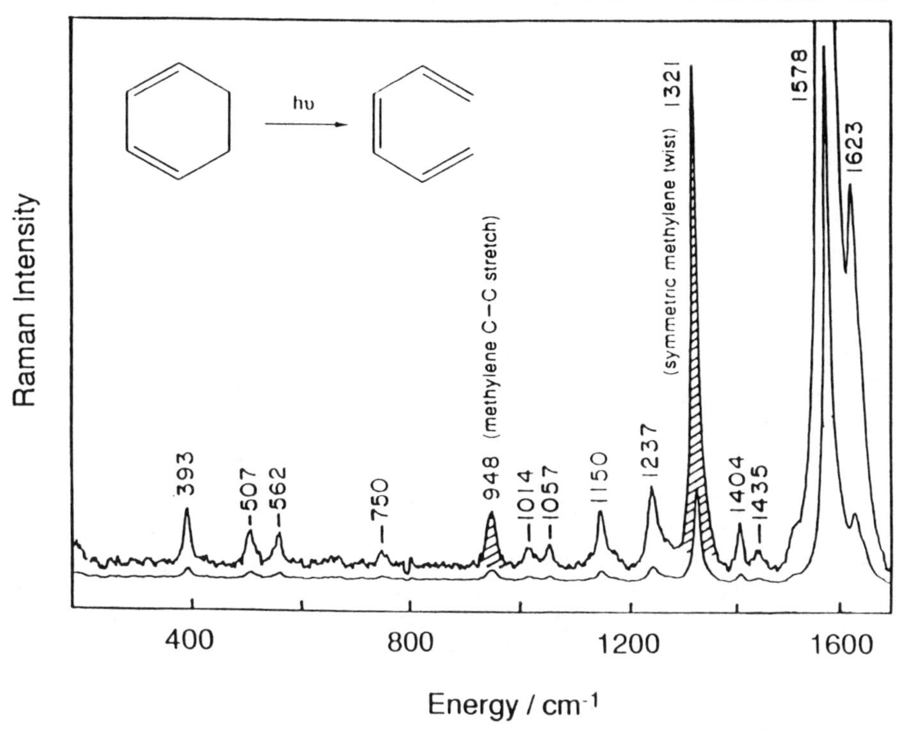

Fig. 4. Resonance Raman spectrum of 1,3-cyclohexadiene, excited at 257 nm. The cyclohexane solvent peaks have been subtracted. The inset is divided by 20. The peaks at 393 and 1623 cm^{-1} are the ethylenic torsion and stretch, respectively, of *cis*-1,3,5-hexatriene formed within the transit time of the sample through the laser beam.

assigned as the methylene C-C single bond stretch. The absence of RR intensity in the CH_2 symmetric rocking mode expected at 1316 cm^{-1} indicates that conrotatory motion of the CH_2 groups does not occur in the Franck-Condon region, as predicted for a disrotatory ring-opening.

Since a disrotatory twist of the methylene groups is a nontotally symmetric motion, (b_1 symmetry for the approximate C_{2v} point group), the corresponding nontotally symmetric C—H rocking fundamental (predicted to lie at 1356 cm^{-1}) is symmetry-forbidden. Therefore, fundamental intensity should not be observed in this mode. Consistent with this expectation, no unusually intense lines are observed in the 1200–1300 cm^{-1} region that can be assigned to either the symmetric or nontotally symmetric rocking modes. If structural evolution corresponding to disrotatory motion was occurring out of the Franck-Condon region, we would expect a decrease in frequency of the nontotally symmetric CH_2 rocking mode upon excitation which would give rise to even-overtone scattering. However, each peak in the corresponding overtone spectral region can be assigned as an overtone or combination of existing fundamentals (see Figure 5). Furthermore, calculations employing Equations (1) and (2) predict

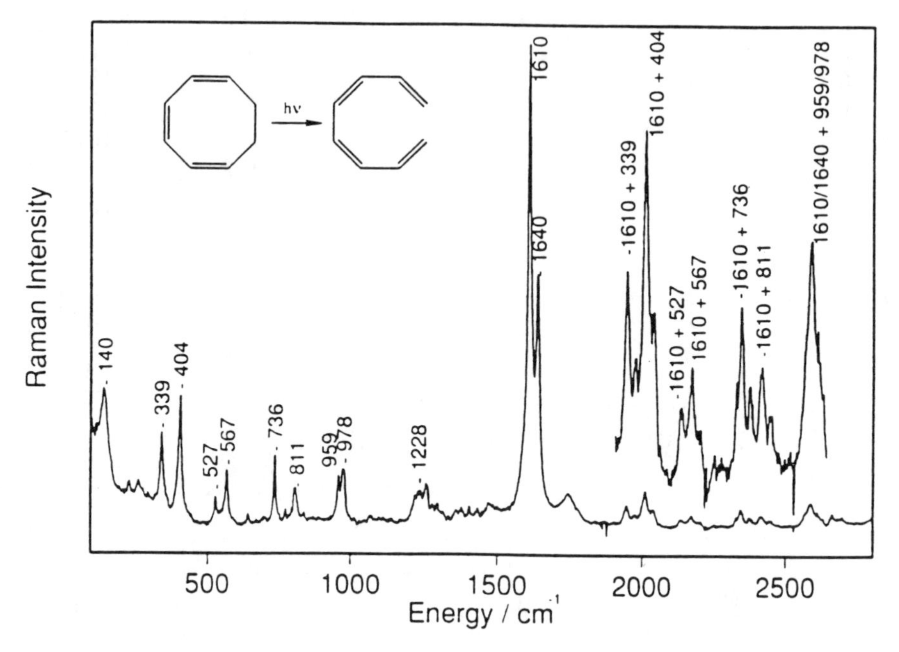

Fig. 5. Resonance Raman spectrum of 1,3,5-cyclooctatriene excited at 257 nm. Large intensities in the low-frequency modes indicate that a global flattening of the ring dominates the early part of the ring-opening reaction coordinate. The peak at 1257 cm^{-1} corresponds to photoproduct scattering formed within the transit time of the sample through the laser beam.

that if a frequency change of more than 200 cm^{-1} occurred, the resulting overtone intensity would be within our detection limits in this region. The absence of the overtone of the 1356 cm^{-1} b_1 twist mode indicates that if a frequency change exists, it is no more than $\sim$ 200 cm^{-1}. It appears that because this system is substantially less planar than CHD in the ground state, the majority of COT's motion out of the Franck–Condon region is along ring-flattening modes; this planarization must evidently occur before any significant evolution along the disrotatory CH_2 twisting coordinate can take place.

Sigmatropic Rearrangements. We have also studied the excited-state dynamics of the photochemical hydrogen migration of 1,3,5-cycloheptatriene (CHT) using RR intensity analysis [9]. This reaction represents a class of pericyclic rearrangements known as sigmatropic shifts in which the breaking and formation of a σ-bond occurs with accompanying rearrangement of the olefinic bonds. The RR spectrum of CHT excited at 257 nm is shown in Figure 6. Similar to COT, many modes exhibit RR intensity, demonstrating the multidimensional nature of the reaction. Also, the electronic transition is accompanied by large Δ's in the ethylenic modes, but the most intrigu-

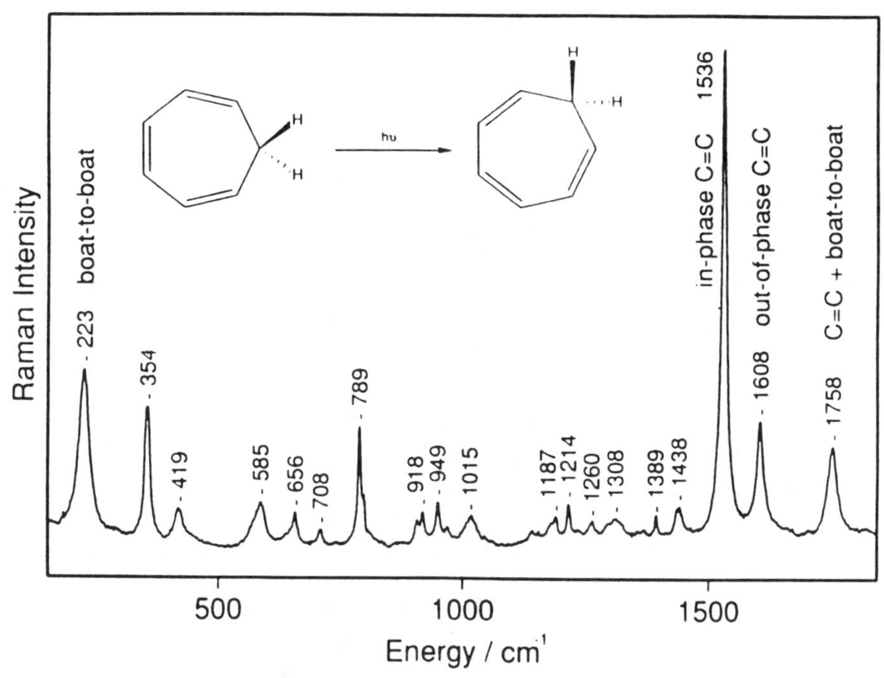

Fig. 6. Resonance Raman spectrum of 1,3,5-cycloheptatriene excited at 257 nm. The large intensity in the low-frequency, boat-to-boat mode demonstrates the planarization of the cycloheptatriene ring immediately upon excitation.

ing observation is the absence of intensity in the methylene C—H stretching mode near $\sim$ 2900 cm^{-1} (not shown). Since the C—H bond breaks in the formation of photoproduct, we might expect to see fundamental RR intensity in this mode. However, the absence of intensity shows that no displacement of the excited state in the Franck–Condon region exists along this coordinate. In contrast, the most significant intensity is in the 223 cm^{-1} boat-to-boat mode, as well as other low-frequency modes indicative of planarization of the CHT ring. This pattern of intensities shows that planarization of the CHT ring occurs before the hydrogen migration can take place. Similar to COT, it is necessary for the saturated portion of the ring to come into the same plane as the π-system before the reaction can occur. Therefore, the [1,7]-sigmatropic shift of CHT is a sequential reaction in which planarization of the CHT ring precedes the sigmatropic shift.

These three examples have given much insight into the use of resonance Raman intensities to study the first stages of a photochemical reaction. The RR intensities of CHD clearly show that the stereochemical preference for the conrotatory reaction coordinate is established immediately after excitation of the reactant. The pattern of RR intensities of COT are characteristic of a more complicated ring-opening reaction coordinate, involving large motions of the

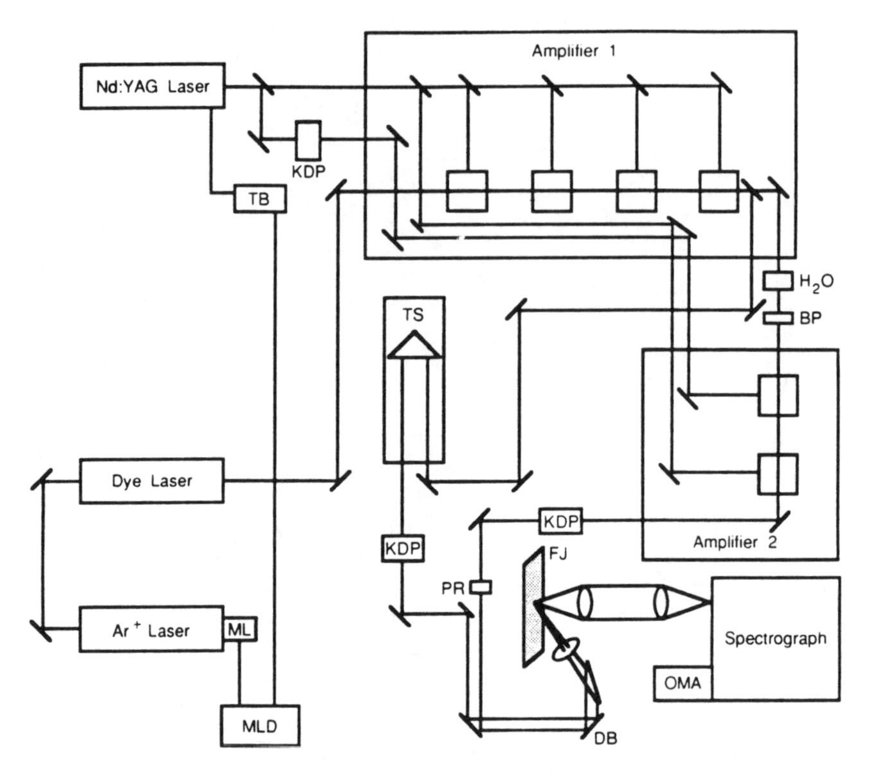

Fig. 7. Schematic of the picosecond, two-color, UV Raman apparatus. BP = band-pass filter, DB = dichroic beam splitter, FJ = flowing jet, H_2O = water continuum cell, KDP = potassium dihydrogen phosphate doubling crystal, ML = mode locker, MLD = mode locker driver, OMA = optical multichannel analyzer, PR = polarization rotator, TB = timing box, TS = translation stage.

ring in the first stages of the reaction and no initial conrotatory motion of the CH_2 groups. Similar intensity patterns are seen in the hydrogen migration reaction of CHT, as planarization of the ring is required before any hydrogen movement occurs. Evolution along these planarization coordinates in both COT and CHT leads to a $\sim$ 10-fs depopulation of the initially prepared excited state, similar to that in CHD [8, 9].

Comparing the three examples given here demonstrates the importance of a planar excited-state in the pericyclic reaction mechanism. The ground-state structure of CHD is close to planar. Thus, limited motion along planarization coordinates is required in the excited state, so we observe the ring-opening reaction right out of the Franck-Condon region. However, COT and CHT are significantly distorted from planarity, and the Raman intensities demonstrate that flattening of both rings must precede the subsequent changes in bonding. Resonance Raman intensity analysis has thus provided unique mode-specific information about the excited-state nuclear dynamics that occur in the first femtoseconds of these pericyclic photochemical reactions.

Picosecond Time-Resolved Resonance Raman Spectroscopy

The determination of the kinetics of product formation is also a central problem in understanding photochemical reactivity. Furthermore, the ability to determine the photoproduct structure is critical in elucidating the nature of the reaction coordinate. The kinetic and structural aspects of product formation can be eloquently determined by time-resolved resonance Raman spectroscopy [30]. By monitoring the temporal evolution of Raman intensities, one can determine the kinetics of photoproduct appearance. Also, since the vibrational spectrum is sensitive to molecular structure, it provides a wealth of information on photoproduct structure and conformational evolution.

In the photochemical pericyclic rearrangements considered here, the electronic absorption of the products and reactants are located in the UV. To study these systems, a source of radiation capable of providing separately tunable UV pulses to preferentially excite the parent and resonantly enhance product scattering is required. Furthermore, the dynamics of these rearrangements are extremely rapid necessitating the use of ultrafast lasers. The two-color, picosecond UV apparatus developed to meet this task is presented in Figure 7 [12, 40]. The heart of the system is a synchronously-pumped dye laser providing pulses of picosecond duration. The energy of these pulses is increased to $\sim$ 1 mJ utilizing a four-stage amplifier at a repetition rate of 50 Hz. A portion of the first amplifier output is removed and frequency doubled to produce a tunable probe beam from 284 to 298 nm. The remainder of the amplifier output is focused into a cell of H_2O for continuum generation. A portion of the continuum centered at 550 nm is removed and subsequently amplified to $\sim$ 600 μJ. Frequency doubling of this light produces the actinic pulse at 275 nm. The polarization of the pump pulse is rotated to 55° relative to that of the probe minimizing the contribution of molecular rotations to the observed spectra. The time resolution of the experiment is dictated by the 2-ps cross-correlation of the pump and probe beams. The arrival of the probe pulse at the sample relative to the pump is adjusted by varying the spatial path of the probe allowing for delays up to 10 ns. The experiment is performed by initiating the chemistry with the actinic pulse and monitoring the vibrational spectrum with the probe as a function of time.

The conrotatory ring-opening of CHD. To illustrate the information available from this technique, Figure 8 presents difference Stokes and anti-Stokes resonance Raman spectra of the photoconversion of 1,3-cyclohexadiene (CHD) to *cis*-hexatriene (c-HT) [10, 11]. These difference spectra are constructed by subtracting the probe-only spectrum from the probe spectrum in the presence of photolysis. In the Stokes data, negative intensity at 1578 and 1323 cm^{-1} is assigned to the depletion of CHD created by the presence of the actinic pulse. At 4 ps, positive intensity is observed at 1610 cm^{-1} corresponding to the appearance of the ground-state c-HT photoproduct. This line increases in intensity for delays up to 100 ps with no further evolution observed out to 10 ns. The intensity of the photoproduct ethylenic line is presented as a function of time in Figure 9A. The best fit to the data by a single exponential results in an appearance time of 6 ± 1 ps which is the time necessary for the completion of a photochemical ring-opening reaction in solution.

The anti-Stokes data (Figure 8) also provide information on photoproduct appearance as well as the kinetics of vibrational relaxation. The observed anti-Stokes intensity allows for the investigation of vibrationally hot molecules. At 0 ps, little anti-Stokes intensity is observed, consistent with the absence of ground-state c-HT at this early time. At longer delays, intensity in vibrational lines characteristic of c-HT increase in intensity and subsequently decay due to vibrational relaxation. A plot of this intensity as a function of time is presented in Figure 9B. The best fit to the data results in an appearance time of 7 ± 2 ps and decay time of 9 ± 2 ps. The good agreement between the appearance of the Stokes and anti-Stokes scattering argues that ground-state c-HT is produced in 6 ps. Furthermore, these molecules are initially vibrationally hot after internal conversion from the excited state, reaching thermal equilibrium presumably via energy exchange with the solvent in 9 ps.

The time-resolved resonance Raman data also provide information on the structure of the ground-state photoproduct. Although it is tempting to assign the observed product scattering to the *all-cis* conformer, c-HT can isomerize about either single bond to form *mono* or *di-s-trans* c-HT. Close inspection of the data reveals that conformational relaxation is indeed occurring. In the Stokes spectra (Figure 8), the most intense ethylenic line of the photoproduct increases in frequency from 1610 to 1625 cm^{-1} and the single bond stretch also shifts from 1236 to 1249 cm^{-1} consistent with conformational relaxation [11, 41]. Furthermore, the 100-ps spectrum reveals the presence of a second ethylenic line at 1572 cm^{-1}. If we consider the symmetry of c-HT, the *all-cis* and *di-s-trans* conformers are of C_{2v} symmetry resulting in predicted resonance Raman intensity in a single, in-phase ethylenic line. However, the lower symmetry of the *mono-s-trans* conformer turns on intensity in both the in- and out-of-phase ethylenic modes. The presence of a double ethylenic line in the 100-ps Stokes spectrum is consistent with the presence of *mono-s-trans* HT only. This result immediately raises the question: Is there any evidence for all-cis HT in this reaction mechanism? Close inspection of the

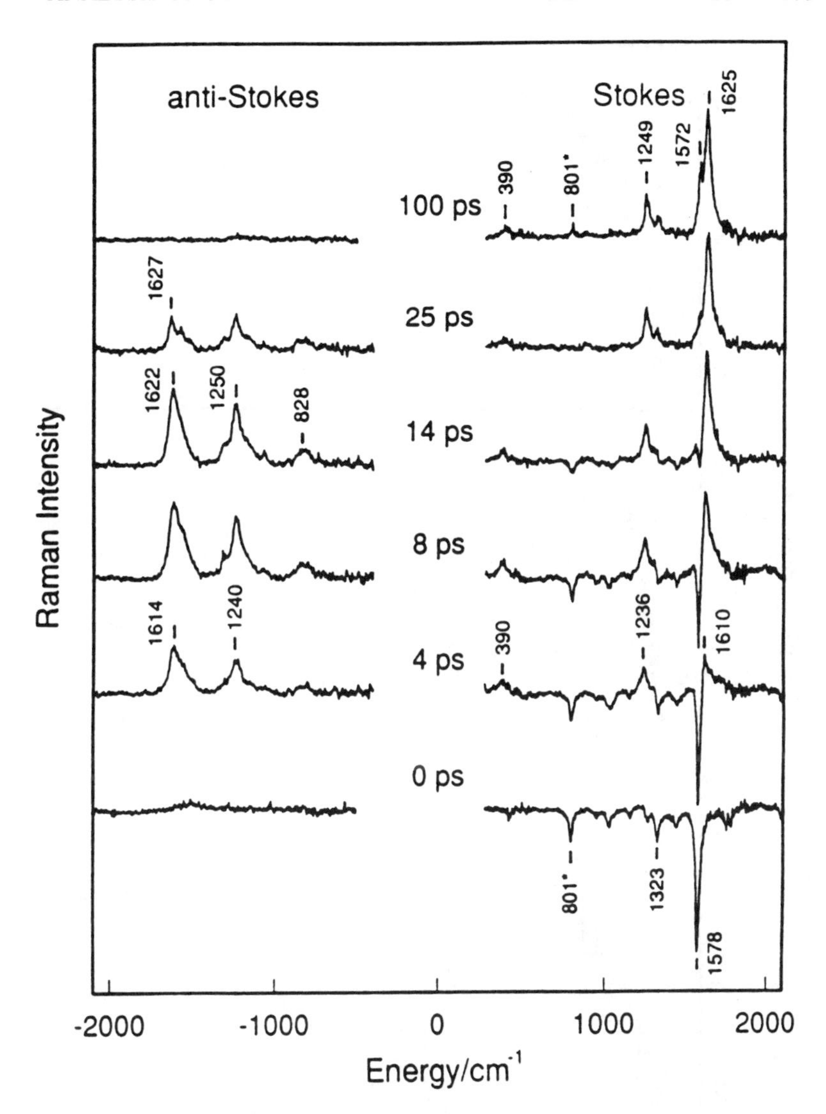

Fig. 8. Resonance Raman anti-Stokes and Stokes difference spectra of the 1,3-cyclohexadiene (CHD) to *cis*-hexatriene (c-HT) photoconversion. Anti-Stokes spectra were obtained using a 1-color format at 284 nm. The two-color Stokes spectra were obtained with a 275-nm pump and a 284-nm probe. The line at 801 cm^{-1} in the Stokes spectra is due to the cyclohexane solvent.

anti-Stokes data reveals a line at 828 cm^{-1} corresponding to the in-plane CH_2 rocking mode of the *all-cis* conformer. This observation coupled with the intensity and frequency evolution in the Stokes data demonstrates that c–HT initially appears on the ground state surface as the *all-cis* conformer and then it undergoes conformational relaxation to form *mono-s-trans*-HT.

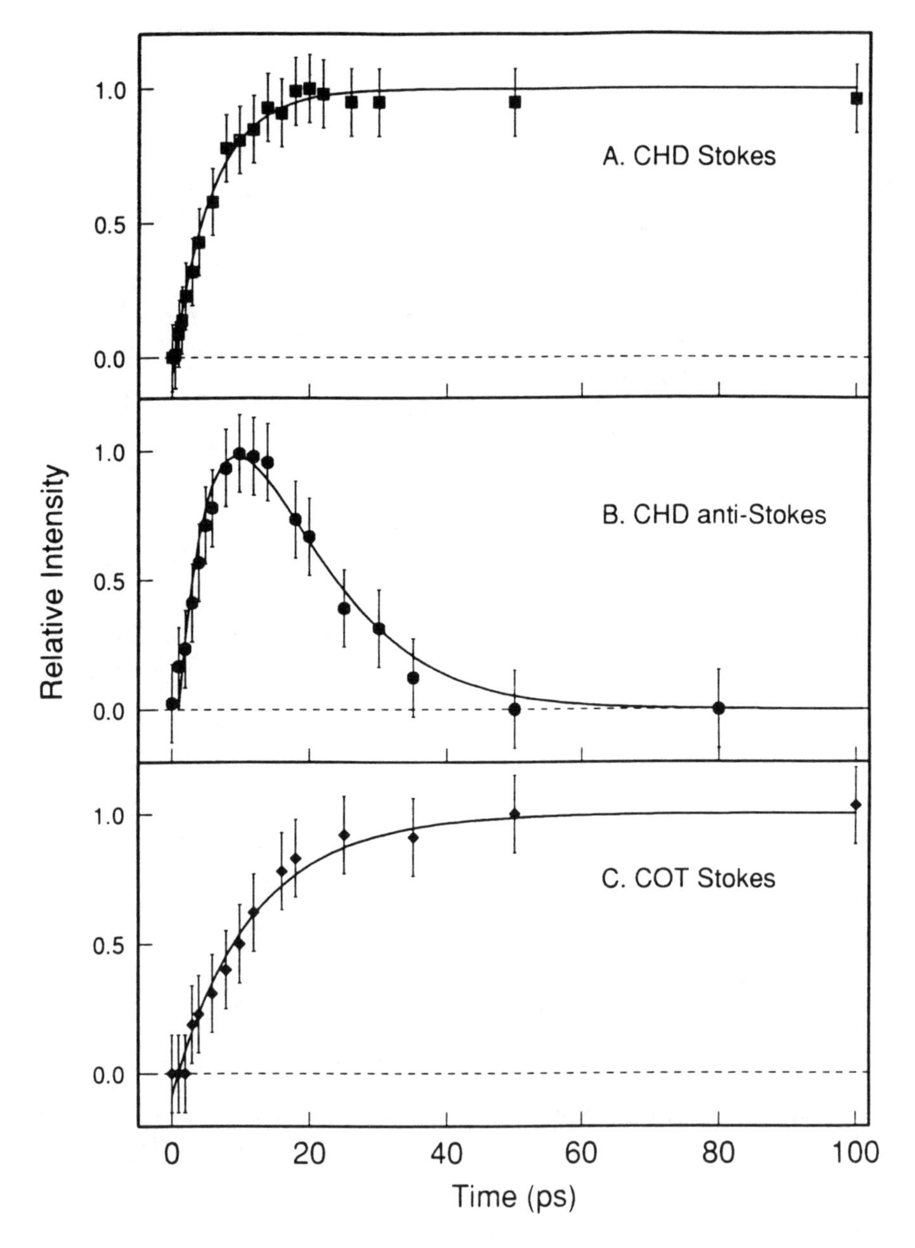

Fig. 9. (A) Intensity of the *cis*-hexatriene photoproduct ethylenic intensity as a function of time after the photolysis of CHD. The best fit to the data by a single exponential resulted in an appearance time of 6 ± 1 ps. (B) Intensity of the *cis*-hexatriene photoproduct ethylenic anti-Stokes intensity as a function of time. The best fit to the data by a double exponential resulted in an appearance time of 8 ± 2 ps and a thermal cooling time of 9 ± 2 ps. (C) Intensity of the *cis*, *cis*-octatetraene ethylenic line produced from the photochemical ring opening of 1,3,5-cyclooctatriene. Best fit to the data by a single exponential resulted in an appearance time of 12 ± 2 ps.

The rate of isomerization is undoubtably increased by the steric interactions present in *all-s-cis*-HT and the fact that the molecule is transiently hot.

The disrotatory ring-opening of COT. We have also examined the disrotatory ring-opening reaction in 1,3,5-cyclooctatriene (COT). The photoproduct formed from the disrotatory ring-opening of COT appears on the ground state in 12 $\pm$ 2 ps (Figure 9C) [12]. The similarity between this time constant and the photoproduct formation time for the conrotatory ring-opening of CHD demonstrates that the rate of photoproduct formation is insensitive to the stereochemistry of the reaction. *The* $\sim$ 10 ps *photoproduct appearance time is a general feature of electrocyclic ring-opening reactions.*

The hydrogen migration in CHT. The pericyclic photochemical hydrogen migration of 1,3,5-cycloheptatriene (CHT) is also complete in picoseconds [13]. Figure 10 presents the transient Raman difference spectra of this photochemical sigmatropic shift. At 0 ps, negative intensity is observed at 1536, 1610, and 1758 cm^{-1} corresponding to ground state depletion of vibrationally relaxed CHT. This intensity, created by the actinic pulse, decays as the probe delay time is increase with recovery complete by 80 ps.

Although it is tempting to simply measure the magnitude of the depletion to determine the rate of ground-state recovery, the presence of positive spectral features to the low-wavenumber side of the 1536 cm^{-1} CHT ethylenic line in the 25-ps spectrum demonstrates the presence of other molecular species. The right panel in Figure 10 presents the difference spectra after elimination of the vibrationally relaxed CHT depletion revealing the presence of two new species with ethylenic frequencies at 1545 cm^{-1} and 1529 cm^{-1}. The kinetic behavior of these species allows for the assignment of their molecular nature. First, the scattering from vibrationally cold CHT at 1536 cm^{-1} (Figure 11A) recovers with a time constant of 26 $\pm$ 4 ps. The temporal evolution of the 1529 cm^{-1} species is presented in Figure 11B. This intensity appears with a time constant of 26 $\pm$ 4 ps and decays in 15 $\pm$ 5 ps. The kinetic behavior and frequency of this scattering indicates that this intensity corresponds to vibrationally-hot CHT. The 7 cm^{-1} frequency decrease to 1529 cm^{-1} is due to population of higher levels of the anharmonic ground-state ethylenic vibrational manifold. The similarity in rise times between vibrationally-relaxed and unrelaxed CHT demonstrates that CHT appears on the ground state in 26 ps. Finally, the ethylenic line at 1545 cm^{-1} rises and decays within the instrumental response (Figure 11C) with similar kinetics observed for the sample optical absorbance. This behavior is consistent with scattering from excited-state CHT formed during the temporal overlap of the actinic and probe pulses. In summary, despite the very different nature of the reaction coordinate, the hydrogen migration in CHT occurs on a timescale that is very similar to that observed in electrocyclic ring-opening reactions.

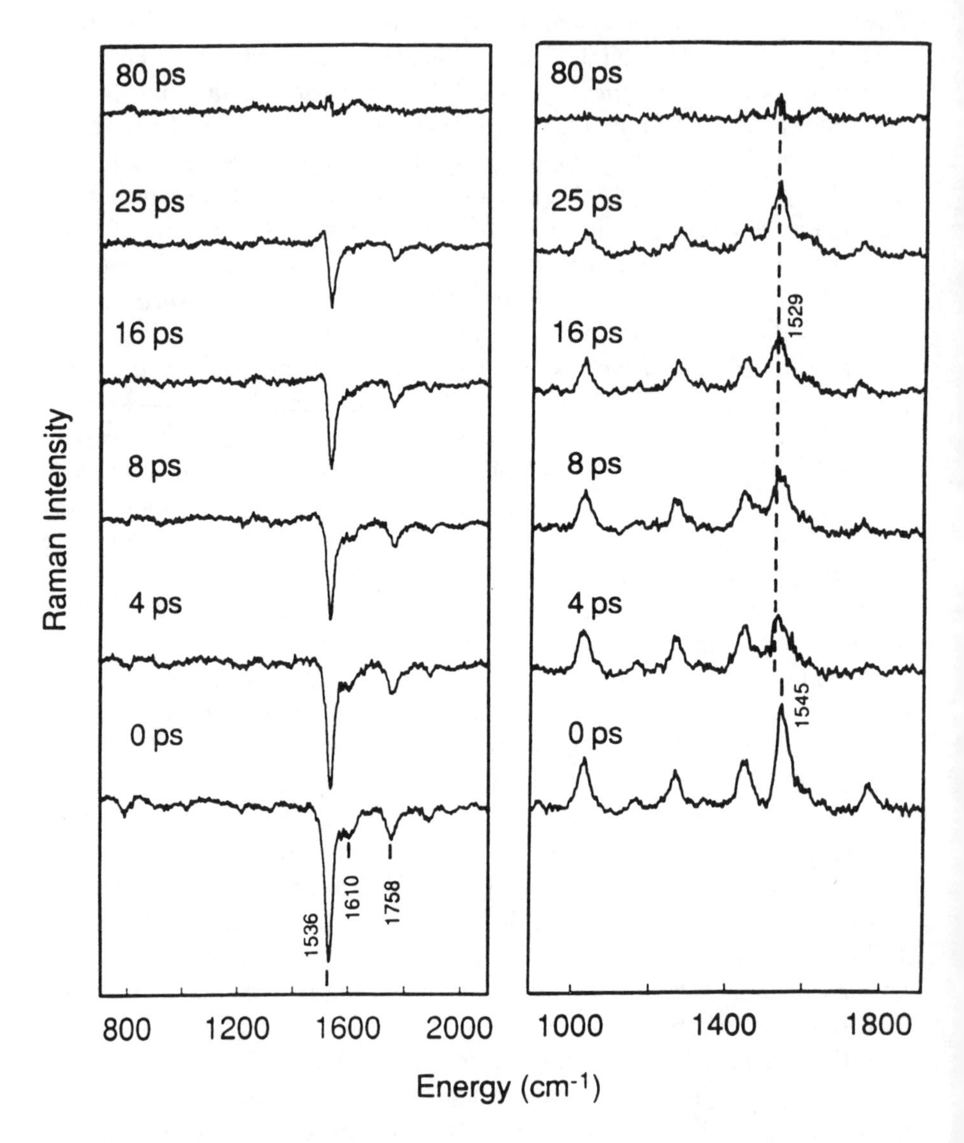

Fig. 10. *Left panel;* Resonance Raman difference spectra of the (1,7)-photochemical sigmatropic rearrangement of 1,3,5-cycloheptatriene (CHT) excited at 284 nm. The negative lines at 1536, 1610 and 1758 cm^{-1} are due to the depletion of ground state CHT. *Right panel:* Resonance Raman difference spectra of molecular intermediates in the ground state appearance of photoexcited CHT. Spectra were produced by the addition of the ground-state CHT spectrum to the difference spectra in A until negative features due to the depletion of ground state CHT were eliminated. The spectra exhibit scattering from two separate intermediates, the first corresponds to excited-state CHT with an ethylenic frequency of 1545 cm^{-1} and the second to hot ground-state CHT with ethylenic frequency of 1529 cm^{-1}.

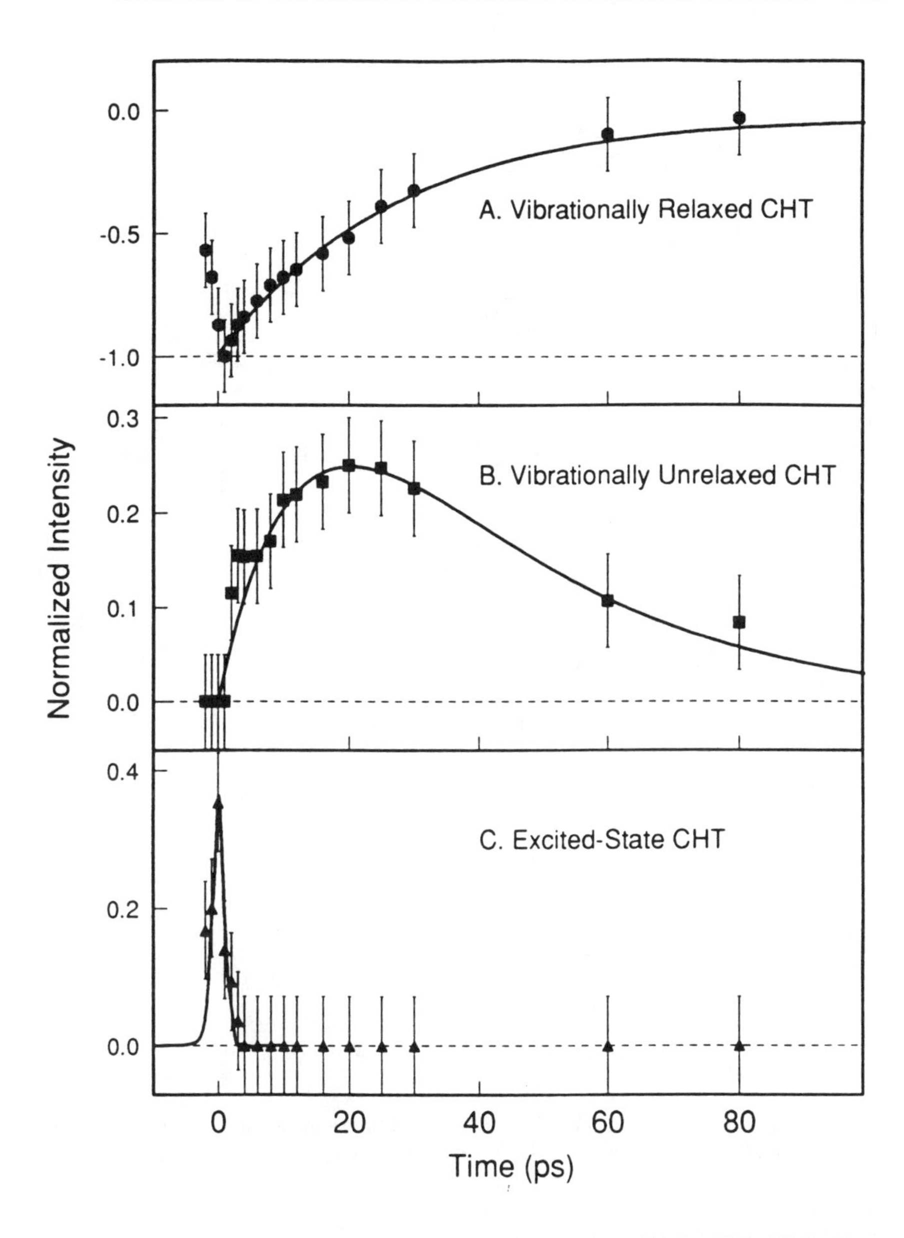

Fig. 11. Picosecond kinetics of the 1,7-sigmatropic hydrogen shift in CHT. (A) Recovery kinetics of vibrationally relaxed CHT scattering at 1536 cm^{-1}. (B) Kinetics of scattering from the hot ground state of CHT at 1529 cm^{-1}. (C) Kinetics of scattering from the excited state of CHT at 1545 cm^{-1}. Intensities were determined from least-squares analysis of the difference spectra in Figure 10. The best fit to the vibrationally unrelaxed CHT data resulted in an appearance time of 26 ± 10 ps and a decay time of 15 ± 5 ps. The best fit to the vibrationally relaxed CHT data resulted in an appearance time of 26 ± 10 ps.

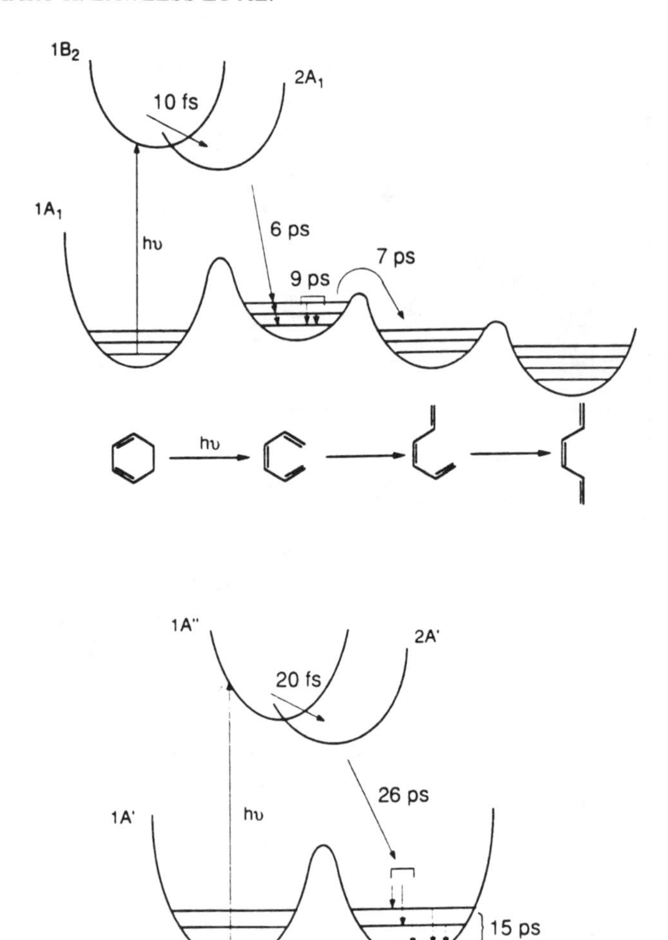

Fig. 12. Schematic reaction coordinates for (A) the photochemical ring-opening of 1,3-cyclohexadiene and (B) the photochemical hydrogen migration in 1,3,5-cycloheptatriene.

Conclusions

The combination of resonance Raman intensity analysis and time-resolved resonance Raman spectroscopy has allowed us to develop a detailed picture of the excited-state dynamics and photoproduct formation kinetics of pericyclic photochemical rearrangements (Figure 12). The photochemical ring-opening reaction in CHD is characterized by rapid evolution along the conrotatory reaction coordinate predicted by orbital-symmetry conservation, while the disrotatory ring-opening reaction in COT requires significant ring planarization before any large changes in methylene orientation take place. However, evolution in the excited state of both systems is followed by the decay of the initially prepared excited state in $\sim$ 10 fs. This extremely rapid internal conversion is due to strong coupling between this surface and a lower-lying electronic excited state. The timescale for internal conversion is much faster than typical vibrational relaxation and dephasing times, suggesting that the internal conversion is occurring through non-stationary vibrational states [42–44]. Decay out of this lower surface along both conrotatory and disrotatory pathways results in population of the ground state with a time constant of $\sim$ 10 ps after which conformational relaxation as well as vibrational cooling occur.

Similarly, the photochemical hydrogen migration in CHT occurs with rapid excited-state evolution towards a planar geometry from which the sigmatropic shift can take place. In this case, the initial dynamics correspond to flattening of the ring, with the hydrogen migration occurring farther along the reaction coordinate. Similar to CHD and COT, the initially prepared excited state of CHT depopulates on the femtosecond timescale. The difference between the 10-fs decay time and the 26-ps ground-state appearance time indicates the participation of a lower-lying excited state in the photochemistry. Finally, the photoproduct dynamics are completed with the $\sim$ 15 ps relaxation of vibrationally hot CHT.

Comparison between the initial dynamics of CHD, COT, and CHT demonstrates the need for a planar excited-state structure as a precursor to pericyclic bonding changes. The RR intensities of CHD show that the majority of the changes (besides bond-order inversion) occur in the methylene portion of the ring, specifically, stretching of the CH_2—CH_2 single bond and twisting of the CH_2 groups. However, the initial excited-state dynamics of COT and CHT correspond to ring planarization and not motions characteristic of their respective pericyclic reactions. These observations suggest that a planar or near-planar intermediate structure is a prerequisite for an electrocyclic rearrangement. The non-planar, ground-state structures of COT and CHT do not allow proper orbital overlap between carbons involved in the pericyclic reaction. The initial evolution along ring-planarization coordinates creates an appropriate intermediate structure from which the predicted bonding changes can take place. Therefore, the reaction coordinates of COT and CHT are sequential, involving a series of structural intermediates. In contrast, the near-planar, ground-state geometry of CHD allows for excited-state evolution along the predicted conrotatory reaction coordinate immediately after excitation, demonstrating that the rearrangement is concerted. This diversity of excited-state dynamics shows that atomic orbital overlap is an important factor in pericyclic photochemistry.

References

1. Dauben, W. G., McInnis, E. L., and Michno, D. M., in *Rearrangements in Ground and Excited States*, Vol. 3, edited by P. de Mayo (Academic Press, New York, 1980), p. 91.
2. Jacobs, H. J. C. and Havinga, E., in *Advances in Photochemistry*, Vol. 11, edited by Pitts, J. N. Jr., Hammond, G. S., and Gollnick, K. (Interscience Publishers, New York, 1979), p. 305.
3. Hoffmann, R. and Woodward, R.B., *Acc. Chem. Res.* **1**, 17 (1968).
4. Fukui, K., *Acc. Chem. Res.*. **4**, 57 (1971).
5. Woodward, R. B. and Hoffmann, R., *The Conservation of Orbital Symmetry* (Verlag Chemie, Weinheim/Deerfield Beach, 1981).
6. Trulson, M. O., Dollinger, G. D., and Mathies, R. A., *J. Am. Chem. Soc.* **109**, 586 (1987).
7. Trulson, M. O., Dollinger, G. D., and Mathies, R. A., *J. Chem. Phys.* **90**, 4274 (1989).
8. Lawless, M. K. and Mathies, R. A., *J. Chem. Phys.*, submitted (1993).
9. Reid, P. J. and Mathies, R. A., *J. Phys. Chem.*, in press (1993).
10. Reid, P. J., Doig, S. J., and Mathies, R. A., *Chem. Phys. Lett.* **156**, 163 (1989).
11. Reid, P. J., Doig, S. J., Wickham, S. D. and Mathies, R. A., *J. Am. Chem. Soc.* **115**, 4754 (1993).
12. Reid, P. J., Doig, S. J. and Mathies, R. A., *J. Phys. Chem.* **94**, 8396 (1990).
13. Reid, P. J., Wickham, S. D. and Mathies, R. A., *J. Phys. Chem.* **96**, 5720 (1992).
14. Lawless, M. K. and Mathies, R. A., *J. Chem. Phys.* **96**, 8037 (1992).
15. Amstrup, B., Langkilde, F. W., Bajdor, K. and Wilbrandt, R., *J. Phys. Chem.* **96**, 4794 (1992).
16. Ci, X. and Myers, A. B., *J. Chem. Phys.* **96**, 6433 (1992).
17. Peteanu, L. A. and Mathies, R. A., *J. Phys. Chem.* **96**, 6910 (1992).
18. Phillips, D. L. and Myers, A. B., *J. Chem. Phys.* **95**, 226 (1991).
19. Morikis, D., Li, P., Bangcharoenpaurpong, O., Sage, J. T. and Champion, P. M., *J. Phys. Chem.* **95**, 3391 (1991).
20. Myers, A. B., *J. Opt. Soc. Am. B* **7**, 1665 (1990).
21. Sweeney, J. A. and Asher, S. A., *J. Phys. Chem.* **94**, 4784 (1990).
22. Trulson, M. O. and Mathies, R. A., *J. Phys. Chem.* **94**, 5741 (1990).
23. Ziegler, L. D., Chung, Y. C., Wang, P. G. and Zhang, Y. P., *J. Phys. Chem.*, **94**, 3394 (1990).
24. Shin, K.-S. K. and Zink, J. I., *Inorg. Chem.* **28**, 4358 (1989).
25. Loppnow, G. R. and Mathies, R. A., *Biophys. J.* **54**, 35 (1988).
26. Sue, J. and Mukamel, S., *J. Chem. Phys.* **88**, 651 (1988).
27. Sension, R. J., Kobayashi, T. and Strauss, H. L., *J. Chem. Phys.* **87**, 6233 (1987).
28. Myers, A. B. and Mathies, R. A., in *Biological Applications of Raman Spectrometry*, Vol. 2, edited by Spiro, T. G. (Wiley-Interscience, New York, 1987), p. 1.
29. Schomacker, K. T. and Champion, P. M., *J. Chem. Phys.* **84**, 5314 (1986).
30. Takahashi, H., ed., *Time-Resolved Vibrational Spectroscopy V* (Springer-Verlag, Berlin, 1992).
31. Phillips, D. L. and Myers, A. B., *J. Phys. Chem.* **95**, 7164 (1991).
32. Yang, T.-S. and Myers, A. B., *J. Chem. Phys.* **95**, 6207 (1991).
33. Lee, S.-Y. and Heller, E. J., *J. Chem. Phys.* **71**, 4777 (1979).
34. Tang, J. and Albrecht, A. C., in *Raman Spectroscopy*, Vol. 2, edited by Szymanski, H. A. (Plenum Press, New York, 1971), p. 33.
35. Heller, E. J., Sundberg, R. L. and Tannor, D., *J. Phys. Chem.* **86**, 1822 (1982).
36. Hudson, B. S., Kohler, B. E., and Schulten, K., in *Excited States*, Vol. 6, edited by Lim, E. C. (Academic Press, New York, 1982), p. 1.
37. Buma, W. J., Kohler, B. E., and Song, K., *J. Chem. Phys.* **92**, 4622 (1990).
38. Buma, W. J., Kohler, B. E., and Song, K., *J. Chem. Phys.* **94**, 4691 (1991).
39. Share, P. E. and Kompa, K. L., *Chem. Phys.* **134**, 429 (1989).
40. Doig, S. J., Reid, P. J., and Mathies, R. A., *J. Phys. Chem.* **95**, 6372 (1991).

41. Yoshida, H., Furukawa, Y., and Tasumi, M., *J. Mol. Struct.* **194**, 279 (1989).
42. Jean, J. M., Fleming, G. R., and Friesner, R. A., *Ber. Bunsenges. Phys. Chem.* **95**, 253 (1991).
43. Schoenlein, R. W., Peteanu, L. A., Mathies, R. A., and Shank, C. V., *Science* **254**, 412 (1991).
44. Onuchic, J. N. and Wolynes, P. G., *J. Phys. Chem.* **92**, 6495 (1988).

11. Chemistry in Clusters: Solvation at the Single Molecule Level

JACK A. SYAGE
The Aerospace Corporation, P.O. Box 92957/M5-754, Los Angeles, CA 90009, U.S.A.

1. Introduction

One of the more familiar statements made about clusters is that they represent a phase of matter than is neither gas phase nor condensed phase, but somewhere in between. If so, then it is reasonable to expect that the physical and chemical behavior of clusters should take on properties that are intermediate between gas phase and condensed phase properties. The impetus to cluster research is driven largely by the potential to learn about the correspondence between these conventional phases of matter and the hope that some unforeseen molecular level properties of the condensed phase will be revealed in the cluster environment.

Research in clusters is pursued in a wide variety of scientific disciplines, including atomic physics (condensation, electron shell structure), laser physics (excimers), material science (crystal growth, phase transitions), chemistry (microsolvation, catalysis), and biochemistry (site-selective bonding, enzyme activity) [1, 2]. Our particular interest is in chemical reactivity in clusters and the dynamical events (energy redistribution, bond rearrangement, solvent reorganization) that accompany chemistry. The chemistry that we will discuss is initiated by a photoexcitation or photoionization pulse and the evolution of cluster properties are interrogated at later times by various probe excitation and detection techniques. The classes of chemistry in clusters reviewed here include ion-molecule reactions (e.g. electron and proton transfer, and sigmatropic rearrangements) [3–11], neutral excited-state proton transfer [11–18], and free radical chemistry [19].

The most common method of producing clusters is by supersonic expansion. This process gives rise to molecular cooling and low-energy collisions that promote cluster growth [20]. The cold clusters are formed nearly exclusively in their vibrationless level with rotational energy that is negligible on a chemical-energy scale. Photoexcitation energy can account for a large fraction of the total cluster internal energy. These experimental conditions allow important properties of the clusters to be selectively controlled, making them ideally suited as molecular-level models of the condensed phase.

J.D. Simon (ed.), Ultrafast Dynamics of Chemical Systems, 289–326.

(1) It is possible to measure properties of individual cluster sizes by optically-selective excitation and mass-resolved detection. A series of measurements as a function of cluster size provides direct information on how properties such as reactivity, electronic-state energies, energy and structural relaxation, etc. are influenced by addition of single molecules. (2) Because the internal energy of the clusters can be made low relative to the photoexcitation energy, the excited-state clusters are essentially isoenergetic and constitute a microcanonical ensemble (i.e., an ensemble of noninteracting subsystems at constant energy). Thermal averaging effects are minimized in photochemical cluster studies. (3) The specific cluster size, composition (e.g., when using mixed solvents), and geometry are often known making it possible to explore specific structural and solvent-core effects that would otherwise be averaged out in condensed-phase studies.

In this article, we review some of our recent work in cluster chemistry. Many of the examples we show can be classified as well-behaved cluster chemistry. These are cases where the observed chemistry corresponds to known reactions in the gas phase or condensed phase and is readily explanable by simple thermochemical models. More interesting are the unusual and unexpected results that can be attributed to some specific cluster environment such as structure and composition of an inner solvent shell. These effects would be obscured in condensed-phase studies, which measure an averaged solvent environment. From cluster work emerges a deeper understanding of the role of solvation and structure on chemical reactivity. We will cover two themes. (1) Chemical reactivity, as measured by rate of reaction, is investigated for specific solvent compositions and size. These studies include the use of mixed solvents to examine impurity effects. Time-resolved methods for measuring dynamics of solvent reorganization (i.e., "solvation") will also be discussed. (2) The correspondence between gas-phase to condensed-phase chemistry is investigated as a function of cluster size. These include studies to observe the onset to condensed-phase chemistry with increasing cluster size as well as studies to show both gas-phase and condensed-phase type chemistry in clusters that occur in relative proportions that are strongly dependent on the excitation process.

2. Experimental Methods

Most of our work employs multiphoton ionization and mass spectrometry. We have recently brought on line a time-of-flight photoelectron spectrometer of unusual design and a free jet chamber employing ellipsoidal reflection optics for detecting laser induced fluorescence. We describe these systems below:

(1) The molecular beam time-of-flight (TOF) mass spectrometer (MS) system pictured in Figure 1a is our work-horse apparatus. It employs a

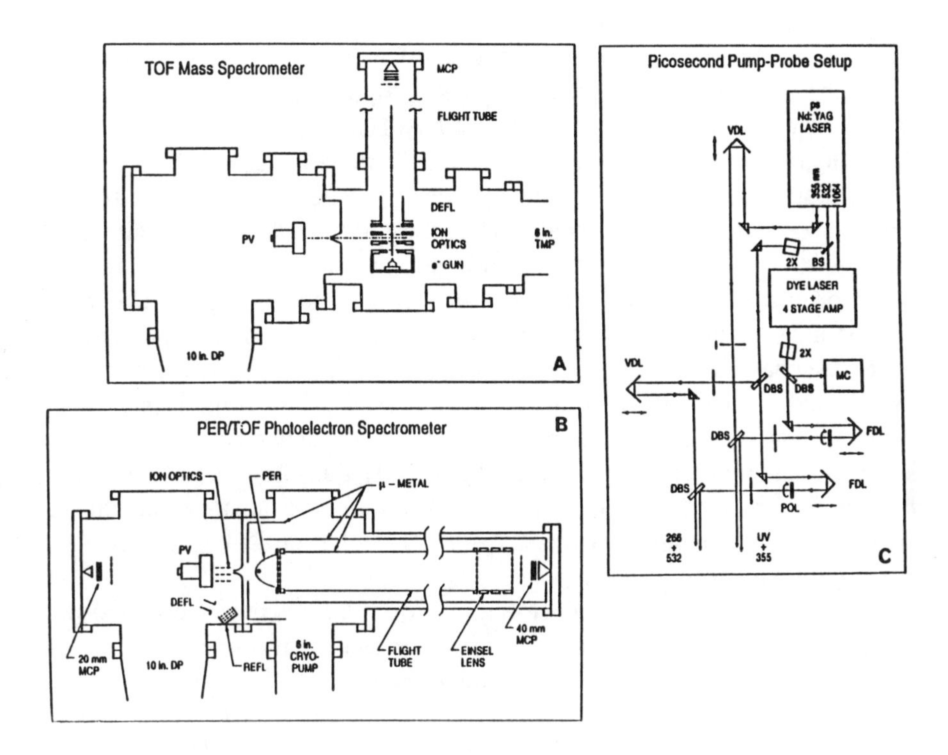

Fig. 1. Experimental schematic: (a) Molecular beam mass spectrometer, (b) molecular beam photoelectron spectrometer, (c) picosecond pump-probe arrangement. Terms are defined as follows: BS – beam splitter; DBS – dichroic beam splitter; DEFL – deflector plates; DP – diffusion pump; FDL – fixed delay line; I – iris; MC – monochromator; MCP – microchannel plate detector; PER – paraboloidal electrostatic reflector; POL – polarization rotator, PV – pulsed valve; REFL – reflecting grids; TMP – turbomolecular pump; VDL – variable delay line, 2X – doubling crystal period.

temperature-controlled pulsed valve that can be remotely positioned along three-axes during experiments. The free-jet expansion is skimmed and the resulting collimated beam is passed through a set of TOF ion optics. Three types of experiments are conducted in this apparatus. Multiphoton ionization is used for time-resolved measurements by picosecond pump-probe excitation [11–14, 19] and for spectroscopic measurements by nanosecond excitation. Electron-impact (EI) ionization is used to measure ionization cross sections in non-cluster experiments [21–23], to monitor cluster distributions as a function of pulsed valve conditions, and to measure thermochemical thresholds for cluster ion fragments. Finally, a combination of EI ionization, mass filtering, and laser photodissociation is used to measure size-specific electronic spectra and fragmentation mechanisms of cluster ions [7, 24, 25].

(2) Molecular beam photoelectron spectroscopy (PES) is carried out in an apparatus shown in Figure 1b. This device is based on a new concept of using a paraboloidal electrostatic reflector (PER) [26] to dramatically increase the collection efficiency of electrons detected by TOF energy analyzers [11, 13, 27]. This feature is crucial for studies using low energy laser pulses (e.g., in ultrafast spectroscopy) and for clusters where the density for any particular cluster size in the molecular beam can be very low. The chamber containing the pulsed valve also includes ion acceleration optics and a reflectron device for recording ion mass spectra. Because PES does not distinguish electrons originating from different cluster masses it is important to choose systems that have well characterized cluster spectroscopy so that specific cluster sizes can be selected by optical excitation. Alternatively, we have some capacity for detecting mass-selective PES spectra by simultaneously recording ion mass spectra (chamber I) and PES spectra (chamber II). Pulsed valve and excitation conditions can be adjusted to enhance or discriminate against certain cluster sizes and these changes in the mass spectrum can be correlated with changes in the photoelectron spectrum.

(3) Laser induced fluorescence of cluster products is detected in another apparatus using an ellipsoidal reflector to collect a large fraction of emission. Emission is dispersed in a monochromator and detected using a fast microchannel plate photomultiplier tube and a scanning 100-ps wide boxcar gate. Future experiments will include time-resolved stimulated emission pumping and fluorescence depletion measurements to monitor the progress of cluster dynamics in excited electronic states.

In order to sort out the variety of cluster processes that can occur simultaneously, it is necessary to employ a battery of excitation and interrogation techniques. The ones we most commonly use are described below:

(1) Picosecond pump-probe is used to initiate (pump) excited neutral reactions and detect (probe) by delayed ionization. A number of possible choices for the pump and probe pulses are available from the Nd:YAG harmonics and the dye laser (the latter employs the usual wavelength extension capabilities, e.g., doubling, mixing, Raman shifting). Two readily available and frequently used pump-probe arrangements are shown in Figure 1c. One arrangement uses the Nd:YAG harmonics, 266 nm as the pump and 532 nm (or 355 nm) as the probe. The other arrangment uses tunable UV output from the dye laser as the pump and 355 nm as the probe. We employ other combinations on occasion for both cluster and non-cluster work (i.e., 266 + Vis, UV + 532, UV + Vis). The number of labs besides our own combining time-resolved methods and cluster studies is small, but growing steadily [17, 18, 28–33].

(2) Nanosecond excitation and ionization is used to study cluster-ion properties [6–8, 11]. In some cases, the cluster ions are photoexcited to initiate chemistry [7, 8]. The cluster ion energies can be controlled to some extent

by controlling laser intensity and wavelength and choosing appropriate intermediate molecular excited states for the resonance ionization. Two-color arrangements are used to resonantly threshold ionize or to ionize and then resonantly photodissociate the ions.

(3) Electron-impact (EI) ionization is used for several purposes. We use low-energy EI to record cluster mass spectra as a diagnostic for adjusting our cluster source conditions to produce some desired cluster distribution. Another fruitful use of EI is as a convenient and non-selective ionization source for cluster-ion photodissociation studies. The method was particularly useful for our work on $(CH_3I)_n^+$ photodissociation spectra and dissociation mechanisms [7, 24, 25]. It is difficult to form $(CH_3I)_n^+$ parent ions by laser excitation because the first excited state is dissociative, which competes with ionization, and because the higher excited bound states require two-photon excitations at pulse energies that typically fragment the cluster ions [34].

The cluster samples were prepared by a variety of procedures. Solid compounds, such as phenol and aniline, were placed in a sample holder in the pulsed valve and heated to 40–60°C. For ammonia solvation, a 2-liter high-pressure cylinder was made up containing 1–10% ammonia in helium or argon. Water and methanol solvation was achieved by flowing helium through a bubbler containing either of these solvents. The CH_3I samples were prepared by flowing Ar (20–30 psia) through a bubbler that was cooled in a bath of dry ice/acetone or ice water, depending on the extent of clustering desired.

3. Excited State Proton Transfer

Aromatic alcohols are known to be more acidic in their S_1 state than in S_0 (i.e., the so-called pH jump) [35]. These reactions have been studied extensively in the condensed phase including several time-resolved experiments [35–37]. Cheshnovsky and Leutwyler pioneered the study of excited-state proton transfer (ESPT) in clusters by studying the system 1-naphthol-$(NH_3)_n$ [15]. The steady-state mass spectra and emission spectra they recorded using nanosecond excitation indicated a cluster size threshold for reaction of $n = 4$. Solgadi *et al.* measured steady-state threshold ionization curves for phenol-$(NH_3)_n$ and also reported an ESPT threshold of $n = 4$ [16], even though phenol is believed to be a weaker acid than naphthol. We have carried out mass-resolved picosecond measurements of the phenol (PhOH) ESPT reaction. The proton transfer, represented by

$$\mathrm{PhOH}^* \cdot \mathrm{B}_n \underset{k}{\longrightarrow} \mathrm{PhO}^{*-}\mathrm{H}^+\mathrm{B}_n$$

corresponds to a conversion from the locally excited S_1 state of the phenol acid to an excited ion-pair state (B_n refers to the solvent cluster consisting

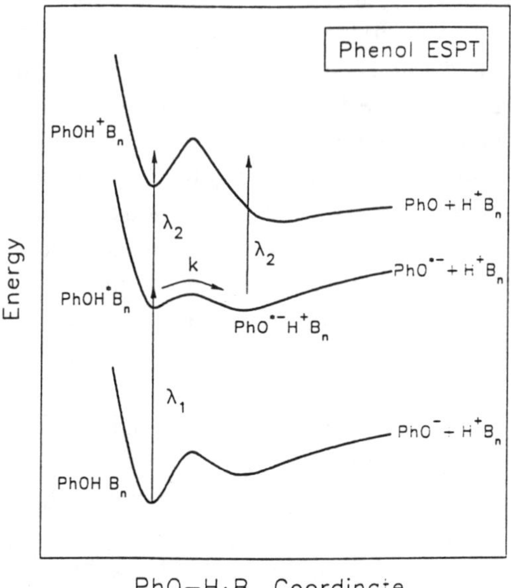

Fig. 2. Energy diagram for PhOH · $(NH_3)_n$ consistent with a cluster size of $n \sim 5$. ESPT produces an excited state ion-pair product that is ionized to the weakly-bound outer potential well of the cluster ion.

of n molecules of base B). Approximate potential energy curves for the relevant ground, excited, and ionic states of the cluster are shown in Figure 2 along with the picosecond pump-probe excitation/detection scheme used to measure the chemical rates. The energy curves are strongly dependent on solvent cluster size and basicity (explained below). The illustration given in Figure 2 approximates the case for phenol-$(NH_3)_5$. Reactant and product are distinguishable by mass spectrometry because each species fragments differently upon ionization (Figure 2). Ionization of reactant clusters $PhOH^* \cdots B_n$ produces the cluster ions $PhOH^+ \cdots B_n$ on the inner potential well that are strongly bound along the O—H coordinate. Whereas, ionization of the product ion-pairs $PhO^{*-} \cdots H^+B_n$ produces cluster ions $PhO \cdots H^+B_n$ on the outer well that dissociate readily to give H^+B_n. These ion signals are convenient probes of the dynamics along the neutral excited S_1/ion-pair reactive surface.

3.1 Stepwise Solvation Properties

Phenol, in a variety of solvent clusters, was excited at 266 nm. The excess energies for NH_3 solvation range from about 1900 cm^{-1} ($n = 1$) to about 2250–2800 cm^{-1} (estimated for $n = 3-7$) [16, 38]. The S_1 origin of uncomplexed phenol is at 274.7 nm [38]. A set of reactant decay measurements are presented in Figure 3 ($\lambda_2 = 532$ nm) as a function of solvent cluster

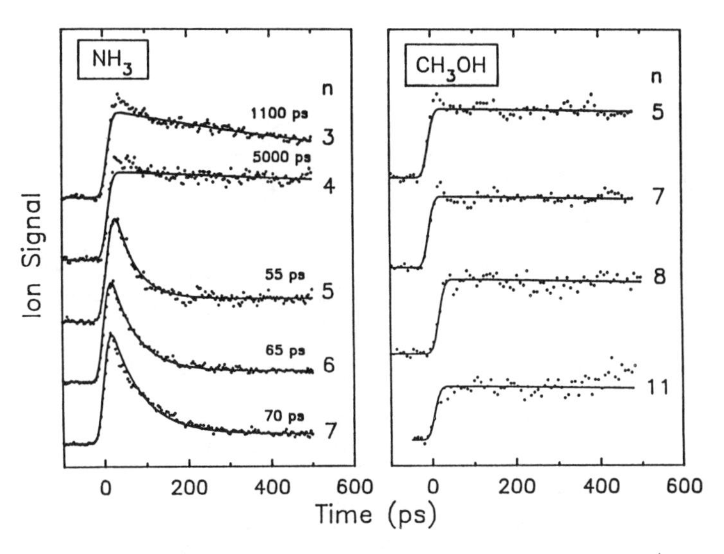

Fig. 3. Observed lifetime of the reactant $PhOH^* \cdot B_n$ (as measured by $PhOH^+ \cdot B_n$ detection) in solvent clusters $(NH_3)_n$ and $(CH_3OH)_n$. The calculated curves are described in [12]. Calculated curves for the $(CH_3OH)_n$ data assume a 10 ns lifetime.

size n for the solvents NH_3 and CH_3OH. A distinct threshold for ESPT was observed for NH_3 solvation at $n = 5$. However, no reaction was observed for CH_3OH solvation (up to $n = 11$, Figure 3). Results for other solvents are given elsewhere [12].

The reaction threshold observed for $(NH_3)_n$ solvation at $n = 5$ is due largely to stabilization of the ion-pair Coulombic potential by the proton affinity of NH_3. The ion-pair or proton-transfer states in clusters (or condensed phase) correspond to gas-phase charge transfer states. These gas-phase states, shown in Figure 4a, occur at very high energy relative to the covalent S_1 state because of the large energy required to form a free proton (the H atom ionization potential is 13.6 eV). States with large proton character, however, are stabilized enormously by complexation to molecules with high proton affinity. The interaction of a single NH_3 to PhOH is shown in Figure 4b to reduce the energy of the charge-transfer or ion-pair state dramatically relative to S_1. (NH_3 has a proton affinity of 8.8 eV) [39]. The stabilization energy for successive NH_3 molecules becomes less and less (the stepwise proton affinity) [40], however, the cumulative stabilization can be enough to lower the energy of the ion-pair state to below that of the S_1 state, thus making ESPT thermodynamically allowed. The experimental measurements indicate that this occurs for five NH_3 molecules and the calculated potentials in Figure 4c are in near agreement with this result.

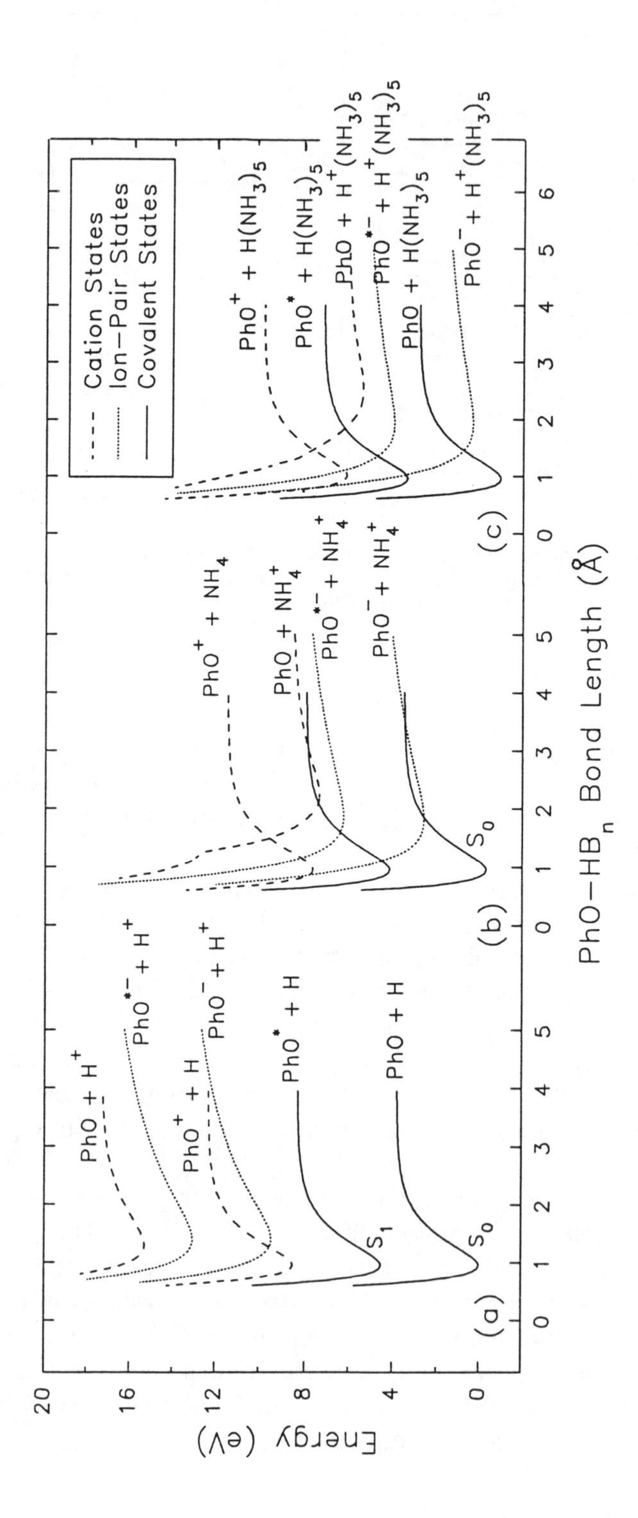

Fig. 4. Approximate PhOH · $(NH_3)_n$ cluster potential energy curves for the reactive PhO—H coordinate. (a) $n = 0$, (b) $n = 1$, and (c) $n = 5$.

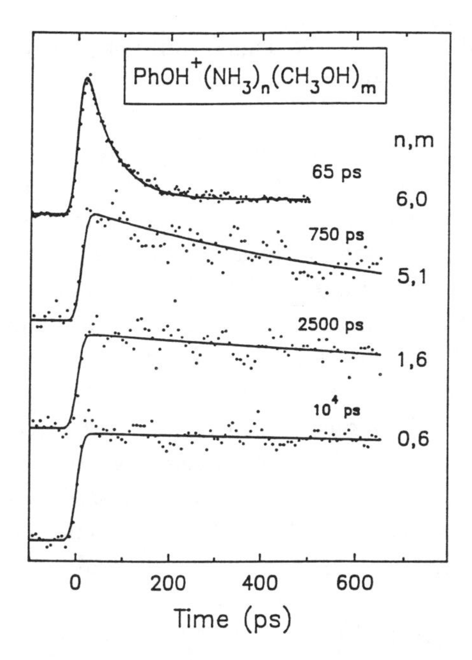

Fig. 5. Measured rates of ESPT for the mixed solvent clusters $PhOH^*(NH_3)_n(CH_3OH)_m$ using 266-nm pump and 532-nm probe excitation.

If the rates of ESPT in Figure 3a are plotted versus n, the resulting curve takes on a shape and inflection that is reminiscent of an acid-base titration curve. The similarity arises because our mass-specific, time-resolved cluster experiments are, in fact, a *single molecule titration* measurement. In similar experiments by Zewail and coworkers [17], and Bernstein, Kelley, and coworkers [18] on 1-naphthol-$(NH_3)_n$, a threshold solvent cluster size of $n = 3$ was observed for ESPT versus $n = 5$ for phenol. This is consistent with the more acidic 1-naphthol molecule [41] requiring fewer base molecules to dissociate the proton.

3.2 DISRUPTION OF THE SOLVENT STRUCTURE

Adding a dissimilar solvent molecule to an otherwise pure solvent cluster can cause a change in the hydrogen bonding network that can disrupt proton transfer. Reaction rates for mixed solvent distributions $PhOH \cdot (NH_3)_n(CH_3OH)_m$ of similar cluster size ($n + m$ = 6 or 7) were measured and are presented in Figure 5. The substitution of a single CH_3OH molecule for an NH_3 molecule causes a substantial decrease in reaction rate. Excited phenol in neat NH_3 solvent cluster $(n, m) = (6, 0)$ reacts in 65 ps whereas the $(5, 1)$ cluster reacts in 750 ps. Because the (5,0) cluster reacts in 60 ps (Figure 3), the (5,1) result indicates that the mere *addition* of a CH_3OH molecule effectively "poisons" the reactivity of the (5,0) cluster. This is quite surprising because one would not expect the proton affinity of the solvent to decrease by simply adding CH_3OH.

An explanation for the slowed reaction is that CH_3OH acts as an intruder that disrupts the hydrogen bonding network of the neat $(NH_3)_n$ solvent structure that is crucial to solvating and stabilizing the dissociated proton [36, 37]. It is known from gas-phase cluster studies that the closed-shell structure $NH_4^+(NH_3)_4$ is very favorable for stabilizing a proton [40, 42]. However, spectroscopic and mass spectrometric evidence for the mixed NH_3/CH_3OH solvation of PhOH indicates that NH_3 forms the stronger bond to PhOH, but that CH_3OH forms the stronger bond to PhOH $\cdot$ NH_3 [14]. The experimental evidence is supported by reports that the NH_3—CH_3OH bond is stronger than the NH_3—NH_3 bond [43]. Consequently, given a chance, CH_3OH will displace NH_3 from the first solvation shell and prevent a pure NH_3 solvent cluster from forming. Also CH_3OH can form a stable hydrogen bond to the phenol oxygen which diminishes excited state acidity, whereas NH_3 avoids that site [38].

The concept of critical solvent structures in ESPT reactions has been investigated in the solution phase by Robinson and coworkers [36]. Their picosecond measurements and theoretical analyses on aromatic acids (e.g., 1- and 2-naphthol) in H_2O and alcohol (CH_3OH and C_2H_5OH) support the notion that a critical solvent cluster core is necessary to act as an efficient proton acceptor. For 2-naphthol in H_2O, the critical solvent-core size is about four [36].

Although, the structure of clusters that involve hydrogen bonds can often be deduced or determined from optical shifts and mass spectral distributions, direct spectroscopic measurement of solvent structure about a reactive solute molecule has not yet been made. Advances in high resolution spectroscopy [44–46] and new applications in mass spectrometry [14, 47] are beginning to yield information on the bonding structure of homogeneous and mixed clusters; however, these efforts are young. Progress in understanding the influence of solvent structure on chemical reactivity and other dynamical processes will benefit greatly by a closer association of dynamicists and spectroscopists.

3.3 Solvent Reorganization in Clusters

As a consequence of proton transfer, the solvent is no longer at equilibrium because of the large dipole formed by the charge separation along the O—H coordinate. This is a substantial perturbation that should give rise to further dynamics in the cluster. We represent the separation of reaction and solvent dynamics as

$$\mathrm{PhOH}^{*}\cdots \mathrm{B}_n \underset{k}{\longrightarrow} [\mathrm{PhO}^{*-}\cdots \mathrm{H}^{+}\mathrm{B}_n]_{ur} \underset{k_s}{\longrightarrow} [\mathrm{PhO}^{*-}\cdots \mathrm{H}^{+}\mathrm{B}_n]_{r}$$

where the subscripts ur and r denote unrelaxed and relaxed solvent configurations and k_s is the solvent relaxation rate constant. The product formation curves in Figure 6a show two time-dependences, a rapid rise that matches the decay of the $PhOH^{+}\cdot (NH_3)_n$ signals (55–70 ps) and a much slower growth component of about 300 ps. The fast time response is due to the ESPT reaction. The slow time response is ascribed to solvent cluster reorganization.

How the potential energy along the O—H coordinate might change with time due to solvent reorganization is illustrated qualitatively in Figure 6b. The solvent network must reorganize to achieve a structure that best accomodates the proton. This reorganization or "solvation of the proton" causes delocalization of charge and lengthening of the O—H bond. The relaxing ion-pair moves closer to a geometry representing the cluster-ion equilibrium configuration. This improves the Franck-Condon (FC) factors for ionization and leads to an increased ionization efficiency with time. The curves in Figure 6c were calculated for different relative ionization efficiencies from the unrelaxed (i_b) and relaxed (i_c) product clusters. The details of this simple kinetic model are given elsewhere [12]. The calculations that best fit the data in Figure 6a are indeed those for which the ionization efficiency increases with time. These results are consistent with Franck-Condon factors for ionization that improve as a result of solvent reorganization.

A more definitive measure of geometry change is to monitor individual FC components to ionization [13]. This is achieved by time-resolved photoelectron spectroscopy. Figure 7 contains photoelectron spectra (PES) for different probe delay times. The spectra are broad because of the distribution of cluster sizes, large density of low frequency modes, rapid vibrational redistribution, and large geometry change upon ionization. The $t = -100$ ps spectrum is identical to the 266 nm only spectrum (not shown) and therefore corresponds to ionization to the $PhOH^{+}\cdot (NH_3)_n$ complex (Figure 2 or 7b). For positive t a low electron energy band appears. Low-energy electrons correspond to high-energy ions and would be the outcome of poor FC factors to ionization. However, in time, the solvent reorganizes to better solvate the proton and evolves to a geometry that is more similar to the ion geometry. Consequently, the FC factors for ionization improve (become more vertical), thus producing lower energy ions (i.e., higher energy electrons). Hence, the PES is expected to shift from low electron energy to high electron energy. This effect is analogous to that which causes the time dependence of emission bands used to measure solvent relaxation in the condensed phase [48–50].

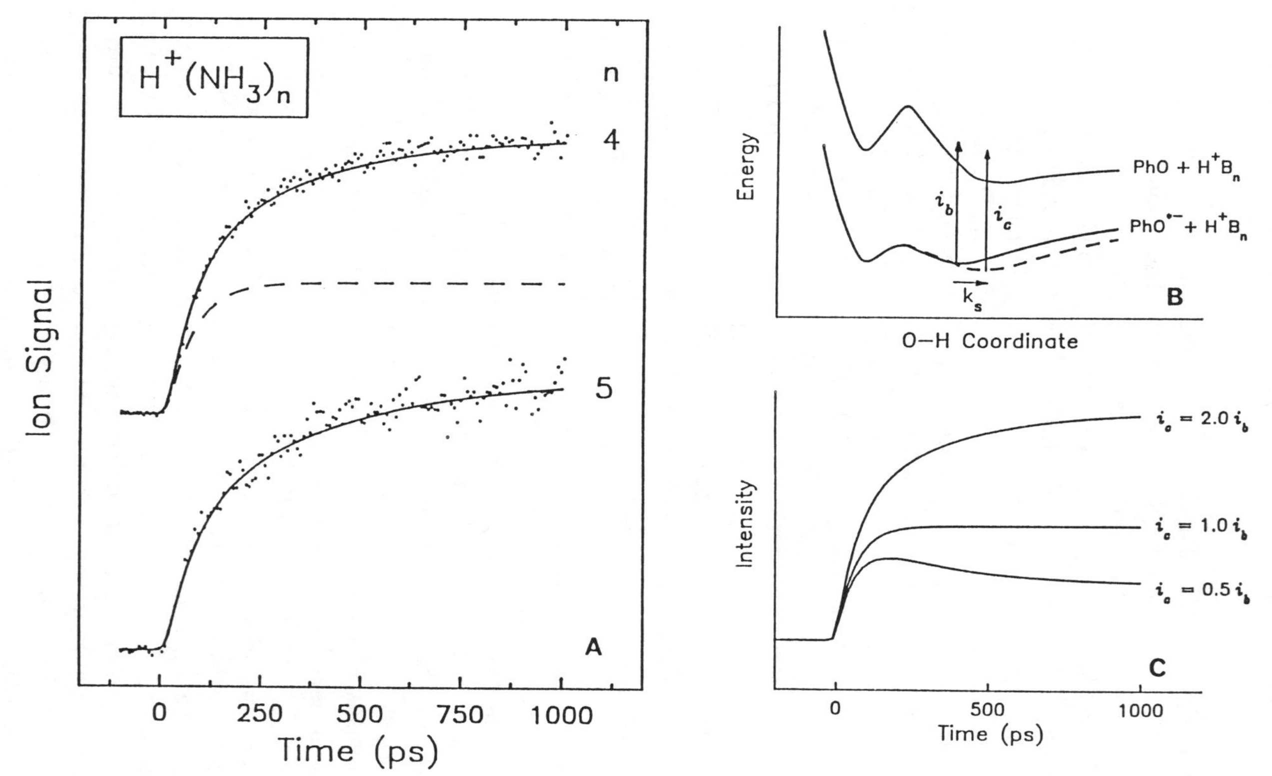

Fig. 6. (a) ESPT product formation kinetics (as measured by H^+B_n detection) for NH_3 solvation using 266-nm pump and 532-nm probe excitation. The data were fitted according to [12] for $k = (65\ ps)^{-1}$, $k_s = (350\ ps)^{-1}$, $i_c/i_b = 2.4$ (for $n = 4$); and $k = (65\ ps)^{-1}$, $k_s = (300\ ps)^{-1}$, $i_c/i_b = 2.0$ (for $n = 5$). Evaporation in the dissociative ionization process accounts for formation kinetics appearing at $n = 4$ despite reaction threshold at $n = 5$. (b) Schematic representation of the evolving solvent-relaxed ion-pair potential. The dotted line is a portrayal of the O—H potential for a relaxed solvent coordinate. (c) Time dependence calculated for $k = (65\ ps)^{-1}$, $k_s = (300\ ps)^{-1}$ and different ratios of unrelaxed (i_b) to relaxed (i_c) ionization efficiencies.

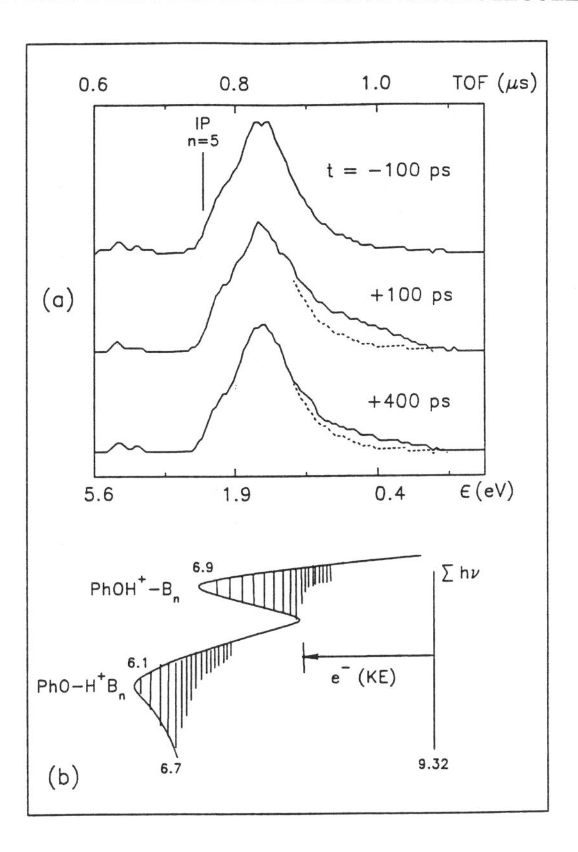

Fig. 7. (a) Time-resolved photoelectron spectra as a function of pump-probe delay time t (low resolution using 5 V acceleration). The $t = -100$ ps spectrum is superimposed on the latter two spectra (dashed line) to emphasize the low energy band. (b) Cluster ion potential energy curve along the O—H coordinate for B_n = $(NH_3)_5$ (energy values in eV, [13]).

The time dependence for the low-energy and high-energy photoelectron bands is plotted in Figure 8. The low-energy band (high-energy ions) increases rapidly in intensity and then decays at a slower rate. The high-energy band (low-energy ions) shows a corresponding slow rate of increase. A number of observations support the interpretation of a Franck-Condon distribution that changes with time due to solvent reorganization. The direction of the shift is from high-energy to low-energy ion excitation (red-shifted transitions to the ion), as expected for solvent relaxation. The low-energy band decays to a plateau level rather than to the baseline suggesting a spectral shift that is less than the width of the band. The high-energy band should grow from a similar non-zero baseline at $t < 0$, however, this level is obscured by the one-color background signal. The calculated curves in Figure 8 are based on a simple model that assumes a Gaussian photoelectron band that shifts to a new

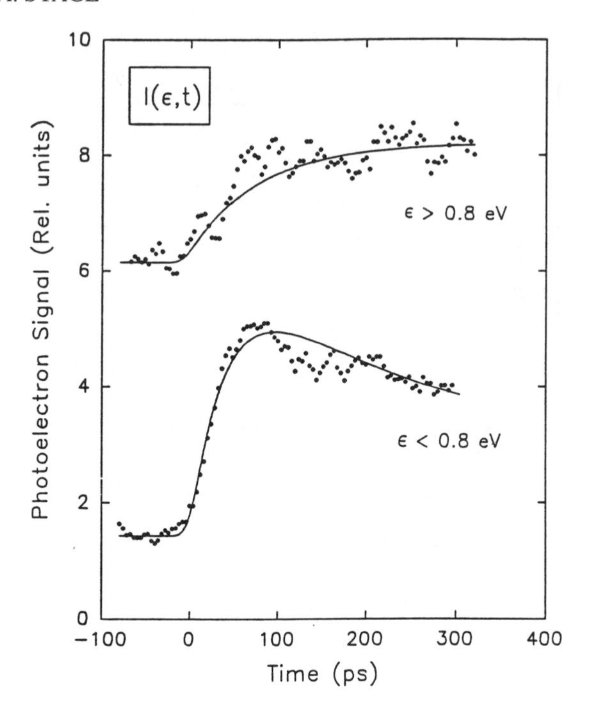

Fig. 8. Time-dependence of low-energy ($0.8 > \varepsilon > 0$ eV) and high-energy ($2.6 > \varepsilon > 0.8$ eV) photoelectron bands. Fitted curves were calculated for rate constants $k = (50 \text{ ps})^{-1}$ and $k_s = (300 \text{ ps})^{-1}$ and spectral parameters discussed in [13].

peak energy by a single-exponential rate constant k_s [13]. At this preliminary stage, the fits are not quantitative; however, the rate constants $k = (50 \text{ ps})^{-1}$ and $k_s = (300 \text{ ps})^{-1}$, used to fit the data, are in excellent agreement with the time-resolved mass spectrometry results in Figure 6a.

4. Free Radical Chemistry in Clusters

Free radical chemistry is usually associated with energetic chemistry (e.g., combustion, propulsion), but it is also important in the natural environment (e.g., atmospheric chemistry). The relationship between free radical chemistry in the gas phase and the condensed phase has attracted intense interest lately since the discovery that Antarctic stratospheric ozone depletion is aggravated by heterogeneous chemistry occuring on ice crystals [51]. Similarly, tropospheric pollution chemistry proceeds differently in the gas phase versus in aqueous aerosols. For example, the reaction of the common pollutant NO_2 with water to form nitric acid is endothermic in the gas-phase, but reacts efficiently in the aqueous phase. In water, the reaction is second order in NO_2 [52]; however, the kinetics may be different in small clusters.

Understanding this difference could explain the uptake of NO_2 in aqueous aerosols (in clouds and humid air) and its role in forming acid rain. We are beginning studies to measure atmospherically relevant chemistry in clusters. Another area where we are applying cluster research is in efforts to improve the efficiency of chemical propulsion. Rocket thrusters typically use cryogenic propellants (e.g., liquid H_2, O_2) or room temperature liquid propellants (e.g., hydrazines, hydrocarbons, N_2O_4) that are injected into a combustion chamber where they are assumed to volatilize before reacting. In fact, for many thruster designs, there is a growing suspicion that substantial chemistry occurs at the gas-liquid interface, which can adversely affect chemical thrust performance. One can find many examples in nature of chemistry occuring at gas-liquid interfaces. Studies of these chemistries in clusters could very well provide crucial understanding of the correspondence between gas-phase and condensed-phase chemical properties.

4.1 METHYL RADICAL REACTION IN $(CH_3I)_n$ CLUSTERS

In this section we give an example of a time-resolved study of radical chemistry in $(CH_3I)_n$ clusters where a combustion process is believed to have been initiated [19]. The chemistry is assumed to be driven by the exothermicity of the reaction

$$CH_3I + CH_3I \longrightarrow C_2H_6 + I_2 \qquad \Delta H = -0.52 \text{ eV}$$

A bottle of CH_3I clearly does not combust so one can safely conclude that the activation energy for the above reaction is very high. However, the reaction can be initiated by a radical mechanism that provides enough heat to sustain thermal chemistry. The cluster photoreactions were initiated by exciting into the directly dissociative $\tilde{A}$ state. The monomer reaction dissociates into CH_3 + I in less than 1 ps [53], however for certain cluster sizes one can expect sufficient caging of the initially formed radicals to induce chemistry within the cluster. Vaida *et al.* suggested that I_2 is produced in $(CH_3I)_n$ clusters on the basis of the detection of I_2^+ following nanosecond resonance ionization through the valence $\tilde{A}$ state [54]. These results, however, did not distinguish between I_2^+ formed by ionization of neutral product I_2 versus dissociation of cluster ions. To definitely establish the neutral cluster reaction channels, we identified the purely ionic fragmentation processes by employing direct variable-energy electron-impact (EI) ionization, and by recording size-specific cluster ion photodissociation mass spectra [7, 19, 25]. Since our work, two other studies of $(CH_3I)_n$ cluster chemistry have confirmed the neutral I_2 mechanism [55, 56].

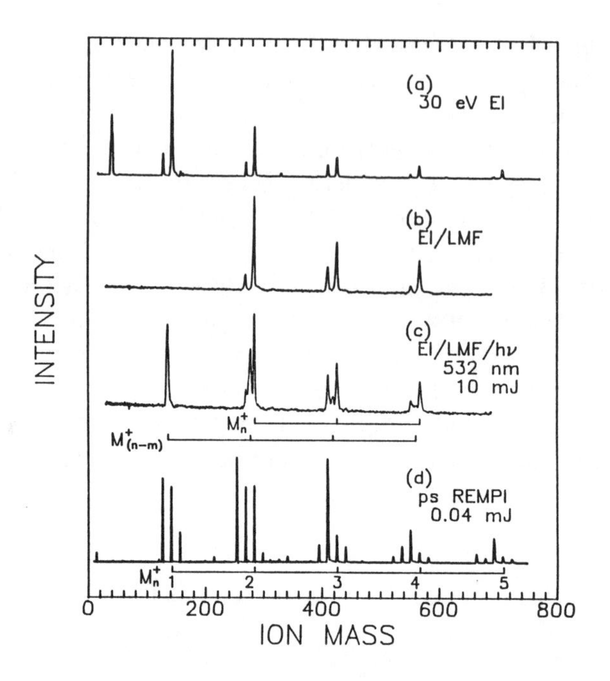

Fig. 9. $(CH_3I)_n$ cluster ionization mass spectra (M = CH_3I). (a) Direct ionization by EI excitation at 30 eV. (b) EI ionization and low mass filter (LMF). (c) Same, but with resonant photodissociation of cluster ions. (d) Picosecond $\tilde{A}$ state REMPI.

4.2 Neutral versus Ionic Cluster Reactions

When using resonance ionization to study photodissociation processes, the question often arises whether the detected fragment ions occur by ionization of neutral products or dissociation of parent ions. This competition is a concern in cluster research because the weak van der Waals (vdW) bonds usually mean that both neutral and ionic dissociation will occur. Picosecond pump-probe measurements help to sort out these competitive processes because they usually occur over different timescales. However, we also take the independent approach of measuring directly the ion dissociation mechanisms by EI ionization and by mass-selective ion photodissociation. The ion photodissociation experiments make use of a crossed electron-laser-molecular beam configuration as described in Section 2. The molecular clusters are first ionized by EI excitation, then mass filtered, and finally resonantly photodissociated using nanosecond laser pulses [7, 24, 25].

The methodology behind using both resonance ionization and direct ionization to identify neutral cluster reactions is illustrated in Figure 9 for $(CH_3I)_n$ clusters. EI ionization is used to ionize the clusters directly in order to ascertain fragment ions that occur exclusively by ion dissociation mechanisms.

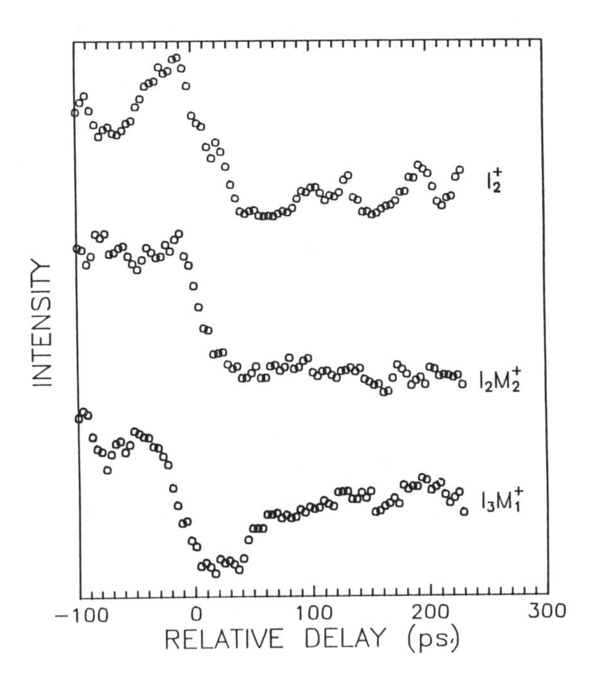

Fig. 10. Picosecond measurements using 266-nm pump and 532-nm probe: M_n ($M = CH_3I$) reaction for loss of two and three CH_3 groups [I_2M_{n-2} ($n = 2, 4$) and I_3M_{n-3} ($n = 4$), respectively].

The electron beam flux is sufficiently low that all excitation events occur by a single electron. The EI mass spectrum in Figure 9a shows that dissociation following cluster ionization consists only of C—I bond breaking or van der Waals (vdW) evaporation. The operation of the TOF low mass rejection filter is illustrated in Figure 9b where all masses less than the dimer mass are rejected from the excitation volume. (The details of this device are described elsewhere.) [7] In Figure 9c, the remaining ions are excited by a laser pulse that is resonant with a predissociative absorption in CH_3I. This excitation, which is known to cleave the C—I bond in monomer CH_3I^+ results only in vdW dissociation in the cluster ions. It is not surprising that the weakest bond breaks. (In Section 5.2 we present an example of a cluster ion dissociation in which a strong bond breaks preferentially to a weak bond.) Finally, Figure 9d contains the picosecond mass spectrum of $(CH_3I)_n$ resulting from excitation of the dissociative $\tilde{A}$ state. A distinctly different spectrum is observed from the ion fragmentation spectra in Figures 9a–c. As a result of the overall procedure encompassed by Figure 9, the additional signals in Figure 9d can be assigned to ionization of *neutral* reaction products.

The excited state neutral clusters undergo very rapid chemistry. Examples of picosecond pump-probe traces for some of the major products are given

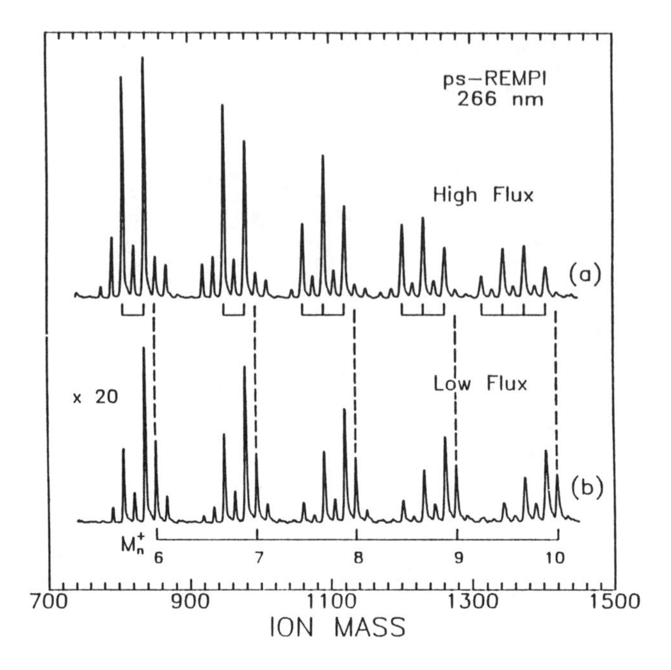

Fig. 11. ps-REMPI mass spectra showing reactions in larger clusters. The 266-nm high and low photon flux conditions differ by about a factor of three (within the range of 10^{27}–10^{28} $cm^{-2}\ s^{-1}$). The scaling factor in (b) shows that the overall signal intensity is steeply dependent on photon flux, however, the relative intensities are much less sensitive.

in Figure 10. A more complete mapping of the product time dependence versus cluster size has been reported elsewhere [19]. The main conclusion is that the chemistry is complex and that sequential processes indicative of a chain reaction appear to be occuring on the timescale of the 20–30 ps laser pulse durations. We explain only a few signals here. The measured rates for the formation of $I_2(CH_3I)_{n-2}$ as a function of cluster size n show a sharp decrease in signal level at $t = 0$ and remains constant to long delay times. The latter behavior suggests that photodissociation takes place in the long-lived ion, presumably due to the I_2 (or I_2^+) component. However the sharp threshold with delay time indicates that the ion, and therefore the neutral product, is formed within the instrument resolution of about 20 ps. Also shown in Figure 10 is the product $I_3(CH_3I)_{n-3}$, corresponding to loss of three CH_3 groups, that opens up for cluster sizes $n \geq 4$. The ion signal undergoes a rapid depletion at $t = 0$ within the instrument response time and then recovers for $t > 0$. Because of the signal recovery, it is unlikely that the dissociation occurs in the long-lived ion, but rather in the short-lived neutral. The $I_3M_{n-3}^+$ response is strong evidence for a neutral reaction mechanism.

4.3 "HOT" RADICAL REACTION MECHANISM

The 266 nm picosecond mass spectra in Figures 9d and 11 show that the photoexcited clusters can lose a substantial fraction of the CH_3 groups allowable by cluster size. Also evident in Figure 11 is an alternation of peak intensities separated by two CH_3 mass units. Because the product distribution is observed to shift to greater demethylation at higher pulse energies, it is believed that sequential photon absorption plays some role in demethylation. However, because demethylation is relatively extensive even at low pulse energies (Figure 11b), it seems unlikely that sequential photon absorption can account for all the products observed. This conclusion is supported by the photon flux and absorption cross section data [19].

The picosecond measurements and mass spectral features can be explained by a simple, but not necessarily unique, reaction mechanism based on the chemistry of a "hot" CH_3 radical. The photodissociation of CH_3I by 266 nm light provides a 4.66 eV photon to break the 2.30 eV C—I bond. The excess energy in the direct dissociation is released mainly as kinetic energy in the CH_3 radical [57], which can lead to the exothermic reaction

$$[CH_3 + (CH_3I)_{n-1}] \longrightarrow C_2H_6 + I(CH_3I)_{n-2} \qquad \Delta H = -1.37 \text{ eV}$$

where we assume that the photodissociated I atom is lost (recombination of I atoms would contribute an additional exothermicity of $\Delta H = -1.58$ eV). An alternative pathway to production of ethane would involve the absorption of two photons by $(CH_3I)_n$ to produce two CH_3 radicals that could recombine to release 3.81 eV of heat. In both cases the heat of reaction, which is augmented by the initial radical kinetic energy as well as further recombination including the dissociated I atoms, could drive additional chemistry based on the exothermicity of ethane formation. The hot cluster product above, for example, could undergo the thermal rearrangement

$$I(CH_3I)_{n-2} \longrightarrow C_2H_6 + I_3(CH_3I)_{n-4} \qquad \Delta H = -0.52 \text{ eV}$$

with the additional heat sustaining further loss of ethane. Again subsequent photodissociation would enhance the rate of C—I bond breaking and exothermic production of C_2H_6.

The near absence of a multiple I atom loss product is consistent with efficient caging of this heavy mass which acquires little kinetic energy (0.15–0.25 eV) by photodissociation and even less by the thermal route. This provides an efficient means for production of the major product I_2. Recent work by Ziegler *et al.* [55] and Fan and Donaldson [56] show that on the nanosecond timescale thermally hot I_2 is observed free of the clusters (although the excitation energy and, therefore, the I_2 energy content differ in these two studies). It will be interesting in a future time-resolved experiment to measure the rate of production of I_2 in the cluster and the subsequent rate of evaporation or separation from the cluster. Perhaps I_2 in different vibrational levels will show different time dependences reflecting some diversity of the local environment in clusters.

5. Ion-Molecule Chemistry

5.1 Deciphering Gas-Phase Reaction Coordinates

Large compilations of data now exist for gas-phase bimolecular ion-molecule reactions. The most common and routine properties measured in gas-phase experiments are product branching ratios, enthalpies, and activation energies. Much less information is obtained regarding the reactive complex in terms of structure, minima energy, and boundness. Rather these properties are inferred from thermodynamic and kinetic data or from theory, and are often represented by a reaction coordinate for the minimum energy path to products. Although, the minimum energy path is the most probable route for product formation, it is not the most likely owing to the random nature of collision-pair impact parameter and orientation. However, determining a multidimensional potential energy surface encompassing a large gamut of collision parameters is not feasible except for the simplest reactions (e.g., atom-diatom reactions). Cluster research, however, can provide crucial information.

Photoexcitation of van der Waals (vdW) complexes is a useful way to study the bound and quasibound reactive complexes associated with gas-phase bimolecular collisions [58, 59]. Ionization of vdW complexes to form reactive complexes of ion-molecule reactions [3–11] is especially favorable because they tend to be stable due to charge-induced dipole interactions. These complexes assume the lowest energy structure and so can be expected to be typical of the minimum energy reaction coordinate of a bimolecular reaction. At any rate, a van der Waals complex represents a single coordinate; in other words an aligned complex.

Ion-molecule kineticists, for many years, have postulated that some reactions (mostly fast) proceed along barrierless single-well reaction coordinates whereas other reactions (mostly slow) involve double-well barrier processes [60]. Included in the latter group were classes of reactions in which the charge distribution in the reactant was delocalized from the eventual charged leaving group (as can happen in electron and proton transfer reactions and S_n2 substitutions [62]). The double-minima barrier mechanism [60, 61] became widely accepted well before the existence of two distinct and stable reactive complexes was experimentally confirmed [10, 11]. $PhOH^+$ is an acid cation that has a ground electronic-state charge distribution delocalized from the poten-

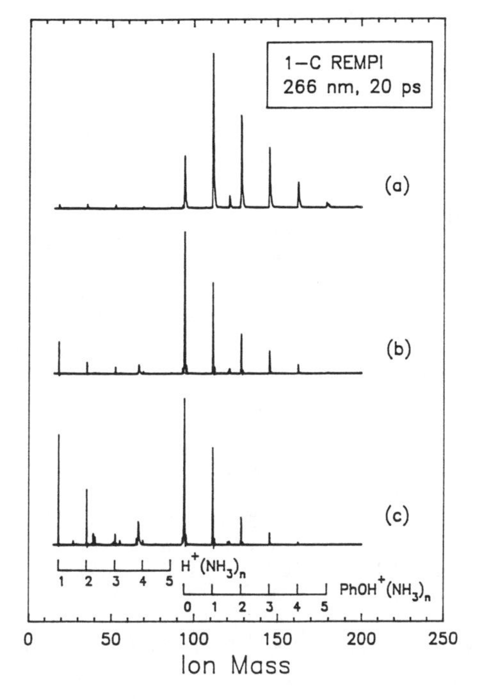

Fig. 12. One-color picosecond REMPI mass spectra as a function of increasing pulse energy. (a) two-photon signals, (b) appearance of $H^+(NH_3)_n$ at the three-photon level, (c) appearance of aromatic fragmentation at the four-plus photon level.

tial proton leaving group, which prompted us to investigate the possibility of detecting two stable complex forms for the reaction

$$PhOH^+ + (NH_3)_n \longrightarrow PhO + H^+(NH_3)_n$$

Using a picosecond delayed-ionization technique, we determined that there are two stable reaction complexes: $PhOH^+ \cdots (NH_3)_n$ and $PhO \cdots H^+(NH_3)_n$. By referring to Figure 2, it can be seen that zero time-delay pump-probe ionization forms complex ions of the first type. These are strongly bound to proton transfer as seen by the lack of $H^+(NH_3)_n$ in the picosecond two-photon ionization mass spectrum in Figure 12a. Only by increasing pulse energy so that the ion complex absorbs a photon was the barrier surmounted to produce $H^+(NH_3)_n$ (Figures 12b and 12c). The product complex can be formed by taking advantage of the ESPT reaction from S_1 phenol and using delayed ionization, as shown in Figure 2. This ion complex, being formed on the product side of the barrier, readily dissociates to $H^+(NH_3)_n$ as seen in Figure 13. For high energy ionization (two 532-nm photons), the second complex was observed to be 95% dissociative [11]. Witness the decay of $PhOH^+(NH_3)_n$ signal to near zero for 532-nm ioniza-

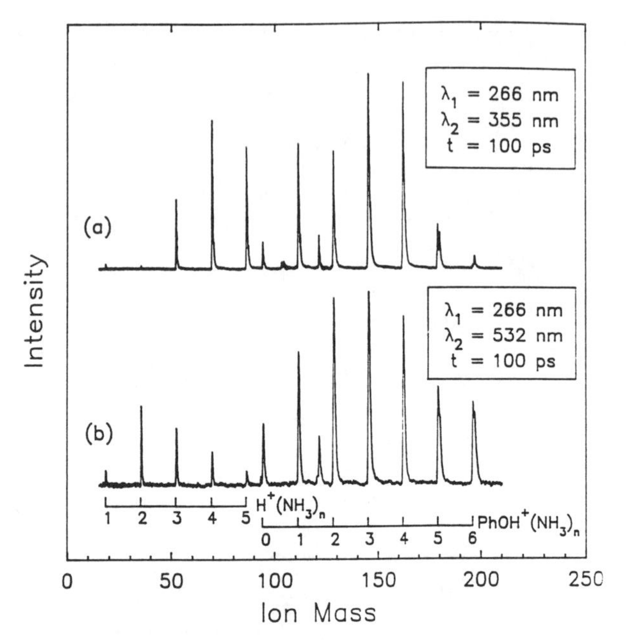

Fig. 13. Two-color picosecond delayed ionization REMPI mass spectra for (a) $\lambda_2 = 355$-nm, one-photon ionization, and (b) $\lambda_2 = 532$-nm, two-photon ionization.

tion (Figures 3 and 14a). For lower energy ionization (one 355-nm photon), the $PhOH^+(NH_3)_n$ signal decays to a plateau level corresponding to 65% dissociation (Figure 14b) [11]. Because at later times the signal is due to the $PhO \cdots H^+(NH_3)_n$ complex, we can conclude that it is bound and stable.

What was uncertain from our earlier work was the height of the barrier because we did not know the distribution of ion energies. This measurement has since been made by photoelectron spectroscopy. Figure 15 shows the PES for different photon energies. A distribution of cluster sizes is ionized, all of which contribute to the PES. The maximum electron energy for each cluster size is marked, based on reported ionization potentials [16]. Figure 15a corresponds to the photoelectron spectrum for cluster sizes $n > 4$, achieved by using an excitation wavelength of 286 nm that is to the red of the $S_1 \leftarrow S_0$ origin transition for the smaller cluster sizes [16]. The energy distribution for these larger complexes extends to relatively high energy. Similar results are seen for the other spectra. The spectra in Figures 15d and 15e demonstrate that the cluster ions for $n = 1$ and 2 are also formed with substantial energy. These spectra compare the results for a large cluster distribution (not greatly exceeding $n = 6$) produced by ionization in the cold central portion of the gas pulse (Figure 15d) and for a small cluster distribution (dominated by $n \leq 3$) produced by ionization resonant with $n = 1$ in the warm leading edge of the

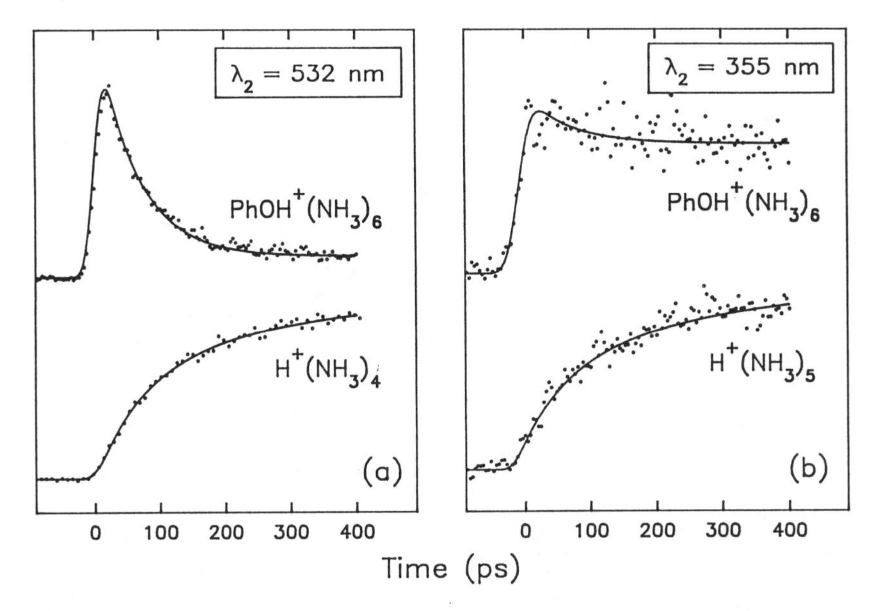

Fig. 14. Time-resolved measurements of reactant and product ionization signals $[PhOH]^+ \cdot (NH_3)_n$ and $H^+(NH_3)_n$, respectively, for cluster size $n = 6$. (a) $\lambda_2 = 532$-nm, two-photon ionization, (b) $\lambda_2 = 355$-nm, one-photon ionization. We show $H^+(NH_3)_4$ traces for $n = 4$ and $n = 5$ because, typically, two and one solvent molecules evaporate at these respective ionization energies following dissociative proton transfer [19b]. The calculated curves were fitted to a model in Ref. 10 for rate constants of $k = (65\ \text{ps})^{-1}$ and $k_s = (350\ \text{ps})^{-1}$ and proton dissociation fractions for the product ion of 100% for 532-nm ionization and 65% for 355-nm ionization. The reactant ion is essentially undissociated.

gas pulse (Figure 15e). The latter spectrum shows that the energy distribution in these cluster ions extends to high energy.

Despite the congested spectra, it is possible to see that a fairly large fraction of ions are formed with energies greater than 1 eV. Yet these complex ions are known to be stable to proton dissociation according to the mass spectra in Figure 12. These results would argue that the ion-molecule proton transfer reaction above (at least for small n) has a barrier of at least 1 eV. The model for the potential energy surface, discussed below, sets an upper limit for the barrier of 1.5 eV.

Potential Energy Curve and Reaction Coordinate: We review the general features of a potential energy model for delocalized charge transfer described in detail elsewhere [10, 11]. The model proposed for the potential energy surface and reaction coordinate is pictured in Figure 16. Potential energy curves along the O—H coordinate are drawn for isolated $[PhOH]^+$ (back panel) and for $[PhOH \cdot NH_3]^+$ (front panel). The N—H coordinate running from back to front represents the approach of reactants in a bimolecular reaction and the O—H coordinate, just mentioned, represents departure of products.

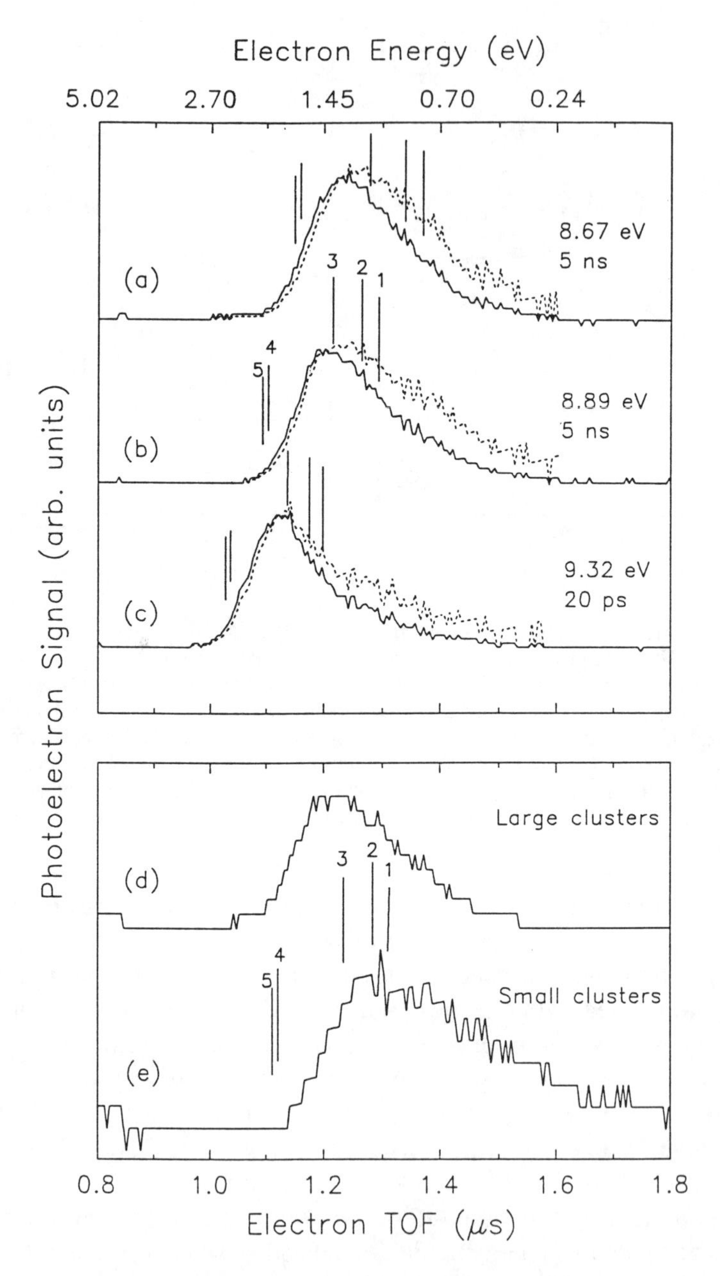

Fig. 15. Photoelectron spectra for resonance two-photon ionization of PhOH · $(NH_3)_n$ clusters. (a) 286-nm excitation, $n_{max} \sim 4 - 5$; (b) 279-nm excitation, $n_{max} \sim 3$; (c) 266-nm excitation, $n_{max} \sim 3 - 4$; (d) 280-nm excitation at center of gas pulse, $n_{max} \sim 4$; (e) 280-nm excitation at edge of gas pulse, $n_{max} \sim 2$. The n_{max} values are estimated from mass spectra and knowledge of cluster $S_0 \leftarrow S_1$ transition energies. The vertical lines represent IPs for clusters of size n. The dotted curves in (a)–(c) are corrected for electron kinetic energy detection efficiency.

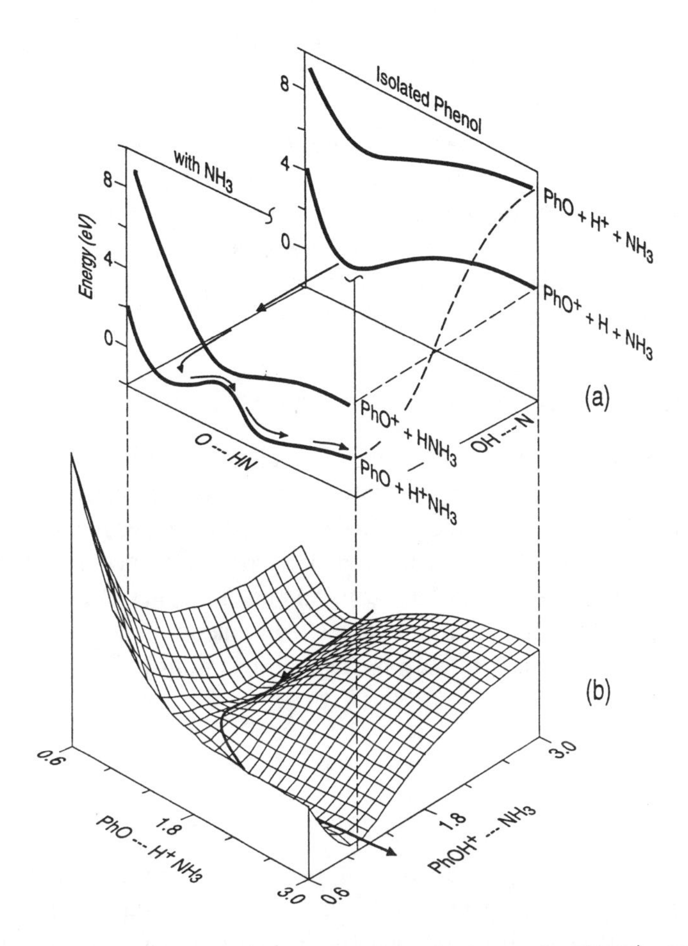

Fig. 16. Model of the potential energy reaction coordinate for the reaction $PhOH^+ + NH_3 \longrightarrow PhO + NH_4^+$. (a) Potential energy curves for uncomplexed and complexed $PhOH^+$ showing the surface crossing that gives rise to the barrier to proton transfer. (b) A rendition of the lower adiabatic potential energy surface for the collinear O· · ·H · · · N coordinate (in Å). Reaction paths are drawn for visualization in (a) and (b).

The reaction coordinate is approximated by the path marked by arrows. The barrier is predicted on the basis of a simple concept. The ground electronic state of $PhOH^+$ dissociates to PhO^+ + H. The dissociative limit for PhO + H^+ is about 5 eV higher in energy (given by the difference in the ionization potentials of PhO and H, 8.56 and 13.60 eV, respectively) [63] and is probably associated with some higher unknown excited electronic state of $PhOH^+$. This dissociative limit is stabilized enormously as NH_3 approaches along the N—H coordinate because of the substantial 8.86 eV proton affinity [39] of NH_3. Because the ground state dissociative limit PhO^+ + H experiences considerably less stabilization from NH_3, the dissociative limits are reversed in energy by the approach of reactant NH_3 (dotted lines, Figure 16a). This causes a surface crossing that creates the barrier to reaction. The surface crossing reaction barrier should be a general property of reactions in which the charged leaving group is not the lowest energy dissociative limit of the isolated ion. And this situation will be typical of ions where charge is delocalized from the leaving group. A visualization of the shape of the lower adiabatic surface for the collinear $O \cdots H \cdots N$ coordinate is given in Figure 16b [10, 11]. The reaction path drawn represents the lowest energy reaction coordinate in the classical limit (ignoring zero-point energy).

5.2 ELECTRON TRANSFER INDUCED BY DISSOCIATION

All chemistry defined by transformation of chemical bonds involves charge transfer in some form of interpretation. Thermodynamically driven reactions seek products that minimize electronic energy. Changes in electronic energy are brought about by redistribution of charges and occur in both the gas phase and solution phase. Many common reactions in solution phase involve transfer of relatively localized charges from one group to another. Two good examples are S_n2 and proton transfer (PT) reactions [64], as shown below

$$X^- + RY \longrightarrow XR + Y^- \qquad (S_n2)$$

$$AH^+ + B \longrightarrow A + BH^+ \qquad (PT)$$

The stability of charge reactions are usually strongly dependent on solvent properties (i.e., polarity, proton affinity). The above classes of reactions also occur in the gas-phase; however, their stability is greatly affected by the absence of solvent. The role of electron or charge transfer is obviously of great importance to chemistry, so we will open up the discussion somewhat to consider non-chemical electron transfer processes. (We define electron transfer (ET) as the specific case of charge transfer involving increments of unit charge.) Two examples of the latter processes in the gas phase are collisional charge transfer (CT) and Penning ionization, shown below

$$A^+ + B \longrightarrow A + B^+ \qquad (CT)$$

$$A^* + B \longrightarrow A + B^+ \qquad (Penning)$$

where A^* is an excited state that exceeds the ionization potential of B (A is often a rare gas atom).

We have already discussed the basis for the very different properties of gas-phase and solution-phase PT reactions by studies of ESPT in clusters.

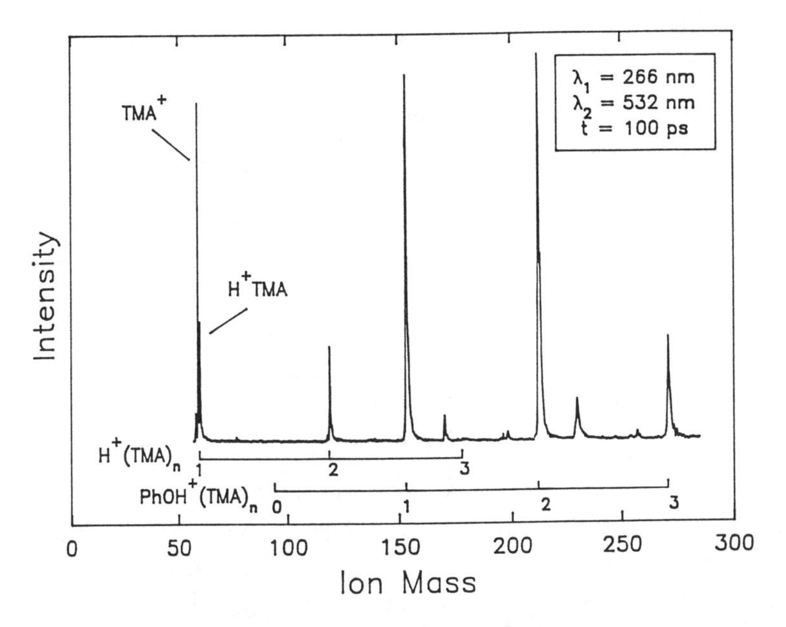

Fig. 17. Two-color picosecond TOF mass spectrum of PhOH · TMA_n clusters. An NH_3 impurity gives rise to the weak $[PhOH]^+NH_3TMA_n$ signals at 170 amu and 229 amu.

Here we present two examples of electron transfer in clusters resulting from dissociation mechanisms, namely

$$\begin{aligned} AB + h\nu &\longrightarrow A^+B \longrightarrow A + B^+ \\ CD + h\nu &\longrightarrow C^+D \longrightarrow C^+ + D \\ &\qquad\qquad\;\; \longrightarrow E + F + D^+ \end{aligned}$$

The first example occurs for ionization of phenol (A) bound to trimethylamine (B). The second example was observed for ionization and subsequent dissociative excitation of aniline (C) to C_5H_6 (E) and HNC (F) in the presence of hydrazine (D).

Phenol-trimethylamine: We have made extensive time-resolved measurements of ESPT involving PhOH with trimethylamine (TMA) cluster solvation. In accord with the higher proton affinity of TMA relative to NH_3 (9.8 vs 8.8 eV, respectively) [39], we observed a threshold solvent cluster size of $n = 3$ vs $n = 5$ for NH_3. Interesting size-dependent chemistry was also observed for the cluster ion $PhOH^+(TMA)_n$ that was not observed for $PhOH^+(NH_3)_n$. Solvent-specific chemistry is evident by comparing the mass spectra in Figure 17 (TMA solvation) with the mass spectra in Figures 12 and 13 (NH_3 solvation). The principal difference is the presence of unprotonated solvent cation (B^+) and the absence of phenol monomer cation ($PhOH^+$) for TMA solvation. The reason for this behavior is that TMA has a lower ionization potential (7.81 eV) than the chromophore phenol (8.51 eV) [63]. Following ionization of phenol, a rapid ET occurs so that dissociation of $[\mathrm{PhOH} \cdot \mathrm{TMA}]^+$ leads only to PhOH + TMA^+, in contrast to the behavior for NH_3 solvation. When more than one TMA is bound to PhOH^+, the chemistry appears more like that of NH_3 solvation in that one observes signal due to H^+B_n and $[\mathrm{PhOH} \cdot \mathrm{TMA}_n]^+$. However, the ET has still occurred in these larger cluster ions. The solvent size-dependent chemistry for TMA is summarized as follows:

$$[\mathrm{PhOH} \cdot \mathrm{TMA}_n]^+ \longrightarrow \begin{cases} \mathrm{PhOH}^+ + \mathrm{TMA} & (X) \\ \mathrm{PhOH} + \mathrm{TMA}^+ & (n = 1) \\ [\mathrm{PhOH}] \cdot \mathrm{TMA}^+_{n-m} + \mathrm{TMA}_m & (n > 1) \\ [\mathrm{PhOH}] \cdot \mathrm{TMA}_{n-m} + \mathrm{TMA}^+_m & (X) \\ \mathrm{PhO} + \mathrm{H}^+\mathrm{TMA}^+_{n-m} + \mathrm{TMA}_m & (n > 2) \end{cases}$$

where X stands for "no reaction". The model assumes a fast electron transfer followed by one of three possible cluster-size-dependent dissociations. For $n = 1$, only one bond can break, thus producing TMA^+. The presence of TMA^+ and absence of PhOH^+ in Figure 17 indicates that ET is much faster than dissociation. Cluster ions for $n > 1$ preferentially dissociate a neutral TMA to preserve the stronger cluster ion-dipole bond(s). Thus, unprotonated TMA^+_n is not observed for $n > 1$ in Figure 17. For $n > 2$, ESPT is presumed to occur, leading to the formation of neutral excited state ion-pairs, which then undergo dissociative ionization to the outer-cluster ion potential well (cf. Figure 2).

These results have important implications toward understanding gas-phase and bulk-phase correspondences. In a bimolecular gas-phase collision, PhOH loses its charge to TMA. However, in clusters or the bulk phase, phenol retains proximity to charge even though it resides on a partner molecule.

Aniline-hydrazine: We undertook a study to understand how single-molecule solvation would affect the well-known ion photodissociation

$$C_6H_5NH_2^+ + h\nu \longrightarrow C_6H_5NH_2^{+*} \longrightarrow C_5H_6^+ + \mathrm{HNC}$$

This reaction is termed metastable because of its slow rate of decay at moderately high ion energies ($k \sim$ 1–2 μ s^{-1} at 5–6 eV ion energy) [65]. This level of ion energy is achieved using 266-nm excitation (two-photon ionization, one-photon ion absorption) [66]. Our cluster work showed that the addition of just one or two NH_3 molecules reduces the dissociation rate by at least an order of magnitude at comparable excitation energies [7]. Yet, the reaction clearly occurs and without other competing chemistry [6].

We employed a mass-selected cluster ion dissociation technique coupled with fragment ion kinetic energy filtering to distinguish $C_5H_6^+$ formed from monomer versus solvated aniline cation [7]. However, our best evidence

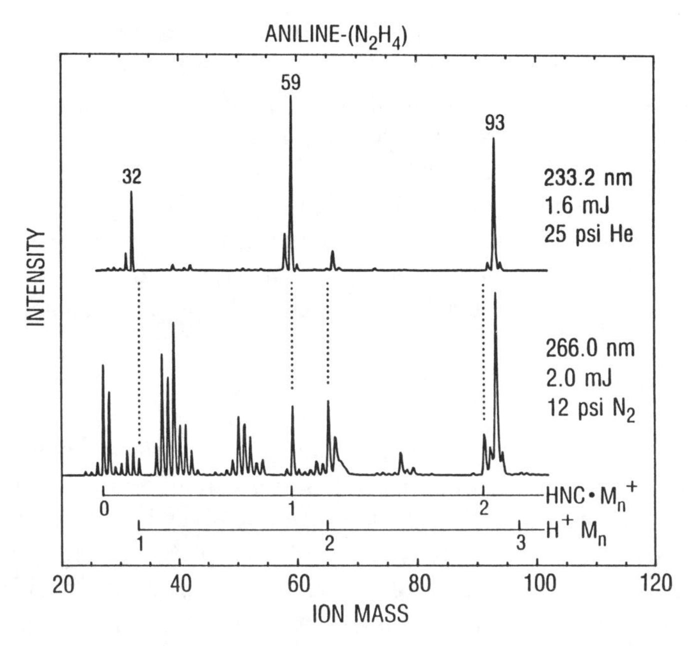

Fig. 18. Resonance ionization mass spectra of aniline-(N_2H_4) for different excitation conditions. The HNC · $(N_2H_4)_n^+$ series is evidence of a dissociation-induced electron transfer mechanism.

for solvated metastable decay was obtained inadvertently when we used hydrazine (N_2H_4) solvation [6]. We investigated a series of hydrogen bonding solvents. The solvents studied and their ionization potentials (IPs) are NH_3 (10.18 eV), N_2H_4 (8.74 eV), H_2O (12.6 eV), and CH_3OH (10.84 eV) [63]. In all cases, except N_2H_4, the mass spectra for 266-nm ionization/excitation consisted of the $C_6H_5NH_2^+ \cdot B_n$ series and $C_5H_6^+$. This result indicates that the solvent binds to the $-NH_2$ group of aniline and not the ring, as expected. However, it could not be determined whether any $C_5H_6^+$ originated from clusters.

For N_2H_4 solvation, a new reaction unfolds involving a long-range electron transfer. The mass spectra in Figure 18 reveal strong signals due to HNC · B_n^+. Solvated metastable dissociation is now evident because the otherwise neutral fragment HNC is now bound to a detectable charged molecule. The reason for the ET from N_2H_4 can be understood by considering the IP of all species involved. The chromophore aniline carries the charge in the cluster ion because of its low IP of 7.7 eV relative to that of all the solvents. Upon aniline cation dissociation, the C_5H_6 radical becomes the species with the lowest IP (8.97 vs 13.6 eV for HNC) [63] in all solvents studied except N_2H_4.

The dissociation-induced electron transfer involving N_2H_4 has several interesting implications worth noting. (1) The ET travels the length of several bonds. Also because it must occur during or after dissociation, but before a large separation takes place between fragments, the rate of ET must be very rapid. (2) The presence of HNC $\cdot$ B_n^+ fragments implies that the weak —NH_2 $\cdot$ B bond (about 1 eV bond energy) remains intact while slow and far more energetic chemistry occurs. This result would appear to violate a statistical mechanism in which energy redistribution should lead to breakage of the weakest bond. We have no explanation for this intriguing result.

5.3 PHOTOCHEMISTRY OF SULFIDE DIMER CATIONS

We have so far discussed cluster photochemistry for a variety of solute excitation conditions (bound and repulsive state excitation) and solute-solvent bonding interactions (van der Waals, hydrogen-bonding, ion-induced dipole attraction). Here we review a study of the photoexcitation of organosulfide (thioether) cluster ions formed by ionization of dimethyl sulfide (DMS) clusters [8]. These systems are unusual because the bond for the dimer ion core is "covalent" due to a three-electron, two-center interaction (3e, 2c) [67–69] represented in Figure 19a. Odd-electron bonds have a prominent role in radical chemistry, yet they are poorly understood compared with conventional two-electron neutral bonds. The three-electron sigma-like bond in the thioether dimer cation forms from the overlap of an open-shell (ion) and closed-shell (neutral) p-orbital and has a bond energy of about 25 kcal/mol ($\sim$ 1 eV) for DMS [68].

The study of reactions in cluster ions has direct correspondence to solution-phase chemistry, particularly nucleophilic displacements for which the formation of carbocations constitutes the central reactive site. Reaction intermediates for nucleophilic displacement with sulfides and halogenated sulfides often involve electropositive sulfur atoms. The stable $S—S^+$ bond and the alkylating nature of sulfides, which we observe in the cluster ions, are responsible for such adverse effects as the blistering action of mustard gas [2,2'–bis(chloroethyl)sulfide], or in a controlled environment are capable of beneficial effects such as the selective destruction of tumor cells in cancer treatment [70]. Sulfer-sulfer cross-linkage bonds have biochemical importance in the activity of enzymes and commercial importance in the manufacture of rubber.

The (3e, 2c) sulfide dimer cation has strong visible absorptions that are well separated from the far ultraviolet absorption of the neutral monomer and the near ultraviolet absorptions of the monomer cation [68, 69]. The origin of these blue-visible absorptions can be seen in the correlation diagram in Figure 19b. The splitting of the $3p_x$ orbitals creates a σ, σ^* transition.

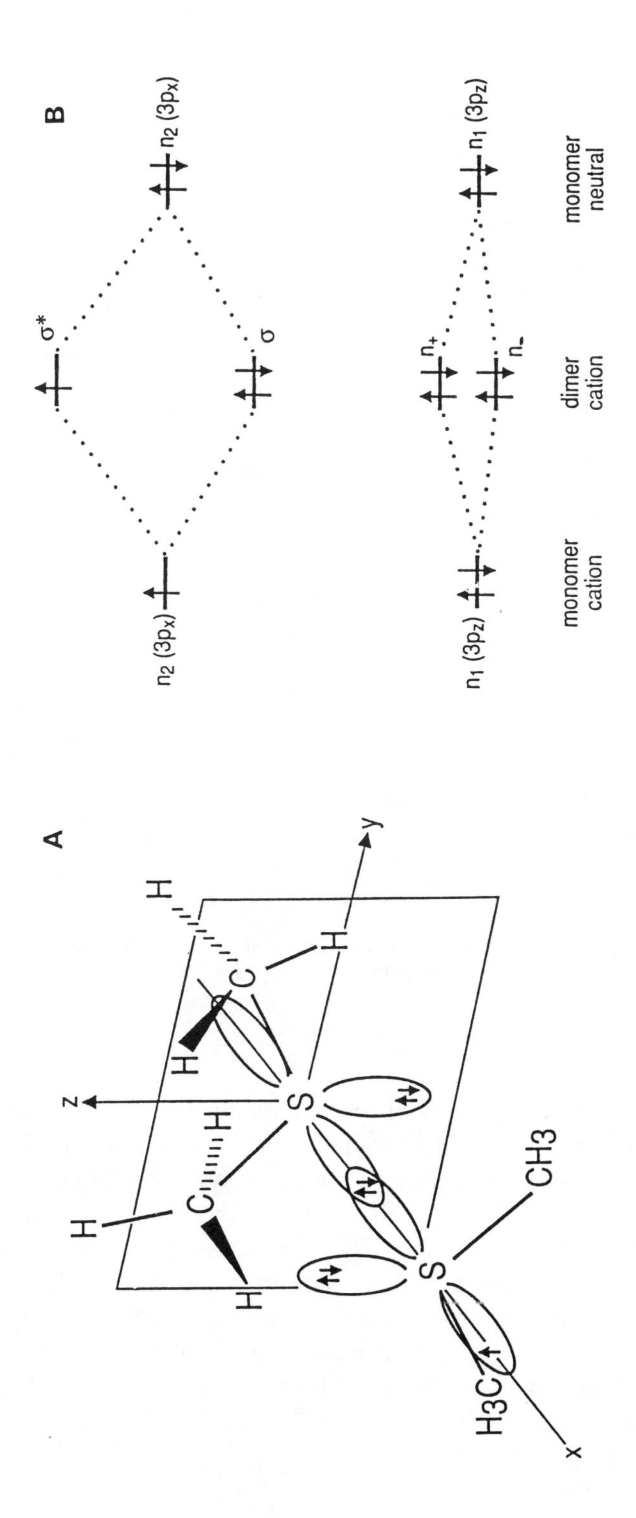

Fig. 19. (a) Geometry and coordinate system for DMS monomer and dimer cation. (b) Energy level diagram for the sulfide dimer cation. The $3p_x$ is a pure p-orbital, whereas, $3p_z$ is actually a hybrid orbital with about 25% $3s$ admixture.

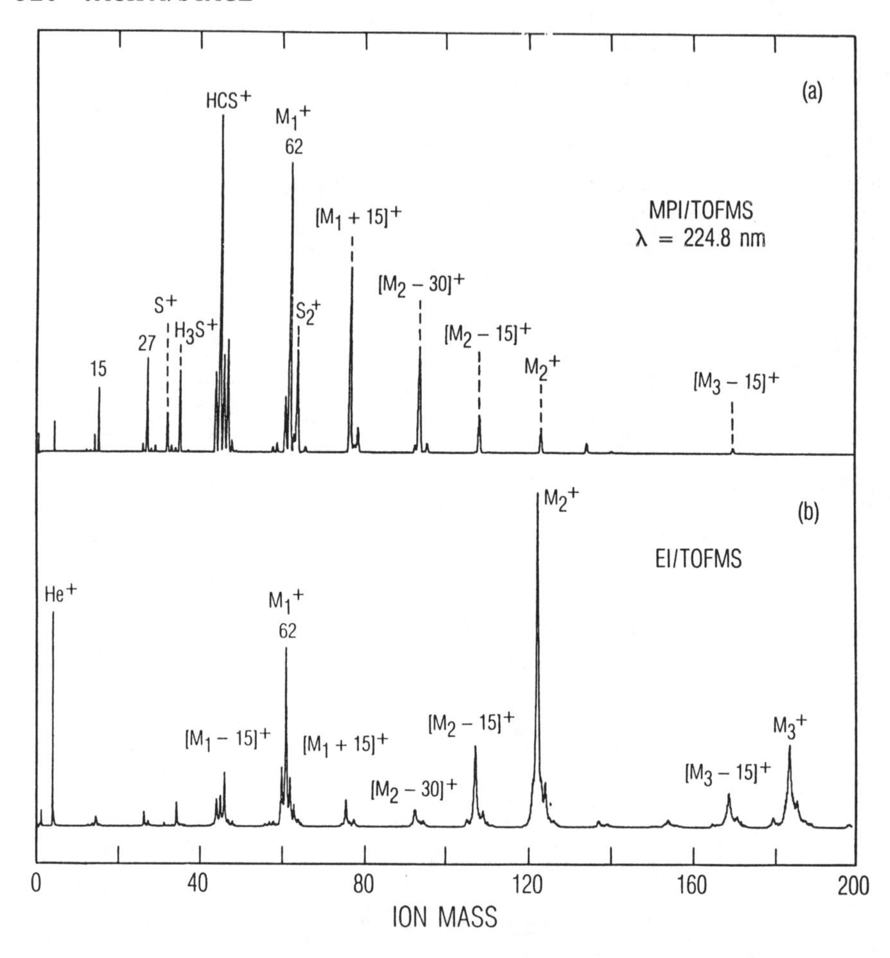

Fig. 20. Mass spectra of DMS clusters: (a) Resonance ionization and ion absorption at 225 nm. (b) EI excitation showing different fragmentation behavior.

The results in Figures 20 and 21 reveal dramatic differences in the cluster ion chemistry induced by EI and by photochemical excitation. The principal cluster ion chemistry in EI ionization/dissociation is S–C bond cleavage leading to demethylation and hydride transfer producing protonated DMS monomer and cluster ions (Figure 20b). These results are similar to what is observed in bimolecular gas-phase reactions involving DMS cation and neutral [67, 71]. Photochemical excitation, however, produces distinct photoproducts that depend strongly on wavelength. The cluster fragmentation in Figure 20a is attributable mostly to cluster ion absorption at 225-nm and leads to photoproducts S_2^+, $(CH_3)_3S^+$, CS^+, and HCS^+ that are either absent or much weaker by EI excitation.

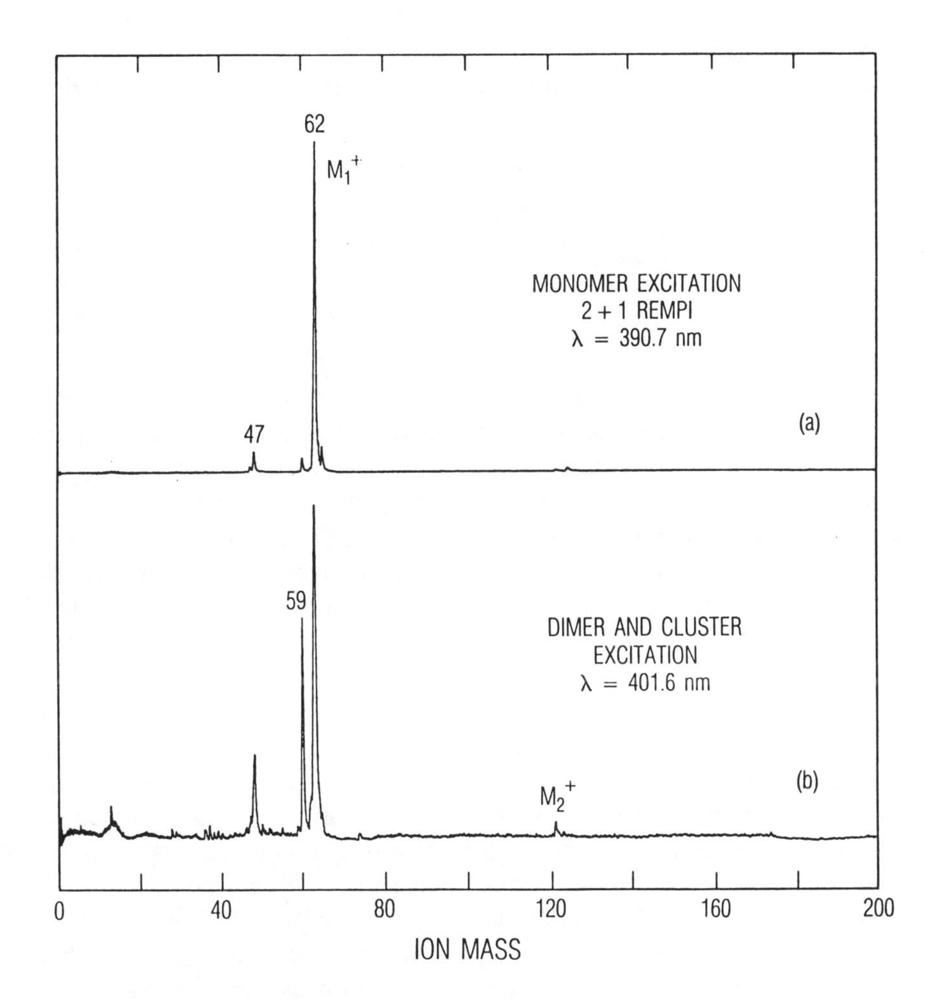

Fig. 21. Two-photon resonance, one-photon ionization mass spectra of DMS monomer and dimer/cluster: (a) monomer ionization through the origin intermediate state at 390.7 nm. (b) Dimer and cluster ionization/excitation using 401.6 nm excitation.

Cluster ion excitation to the dimer cation σ^* state at 400 nm generates entirely different chemistry than for the cases above. Cluster ions are first produced just above the ionization threshold by a two-photon resonance ($4p \leftarrow 3p_x$), one-photon ionization at wavelengths around 400 nm. Photochemistry is then promoted by subsequent ion absorption. Pulse energies were reduced so that excitation through the origin transition of DMS monomer

(51200 cm^{-1}) [72] produced little ion dissociation (Figure 21a). The excitation wavelength was shifted to the red of the monomer origin to excite primarily DMS clusters (Figure 21b). A distinctly singular peak at mass 59 was observed. The structure of mass 59 could either be an open β-thiovinyl cation or a closed cyclic thiirenium cation, namely

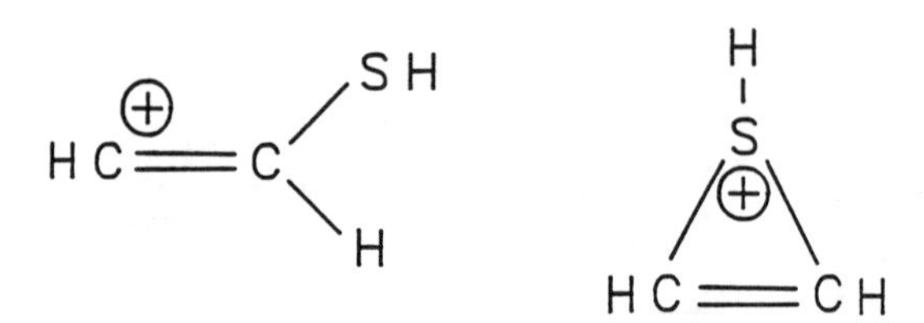

Molecular orbital calculations indicate that the cyclic form is more stable by 1–14 kcal/mol with a barrier to interconversion of about 13 kcal/mol [73].

The cyclic thiirenium structure has been postulated as a critical intermediate or transition state in certain solution-phase rearrangement and addition reactions involving organosulfide molecules [70]; however, its previous detection has been elusive. Trimethyl thiirenium has been observed by low temperature NMR [74]. Mass 59 has only been observed as a trace component in gas-phase mass spectrometry of organosulfide molecules [75].

6. Summary and Conclusions

Studies of reactivity in isolated molecular clusters are providing valuable information on the interactions of individual solvent or lattice molecules on chemical dynamics. The two main objectives of our cluster work are to gain understanding of (1) the correspondence between gas-phase and bulk-phase chemistry and (2) the dependence of chemical properties on solvent environment. Clusters are uniquely suited for these studies because their size quite naturally spans the limits of the gas phase and the bulk phase and because their specific composition and geometry are usually fairly well known. Much of our work is based on proton transfer from the molecule phenol to a hydrogen-bonding solvent cluster. This prototype system has proved valuable in addressing many of the issues above. We observed distinct cluster size thresholds for excited state proton transfer and explained the results using an energy diagram showing how the isolated gas-phase phenol excited state potentials are modified by stepwise addition of solvent molecules. Though stepwise solvent binding energies are a strong driving force to reactivity, simple additivity arguments are insufficient for describing the role of first- and second-shell solvent molecules. Mixed solvent studies with phenol showed that addition of a single dissimilar molecule to an otherwise reactive cluster can seriously suppress the rate of ESPT. We also investigated the cluster ions as models of gas-phase reactive complexes and resolved a longstanding issue regarding double-minima potentials for ion-molecule reactions.

Our studies have yielded predictable results and surprising results. The former includes chemistry that is directly related to gas-phase or condensed-phase reactions and easily explained by simple energy arguments. The surprising results are those that are unique to the cluster environment and highlight that there is still much to be learned about chemistry in general. For example, we are baffled by how a weak solvent bond can survive an energetic reaction involving multiple breaking of much stronger bonds (aniline cation fragmention bound to N_2H_4). In another example, ionization and excitation of dimethyl sulfide clusters yielded primarily products associated with gas-phase ion molecule reactions. However, excitation of a strong dimer cation σ^* transition yielded almost exclusively the elusive cyclic thiirenium ion, implicated as a transition state immediate in a number of condensed-phase sulfide reactions.

Cluster research is a large field, but a young one. The field is still full of surprises (witness C_{60} developments). Clusters are just beginning to make serious inroads in our understanding of matter. There are broad classes of chemistry that have not yet been investigated from the perspective of clusters (photochemical rearrangements, nucleophilic displacements, stereochemistry, to name a few). Development of new research techniques involving clusters shows no sign of abating. One might expect cluster techniques to be useful for engineering nanotechnologies. Clusters are also ideally suited for studying physical properties of gas-liquid interfaces. It is certain that more application will evolve and that cluster research will remain exciting for many years to come.

Acknowledgements

I am particularly indebted to Jhobe Steadman who made large contributions to the cluster work reviewed here. These studies were supported by the Aerospace Sponsored Research program.

References

1. Bernstein, E. R., ed. *Atomic and Molecular Clusters*, (Elsevier, New York, 1990).
2. Halberstadt N. and Janda, K. C., eds.; *Dynamics of Polyatomic van der Waals Complexes*, NATO ASI Series, Vol. 227, 1990.
3. Breen, J. J., Tzeng, W.-B., Kilgore, K., Keesee, R. G., and Castleman Jr., A. W., *J. Chem. Phys.* **90**, 19 (1989); O. Echt, O., Dao, P. D., Morgan, S., and Castleman Jr., A. W., *J. Chem. Phys.* **82** 4076 (1985).
4. Garvey J. F. and Bernstein, R. B., *J. Am. Chem. Soc.* **109**, 1921 (1987); Garvey J. F. and Bernstein, R. B., *J. Phys. Chem.* **90**, 3577 (1986).

5. Bernstein, E. R., *J. Phys. Chem.* **96**, (1992).
Brutschy, B., *J. Phys. Chem.* **94**, 8637 (1990).
6. Syage, J. A., *J. Phys. Chem.* **93**, 107 (1989).
7. Syage, J. A., *J. Chem. Phys.* **92**, 1804 (1990); Syage J. A., and Steadman, J., *Rev. Sci. Instrum.* **61**, 1204 (1990).
8. Syage, J. A., Pollard, J. E., and Cohen, R. B., *J. Phys. Chem.* **95**, 8560 (1991).
9. Mikami, N., Okabe, A., and Suzuki, I., *J. Phys. Chem.* **92**, 1858 (1988); Mikami, N., Suzuki, I., and A. Okabe, A., *J. Phys. Chem.* **91**, 5242 (1987).
10. Steadman J. and Syage, J. A., *J. Am. Chem. Soc.* **113**, 6786 (1991).
11. Syage J. A., and Steadman, J., *J. Phys. Chem.* **96**, 9606 (1992).
12. Syage J. A., and Steadman, J., *J. Chem. Phys.* **95**, 2497 (1991); Steadman J. and Syage, J. A., *J. Chem. Phys.* **92**, 1630 (1990).
13. Syage, J. A., *Chem. Phys. Lett.* **202**, 227 (1993).
14. Steadman J. and Syage, J. A., *J. Phys. Chem.* **95**, 10326 (1991).
15. Cheshnovsky, O. and Leutwyler, S., *Chem. Phys. Lett.* **121**, 1 (1985); Cheshnovsky, O. and Leutwyler, S., *J. Chem. Phys.* **88**, 4127 (1988); Droz, T., Knochenmuss, R., and Leutwyler, S., *J. Chem. Phys.* **93**, 4520 (1990).
16. Solgadi, D., Jouvet, C., and Tramer, A., *J. Phys. Chem.* **92**, 3313 (1988); Jouvet, C., Lardeux-Dedonder, C., Richard-Viard, M., Solgadi, D., and Tramer, A., *J. Phys. Chem.* **94**, 5041 (1990).
17. Breen, J. J., Peng, L. W., Willberg, D. M., Heikal, A., Cong, P., and Zewail, A. H., *J. Chem. Phys.* **92**, 805 (1990).
18. Hineman, M. F., Bruker, G. A., Kelley, D. F., and Bernstein, E. R., *J. Chem. Phys.* **97**, 3341 (1992); Kim, S. K., Li, S., and Bernstein, E. R., *J. Chem. Phys.* **95**, 3119 (1991).
19. Syage J. A. and Steadman, J., *Chem. Phys. Lett.* **166**, 159 (1990); Steadman, J., Fournier, E. W., and Syage, J. A., *Appl. Opt.* **29**, 4962 (1990).
20. Kappes M. and Leutwyler, S., in *Atomic and Molecular Beam Methods*, Vol. I, edited by G. Scoles, (Oxford University Press, New York, 1988), p. 380.
21. Syage, J. A., *Phys. Rev. A* **46**, 5666 (1992).
22. Syage, J. A., *J. Phys. B.* **24**, L527 (1991).
23. Syage, J. A., *J. Chem. Phys.* **97**, 6085 (1992).
24. Syage, J. A., Pollard, J. E., and Steadman, J., *Chem. Phys. Lett.* **161**, 103 (1989).
25. Syage J. A. and Steadman, J., *J. Chem. Phys.* manuscript in preparation.
26. Trevor, D. J., Van Woerkom, L. D., and Freeman, R. R., *Rev. Sci. Instrum.* **60**, 1051 (1989); Liu, G., Barton, J. J., Bahr, C. C., and Shirley, D. A., *Nucl. Instr. and Meth.*, **A246**, 504 (1986).
27. Steadman J. and Syage, J. A., *Rev. Sci. Instrum.*, in press.
28. Ray, D., Levinger, N. E., Papanikolas, J. M., and Lineberger, W. C., *J. Chem. Phys.* **91**, 6533 (1989); Papanikolas, J. M., Gord, J. R., Levinger, N. E., Ray, D., Vorsa, V., and Lineberger, W. C., *J. Phys. Chem.* **95**, 8028 (1991).
29. Kazishka, A. J., Shchuka, M. I., Wittmeyer, A. S., and Topp, M. R., *J. Phys. Chem.* **95**, 3663 (1991).
30. Felker P. M. and Zewail, A. H., *Chem. Phys. Lett.* **94**, 448, 454 (1983); Connell, L. L., Ohline, S. M., Joireman, P. W., Corcoran, T. C., and Felker, P. M., *J. Chem. Phys.* **96**, 2585 (1992).
31. Cassasa, M. P., Stephenson, J. C., and King, D. S., *J. Chem. Phys.* **89**, 1966 (1988).
32. Sipior J. and Sulkes, M., *J. Chem. Phys.* **88**, 6146 (1988); Teh, C. K., Sipior, J., and Sulkes, M., *J. Phys. Chem.* **93**, 5393 (1989); Arnold S. and Sulkes, M., *J. Phys. Chem.* **96**, 4768 (1992).
33. Fuke, K., Keizo, K., Misaizu, F., and Kaya, K., *J. Chem. Phys.* **95**, 4074 (1991).
34. Woodward, A. M., Colson, S. D., Chupka, W. A., and White, M. G., *J. Phys. Chem.* **90**, 274 (1986); Walter, K., Weinkauf, R., Boesl, U., and Schlag, E. W., *J. Chem. Phys.* **89**, 1914 (1988).
35. Kosower E. M., and Huppert, D., *Annu. Rev. Phys. Chem.* **37**, 127 (1986).

36. Lee, J., Griffin, R. D., and Robinson, G. W., *J. Chem. Phys.* **82**, 4920 (1985); Lee, J., Robinson, G. W., Webb, S. P., Philips, L. A., and Clark, J. H., *J. Am. Chem. Soc.* **108**, 6538 (1986)
37. Huppert, D. H., Jayaraman, A., Maines Sr., R. G., Steyert, D. W., and Rentzepis, P. M., *J. Chem. Phys.* **81**, 5596 (1984).
38. Gonohe, N., Abe, H., Mikami, N., and Ito, M., *J. Phys. Chem.* **89**, 3642 (1985); Oikawa, A., Abe, H., Mikami, N., and Ito, M., *J. Phys. Chem.* **87**, 5083 (1983).
39. Lias, S. G., Liebman, J. F., and Levin, R. D., *J. Phys. Chem. Ref. Data* **13**, 695 (1984).
40. Kebarle, P., *Annu. Rev. Phys. Chem.* **28**, 445 (1977); Grimsrud E. P. and Kebarle, P., *J. Am. Chem. Soc.* **95**, 7939 (1973); Mautner M. J. and Speller, C. V., *J. Phys. Chem.* **90**, 6616 (1986).
41. Ireland J. F. and Wyatt, P. A. H., *Adv. Phys. Org. Chem.* **12**, 131 (1976).
42. Price, J. M., Crofton, M. W., and Lee, Y. T., *J. Phys. Chem.* **95**, 2182 (1991).
43. Pimentel G. C. and McClellan, A. L., *The Hydrogen Bond*, (W. H. Freeman, San Francisco, 1960); Dykstra, C. E., *J. Phys. Chem.* **94**, 6948 (1990).
44. Peteanu L. A. and Levy, D. H., *J. Phys. Chem.* **92**, 6554 (1988); Champagne, B. B., Pfanstiel, J. F., Plusquellic, D. F., Pratt, D. W., van Herpen, W. M., and Meerts, W. L., *J. Phys. Chem.* **94**, 6 (1990).
45. Venturo, V. A., Maxton, P. M., and Felker, P. M., *J. Phys. Chem.* **96**, 5234 (1992); Connell, L. L., Corcoran, T. C., Joireman, P. W., and Felker, P. M., *J. Phys. Chem.* **94**, 1229 (1990).
46. Wessel J. E. and Syage, J. A., *J. Phys. Chem.* **94**, 737 (1990); Syage J. A. and Wessel, J. E., *J. Chem. Phys.* **89**, 5962 (1988).
47. Wei, S., Shi, Z., and Castleman Jr. A. W., *J. Chem. Phys.* **94**, 3268 (1991).
48. Maroncelli, M., MacInnis, J., and Fleming, G. R., *Science* **243**, 1674 (1989); Rothenberger, G., Negus, D. K., and Hochstrasser, R. M., *J. Chem. Phys.* **79**, 5360 (1983).
49. Simon, J. D., *Acc. Chem. Res.* **21**, 128 (1988); Su S.-G. and Simon, J. D., *J. Chem. Phys.* **89**, 908 (1988).
50. Barbara, P. F., Walker, G. C., and Smith, T. P., *Science,* **256**, 975 (1992); Jarzeba, W., Walker, G. C., Johnson, A. E., Kahlow, M. A., and Barbara, P. F., *J. Phys. Chem.* **92**, 7039 (1988); Kozik, M., Sutin, N., and Winkler, J. R., *Coord. Chem. Rev.* **97**, 23 (1990).
51. Anderson, J. G., Toohey, D. W., and Brune, W. H., *Science* **251**, 39 (1991).
52. Schwartz S. E., and White, W. H., in *Trace Atmospheric Constituents: Properties, Transformations, and Fates*, (John Wiley and Sons, New York, 1983).
53. Knee, J. L., Khundkar, L. R., and Zewail, A. H., *J. Chem. Phys.* **83** 1996 (1985).
54. Sapers, S. P., Vaida, V., and Naaman, R., *J. Chem. Phys.* **88** 3638 (1988).
55. Wang, P. G., Zhang, Y. P., Ruggles, C. J., and Ziegler, L. D., *J. Chem. Phys.* **92**, 2806 (1990).
56. Fan and Y. B., Donaldson, D. J., *J. Phys. Chem.* **96**, 19 (1992).
57. Sparks, R. K., Shobatake, K., Carlson, L. R., and Lee, Y. T., *J. Chem. Phys.* **75** 3838 (1981).
58. Wittig, C., Sharpe, S., and Beaudet, R. A., *Acc. Chem. Res.* **21**, 341 (1988); Radhakrishnan, G., Buelow, S., and Wittig, C., *J. Chem. Phys.* **84**, 727 (1986).
59. Halberstadt N. and Soep, B., *J. Chem. Phys.* **80**, 2340 (1984).
60. Moylan C. R. and Brauman, J. I., *Annu. Rev. Phys. Chem.* **34**, 187 (1983); (b) Comita P. B. and Brauman, J. I., *Science* **227**, 863 (1985); Bowers, M. T., ed. *Gas Phase Ion Chemistry*, Vols. I and II, (Academic Press, New York, 1979).
61. Farneth W. E. and Brauman, J. I., *J. Am. Chem. Soc.* **98**, 7891 (1976); Jasinski J. M. and Brauman, J. I., *J. Am. Chem. Soc.* **102**, 2906 (1980).
62. Cyr, D. M., Bishea, G. A., Scarton, M. G., and Johnson, M. A., *J. Chem. Phys.* **97**, 5911 (1992).
    Cyr, D. M., Posey, L. A., Bishea, G. A., Han, C.-C., and Johnson, M. A., *J. Am. Chem. Soc.* **113**, 9697 (1991).
63. Lias, S. G., Bartmess, J. E., Liebman, J. F., Holmes, J. L., Levin, R. D., and Mallard, W. G., *J. Phys. Chem. Ref. Data* **17**, 1 (1988).

64. Lowry T. H. and Richardson, K. S., *Mechanism and Theory in Organic Chemistry*, (Harper and Row, New York, 1976).
65. Baer T. and Carney, T. E., *J. Chem. Phys.* **76**, 1304 (1982).
66. Proch, D., Rider, D. M., and Zare, R. N., *Chem. Phys. Lett.* **81**, 430 (1981); Kühlewind, H., Neusser, H. J., and Schlag, E. W., *J. Chem. Phys.* **82**, 5452 (1985).
67. Clark, T., *J. Am. Chem. Soc.* **110**, 1672 (1988); Illies, A. J., Livant, P., and McKee, M. L., *J. Am. Chem. Soc.* **110**, 7980 (1988).
68. Asmus, K.-D., *Acct. Chem. Res.* **12**, 436 (1979).
69. Naito, A., Akasaka, K., and Hatano, H., *Mol. Phys.* **44**, 427 (1981).
70. Block, E., *Reactions of Organosulfur Compounds*, (Academic Press, New York, 1978), Chap. 4.
71. Drewello, T., Lebrilla, C. B., Schwarz, H., de Koning, L. J., Fokkens, R. H., Nibbering, N. M. M., Anklam, E., and Asmus, K.-D., *J. Chem. Soc. Chem. Commun.*, 1381 (1987); Musker, W. K., Gorewit, B. V., Roush, P. B., and Wolford, T. L., *J. Org. Chem.* **43**, 3235 (1978).
72. Syage, J. A., Pollard, J. E., and Cohen, R. B., *Appl. Opt.* **26**, 3516 (1987).
73. Csizmadia, I. G., Duke, A. J., Lucchini, V., and Modena, G., *J. Chem. Soc. Perkin Trans. II*, 1808 (1974).
74. Capozzi, G., DeLucchi, O., Lucchini, V., and Modena, G., *Chem. Commun.*, 248 (1975).
75. Möckel, H. J., *Fresenius Z. Anal. Chem.* **295**, 241 (1979).

12. Quantum Brownian Oscillator Analysis of Pump-Probe Spectroscopy in the Condensed Phase

YOSHITAKA TANIMURA and SHAUL MUKAMEL
Department of Chemistry, University of Rochester, Rochester, New York 14627, U.S.A.

1. Introduction

Nuclear motions and relaxation play an important role in determining the rates and outcomes of chemical processes in the condensed phase [1–12]. Electron transfer, isomerization and bimolecular reactions are directly effected by intramolecular vibrations as well as solvent motions.

Femtosecond spectroscopy allows the direct probe of elementary nuclear motions [11, 13]. Much physical insight can be gained by formulating nonlinear spectroscopy in terms of nonlinear response functions which are given as sums of contributions of Liouville space paths [14]. Each of these paths may be represented using a nuclear wavepacket in phase space. Comparison of the wavepackets with the experimental observable allows the development of a powerful semiclassical representation of nonlinear spectroscopy. The nonlinear response functions obtained from nonlinear spectroscopies can then be used to calculate other processes including curve crossing and electron transfer [15, 16, 17].

The multimode Brownian oscillator model provides a convenient means for incorporating nuclear degrees of freedom in the response function [14]. We have recently used a path integral approach [18] to develop exact closed expressions for the nuclear wavepackets in phase space and nonlinear response functions for this model [19] (hereafter denoted TM). In this paper, we apply these results to the analysis of pump-probe spectroscopy. In Section 2, we introduce the response functions and present expressions for the wavepackets using the Condon approximation which neglects the variation of the transition dipole moment with nuclear coordinates. More general expressions which do not involve the Condon approximation are given in the Appendix. In Section 3, we apply these results to impulsive pump-probe spectroscopy and analyze the roles of high frequency (underdamped) modes as well as overdamped solvation modes.

J.D. Simon (ed.), Ultrafast Dynamics of Chemical Systems, 327–343.

2. Phase Space Wavepacket Representation for the Optical Response

We consider a two electronic level system with a ground state $|g>$ and an excited state $|e>$ interacting with an external electromagnetic field $E(t)$;

$$H_s = H_0 - E(t)V, \tag{2.1}$$

where

$$H_0 = |g> H_g < g| + |e> H_e < e|, \tag{2.2}$$

with

$$\begin{cases} H_g = \frac{p^2}{2M} + U_g(q), \\ H_e = \frac{p^2}{2M} + U_e(q), \end{cases} \tag{2.3}$$

and p, q, and M represent the momentum, the coordinate and the mass of a nuclear coordinate, respectively. V is the dipole operator which is given by

$$V = |g> \mu(q) < e| + |e> \mu(q) < g|. \tag{2.4}$$

$\mu(q)$ is the dipole matrix element between the two states which depends on the nuclear coordinate. The potentials of the excited and the ground states are assumed to be harmonic:

$$\begin{cases} U_g(q) = \frac{1}{2} M\omega_0^2 q^2, \\ U_e(q) = \frac{1}{2} M\omega_0^2 (q+D)^2 + \hbar\omega_{eg}^0, \end{cases} \tag{2.5}$$

where ω_{eg}^0 is the electronic energy gap and D is the displacement of the potential. This system is embedded in a solvent, which is modelled as a set of harmonic oscillators with coordinates x_n and momenta p_n. The interaction between the system and the n-th oscillator is assumed to be linear with a coupling strength c_n. The total Hamiltonian is then given by [20, 21]

$$H = H_s + H', \tag{2.6}$$

where

$$H' = \sum_n \left[\frac{p_n^2}{2m_n} + \frac{m_n \omega_n^2}{2} \left(x_n - \frac{c_n q}{m_n \omega_n^2} \right)^2 \right]. \tag{2.7}$$

We assume the entire system is initially at equilibrium in the ground electronic state:

$$\rho_g = |g><g| \exp[-\beta(H_g + H')]/\mathrm{Tr}\{\exp[-\beta(H_g + H')]\}, \tag{2.8}$$

where $\beta \equiv 1/k_B T$ is the inverse temperature. All effects of the heat bath on the system are determined by a spectral distribution of coupling strength defined by

$$J(\omega) = \pi \sum_n \frac{c_n^2}{2m_n\omega_n}\delta(\omega - \omega_n). \tag{2.9}$$

By introducing a frequency dependent friction, $\tilde{\gamma}(\omega) \equiv J(\omega)/\omega$, the anti-symmetric and a symmetric equilibrium correlation functions of the nuclear coordinate are expressed as

$$\begin{aligned}\chi(t) &\equiv \frac{i}{\hbar}\langle q(t)q - qq(t)\rangle_g \\ &= \frac{1}{M}\int_{-\infty}^{\infty} \frac{d\omega}{\pi} \frac{\omega\tilde{\gamma}(\omega)}{(\omega_0^2 - \omega^2)^2 + \omega^2\tilde{\gamma}^2(\omega)} \sin(\omega t),\end{aligned} \tag{2.10}$$

and

$$\begin{aligned}S(t) &\equiv \frac{1}{2}\langle q(t)q + qq(t)\rangle_g \\ &= \frac{\hbar}{M}\int_{-\infty}^{\infty} \frac{d\omega}{2\pi} \frac{\omega\tilde{\gamma}(\omega)}{(\omega_0^2 - \omega^2)^2 + \omega^2\tilde{\gamma}^2(\omega)} \coth\left(\frac{\beta\hbar\omega}{2}\right) \cos(\omega t).\end{aligned} \tag{2.11}$$

We further introduce the auxiliary function

$$g_\pm(t) \equiv \xi^2 \int_0^t dt' \int_0^{t'} dt'' \left[S(t'') \pm \frac{i\hbar}{2}\chi(t'')\right], \tag{2.12}$$

where

$$\xi \equiv \frac{MD\omega_0^2}{\hbar} = d\sqrt{\frac{M\omega_0^3}{\hbar}} = \frac{2\lambda}{d}\sqrt{\frac{\hbar}{M\omega_0}}. \tag{2.13}$$

Here, we defined the dimensionless nuclear displacement parameter, $d \equiv D\sqrt{M\omega_0/\hbar}$, and the Stokes shift parameter

$$\lambda \equiv \frac{MD^2\omega_0^2}{2\hbar} = \frac{d^2\omega_0}{2}. \tag{2.14}$$

The optical response can be expressed in terms of the optical polarization

$$P(t) \equiv \mathrm{Tr}\left\{(|e><g| + |g><e|)\int dp \int dq\mu(q)W(p,q,t)\right\}, \tag{2.15}$$

where $W(p, q, t)$ is a phase space wavepacket which depends on the interaction between the driving field and the system. Denoting the nuclear wavepacket to n-th order in the field by $W^{(n)}(p, q, t)$, we have

$$P^{(n)}(t) = \int dp \int dq \mu(q) W^{(n)}(p, q, t), \tag{2.16}$$

and

$$W(p, q, t) = \sum_n \sum_\alpha W_\alpha^{(n)}(p, q, t), \tag{2.17}$$

where $W_\alpha^{(n)}$ is the contribution of the α-th path to the n-th order polarization.

Since the equations are simpler in the Condon approximation, $\mu(q) = \mu$, we first give the expressions for this case. From TM, the first order distribution function is given by

$$\begin{aligned} W_1^{(1)}(p, q, t) = &\int_0^\infty dt_1 E(t - t_1) \mu^2 (4\pi^2 \langle p^2 \rangle_g \langle q^2 \rangle_g)^{-1/2} \\ &\times \exp\left[-\frac{1}{2\langle q^2 \rangle_g} (q - \bar{q}_1^{(1)}(t_1))^2 - \frac{1}{2\langle p^2 \rangle_g} (p - \bar{p}_1^{(1)}(t_1))^2 \right] \\ &\times R_1^{(1)}(t_1) + c.c., \end{aligned} \tag{2.18}$$

where

$$\begin{aligned} \langle q^2 \rangle_g &= \xi^2 \ddot{g}_+(0), \quad \langle p^2 \rangle_g = \xi^{-2} M^2 d^4 g_+(t)/dt^4 \,|_{t=0}, \\ \omega_{eg} &\equiv \omega_{eg}^0 + \lambda, \end{aligned} \tag{2.19}$$

and the center of the coordinate and the momentum of distribution function are given by

$$\bar{q}_1^{(1)}(t_1) = -i\xi^{-1} \dot{g}_-(t_1), \quad \bar{p}_1^{(1)}(t_1) = -iM\xi^{-1} \ddot{g}_-(t_1). \tag{2.20}$$

The first order response functions is given by

$$R_1^{(1)}(t_1) \exp[Q_1^{(2)}(t_1)], \tag{2.21}$$

where

$$Q_1^{(2)}(t_1) = -i\omega_{eg} t_1 - g_-(t_1). \tag{2.22}$$

The linear polarization is then expressed by using the response function, Equation (2.21), as

$$P^{(1)}(t) = -i\mu^2 \int_0^\infty dt_1 E(t - t_1) R_1^{(1)}(t_1) + c.c. \tag{2.23}$$

For the third order polarization, we need distribution functions for four Liouville space paths [11], $eg \to ee \to eg$, $ge \to ee \to eg$, $ge \to gg \to eg$ and $eg \to gg \to eg$ denoted by α =1–4, respectively. The corresponding phase space wavepackets are then given by

$$\begin{aligned} W^{(3)}(p,q,t) &= \int_0^\infty dt_1 \int_0^\infty dt_2 \int_0^\infty dt_3 E(t-t_3)E(t-t_2-t_3) \\ & E(t-t_1-t_2-t_3)\mu^3(4\pi^2\langle p^2\rangle_g\langle q^2\rangle_g)^{-1/2} \\ & \times \sum_{\alpha=1}^{4} \exp\left[-\frac{1}{2\langle q^2\rangle_g}(q-\bar{q}_\alpha^{(3)}(t_3,t_2,t_1))^2\right. \\ & \left. -\frac{1}{2\langle p^2\rangle_g}(p-\bar{p}_\alpha^{(3)}(t_3,t_2,t_1))^2\right] \\ & \times R_\alpha^{(3)}(t_3,t_2,t_1) + c.c., \end{aligned} \tag{2.24}$$

where

$$\begin{cases} \bar{q}_1^{(3)}(t_3,t_2,t_1) = -i\xi^{-1}[\dot{g}_-(t_1+t_2+t_3) - \dot{g}_+(t_2+t_3) - \dot{g}_+(t_3)], \\ \bar{q}_2^{(3)}(t_3,t_2,t_1) = -i\xi^{-1}[-\dot{g}_+(t_1+t_2+t_3) + \dot{g}_-(t_2+t_3) + \dot{g}_+(t_3)], \\ \bar{q}_3^{(3)}(t_3,t_2,t_1) = -i\xi^{-1}[-\dot{g}_+(t_1+t_2+t_3) + \dot{g}_+(t_2+t_3) + \dot{g}_-(t_3)], \\ \bar{q}_4^{(3)}(t_3,t_2,t_1) = -i\xi^{-1}[\dot{g}_-(t_1+t_2+t_3) - \dot{g}_-(t_2+t_3) + \dot{g}_-(t_3)], \end{cases} \tag{2.25}$$

and

$$\begin{cases} \bar{p}_1^{(3)}(t_3,t_2,t_1) = -iM\xi^{-1}[\ddot{g}_-(t_1+t_2+t_3) - \ddot{g}_+(t_2+t_3) + \ddot{g}_+(t_3)], \\ \bar{p}_2^{(3)}(t_3,t_2,t_1) = -iM\xi^{-1}[-\ddot{g}_+(t_1+t_2+t_3) + \ddot{g}_-(t_2+t_3) + \ddot{g}_+(t_3)], \\ \bar{p}_3^{(3)}(t_3,t_2,t_1) = -iM\xi^{-1}[-\ddot{g}_+(t_1+t_2+t_3) + \ddot{g}_+(t_2+t_3) + \ddot{g}_-(t_3)], \\ \bar{p}_4^{(3)}(t_3,t_2,t_1) = -iM\xi^{-1}[\ddot{g}_-(t_1+t_2+t_3) - \ddot{g}_-(t_2+t_3) + \ddot{g}_-(t_3)]. \end{cases} \tag{2.26}$$

The third order response function is now given by

$$R_\alpha^{(3)}(t_3,t_2,t_1) = \exp[Q_\alpha^{(4)}(t_3,t_2,t_1)], \tag{2.27}$$

where

$$\begin{cases} Q_1^{(4)}(t_3,t_2,t_1) = -i\omega_{eg}(t_1+t_3) - g_-(t_1) - g_+(t_3) \\ \quad -[g_+(t_2) - g_+(t_2+t_3) - g_-(t_1+t_2) + g_-(t_1+t_2+t_3)], \\ Q_2^{(4)}(t_3,t_2,t_1) = -i\omega_{eg}(-t_1+t_3) - g_+(t_1) - g_+(t_3) \\ \quad +[g_-(t_2) - g_-(t_2+t_3) - g_+(t_1+t_2) + g_+(t_1+t_2+t_3)], \\ Q_3^{(4)}(t_3,t_2,t_1) = -i\omega_{eg}(-t_1+t_3) - g_+(t_1) - g_-(t_3) \\ \quad +[g_+(t_2) - g_+(t_2+t_3) - g_+(t_1+t_2) + g_+(t_1+t_2+t_3)], \\ Q_4^{(4)}(t_3,t_2,t_1) = -i\omega_{eg}(t_1+t_3) - g_-(t_1) - g_-(t_3) \\ \quad -[g_-(t_2) - g_-(t_2+t_3) - g_-(t_1+t_2) + g_-(t_1+t_2+t_3)]. \end{cases} \tag{2.28}$$

Then, using the response function Equation (2.27), we have the third-order polarization in the form;

$$P^{(3)}(t) = i\mu^4 \sum_{\alpha=1}^{4} \int_0^\infty dt_3 \int_0^\infty dt_2 \int_0^\infty dt_1 E(t-t_3)E(t-t_2-t_3)$$
$$E(t-t_1-t_2-t_3)R_\alpha^{(3)}(t_3,t_2,t_1) + c.c. \tag{2.29}$$

These quantities are generalized to the non-Condon case [19] and the results are given in the Appendix.

The Brownian oscillator model provides a picture in terms of wave packets in phase space which can be calculated semiclassically. Using a classical Langevin equation, Yan and Mukamel derived closed expressions for the wavepackets [22]. The equations presented here generalize these results in two aspects. (i) We use a microscopic description of the bath which provides a consistent treatment of relaxation and dephasing at all temperatures. The Langevin equation used earlier is valid at high temperatures. Yan and Mukamel have shown how the exact expression for the response function can be obtained from the Langevin equation by further assuming the cumulant expansion and including the fluctuation dissipation theorem. However, the expression of the wavepackets was given only in the high temperature limit, since the semi-classical Langevin equation approach cannot keep track of the quantum coherence between the system and the noise source (the heat bath). This coherence is less important at high temperatures due to the fast dephasing, but becomes dominant at low temperatures. (ii) We allow for an arbitrary dependence of the transition dipole moment on nuclear coordinates and thus relax the Condon approximation.

3. Impulsive Pump-Probe Spectroscopy with Non-Condon Dipole Moment

We assume the variation of the transition dipole with the nuclear coordinate (non-Condon effects) is given in the form

$$\mu(q) = \mu_0 \exp(cq). \tag{3.1}$$

We may calculate the response functions and the phase space distribution functions for this model by replacing all c_j in the Appendix by c.

In a pump-probe experiment, the system is first subjected to a short pump pulse, then after a delay τ, a second probe pulse interacts with the system. The external electric field is given by

$$E(t) = E_1(t+\tau)\exp(-i\Omega_1 t) + E_1^*(t+\tau)\exp(i\Omega_1 t)$$
$$+ E_2(t)\exp(-i\Omega_2 t) + E_2^*(t)\exp(i\Omega_2) \tag{3.2}$$

where $E_1(t)$ and $E_2(t)$ are the temporal envelopes, and Ω_1 and Ω_2 are the center frequency of the pump and the probe field, respectively.

The probe absorption spectrum is [14]

$$S(\Omega_1, \Omega_2; \omega_2, \tau) = -2 \,\mathrm{Im}\, E_2[\omega_2] P^{(3)}[\omega_2], \tag{3.3}$$

where

$$E_2[\omega_2] = \frac{1}{\sqrt{2\pi}} \int_{-\infty}^{\infty} dt \, \exp[i(\omega_2 - \Omega_2)t] E_2(t), \tag{3.4}$$

and

$$P^{(3)}[\omega_2] = \frac{1}{\sqrt{2\pi}} \int_{-\infty}^{\infty} dt \exp[i(\omega_2 - \Omega_2)t] P^{(3)}(t). \tag{3.5}$$

We shall assume impulsive pump and probe pulses [23]

$$E_1(t) = \theta_1 \delta(t + \tau), \quad E_2(t) = \theta_2 \delta(t), \tag{3.6}$$

where θ_1 and θ_2 are the pump and pulse areas, respectively. We shall calculate separately the particle contribution corresponding to the Liouville paths $\alpha = 1$ ($eg \to ee \to eg$) and $\alpha = 2$ ($ge \to ee \to eg$) and the hole contribution corresponding to the Liouville path $\alpha = 3$ ($ge \to gg \to eg$) and $\alpha = 4$ ($eg \to gg \to eg$). For the impulsive pump case, $R_1^{(3)}(t, \tau, 0) = R_2^{(3)}(t, \tau, 0)$ and $R_3^{(3)}(t, \tau, 0) = R_4^{(3)}(t, \tau, 0)$, and we have

$$S(\omega_2 - \Omega_2) = S_{ee}(\omega_2 - \Omega_2; \tau) + S_{gg}(\omega_2 - \Omega_2; \tau). \tag{3.7}$$

Here, S_{ee} and S_{gg} are the contributions of the particle

$$S_{ee}(\omega_2 - \Omega_2) = 2 \,\mathrm{Re} \int_0^{\infty} dt \, \exp[i(\omega_2 - \Omega_2)t] R_1^{(3)}(t, \tau, 0), \tag{3.8}$$

and the hole, given by

$$S_{gg}(\omega_2 - \Omega_2) = 2 \,\mathrm{Re} \int_0^{\infty} dt \, \exp[i(\omega_2 - \Omega_2)t] R_3^{(3)}(t, \tau, 0), \tag{3.9}$$

and we set $\mu_0 \theta_1 = \mu_0 \theta_2 = 1$.

We have also calculated the linear absorption spectra defined by

$$\sigma(\omega) = \int_0^{\infty} dt \, R_1^{(1)}(t) \, \exp(i\omega t) + c.c. \tag{3.10}$$

Hereafter we assume a frequency independent damping, where $\gamma(\omega) = \gamma$, analytical expressions for the symmetric and antisymmetric correlation functions are known [21]. The auxiliary function is then given by

$$g_{\pm}(t) = g'(t) \pm i g''(t), \tag{3.11}$$

where

$$g'(t) = \lambda \left\{ \left[\frac{\lambda_1^2}{2\zeta\omega_0^2}(e^{-\lambda_2 t} + \lambda_2 t - 1) \coth\left(\frac{i\beta\hbar\lambda_2}{2}\right) \right.\right.$$
$$\left. - \frac{\lambda_2^2}{2\zeta\omega_0^2}(e^{-\lambda_1 t} + \lambda_1 t - 1) \coth\left(\frac{i\beta\hbar\lambda_1}{2}\right) \right]$$
$$\left. - \frac{4\gamma\omega_0^2}{\beta\hbar} \sum_{n=1}^{\infty} \frac{1}{\nu_n} \frac{e^{-\nu_n t} + \nu_n t - 1}{(\omega_0^2 + \nu_n^2)^2 - \gamma^2\nu_n^2} \right\}, \tag{3.12}$$

and

$$ig''(t) = i\lambda \left[e^{-\gamma t/2} \left(\frac{\gamma^2/2 - \omega_0^2}{\zeta\omega_0^2} \sin(\zeta t) + \frac{\gamma}{\omega_0^2} \cos(\zeta t) \right) + t - \frac{\gamma}{\omega_0^2} \right]. \tag{3.13}$$

Here, we defined $\nu_n = 2\pi n/\hbar\beta$ and

$$\lambda_1 = \frac{\gamma}{2} + i\zeta, \quad \lambda_2 = \frac{\gamma}{2} - i\zeta, \quad \zeta = \sqrt{\omega_0^2 - \gamma^2/4}. \tag{3.14}$$

In the following, we present the linear absorption spectrum $\sigma(\omega)$, the pump-probe spectrum $S(\omega)$, and the second order wavepackets $W^{(2)}(t)$. The second order wavepackets with the Condon and the non-Condon interactions are obtained from Equations (2.24) and (A11), respectively, by putting $t_3 = 0$, $t_2 = t$, *i.e.* $W^{(2)}(t_1, t) = W^{(3)}(t_1, t, 0)$. For the results of the pump-probe spectrum and the wavepackets, we separately present the contribution from the particle and the hole. Calculations are performed for the Condon case ($c = 0$) and the non-Condon case ($c = 0.1$), and are denoted by a) and b) in all figures. The frequency and the dimensionless displacement are taken to be $\omega_0 = 600$ [cm^{-1}], $d = 1.0$, and $T = 100$ [K].

Figure 1 shows the linear absorption spectra calculated using Equation (3.10) for different choices of γ: 1) the underdamped case, $\gamma = 40$ [cm^{-1}]; 2) the intermediate case $\gamma = 400$ [cm^{-1}]; and 3) the overdamped case $\gamma = 2000$ [cm^{-1}]. The non-Condon spectra are slightly shifted to the blue compared with the Condon spectra. This shift becomes larger as the temperature is increased. The reason is as follows; prior to the pump excitation, the system is in the ground equilibrium state and the nuclear wavepacket, which is well localized if the temperature is low, broadens at high temperatures. As seen from Equation (3.1) the dipole element $\mu(q)$ is a linear function of q for small c. The probability of the system to have large nuclear displacements q increases with temperature, and the non-Condon effect becomes therefore larger at higher temperatures. Since $\mu(q)$ becomes larger where the excitation energy between the ground state and the excitation state is high, then it helps the excitation of the pump and, thus, the spectrum shifts to the blue.

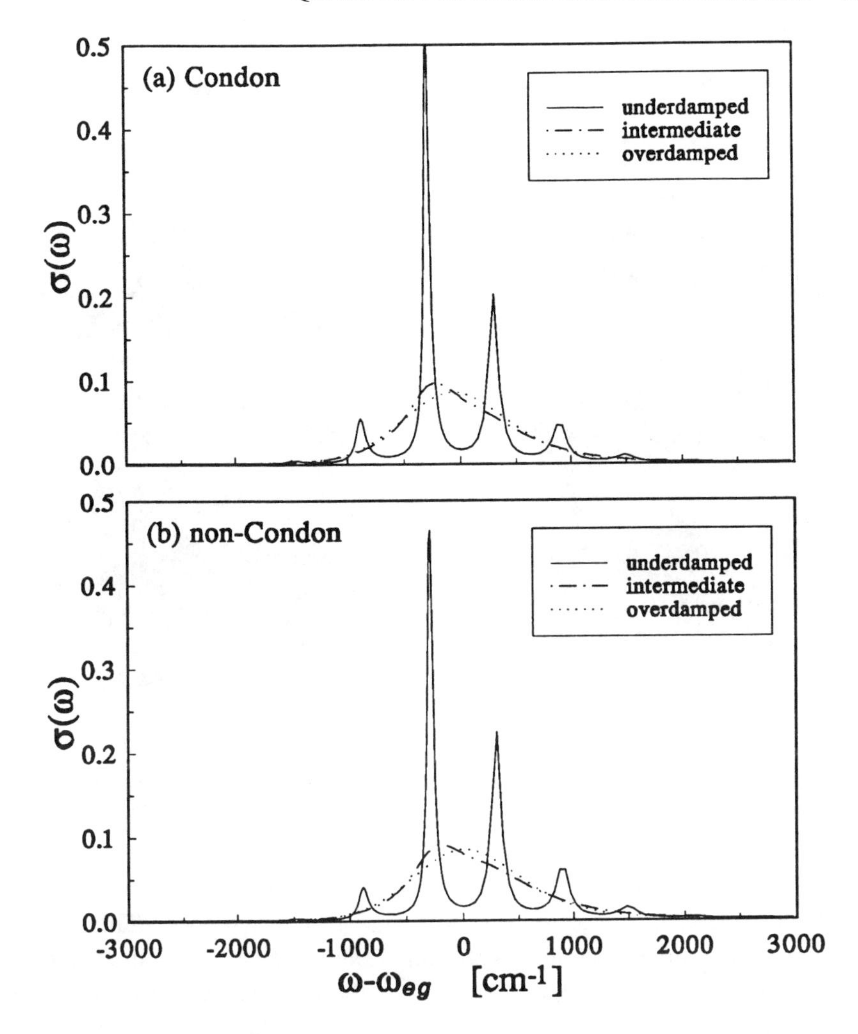

Fig. 1. Absorption spectrum with Condon dipole interaction ($c = 0$) for: 1) the underdamped case, $\gamma = 40$ [cm^{-1}]; 2) the intermediate case $\gamma = 400$ [cm^{-1}]; and 3) the overdamped case $\gamma = 2000$ [cm^{-1}] at different temperatures $T = 100$ [K]. In a) indicate results with Condon approximation ($c = 0$), whereas b) without Condon approximation ($c = 0.1$).

These effects become larger in nonlinear experiments such as pump-probe experiments, since they are higher orders in the dipole interaction.

Figure 2 shows the impulsive pump-probe spectrum for the underdamped case $\gamma = 40$ [cm^{-1}]. Here, we define $\Delta\omega = \omega_2 - \Omega_2 - \omega_{eg}$. Figure 2a is for the Condon approximation whereas 2b is for the non-Condon interaction. In each case, we display separately the contributions of the hole, the particle, and their sum. In Figures 2a and 2b, the particle spectra are rapidly changing in both the Condon and the non-Condon cases; however the hole for the

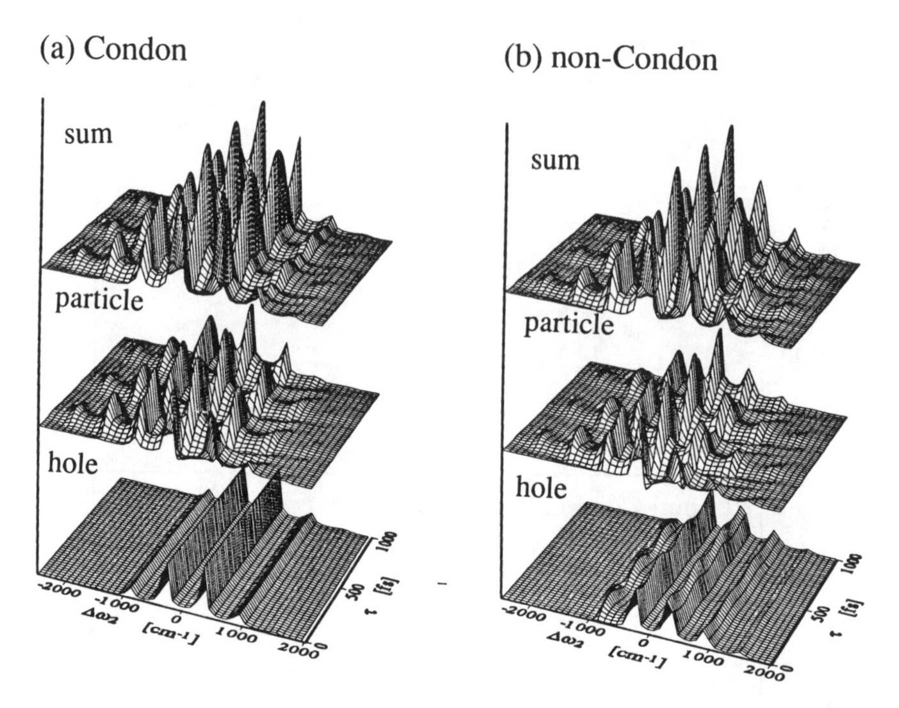

Fig. 2. The impulsive pump-probe spectrum for the underdamped case $\gamma = 40$ [cm^{-1}] at the low temperature $T = 100$ [K]. Here, we define $\Delta\omega = \omega_2 - \Omega_2 - \omega_{eg}$. Figure 2a is for the Condon approximation whereas 2b is for the non-Condon interaction. In each of these, we display separately the contributions of the hole, the particle, and their sum.

Condon case (the bottom of 2a) does not change at all with time. This can be understood by plotting the time evolution of the wavepacket. Figure 3 shows the wavepacket corresponding to Figure 2 (the unit of r is $\sqrt{\hbar/M\omega_0}$). We plotted three wavepackets (the hole, the particle and their sum) as a function of the coordinate and the time. In this underdamped mode, the particle moves from the ground state position $q = 0$ to the equilibrium state $q = d = -1$ (the bottom of the excited state potential) with a coherent oscillation, both for the Condon and the non-Condon cases. However, the hole, which in the Condon case does not change its position and shape, slightly oscillates in the non-Condon case. This is due to the impulsive pump [23]. Under the Condon approximation, the impulsive pump pulse creates a particle in the excited state without changing the Gaussian shape of the wavepacket in the ground-state. Then, the shape of the hole wavepacket is also Gaussian and cannot move in the harmonic potential. However, in the non-Condon case, the coordinate dependent dipole operator affects the shape of the ground equilibrium state.

Figures 4 and 5 show the spectra for the intermediate damping case $\gamma = 400$ [cm^{-1}]. As seen from the middle of Figures of 5a and 5b, the motion

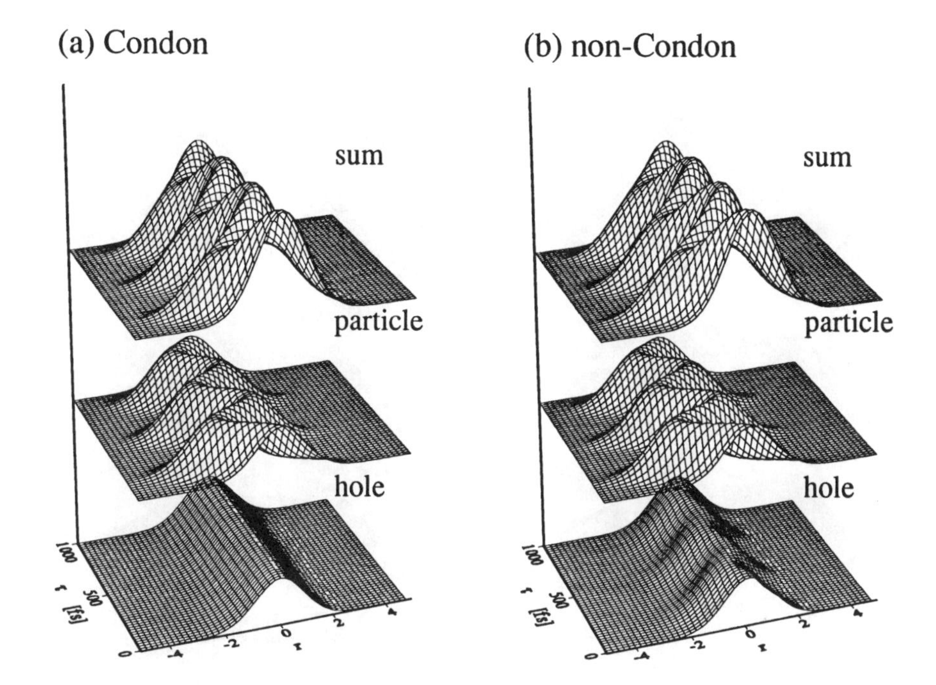

Fig. 3. The time-evolutions of the wavepacket for the underdamped case corresponds to Figure 2 (the unit of r is $\sqrt{\hbar/M\omega_0}$). We plotted three wavepacket (the hole, the particle and the total) as the function of the coordinate and the time.

of the particle is critically damped and the particle quickly moves from the equilibrium position of the ground state ($r = 0$) to the bottom of the excited potential ($r = -1$). These motions are clearly reflected to the spectra Figure 4.

Figures 6 and 7 are for the overdamped damping case $\gamma = 2000$ [cm^{-1}]. In this case, the particle motion is strongly suppressed by the heat bath and particle quickly reaches its equilibrium distribution. Under strong damping, even though with non-Condon interaction, the hole cannot move and shows similar behavior to the Condon case.

Acknowledgements

The support of the National Science Foundation is gratefully acknowledged.

Appendix: Optical Polarization and Wavepackets with Non-Condon Dipole Moment

In order to express the wavepacket element and the optical polarization in compact way, it is convenient to introduce a sign parameter, ε_1, ε_2, and ε_3,

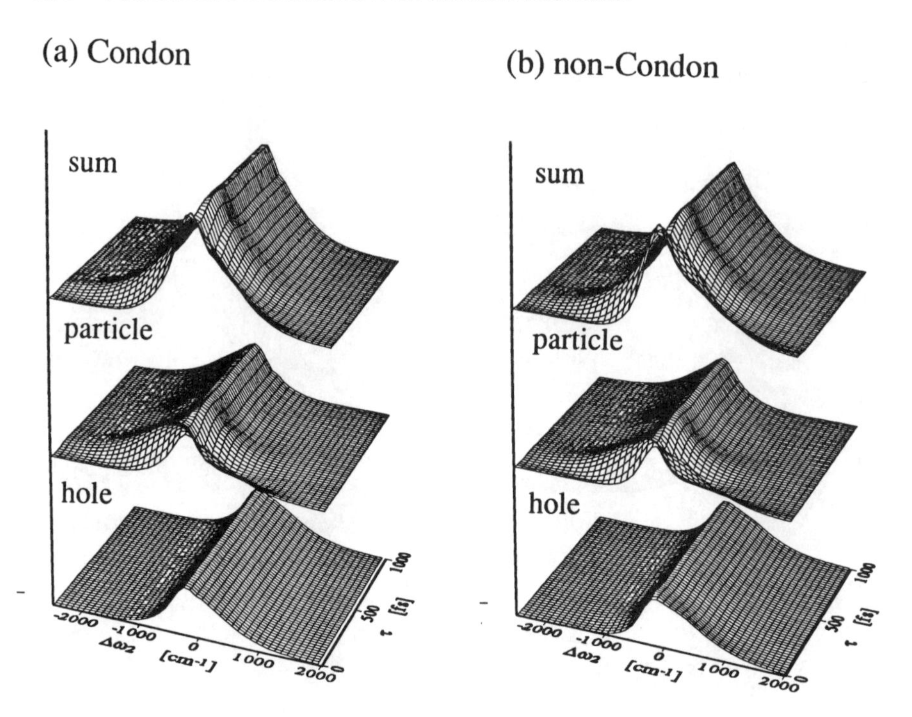

Fig. 4. The impulsive pump-probe spectrum for the intermediate damped case $\gamma = 400$ [cm^{-1}]. The other parameters are the same as the case of Figure 2.

TABLE I

Auxiliary parameters for Equations (A2)–(A15).

Liouville space path α	ε_1	ε_2	ε_3
1	+	+	+
2	−	+	+
3	−	−	+
4	+	−	+

Each Liouville space path is characterized by 3 parameters, as given in Table I. Using these, the first and the third order polarization with the non-Condon

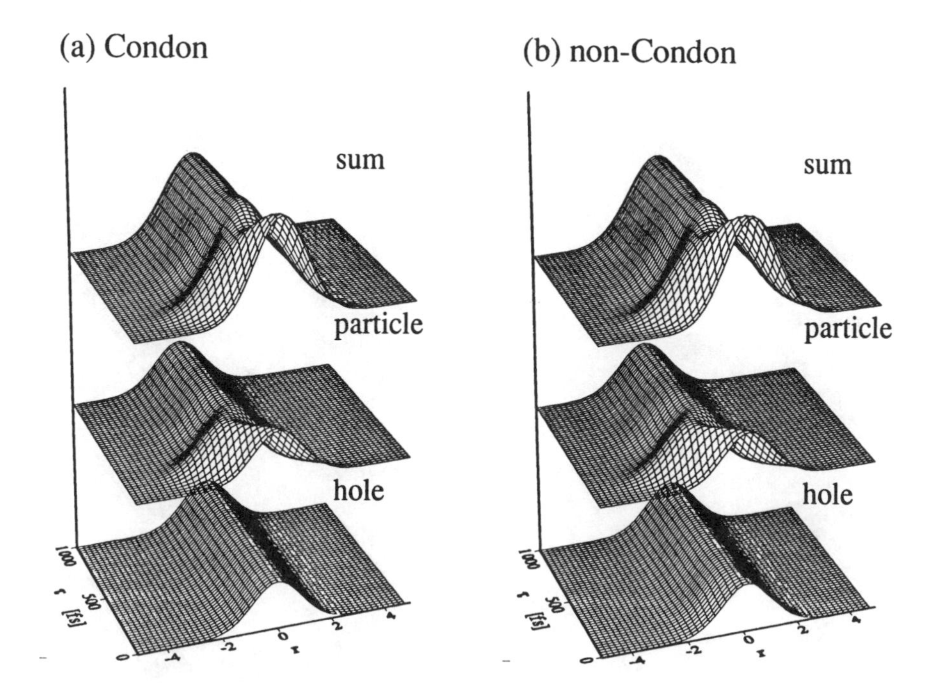

Fig. 5. The time-evolutions of the wavepacket for the intermediate damped case corresponds to Figure 4.

interaction are expressed as [19]

$$P^{(1)}(t) = -i \int_0^\infty dt_1 E(t-t_1)\{\mu(\partial/\partial c_1)\mu(\partial/\partial c_2) \\ R_+^{(1)}(t_1, c_1, c_2)\}|_{c_1=c_2=0} + c.c., \tag{A1}$$

$$P^{(3)}(t) = i \int_0^\infty dt_3 \int_0^\infty dt_2 \int_0^\infty dt_1 E(t-t_3) \\ E(t-t_2-t_3)E(t-t_1-t_2-t_3) \\ \times \sum_{\varepsilon_1\varepsilon_2=\pm} \{\mu(\partial/\partial c_4)\mu(\partial/\partial c_3)\mu(\partial/\partial c_2)\mu(\partial/\partial c_1) \\ R^{(3)}_{\varepsilon_1\varepsilon_2+}(t_3, t_2, t_1, \{c_i\})\}|_{\{c\}=0} + c.c. \tag{A2}$$

Here, the generating function of the non-Condon response functions are defined by

$$R^{(1)}_{\varepsilon_1}(t_1, c_1, c_2) = \exp[Q_{\varepsilon_1}(t_1) + X_{\varepsilon_1}(t_1, c_1, c_2)], \tag{A3}$$

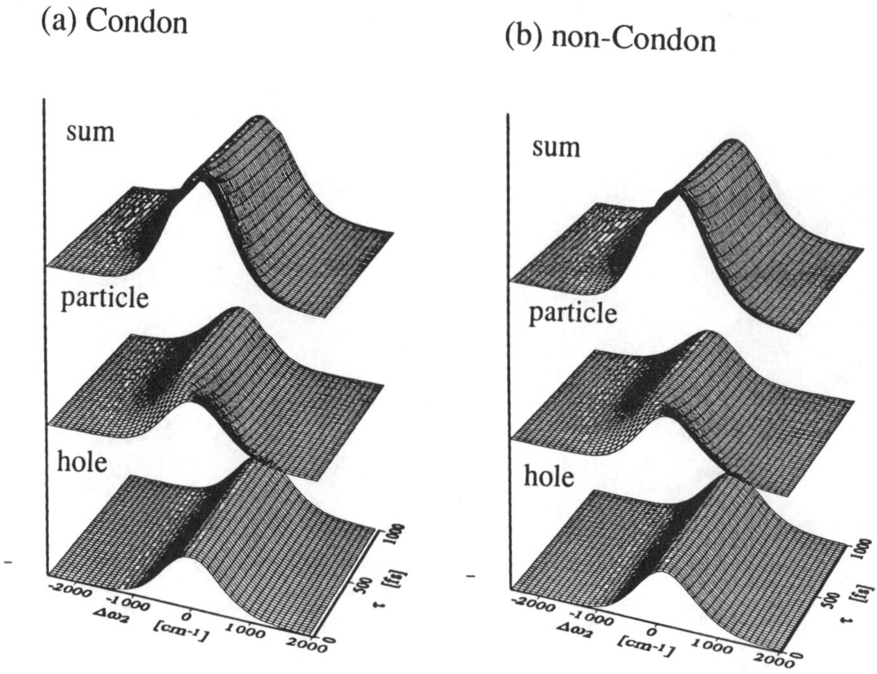

(a) Condon

(b) non-Condon

Fig. 6. The impulsive pump-probe spectrum for the overdamped case $\gamma = 2000$ [cm^{-1}]. The other parameters are the same as the case of Figure 2.

$$R^{(3)}_{\varepsilon_1\varepsilon_2\varepsilon_3}(t_3, t_2, t_1; \{c_j\}) = \exp[Q_{\varepsilon_1\varepsilon_2\varepsilon_3}(t_3, t_2, t_1) + X_{\varepsilon_1\varepsilon_2\varepsilon_3}(t_3, t_2, t_1; \{c_j\})], \tag{A4}$$

where

$$Q_{\varepsilon_1}(t_1) = -i\varepsilon_1\omega_{eg}t_1 - g_{-\varepsilon_1}(t_1), \tag{A5}$$

$$\begin{aligned} Q_{\varepsilon_1\varepsilon_2\varepsilon_3}&(t_3, t_2, t_1) \\ = &-i\omega_{eg}(\varepsilon_1 t_1 + \varepsilon_3 t_3) - g_{-\varepsilon_1}(t_1) - g_{\varepsilon_2\varepsilon_3}(t_3) \\ &-\varepsilon_1\varepsilon_3[g_{\varepsilon_1\varepsilon_2}(t_2) - g_{\varepsilon_1\varepsilon_2}(t_2 + t_3) \\ &-g_{-\varepsilon_1}(t_1 + t_2) + g_{-\varepsilon_1}(t_1 + t_2 + t_3)], \end{aligned} \tag{A6}$$

$$X_{\varepsilon_1}(t_1, c_1, c_2) = -i\xi^{-1}\dot{g}_{-\varepsilon_1}(t_1)(c_1 + c_2) + \xi^{-2}\left[\ddot{g}_{-\varepsilon_1}(t_1)c_1c_2 + \frac{1}{2}(c_1^2 + c_2^2)\ddot{g}(0)\right], \tag{A7}$$

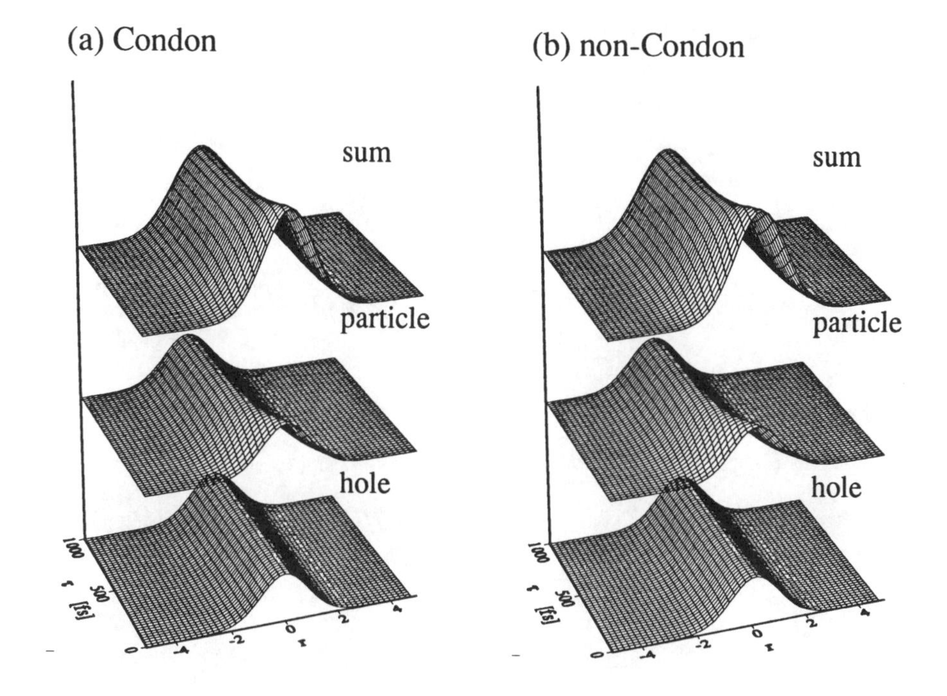

Fig. 7. The time-evolutions of the wavepacket corresponds to Figure 6.

$$\begin{aligned}
X_{\varepsilon_1\varepsilon_2\varepsilon_3}(t_3, t_2, t_1, \{c_j\}) &= c_1\langle\bar{q}_1(\{t\})\rangle_{\varepsilon_1\varepsilon_2\varepsilon_3} + c_2\langle\bar{q}_2(\{t\})\rangle_{\varepsilon_1\varepsilon_2\varepsilon_3} \\
&\quad + c_3\langle\bar{q}_3(\{t\})\rangle_{\varepsilon_1\varepsilon_2\varepsilon_3} + c_4\langle\bar{q}_4(\{t\})\rangle_{\varepsilon_1\varepsilon_2\varepsilon_3} \\
&\quad + \xi^{-2}\left[c_1c_2\ddot{g}_{-\varepsilon_1}(t_1) + c_1c_3\ddot{g}_{-\varepsilon_1}(t_1+t_2)\right. \\
&\quad + c_2c_3\ddot{g}_{\varepsilon_1\varepsilon_2}(t_2) + c_1c_4\ddot{g}_{-\varepsilon_1}(t_1+t_2+t_3) \\
&\quad + c_2c_4\ddot{g}_{\varepsilon_1\varepsilon_2}(t_2+t_3) + c_3c_4\ddot{g}_{\varepsilon_2\varepsilon_3}(t_3) \\
&\quad \left. + \frac{1}{2}(c_1^2 + c_2^2 + c_3^2 + c_4^2)\ddot{g}(0)\right], \qquad \text{(A8)}
\end{aligned}$$

with

$$\left\{\begin{aligned}
\langle\bar{q}_1(\{t\})\rangle_{\varepsilon_1\varepsilon_2\varepsilon_3} &= -i\xi^{-1}[\varepsilon_1\dot{g}_{-\varepsilon_1}(t_1) - \varepsilon_3\dot{g}_{-\varepsilon_1}(t_1+t_2) \\
&\quad + \varepsilon_3\dot{g}_{-\varepsilon_1}(t_1+t_2+t_3)], \\
\langle\bar{q}_2(\{t\})\rangle_{\varepsilon_1\varepsilon_2\varepsilon_3} &= -i\xi^{-1}[\varepsilon_1\dot{g}_{-\varepsilon_1}(t_1) + \varepsilon_3\dot{g}_{\varepsilon_1\varepsilon_2}(t_2+t_3) \\
&\quad - \varepsilon_3\dot{g}_{\varepsilon_1\varepsilon_2}(t_2)], \\
\langle\bar{q}_3(\{t\})\rangle_{\varepsilon_1\varepsilon_2\varepsilon_3} &= -i\xi^{-1}[\varepsilon_1\dot{g}_{-\varepsilon_1}(t_1+t_2) - \varepsilon_1\dot{g}_{\varepsilon_1\varepsilon_2}(t_2) \\
&\quad + \varepsilon_3\dot{g}_{\varepsilon_2\varepsilon_3}(t_3)], \\
\langle\bar{q}_4(\{t\})\rangle_{\varepsilon_1\varepsilon_2\varepsilon_3} &= -i\xi^{-1}[\varepsilon_1\dot{g}_{-\varepsilon_1}(t_1+t_2+t_3) - \varepsilon_1\dot{g}_{\varepsilon_1\varepsilon_2}(t_2+t_3) \\
&\quad + \varepsilon_3\dot{g}_{\varepsilon_2\varepsilon_3}(t_3)].
\end{aligned}\right. \qquad \text{(A9)}$$

Note that Equations (A5) and (A6) are just a compact expression of Equations (2.22) and (2.28), and are very convenient for numerical evaluations. The first and the third order wavepackets are given by

$$\begin{aligned} &W^{(1)}(p,q,t) \\ &= \int_0^\infty dt_1 E(t-t_1)\mu(\partial/\partial c_1)\left(4\pi^2\langle p^2\rangle_g\langle q^2\rangle_g\right)^{-1/2} \\ &\times \exp\left[-\frac{1}{2\langle q^2\rangle_g}(q-\bar{q}_+(t_1,c_1))^2 - \frac{1}{2\langle p^2\rangle_g}(p-\bar{p}_+(t_1,c_1))^2\right] \\ &R_+^{(1)}(t_1,c_1,c_2)|_{c_1=c_2=0} + c.c., \end{aligned} \tag{A10}$$

and

$$\begin{aligned} W^{(3)}(p,q,t) = &\int_0^\infty dt_1 \int_0^\infty dt_2 \int_0^\infty dt_3 E(t-t_3)E(t-t_2-t_3) \\ &E(t-t_1-t_2-t_3)\mu(\partial/\partial c_1)\mu(\partial/\partial c_2)\mu(\partial/\partial c_3) \\ &\times(4\pi^2\langle p^2\rangle_g\langle q^2\rangle_g)^{-1/2} \sum_{\varepsilon_1,\varepsilon_2=\pm} \\ &\times \exp\left[-\frac{1}{2\langle q^2\rangle_g}(q-\bar{q}_{\varepsilon_1\varepsilon_2+}(\{t\},\{c\}))^2\right. \\ &\left.-\frac{1}{2\langle p^2\rangle_g}(p-\bar{p}_{\varepsilon_1\varepsilon_2+}(\{t\},\{c\}))^2\right] \\ &\times R^{(3)}_{\varepsilon_1\varepsilon_2+}(t_3,t_2,t_1;\{c\})|_{\{c\}=0} + c.c. \end{aligned} \tag{A11}$$

where the constants are given in Equation (2.19) and

$$\bar{q}_{\varepsilon_1}(t_1,c_1) = -i\xi^{-1}\dot{g}_{-\varepsilon_1}(t_1) + c_1\xi^{-2}\ddot{g}_{-\varepsilon_1}(t_1), \tag{A12}$$

$$\bar{p}_{\varepsilon_1}(t_1,c_1) \equiv -iM\xi^{-1}\ddot{g}_{-\varepsilon_1}(t_1) + c_1 M\xi^{-2}\dddot{g}_{-\varepsilon_1}(t_1), \tag{A13}$$

and

$$\begin{aligned} \bar{q}_{\varepsilon_1\varepsilon_2\varepsilon_3}(\{t\},\{c\}) = &-i\xi^{-1}[\varepsilon_1\dot{g}_{-\varepsilon_1}(t_1+t_2+t_3) - \varepsilon_1\dot{g}_{\varepsilon_2\varepsilon_1}(t_2+t_3) \\ &+\varepsilon_3\dot{g}_{\varepsilon_2\varepsilon_3}(t_3)] + \xi^{-2}[c_1\ddot{g}_{-\varepsilon_1}(t_1+t_2+t_3) \\ &+c_2\ddot{g}_{\varepsilon_2\varepsilon_1}(t_2+t_3) + c_3\ddot{g}_{\varepsilon_2\varepsilon_3}(t_3)], \end{aligned} \tag{A14}$$

$$\begin{aligned} \bar{p}_{\varepsilon_1\varepsilon_2\varepsilon_3}(\{t\},\{c\}) = &-iM\xi^{-1}[\varepsilon_1\ddot{g}_{-\varepsilon_1}(t_1+t_2+t_3) - \varepsilon_1\ddot{g}_{\varepsilon_2\varepsilon_1}(t_2+t_3) \\ &+\varepsilon_3\ddot{g}_{\varepsilon_2\varepsilon_3}(t_3)] + M\xi^{-2}[c_1\dddot{g}_{-\varepsilon_1}(t_1+t_2+t_3) \\ &+c_2\dddot{g}_{\varepsilon_2\varepsilon_1}(t_2+t_3) + c_3\dddot{g}_{\varepsilon_2\varepsilon_3}(t_3)]. \end{aligned} \tag{A15}$$

The second order wavepackets can be obtained from the third order one by simply setting $t_3 = 0$.

References

1. Maroncelli, M., Maclnnis, J., and Fleming, G. R., *Science* **234**, 1674 (1989).
2. Barbara, P. F. and Jarzeba, W., *Adv. in Photochem.* **15**, 1 (1990).
3. See "Electron Transfer" Special Issue of *Chem. Rev.* **92** (1992) and references therein.
4. Anfinrud, P. A., Han, C., Lian, T., and Hochstrasser, R. M., *J. Phys. Chem.* **95**, 574 (1991); Owrutsky, J. C., Kin, Y. R., Li, M., Sarisky. M. J., and Hochstrasser, R. M., *Chem. Phys. Lett.* **184**, 368 (1991); Rips, I., Klafter, J., and Jortner, J., *J. Phys. Chem.* **94**, 8557 (1990).
5. Xie, X. and Simon, J. D., *Rev. Sci. Instrumen.* **60**, 2614 (1989).
6. Fourkas, J. T. and Fayer, M. D., *Acc. Chem. Res.* **25**, 227 (1992).
7. Scherer, N. F., Carlson, R. J., Matio, A., Du, M., Ruggiero, A. J., Romero-Rochin, V., Cina, J. A., Fleming, G. R., and Rice, S. A., *J. Chem. Phys.* **95**, 1487 (1991).
8. Nelson, K. A. and Ippen, E. P., *Adv. Chem. Phys.* **75**, 1 (1989).
9. Marcus, R. A., *J. Chem. Phys.* **24**, 966 (1956); **24**, 979 (1956); *Ann. Phys. Chem.* **15**, 155 (1964).
10. Zusman, L. D., *Chem. Phys.* **49**, 295 (1980).
11. Mukamel, S., *Adv. Chem. Phys.* **70**, 165 (1988); Mukamel, S., *Ann. Rev. Phys. Chem.* **41**, 647 (1990).
12. Garg, A., Onuchic, J. N., and Ambegaokar, V., *J. Chem. Phys.* **83**, 4491 (1985).
13. Gruebele, M. and Zewail, A. H., *Phys. Today* **43**, 24 (1990).
14. Yan, Y. J. and Mukamel, S., *J. Chem. Phys.* **94**, 997 (1991); *ibid, Phys. Rev.* **A41**, 6485 (1990).
15. Webster, F., Rossky, P. J., and Friesner, R. A., *Comp. Phys.Comm.* **63,** 494 (1991).
16. Tully, J., *J. Chem. Phys.* **93**, 1061 (1990).
17. Sparpaglione, M. and Mukamel, S., *J. Chem. Phys.* **88**, 3263 (1988); *J. Chem. Phys.* **88**, 4300 (1988); Mukamel, S. and Yan, A. J., *Acc. Chem. Res.* **22**, 301 (1989).
18. Feynman, R. P. and Vernon, F. L., *Ann. Phys.* **24**, 118 (1963).
19. Tanimura, Y. and Mukamel, S., *Phys. Rev. E* **47**, 118 (1993); *ibid., J. Opt. Soc. Am. B* (in press).
20. Gehlen, J. N. and Chandler, D., to be published in *J. Chem. Phys.*
21. Grabert, H., Schramm, P., and Ingold, G.-L., *Phys. Rep.* **168,** 115 (1988).
22. Yan, Y. J. and Mukamel, S., *J. Chem. Phys.* **88**, 5735 (1988); **89**, 5160 (1988).
23. Bosma, W. B., Yan, Y. J., and Mukamel, S., *Phys. Rev.* **A42**, 6920 (1990).

13. Charge Transfer Reactions and Solvation Dynamics

JAMES T. HYNES
Department of Chemistry and Biochemistry, University of Colorado, Boulder, CO 80309–0215, U.S.A.

1. Introduction

In this article, we offer a perspective on solvation dynamics from the point of view of chemical reactions, and in particular from an initial viewpoint of heavy particle charge transfer reactions. These reactions include S_N2 nucleophilic displacements – $X^- + RY \rightarrow XR + Y^-$, S_N1 unimolecular ionizations – $RX \rightarrow R^+ + X^-$, dipolar isomerizations, and (with a certain elasticity of definition) ion pair interconversion. Here the solvation dynamics act to influence the motion of the reacting solute nuclear coordinate(s), and thus to influence the reaction rate. The primary reactive coordinate is a nuclear coordinate of the solute.

The related but rather more special case of (outer sphere) electron transfer does not involve the nuclear coordinates of the reactive solute – no bonds are broken or made – and the primary reactive coordinate is a collective coordinate of the solvent.

The outline of this contribution is as follows. In Section 2, we sketch the basic theory for the influence of solvation dynamics on reaction rates and some aspects of the dynamics themselves, as by now conventionally probed by time dependent fluorescence. We then pause to present in Section 3 a model Hamiltonian which attempts to provide a unifying format for the influence of solvation dynamics for reactions. Section 4 is concerned with several case studies, largely via Molecular Dynamics computer simulation, of assorted issues raised previously. Section 5 is devoted to a more detailed examination of the connections between all of the above. Concluding remarks are offered in Section 6.

We pay little attention here, except by way of contrast and illustration, to predictions of dielectric continuum theories. This is not to say that these are not valuable guides, especially for experiment, and indeed we have often employed such models ourselves. Rather, we focus in the limited space available on a more microscopic viewpoint which attempts to clarify the areas where further progress is required. In addition, we focus on the dynamical

J.D. Simon (ed.), Ultrafast Dynamics of Chemical Systems, 345–381.

correction to the Transition State Theory rate constant; it is in this correction that the influence of the solvent dynamics resides. This solvent effect is over and above potentially important solvent effects on the reaction activation free energy barrier. For some general reviews on topics allied to the present chapter, the reader should consult [1–4].

2. Reaction Rate Formulation and Solvation

2.1 HEAVY PARTICLE CHARGE TRANSFER – GROTE–HYNES THEORY

We begin by giving a brief account of Grote-Hynes Theory [1, 5] for reaction rate constants. This theory has been verified for a wide range of solution reactions via computer simulation [3, 6–12] and has also proved useful in comparison with experimental rate studies [12–16].

GH Theory was originally developed to describe chemical reactions in solution involving a nuclear solute reactive coordinate x. The identify of x will depend on the reaction type, i.e., it will be an asymmetric stretch in a bimolecular S_N2 displacement and a separation coordinate in an S_N1 unimolecular ionization. To set the stage, we can picture a reaction free energy profile in the solute reactive coordinate x calculated via the potential of mean force $G_{eq}(x)$ – the system free energy when the system is equilibrated at each fixed value of x. Attention then focusses on the barrier top in this profile, located at $x^\ddagger$ (cf. Figure 1).

In the GH theory, it is assumed that the reaction barrier is parabolic in the neighborhood of $x^\ddagger$ and that the solute reactive coordinate satisfies a generalized Langevin equation (GLE),

$$\ddot{x}(t) = \omega_{beq}^2 x(t) - \int_0^t \mathrm{d}\tau \zeta(t-\tau)\dot{x}(\tau), \tag{2.1}$$

where a random force term inessential for our purposes has been suppressed. It is convenient – and we have assumed this in Equation (2.1) – to take the barrier top location $x^\ddagger$ to be $x^\ddagger = 0$. The square equilibrium barrier frequency is governed by the mean potential curvature

$$\omega_{beq}^2 = -\mu^{-1} \frac{\partial^2 G_{eq}(x)}{\partial x^2} \mid_{x^\ddagger}, \tag{2.2}$$

while the time dependent friction coefficient, per solute mass μ, is governed by the fluctuating forces on the coordinate x through their time correlation function:

$$\zeta(t) = (\mu k_B T)^{-1} \langle \delta F \delta F(t) \rangle \mid_{x^\ddagger} . \tag{2.3}$$

The GLE accounts for the fact that the time scale of these forces is finite, and in particular, finite on the relevant time scale for the barrier crossing, a point to which we return below.

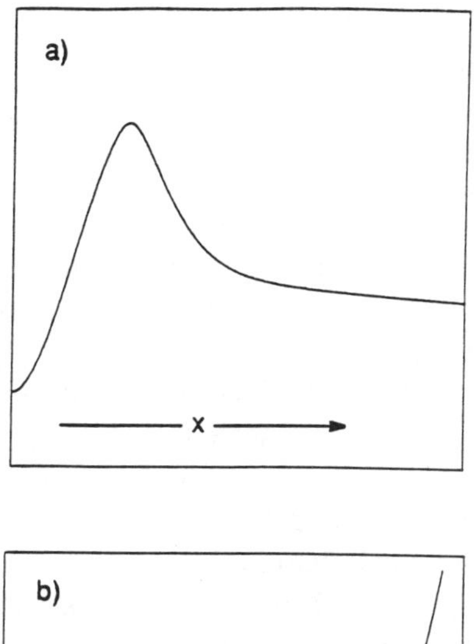

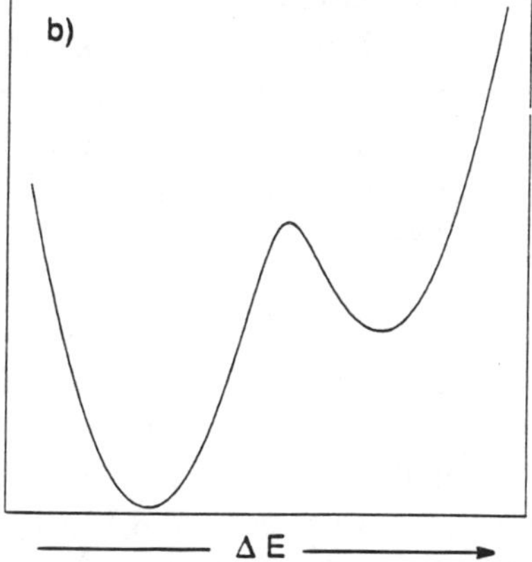

Fig. 1. Schematic free energy curves for (a) a charge transfer reaction in solution, along a solute reactive coordinate x; (b) an electron transfer reaction along a reactive solvent coordinate ΔE.

It should be stressed that the friction in Equation (2.3) is that relevant to the barrier top vicinity, and indeed the entire description is assumed to be valid in just that vicinity [1, 5, 7c]. This is not inconsistent with, and indeed recognizes, the fact that this friction can be different from that relevant to reactants and products [5].

With the above assumptions, the reaction transmission coefficient,

$$\kappa = \frac{k}{k^{TST}}, \tag{2.4}$$

i.e., the ratio of the actual rate constant to its Transition State Theory (TST) value, is found to be the ratio of the reactive frequency λ to the mean barrier frequency [1, 5]

$$\begin{aligned}\kappa &= \lambda/\omega_{beq};\\ \lambda &= \omega_{beq}^2[\lambda + \hat{\zeta}(\lambda)]^{-1};\\ \hat{\zeta}(\lambda) &= \int_0^\infty dt \, \exp(-\lambda t)\zeta(t).\end{aligned} \tag{2.5}$$

(There are of course important *static* solvent effects via the activation free energy in k^{TST} [1], but these are not our current concern.) This self-consistent equation shows that the relevant friction for the reaction is determined by the (Laplace) frequency component of the friction at the reactive frequency λ. This frequency sets the basic time scale for the microscopic events affecting κ. This is a convenient entrée to the discussion of GH theory in terms of solvation dynamics, now pursued.

The TST rate constant in Equation (2.4) refers to the conventional solution application of TST in which it is imagined that *equilibrium* solvation conditions hold [1, 5]. This amounts to picturing the reaction as passage over the equilibrated barrier in x, without any recrossing. Equation (2.1) indicates that these equilibrium conditions will apply when the x coordinate velocity is so low, and the generalized friction due to the coupling of x to the solvent is so small, that the solvent adjusts sufficiently rapidly, i.e., adiabatically, to the x motion to provide the equilibrium solvation incorporated in the equilibrium barrier frequency, but with *no* further effect. In the language of Equation (2.5), the reactive frequency λ would then be the equilibrium barrier frequency ω_{beq}. Whether these conditions are actually met, so that $k = k^{TST}$ and thus $\kappa = 1$, will depend on the time scale of the solvation dynamics and the coupling of the solvent to the reactive solute coordinate x.

To appreciate this latter point, we consider three important limits for the GH theory [17, 18] (although other limits are possible [1, 17, 18]). First, if the solvent is rapid on the time scale of λ^{-1}, then the frequency dependence of $\hat{\zeta}(\lambda)$ can be ignored, and the GH equations reduce to the Kramers Theory result [19]

$$\kappa_{GH} \rightarrow \kappa_{KR} = \frac{\omega_{beq}}{\omega_{beq}\kappa_{KR} + \zeta}, \tag{2.6}$$

in which the zero frequency friction, or friction constant, is

$$\hat{\zeta}(0) = \zeta = (\mu k_B T)^{-1} \int_0^\infty dt \langle \delta F \delta F(t) \rangle. \tag{2.7}$$

If the coupling to the solvent is large, here gauged by $\zeta/\omega_{beq} \gg 1$, the transmission coefficient will be low – $\kappa \sim \omega_{beq}/\zeta$. This is due to extensive recrossing of the barrier induced by the friction. Equivalently, there is extensive spatial nonequilibrium departure from equilibrium solvation conditions.

In this limit, the reaction can be viewed as a diffusion-controlled passage over the barrier, and the full zero frequency solvent dynamics are required for its description.

A second limit is attained if the solvent dynamics are slow on the reactive time scale λ^{-1}, i.e., there is *nonadiabatic* solvation. Then we can set $\hat{\zeta}(\lambda) = \lambda^{-1}\zeta(t=0)$ in Equation (2.5), and the GH equations reduce to [17. 18]

$$\begin{aligned} \kappa_{GH} &\rightarrow \kappa_{na} = \frac{\omega_{bna}^2}{\omega_{beq}^2}; \\ \omega_{bna}^2 &= \omega_{beq}^2 - \zeta(t=0) > 0. \end{aligned} \tag{2.8}$$

In this limit – which is favored by a sharp reaction barrier – the reaction can be viewed as x motion on a nonadiabatic *barrier*, whose frequency is the nonadiabatic value ω_{bna}. (What happens when ω_{bna}^2 is negative constitutes another limit, described below.) This can be seen from the GLE Equation (2.1) by ignoring the time dependence of $\zeta(t)$. This frequency ω_{bna} is less then the corresponding equilibrium value ω_{beq}, since the passage is occuring for fixed solvent configurations, and the solvent cannot respond to provide the equilibrium solvation incorporated in ω_{beq}. Again, there is nonequilibrium solvation and $\kappa < 1$. Near, but not exactly in, this limit, some information about the solvation dynamics is required, but is limited to the early time behavior of $\zeta(t)$.

Note especially that in the nonadiabatic solvatuion, or "frozen solvent" limit, it is the *absence* of solvation dynamics that is important. But it is just this lack that is responsible for the deviation from equilibrium solvation, which assumes the dynamics are effective in always maintaining equilibrium.

We now consider the final, *polarization cage*, limit favored by a more broad reaction barrier [1, 17, 18]

$$\begin{aligned} \omega_{bna}^2 &< 0; \\ \omega_{beq}^2 &< \zeta(t=0), \end{aligned} \tag{2.9}$$

whose consequence is the following. On short time scales where the solvent has not moved while the solute coordinate x crosses the barrier, Equation (2.9) indicates, with Equation (2.1), that the x motion is temporarily trapped, or "caged". The initial friction $\zeta(t=0)$ is too great to allow reaction. No net passage on to the product side of the barrier can occur unless and until there is solvent motion to relax this cage. Thus in this limit the solvent dynamics is *essential* for the reaction to occur; the transmission coefficient reflects this, decreasing more and more as the solvent relaxation time lengthens [17]. The associated multiple recrossing of the barrier is particularly pronounced when the x coordinate motion is overdamped as well [18]: an average trajectory can oscillate several times before ultimately passing to the product region.

The polarization cage limit will apply when there is a relatively broad barrier in the mean potential (small ω_{beq}) and there is strong solute-solvent electrostatic coupling [(large $\zeta(t = 0)$)]. If the reactive frequency is so low as to be small compared to the time scale of the friction, then the polarization cage regime predictions for κ will coincide with the Kramers limit. Otherwise, one just have to solve the GH Equation (2.5). Here something more than the initial time behavior of $\zeta(t)$ will be required to characterize the influence of the solvation dynamics.

One should note that in all of these limits (and in others [1, 17, 18]), both the solvent time scale *and* the magnitude of the solvent coupling to the x coordinate are key in the consequences for the rate of any nonequilibrium solvation conditions. The former will not lead to any departure from the TST equilibrium solvation rate if the solvent coupling is weak. For example, in the nonadiabatic limit, if the coupling is weak – as gauged by $\zeta(t = 0)/\omega_{beq}^2 \ll 1$, then $\kappa \to 1$ and $k \to k^{TST}$. It is not the case that the equilibrium solvation conditions are satisfied *per se* – indeed they are not, since the slow solvent cannot equilibrate to the reactive solute motion. Instead, it is the case that the coupling is so weak that the barrier crossing in x and thus the transmission coefficient are largely impervious to the surrounding solvent.

2.2 Electron Transfer Reactions

We now turn to outer sphere transfer (ET) reactions, e.g. $D^- + A \to D + A^-$, a donor-acceptor electron transfer without significant coupled internal reorganization of the D and A species [20, 21]. A hallmark of such reactions is that the reactive coordinate is itself a many-body collective solvent variable. In particular, if R and P denote the reactant and product, then the reactive coordinate is

$$\Delta E = V_{solv-R} - V_{solv-P}, \tag{2.10}$$

which is the difference in interaction energy, for the solvent molecules in given positions, of the solvent with the reactant and product [22]. In the simplest case of no geometric size changes accompanying the ET, ΔE will be determined exclusively by the Coulombic interactions. We will assume this to be the case in all that follows, and further that the solute intramolecular vibrations play no key role.

For electronically adiabatic ET, the electronic coupling between the R and P diabatic states is sufficiently large that there is a continuous change in electronic character in passage over a reaction barrier in ΔE. Again, one can apply GH Theory to this by assuming [23] a GLE in the barrier top neighborhood, but now for the solvent coordinate $\delta\Delta E(t) = \Delta E(t) - \Delta E^{\ddagger}$, where $\Delta E^{\ddagger}$ locates the barrier top in the equilibrium free energy $G_{eq}(\Delta E)$ (cf. Figure 1b):

$$\delta\Delta\ddot{E}(t) = (\omega_{beq}^{ET})^2 \delta\Delta E(t) - \int_0^t \mathrm{d}\tau \zeta_{ET}^{\ddagger}(t-\tau)\delta\Delta\dot{E}(\tau). \qquad (2.11)$$

Here ω_{beq}^{ET} and $\zeta_{ET}^{\ddagger}(t)$ refer to the equilibrium barrier frequency and the time dependent friction for the *solvent* coordinate $\delta\Delta E$; by contrast, Equation (2.1) refers to the corresponding quantities for a *solute* reactive coordinate.

The TST rate constant in the electronically adiabatic ET context is the well-known Marcus rate constant k_{ET}^{TST} [24, 24]. In our language, solvent dynamical effects can modify the actual rate from this limit due to the friction ζ_{ET} influence. The corresponding GH equations for $\kappa_{ET} = k_{ET}/k_{ET}^{TST}$ are strictly analogous to Equations (2.4) and (2.5) and we do not write them out here.

Although not our main focus, we briefly mention the electronically nonadiabatic ET situation. Here the electronic coupling is sufficiently weak that the intrinsic electronic passage from R to P is slow, even when the isoenergetic conditions in the solvent allow the ET via the Franck–Condon principle. The TST rate for this case contains in its prefactor an *electronic* transmission coefficient κ_{el}, which is proportional to the square of the electronic coupling [20]. But as first described by Zusman [25], if the solvation dynamics are sufficiently slow, the passage up to (and down from [23]) the nonadiabatic curve intersection can influence the rate. This has to do with solvent dynamics in the solvent wells (as opposed to the barrier top description given above) and is best discussed – in Section 4.3 – after we say something about the solvation dynamics involved in time dependent fluorescence. For the moment, we simply note that, even for electronically adiabatic reactions, the conventional picture has been that the well solvation dynamics govern the rate.

2.3 Solvation Dynamics

We now come to the direct probe of solvation dynamics accessible via time dependent fluorescence (TDF) [4]. We focus on the case of a change in the finite dipole moment of a solute upon electronic excitation; other cases may be handled similarly.

Again, the relevant variable is ΔE, as in Equation (2.10), where now R refers to the ground electronic state and P refers to the excited state, which has a different charge distribution. Within a linear response theory approximation (about which we say something more below), the normalized transient nonequilibrium Stokes fluorescence frequency shift is (cf. Figure 2).

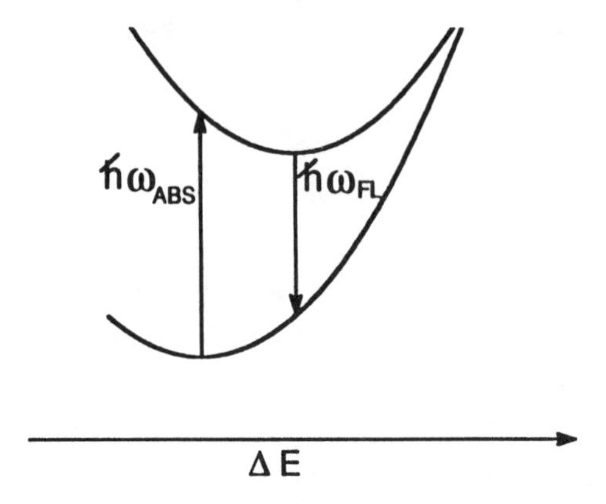

Fig. 2. Schematic free energy diagram for time dependent fluorescence.

$$\frac{\overline{\Delta E}(t) - \overline{\Delta E}(\infty)}{\overline{\Delta E}(0) - \overline{\Delta E}(\infty)} = \Delta_P(t);$$
$$\Delta_P(t) = \frac{\langle \delta\Delta E \delta\Delta E(t)\rangle_{\mathcal{P}}}{\langle \delta\Delta E^2\rangle_P}. \tag{2.12}$$

Here the overbars indicate a nonequilibrium average and $\Delta_P(t)$ is the normalized equilibrium tcf of the fluctuation in ΔE, $\delta\Delta E = \Delta E - \langle\Delta E\rangle_P$, where the averaging and dynamics refer to the solvent in the presence of the solute with the excited electronic state charge distribution, i.e. P.

In fact, a formally exact GLE equation of motion applies to $\delta\Delta E$. With the aid of the projection operator (we drop the subscript P for ease of notation)

$$\mathcal{P}\theta = \delta\Delta E\langle(\delta\Delta E)^2\rangle^{-1}\langle\delta\Delta E\theta\rangle + \delta\Delta\dot{E}\langle(\delta\Delta\dot{E})^2\rangle^{-1}\langle\delta\Delta\dot{E}\theta\rangle, \tag{2.13}$$

one derives the GLE [23, 26]

$$\delta\Delta\ddot{E}(t) = -\omega^2_{\delta\Delta E}\delta\Delta E(t) - \int_0^t \mathrm{d}\tau\, \zeta_{\delta\Delta E}(t-\tau)\delta\Delta\dot{E}(\tau), \tag{2.14}$$

where the time dependent friction for $\delta\Delta E$ is

$$\zeta_{\delta\Delta E}(t) = (\mu_{\delta\Delta E}k_BT)^{-1}\langle\delta F_{\delta\Delta E}\delta F_{\delta\Delta E}(t)\rangle. \tag{2.15}$$

Here the fluctuating "random force" for $\delta\Delta E$ is

$$\begin{aligned}\delta F_{\delta\Delta E} &= \mu_{\delta\Delta E}(1-\mathcal{P})\Delta\ddot{E} = \mu_{\delta\Delta E}\delta\Delta\ddot{E} + k_{\delta\Delta E}\delta\Delta E;\\ \delta F_{\delta\Delta E}(t) &= \exp[(1-\mathcal{P})i\,Lt]\delta F_{\delta\Delta E},\end{aligned} \tag{2.16}$$

where L is the Liouville operator [27] and where the solvent force constant, mass and frequency are

$$k_{\delta\Delta E} = \frac{k_B T}{\langle(\delta\Delta E)^2\rangle}; \qquad \mu_{\delta\Delta E} = \frac{k_B T}{\langle(\delta\Delta \dot{E})^2\rangle};$$

$$\omega^2_{\delta\Delta E} = \frac{k_{\delta\Delta E}}{\mu_{\delta\Delta E}}. \tag{2.17}$$

The double-membered projection operator in Equation (2.13) is not usual [27]; it is motivated by the feature that the $\delta\Delta E$ acceleration should exhibit a restoring force even when the velocity $\delta\Delta\dot{E}$ is negligible. Note that this average restoring force is subtracted off in the definition of the fluctuating force Equation (2.16).

The dynamics of the $\delta\Delta E\ tcf$ are expressed most conveniently via the Laplace transform relation, which with the GLE Equation (2.14), is

$$\hat{\Delta}(\varepsilon) = \{\varepsilon + \omega^2_{\delta\Delta E}[\varepsilon + \hat{\zeta}_{\delta\Delta E}(\varepsilon)]^{-1}\}^{-1}, \tag{2.18}$$

where ε is the Laplace transform variable.

Let us immediately connect to all of this the simplest standard description of TDF [4, 28], in which there is an exponential decay in time for $\Delta(t)$. This amounts to ignoring the inertia and the frequency dependence of the friction to give

$$\hat{\Delta}(\varepsilon) = \left[\varepsilon + \frac{\omega^2_{\delta\Delta E}}{\zeta_{\delta\Delta E}}\right]^{-1}. \tag{2.19}$$

or the exponential time decay

$$\Delta(t) = \exp\left(-\frac{t}{\tau_{\delta\Delta E}}\right), \tag{2.20}$$

with the solvation relaxation time

$$\tau_{\delta\Delta E} = \frac{\zeta_{\delta\Delta E}}{\omega_{\delta\Delta E}}, \tag{2.21}$$

which, it is useful to note from Equation (2.18), is just the area of $\Delta(t)$ without any approximation:

$$\tau_{\delta\Delta E} = \int_0^\infty dt\Delta(t) = \hat{\Delta}(0). \tag{2.22}$$

Thus the "solvation time" $\tau_{\delta\Delta E}$ measures the ratio of the friction constant to the square frequency associated with $\delta\Delta E$. A simple Debye dielectric continuum model for solvation dynamics in fact gives precisely this sort of exponential decay [4, 28]. In this case, the relaxation time is equal to, or simply related to, the longitudinal dielectric relaxation time τ_L, which itself is the Debye relaxation time, divided by the ratio of the static and high-frequency dielectric constants (or a function of this ratio). It should be noted that this continuum prediction time is independent of the solute charge distribution.

But even within the context of a continuum description, there are possibilities for nonexponential behavior. One of these arises from the retention of the frequency $\omega_{\delta\Delta E}$ in Equation (2.18) (cf. Section 4.2), i.e., the incorporation of inertial effects [28, 29]. Another is the case where the solvent exhibits a marked non-Debye dielectric behavior [23, 30, 31], a common occurrence in hydrogen-bonded solvents. Yet another is the inclusion of dielectric relaxation via translational diffusion of the solvent molecules in addition to the reorientational diffusion incorporated in the Debye description [28]. These last two sources of nonexponential solvation dynamics are simple illustrations in which the frequency dependence of the friction $\hat{\zeta}_{\delta\Delta E}$ is critical.

To return briefly to the issue of the linear response theory (LRT) approximation leading to Equation (2.12), there are actually two distinct LRT approximations [32]. The first, upon which Equation (2.12) is based, uses the dynamics of the solvent in the presence of the excited P state, but approximates the solvent initial conditions (which are – due to the Franck–Condon principle – actually those appropriate for the solvent in the presence of the ground R state). The second reverses this procedure, taking the correct initial distribution, but approximating the dynamics as those in the presence of R. (This leads to a tcf formula analogous to Equation (2.12), but in terms of an R, rather than a P, tcf.) The first seems to be decidedly the better approximation [32–34].

However, it is possible that there can be a serious breakdown of any LRT description of the TDF-solvation dynamics connection. This has been discovered and explained in microscopic terms for solvation in methanol in [34]. It seems that one must be very careful in the standard (and even continuing) practice of simply computing tcf's and assuming that these give the TDF response.

In Summary, the major dynamic quantities required as input for predicting and assessing the influence of the solvent on the reaction transmission coefficient is the time dependent friction. For charge transfer reaction involving solute nuclear coordinates, this is the friction $\zeta(t)$ for the solute reactive coordinate x. For any outer sphere ET, it is instead the friction $\zeta_{\delta\Delta E}(t)$ for the solvent coordinate $\delta\Delta E$. Ancillary quantities which are related to these, and perhaps to each other, are the initial value $\zeta(t = 0)$, the square frequencies $\omega_{b\ na}^2$ and $\omega_{\delta\Delta E}^2$, the friction constants ζ and $\zeta_{\delta\Delta E}$ and the time $\tau_{\delta\Delta E} = \zeta_{\delta\Delta E}/\omega_{\delta\Delta E}^2$.

3. Analytical Model

Before addressing the results of simulations and experiments on the issues of the last section, we will now develop in this section a simple model perspective [7c, 17, 18, 35]. This is designed both to shed light on the interpretation in terms of solvation of those results and to emphasize the interconnections (and differences) that may exist. The development below is intended to apply to charge transfer reaction systems with pronounced solute-solvent electrostatic coupling; it is not appropriate for, e.g., neutral reactions in which the solvent influence is mainly collisional [6, 11(a), (b), 36].

We consider the reactive solute system with coordinate x and associated mass μ, in the neighborhood of the barrier top, located at $x = x^{\ddagger} = 0$, and in the presence of the solvent. We characterize the latter by the single coordinate s, with an associated mass μ_s. If the solvent were equilibrated to x in the barrier passage, so that $s = s_{eq}(x)$, the potential for x is just $-1/2\mu\omega_{beq}^2 x^2$, where ω_{beq} is the equilibrium barrier frequency [cf. Equation (2.2)]. To this we add a locally harmonic restoring potential for the solvent to account for deviations from this equilibrium:

$$\frac{1}{2}\mu_s\omega_s^2[s - s_{eq}(x)]^2 = \left(\frac{1}{2}\right)\mu_s\omega_s^2(s - gx)^2. \tag{3.1}$$

Here ω_s is the solvent frequency and we have assumed that the solvent's equilibrium position varies linearly with the solute coordinate x in the barrier top neighborhood:

$$s_{eq}(x) = gx; \qquad g = \frac{ds_{eq}(x)}{dx}\Big|_{\ddagger}. \tag{3.2}$$

This dependence of $s_{eq}(x)$ on x arises most importantly in charge transfer problems from the variation of the charge distribution with x, e.g., in an S_N1 ionization of RX, the ionic character of RX depends on the internuclear RX separation since the solute electronic structure is changing [9, 35]. Even in a simple dipolar isomerization involving no electronic structure change, there is such a dependence because the dipole moment of the solute is changing [18]. Note that at this stage, the precise identity of the solvent coordinate has not been specified. We will return to this question in Section 5.

The Hamiltonian then becomes

$$H = H^{\ddagger} + \frac{1}{2}\mu\dot{x}^2 + \frac{1}{2}\mu_s\dot{s}^2 - \frac{1}{2}\mu\omega_{beq}^2x^2 + \frac{1}{2}\mu_s\omega_s^2(s - gx)^2. \tag{3.3}$$

If we now convert to mass-weighted coordinates $\bar{x} = \mu^{1/2}x$ and $\bar{s} = \mu_s^{1/2}s$ (and then drop the overbars hereafter), the Hamiltonian is

$$H = H^{\ddagger} + \frac{1}{2}\dot{x}^2 + \frac{1}{2}\dot{s}^2 - \frac{1}{2}\omega_{bna}^2x^2 + \frac{1}{2}\omega_s^2s^2 - w_c^2xs. \tag{3.4}$$

Here the square nonadiabatic frequency given by

$$\omega_{bna}^2 = \omega_{beq}^2 - \frac{\omega_c^4}{\omega_s^2}, \tag{3.5}$$

where the $x - s$ coupling square frequency is

$$\omega_c^2 = \left(\frac{\mu_s}{\mu}\right)^{1/2} \omega_s^2 g, \tag{3.6}$$

governs the initial force on x when the solvent coordinate is frozen at its saddle point value $s = s_{eq}(x^\ddagger) = 0$. In general, the coupled equations of motion are

$$\begin{aligned} \ddot{x} &= \omega_{bna}^2 x + \omega_c^2 s; \\ \ddot{s} &= -\omega_s^2 s + \omega_c^2 x. \end{aligned} \tag{3.7}$$

If the solvent adjust rapidly and equilibrates to the solute coordinate x so that there is no force on s, then we have the equilibrium condition

$$s = s_{eq}(x) = \left(\frac{\omega_c^2}{\omega_s^2}\right) x. \tag{3.8}$$

When this is inserted into the first member of Equation (3.7), it gives $\ddot{x} = \omega_{beq}^2 x$, where we have used Equation (3.5); this is just x motion on an equilibrated barrier — i.e., equilibrium solvation.

We will have occasion to further analyze the members of Equation (3.7) as they stand, but it is useful for our subsequent discussion to now simply add a generalized dissipative term to the solvent equation of motion to obtain the set

$$\begin{aligned} \ddot{x} &= \omega_{bna}^2 x + \omega_c^2 s; \\ \ddot{s} &= -\omega_s^2 s - \int_0^t d\tau \zeta_s(t-\tau)\dot{s}(\tau) + \omega_c^2 x, \end{aligned} \tag{3.9}$$

where $\zeta_s(t)$ is the *solvent* time dependent friction coefficient (per unit mass μ_s). These two equations can be cast into the form of a GLE Equation (2.1) for the reactive solute coordinate by Laplace transformation and insertion of the formal solution of the second equation into the first. The Laplace transform of the *solute* time dependent friction coefficient (per unit mass μ) is found to be [18]

$$\hat{\zeta}(\varepsilon) = (\mu k_B T)^{-1} \langle \delta F \delta \hat{F}(\varepsilon) \rangle = \zeta(t=0) \frac{(\varepsilon + \hat{\zeta}_s)}{\varepsilon^2 + \omega_s^2 + \varepsilon \hat{\zeta}_s}, \tag{3.10}$$

where the initial time value is related to the coupling and solvent frequencies

$$\zeta(t=0) = \frac{\omega_c^4}{\omega_s^2}. \tag{3.11}$$

The content of Equation (3.10) can be elaborated by considering the time correlation function of the solvent coordinate itself, when the solute coordinate is fixed at its transition state value $x = 0$. It is a straightforward exercise to show from Equation (3.9) that

$$\begin{aligned} \frac{\langle s\hat{s}(\varepsilon)\rangle_\ddagger}{\langle s^2\rangle_\ddagger} &= \frac{\varepsilon + \hat{\zeta}_s}{\varepsilon^2 + \omega_s^2 + \varepsilon\hat{\zeta}_s}; \\ \langle s^2\rangle_\ddagger &= \frac{k_B T}{\omega_s^2}, \end{aligned} \tag{3.12}$$

(recall that s is mass-weighted) from which we deduce the relationship

$$\zeta(t) = \zeta(t=0)\Delta_\ddagger(t); \quad \zeta(t) = \zeta(t=0)\hat{\Delta}_\ddagger(0) = \zeta(t=0)\tau_s, \tag{3.13}$$

where $\Delta_\ddagger(t)$ is the normalized equilibrium tcf of the solvent coordinate

$$\Delta_\ddagger(t) = \frac{\langle ss(t)\rangle_\ddagger}{\langle s^2\rangle_\ddagger}. \tag{3.14}$$

This is a key relation connecting the time dependent friction on the reactive solute coordinate to the solvation dynamics.

It is useful, for later reference, to consider the friction Equation (3.13) in a few limits. The first example we consider is if the frequency dependence of $\hat{\zeta}_s$ is ignored – $\hat{\zeta}_s(\varepsilon) \approx \zeta_s$ and the solvent acceleration is ignored – i.e., an overdamped solvent; then

$$\hat{\Delta}_\ddagger(\varepsilon) \approx \left[\varepsilon + \frac{\omega_s^2}{\zeta_s}\right]^{-1} \tag{3.15}$$

or in time language there is exponential decay of the solute friction

$$\begin{aligned} \zeta(t) &= \zeta(t=0)\exp\left(-\frac{t}{\tau_s}\right); \\ \tau_s &= \frac{\zeta_s}{\omega_s^2}. \end{aligned} \tag{3.16}$$

This approximation requires that $\zeta_s \gg \omega_s$. This behavior in fact follows from a Debye dielectric continuum model of the solvent when coupled to the solute nuclear motion [17, 18]; indeed, as noted in Section 2.3, in the context of TDF, the Debye model leads to such an exponential dependence of the analogue there of $\Delta_{\ddagger}(t)$. While Equation (3.16) is a useful broad characterization, we will see in Section 4 that it often does not capture the critical microscopic aspects of the friction revealed in MD studies, just as its analogue Equation (2.20) is similarly deficient.

Another limit of interest is if only short times are of importance. In this case, the time dependence of the solvent coordinate tcf $\Delta_{\ddagger}(t)$ can be ignored. Then $\zeta(t) = \zeta(t = 0)$, which by Equation (3.11), is a measure of the solute-solvent coupling frequency. This is particularly relevant for the nonadiabatic limit Equation (2.8), where it is only the initial friction value that counts for the reaction.

(Although we do not address the issue in this chapter, this model perspective can also be used to characterize the reaction coordinate, i.e., the appropriate combination of solute and solvent coordinates along which reactive trajectories move [1, 7c, 18, 35].)

Again we stress that the solvent coordinate s has not yet been specified. However, one can see that if s is the microscopic solvent coordinate $\delta\Delta E$ discussed in Section 2.3 or something like it, then there is a direct connection between solvent dynamics probed in time dependent fluorescence and the time dependent friction on a reacting solute coordinate. We return to this issue in Section 5.

4. Case Studies

In this section, we examine a few case studies for the connection of solvation dynamics and chemical reactions. We begin with a brief survey of heavy particle charge transfer reactions, and then turn to solvation dynamics *per se*, followed by electron transfer reactions.

4.1 Heavy Particle Change Transfer Reactions

For many reactions with high sharp barriers, the time dependent friction on the reactive coordinate can be usefully approximated as the tcf of the force with the reacting solute *fixed* at the transition state. That is to say, no motion of the reactive solute is allowed in the evaluation of Equation (2.3). This condition has its rationale in the physical idea [1, 5] that recrossing trajectories which influence the rate and the transmission coefficient occur on a short time scale. Judging from the results of many MD simulations for a host of different reaction types [6–12], this condition is satisfied; it can be valid even where it is most *a priori* suspect, i.e., for low barrier reactions of the ion pair interconversion class [8].

What does this time dependent friction look like? Figure 3 displays $\zeta(t)$ for the reactive asymmetric stretch coordinate for the $Cl^- + CH_3Cl$ S_N2 sys-

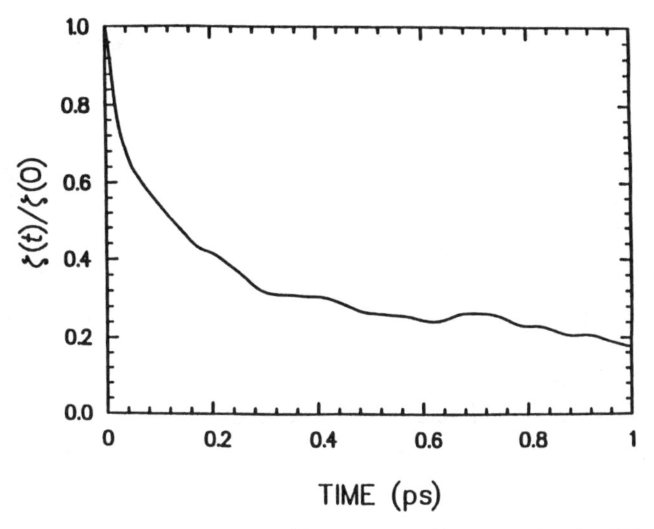

Fig. 3. The time dependent friction $\zeta(t)$ at the transition state for the $Cl^- + CH_3Cl S_N2$ reaction in water [7].

tem in H_2O solvent, while Figure 4 displays its Fourier spectrum. The latter is particularly illuminating, since it displays peaks identifiable from the spectrum of the same pure H_2O liquid [7c]. Thus contributions from the H_2O bends and symmetric and antisymmetric stretches are apparent at frequencies above $\sim$ 1,500 cm^{-1}. At the lower frequency range 300 cm^{-1}–1,000 cm^{-1} are visible the hindered rotational, i.e, librational, contributions of water. A further point of note is that the higher frequency motions (bends, stretches) have a diminished amplitude in the friction as compared to the spectrum, i.e., they are not so strongly *coupled* to the reactive solute as are the water librations. In the time perspective, a fairly rapid initial decay is apparent, followed by a substantial and much longer lived tail. The latter makes a substantial contribution to the large friction constant, i.e., the zero frequency friction [cf. Equation (2.7)], as is evident in Figure 3. Much the same sort of behavior of $\zeta(t)$, in time and in frequency, is observed in an MD simulation for a model of the S_N1 ionization of tert-butyl chloride in water [9]. If we employ the perspective of the model Equation (3.13), it is apparent that the solvation dynamics is decidedly nonexponential for these problems.

Another example of $\zeta(t)$ is shown in Figure 5 for an ion pair fixed at the transition state separation in a model dipolar solvent [8a]. Here the striking feature is a rapid initial drop to negative values, followed by a long positive tail. Such quasi-oscillatory behavior is suggestive of some sort of collective solvent cage motion. Again, the long tail is a significant contributor to the friction constant. This same overall behavior has been observed for various ion pair combinations for Na^+ and Cl^- in water solvent [8b], and it seems to

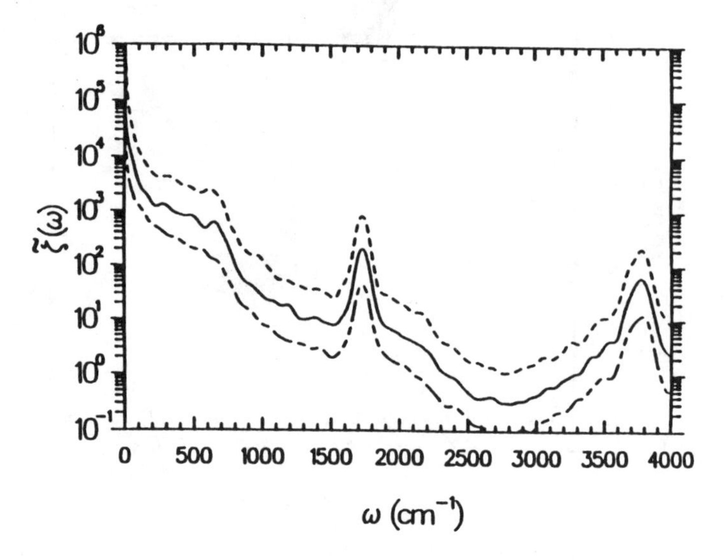

Fig. 4. The friction spectrum, defined as the Fourier transform of $\zeta(t)$ in Figure 3 [7c].

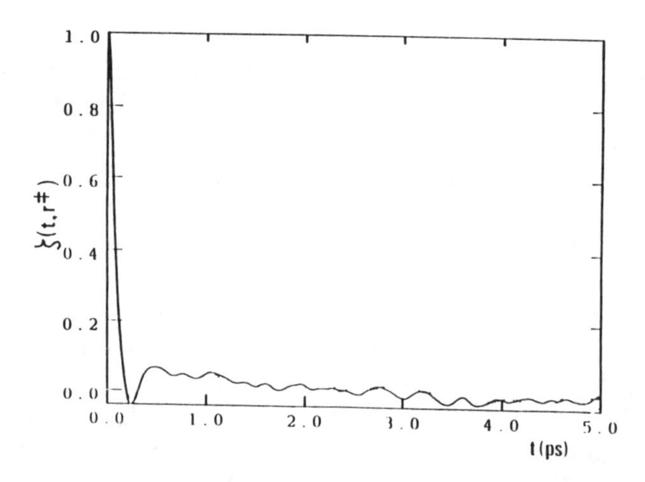

Fig. 5. The time dependent friction $\zeta(t)$ for the ion pair transition state described in Section 4.1 [8a].

be characteristic for ion pair systems. Again, if we exploit the perspective of Equation (3.13), the solvation dynamics is decidedly nonexponential.

The polarization cage limit described in Section 2.1 will apply when there is a relatively broad barrier in the mean potential (small ω_{beq}) and there is strong solute-solvent electrostatic coupling [large $\zeta(t = 0)$]. It has been observed in an MD simulation of ion pair recombination dynamics in a polar solvent [8a], whose friction was just described. This reaction class is especially interesting, in that the reaction barrier is solely solvent-induced; it does not exist in the vacuum. The character of the dynamics is instructive, and we devote some discussion to it. The contact ion pair (CIP), located at a separation of $\sim$ 3.3 Å, is characterized by an average solvation shell whose key feature is a ring of four solvent molecules, at $\sim$ 4 Å from the CIP axis with their dipoles antiparallel to that of the CIP. For the solvent separated ion pair (SSIP), at separation $\sim$ 7.5 Å, the average solvation shell is contracted to $\sim$ 2.5 Å, due to the increased dipole moment of the SSIP, and contains only three solvent molecules. The solvation shell at the transition state, which is located at an ion pair separation of $\sim$ 5.5 Å, has an intermediate structure. The solvation dynamics at the transition state which allow, e.g., the formation of the SSIP, consist of a contraction of the solvation shell, with inward motion of three of its members. For passage to the CIP on the other hand, an expansion of the solvation shell is required. Without these motions, the system is trapped in the transition state neighborhood. It is also instructive to note that the solvent molecule motions seem to be much more translational than reorientational in character. It is also difficult to imagine that the dielectric relaxation time of the solvent has much relevance for such dynamics.

More typical chemical reactions, however, are characterized by sharper reaction barriers. Thus, even though the magnitude of the reactive solute-solvent coupling is strong [large $\zeta(t = 0)$], the intrinsic barrier is of such high frequency that the *nonadiabatic solvation* limit [1, 17, 18]

$$\begin{aligned} \omega_{beq}^2 &> \zeta(t=0); \\ \omega_{bna}^2 &> 0, \end{aligned} \tag{4.1}$$

discussed in Section 2.1 is often a faithful guide to the microscopic mechanism of the barrier passage. The $Cl^- - CH_3Cl$ S_N2 system in H_2O provides a clear illustration [7], for which the nonadiabatic frequency is very high, $\sim$ 500 cm^{-1}, along the antisymmetric stretch solute reactive coordinate. The reaction transmission coefficient is found to be accurately given by the nonadiabatic value Equation (2.8). As alluded to Section 2.1, in the nonadiabatic solvation limit the solvent is effectively frozen during the short time scale during which the fate of a trajectory, reactive or otherwise, is decided – this time is $\sim$ 20 fs in the S_N2 example. The solvation dynamics *per se* are irrelevant. Instead, one can picture a static distribution of barriers in x faced by the reactive solute coordinate x, which depend on the solvent coordinate s. These barriers arise from the different solvation patterns that exist in equilibrium with the solute at its transition state configuration $x^{\ddagger} = 0$. The S_N2 solute transition state structure is symmetric – $Cl^{\delta-}CH_3^{\delta+}Cl^{\delta-}$, but the distribution of solvent configurations has contributions in which there is asymmetric solvation, i.e., better solvation of one Cl or another by the water molecules. Indeed, the model Hamiltonian Equation (3.4) can be rewritten [7b] in the form

$$\begin{aligned} H &= \frac{1}{2}\dot{s}^2 + \frac{\omega_s^2 s^2}{2} + H_x^\ddagger(s); \\ H_x^\ddagger(s) &= G_T^\ddagger + \frac{\dot{x}^2}{2} - \frac{1}{2}\omega_{bna}^2[x - \Delta x(s)]^2 + \Delta V(s); \\ \Delta V(s) &= (\omega_c^4/2\omega_{bna}^2)s^2; \\ \Delta x(s) &= -(\omega_c^2/\omega_{bna}^2)s, \end{aligned} \tag{4.2}$$

which is quite instructive. This represents a sequence of solvation dependent barriers $\Delta V(s)$ for the x motion, located along the line $\Delta x(s)$. These barriers arise from precisely the asymmetric solvation patterns mentioned above [7a]. Calculation of the rate constant from this perspective gives [7b] just the nonadiabatic solvation result Equation (2.8). Indeed, this model succeeds even in accounting for whether trajectories are or are not reactive, depending on the kinetic energy in the solute S_N2 coordinate. For given H_2O solvation configurations, there is a solvent barrier $\Delta V(x)$ along x, and a trajectory will be unsuccessful if the kinetic energy K_x in the antisymmetric stretch – the solute reactive coordinate – is less that $\Delta V(x)$, and successful if it exceeds it. This kind of detail begins to approach that considered in state-resolved gas phase chemistry. Finally, the solvent coordinate s can in fact be identified for this system in terms of a microscopic force [7b].

In this nonadiabatic limit, the transmission coefficient is determined, via Equation (2.8), by the ratio of the nonadiabatic and equilibrium barrier frequencies, and is in full agreement with the MD results [7a–7c]. (By contrast, the Kramers theory prediction based on the zero frequency friction constant is far too low. In the language of Equation (3.13), the solvation time τ_s is not directly relevant in determining κ.) In favorable cases, these frequencies can be calculated by solution phase integral equation technology [37]. This is important, since it indicates a prediction route, via GH theory, which avoids an MD simulation.

Even for cases where the reaction transmission coefficient is independent of the solvent dynamics *per se*, those dynamics can still (and must) play a key role in the overall reactive process. This can be illustrated by the results of an MD simulation of how the S_N2 system $Cl^- + CH_3Cl$ in H_2O actually reaches (and leaves) the transition state [7d]. Recall, from the discussion above, that in the transition state the solute has a symmetric charge distribution $Cl^{\delta-}CH_3^{\delta+}Cl^{\delta-}$. Examination of the barrier climbing process, starting from an ion-dipole complex $Cl^- \cdot CH_3Cl$, during the 500 fs of its duration indicates the following. The changes in the solvent energy and the solvent-reagent interchange potential energy are gradual over most of the 500 fs time span of relevance to the climb. These changes are smooth, implying that the corresponding change in the solvation of the reaction complex is also smooth and gradual. But the change of the charge distribution in the reagents is not at all gradual over this 500 fs. Instead, the transition from ion-dipole to symmetric charge distribution occurs almost entirely over the last 50 fs before the system reaches the transition state [7d].

These results indicate that a major portion of the solvent reorganization to a state appropriate to solvating the symmetric charge distribution of the reagents at the barrier top takes place *well before* the reagent charge distribution begins to change. The solvation cannot adiabatically follow the rapid change in the distribution of the negative charge among the reagents [7d]. This prior solvent reorganization is *required* for the reagents to reach the transition state; the requisite solvation at the transition state has begun to develop well before any change in the charge distribution in the reagents. Together with the results of [7a–7c], this shows very clearly for this S_N2 system that one cannot picture the progress of a chemical reaction as stately motion along the potential of mean force curve – a chemical reaction is a dynamic, and not an equilibrium event.

Of related interest are results for response of water to an instantaneous change in the dipole of a solute [33], for the time scale of the solvent response for several charge-transfer reactions in water, including the S_N2 reaction [38], and for a similar response for $Fe^{2+} - Fe^{3+}$ in water [39]. The time scales found in those studies for the water solvent relaxation under these conditions – and that originally found in [7] for time-dependent friction on the S_N2 transition state – are similar to those found for the prior reorganization of the solvent H_2O.

4.2 SOLVATION DYNAMICS

There are by now a number of computer simulations of solvation dynamics in connection with TDF [32–34, 39–42]. One early example [33] on solvation dynamics in water is especially important in showing that continuum descriptions are particularly inept in capturing the microscopic character of the dynamics. Another noteworthy feature of that study includes the marked importance of first solvent shell motions, which is just the opposite of what a continuum picture might suggest. Here we focus on only a few examples, to illustrate several key points and to make connections to the reaction problem.

The major systems of focus [32] in our discussion are three different diatomic solutes – a neutral pair (NP), a half-ion pair (HIP) with charges $\pm 1/2e$ on the atomic members, and an ion pair (IP), with individual charges $\pm e$. In each case, the solute is immersed in a solvent of model dipolar diatomics, which is approximately similar to CH_3Cl [8a, 10]. The results for the normalized $\delta\Delta E\ tcf$, $\Delta(t)$, are displayed in Figure 6. There are several features to note. First, the significant initial drop in $\Delta(t)$ is very similar for the NP, HIP, and IP solutes. This initial decay is well described by a Gaussian time dependence [32]

$$\Delta_G(t) = \exp\left(-\frac{\omega_{SP}^2 t^2}{2}\right), \tag{4.3}$$

where ω_{SP} is the solvent frequency for the particular solute pair [cf. Equation (2.17)]

$$\omega_{\delta\Delta E}^2 = \omega_{SP}^2 = \frac{\langle(\delta\Delta\dot{E})^2\rangle_{SP}}{\langle(\delta\Delta E)^2\rangle_{SP}}. \tag{4.4}$$

As described in [32], this Gaussian behavior is related to the inertial free streaming of the solvent molecules, whose ensuing coordinate changes alter ΔE. These solvent translational and orientational coordinate changes are independent of the interaction potentials. Such Gaussian behavior has now been found in a number of studies [34, 41, 42], and especially clearly in the case of CH_3CN solvent [41]. For the current, different, solute pairs, the solvent frequency has similar values. The relative insensitivity of ω_{SP} to the solute identify is discussed in [32] and [43a] in terms of its relationship via

$$\omega_{SP} = \left(\frac{k_{SP}}{m_{SP}}\right)^{1/2}$$

to the solvent force constant and mass [cf. Equation (2.17)]. Assorted MD studies of the solvent force constant itself, especially its sometimes marked dependence on the solute charge distribution, may be found in [43].

The longer-time behavior of $\Delta(t)$ depends on the intermolecular interactions [32, 41] and differs noticeably [32] for the various SP cases, with the slowest relaxation for the full IP solute. It is this nonlinear feature that lengthens the solvent relaxation time $\tau_{\Delta E}$ as the dipole moment of the SP increases: $\tau_{\Delta E} = 0.35$ ps for the NP and = 0.54 ps for the IP. It is absent in continuum models of solvent dynamics, as noted in Section 2.3.

The meaning of "inertial" requires some clarification. Clearly $\delta\Delta E$ itself depends on the interaction between the solvent molecules and the solute. What Equation (4.3) says [32] is that the *change* in $\delta\Delta E$ is to be calculated by the change in the molecular positions due solely to free streaming motion, in the *absence* of any forces. The solvent frequency $\omega_{\Delta E}$ encompasses an average inertial dephasing effect, since this free streaming depends on the molecular momenta, subject to the initial Maxwellian distribution. No solvent molecule oscillates with this frequency. Further, one can separately picture a transiently bound "librational" motion for the solvent molecules in the presence of the solute. The frequencies for these motions depend on the motions of the

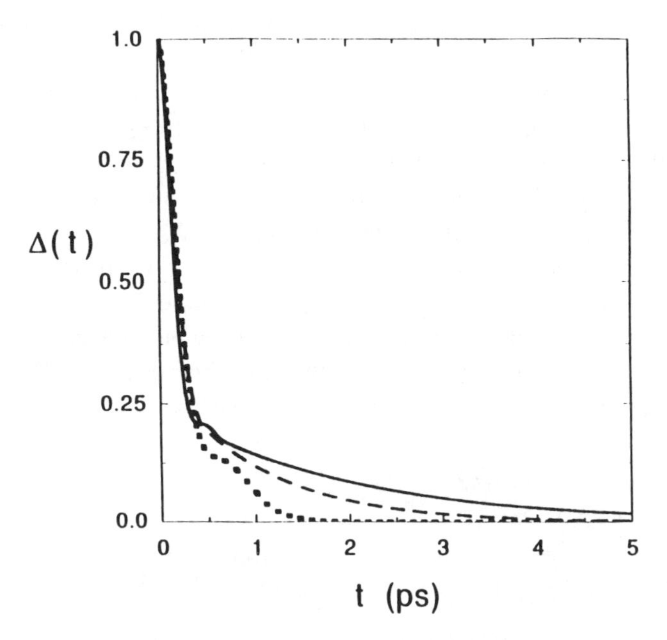

Fig. 6. The solvent coordinate relaxation function $\Delta(t)$ versus time in ps for an IP (–), HIP (- - -) and NP (· · ·). The initial Gaussian Equation (4.3) accounts for 29%, 45% and 66% of the total time areas, respectively [42].

solvent molecules in each other's force fields and in that of the solute. Those frequencies are therefore different from ω_{SP}.

It is certainly the case that all time correlation functions (with continuous forces and torques) must exhibit an initial Gaussian time behavior, and one can find many examples of Gaussian models for associated tcf's in the literature for a variety of problems, e.g., translational motion of a tagged particle. The remarkable thing in the solvation dynamics arena is that this Gaussian behavior is responsible for so *much* of the decay of $\Delta(t)$. As noted above, a 50–70% contribution to the decay has been observed in a number of MD simulations [32, 41]. An important exception is the case of methanol [34], in which the rapid motion of the solvent OH groups can begin to contribute quite early, thereby quickly taking over from the initial inertial Gaussian behavior and reducing its importance. Something similar can be seen in the solvation relaxation in H_2O [33]. These observations also reinforce the distinction, made above, between the solvent frequency ω_{SP} and, e.g., solvent librational frequencics.

To date, almost all the evidence for the importance of the inertial free streaming Gaussian behavior in the solvation dynamics is from MD simulation. However, recently this behavior has evidently been found experimentally in CH_3CN solvent [44] with a time constant similar to that found in a simulation involving the same solvent [41]. It is also important to remark that due to technical difficulties many earlier TDF experiments may have missed substantial contributions to the solvation dynamics. It seems to us that, unless there are rapidly intervening high frequency motions involving, e.g., OH, NH, in the solvent that are electrostatically coupled to the solute, the initial Gaussian decay will always be a quite significant component of the solvation dynamics, at least for solutes of modest size.

The inertial contribution is completely ignored in most available treatments of solvation dynamics. These instead assume that the solvent is overdamped. In the language of Equation (2.14), this would mean not only that the solvent time dependent friction's memory is ignored $[\zeta_{\delta\Delta E}(t) = \zeta_{\delta\Delta E}\delta(t)]$, but also that the acceleration is ignored, to give

$$\delta\Delta\dot{E}(t) = -\frac{\omega_{\delta\Delta E}^2}{\zeta_{\delta\Delta E}}\delta\Delta E(t). \tag{4.5}$$

In this overdamped, diffusive approximation, $\delta\Delta E(t)$ on the average, and thus the tcf $\Delta_{\delta\Delta E}(t)$ decays exponentially [see also Equation (2.20)]:

$$\begin{aligned} \Delta_{\delta\Delta E}(t) &= e^{-(t/\tau_{\delta\Delta E})}; \\ \tau_{\delta\Delta E} &= \zeta_{\delta\Delta E}/\omega_{\delta\Delta E}^2. \end{aligned} \tag{4.6}$$

The Debye continuum behavior is a special case of this. It is evident from Figure 6 that this is not a good description of the dynamics. In fact, we are unaware of any MD simulations for which the popular exponential decay applies.

An improvement would be to retain the acceleration, but continue to ignore the memory in $\zeta_{\delta\Delta E}(t)$. This is then a simple Langevin equation (LE) approach,

$$\delta\Delta\ddot{E}(t) = -\omega_{\delta\Delta E}^2\delta\Delta E(t) - \zeta_{\delta\Delta E}\delta\Delta\dot{E}(t), \tag{4.7}$$

so that the solvation function $\Delta_{\delta\Delta E}(t)$ is given by [28]

$$\begin{aligned} \Delta_{\delta\Delta E}(t) &= \left\{\exp\left(-\omega_{\delta\Delta E}^2\tau_{\delta\Delta E}\frac{t}{2}\right)\right\} \\ &\quad \left\{\cos(\omega_{\delta\Delta E}\gamma t) + \left(\frac{\omega_{\delta\Delta E}\tau_{\delta\Delta E}}{2\gamma}\right)\sin(\omega_{\delta\Delta E}\gamma t)\right\}; \\ \gamma &= \left[1 - \left(\omega_{\delta\Delta E}\frac{\tau_{\delta\Delta E}}{2}\right)^2\right]. \end{aligned} \tag{4.8}$$

(This is a microscopic analogue of the retention of inertia in the continuum description mentioned in Section 2.3.) But, when compared with MD data [32, 45], Equation (4.8) is still unsatisfactory (cf. Figure 6). Although there

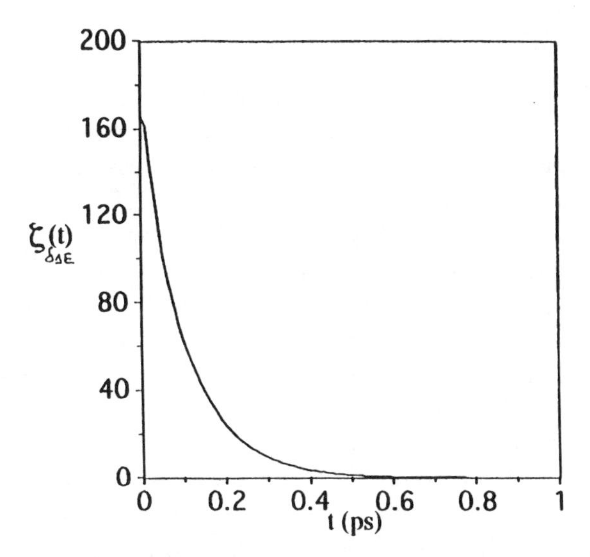

Fig. 7. The time dependent friction $\zeta_{\delta\Delta E}(t)$ for an HIP [45a].

is a certain overall coherence – the enforcement that the average relaxation time

$$\tau_{\delta\Delta E} = \int_0^\infty \mathrm{d}t \Delta_{\delta\Delta E}(t) \tag{4.9}$$

is correctly given leads to agreement in an average sense – the MD shape is in fact poorly reproduced. In particular, the initial Gaussian behavior [Equation (4.3)] is quickly and inappropriately quenched. This is a consequence of assuming that the friction $\zeta_{\delta\Delta E}(t)$ acts instantly to bring on the influence of the generalized forces on $\delta\Delta E(t)$. Comparison of this time dependent solvent friction in Figure 7 with $\Delta_{\delta\Delta E}(t)$ in Figure 6 shows that the former's time development is simply not adequately regarded as instantaneous on the latter's time scale. One must conclude that a full GLE description is required to account for the solvation dynamics. This is a challenging proposition. There are ways to estimate the frequency $\omega_{\delta\Delta E}$, either through dielectric continuum models [28] or, preferably, through integral equation techniques [43b] – this frees us from MD. However, obtaining a handle theoretically on the solvent coordinate friction $\zeta_{\delta\Delta E}(t)$ requires further efforts. The requisite theory for this would, in our opinion, require new insights on the key features of the dynamics of solvent molecules in the first solvent shell or so, *in interaction with* the solute. We return to this in Section 5.

4.3 ELECTRON TRANSFER

We now turn to the electronically adiabatic ET reaction problem (cf. Section 2.2). There has been a plethora of theoretical papers [10, 20, 23, 45, 46, 48] dealing with the possible role of solvent dynamics in causing departures from the standard Marcus TST rate theory [20, 24]. (Related issues can arise in proton transfer reactions [47].) The first of these is due to Zusman [25] and for the most part, further works are variants on this. Zusman focused on the solvent dynamics within each of the R and P wells, in the diffusive approximation so that the exponential decay is assumed to hold in each well. The simples form of Zusman Theory is for a cusped, i.e., sharply peaked, barrier reaction, for which the transmission coefficient $\kappa = k_{ET}/k_{ET}^{TST}$ measuring the departure of the rate constant k from its TST Marcus Theory approximation is, in the symmetric reaction case, given by

$$\kappa_Z = (\omega_R \tau_R)^{-1} \left(\frac{\pi \Delta G^{\ddagger}_{na}}{k_B T} \right)^{1/2} \propto \left[\int_0^{\infty} \mathrm{d}t \Delta_R(t) \right]^{-1} . \qquad (4.10)$$

Here $\Delta G^{\ddagger}_{na}$ is the nonadiabatic barrier height, i.e., the free energy at the intersection of the nonadiabatic R and P curves minus the free energy at the minimum of the R well, which has frequency ω_R. The second form emphasizes the connection with the solvent dynamics tcf $\Delta(t)$ discussed in Section 2.3. The picture here is that the rate limiting steps for the reaction are the diffusive climbing of the R well wall and the diffusive slide down the P well wall (cf. Figure 1b). The passage over the barrier top itself is supposed to be rapid and direct, i.e., there is no solvent-induced recrossing associated with the barrier top *per se*; instead, recrossings of the barrier supposedly arise from "reapproach" from the region of the wells before the bottom of either of the R or P wells is reached. Figure 8a illustrates the concept.

The ET reaction that has been considered to examine these issues is a simplified symmetric model, $A^{-1/2}A^{1/2} \rightarrow A^{1/2}A^{-1/2}$, in the solvent described in Section 4.2. The technical and computational rationales for this somewhat artificial fractional charge model are given in [10]; however, the model is sufficiently realistic to explicitly address the key dynamical issues.

The MD simulation of the reaction is effected via the electronically adiabatic Hamiltonian [10]

$$H_{ad} = \frac{H_R + H_P}{2} - \frac{1}{2}\sqrt{(\Delta E)^2 + 4\beta^2}; \qquad \Delta E = H_R - H_P, \quad (4.11)$$

where $H_{R(P)}$ is the system Hamiltonian when the solute has the $R(P)$ charge distribution and β is the invariant electronic coupling. This is appropriate as representative for many adiabatic ET reactions; in addition, solvent dynamical effects are expected to be most pronounced in the electronically adiabatic limit

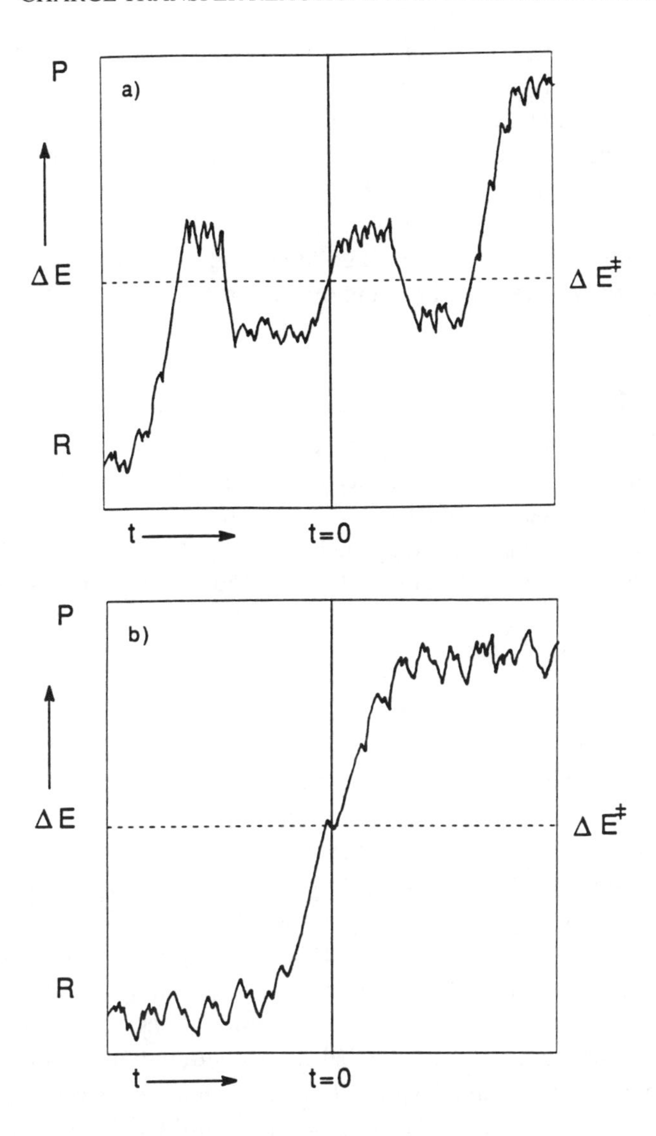

Fig. 8. Schematic contrast of the ET reaction trajectories (a) consistent with the well relaxation picture and (b) consistent with the barrier top dynamics picture.

[23]. With this Hamiltonian, the solute is always in its ground electronic state whatever the solvent configuration may be. The barrier is traversed as ΔE progresses from values appropriate to the neighborhood of equilibrium with R to those similarly appropriate to P. The electronic charge evolves smoothly from that of R, through the transition state distribution $A^0 A^0$ – which is a neutral pair, on to the P charge distribution.

The transmission coefficient κ can be directly computed, for different choices of the electronic coupling β, in an MD simulation for the ET reaction, as can the various quantities in κ_Z [10]. The first important point is that for $\beta = 1$ kcal/mol, κ is quite close to unity; there are few recrossings of the barrier and the Marcus TST Theory is thus an excellent approximation. How does the Zusman Equation (4.10) describe the results? The necessary ingredients for its evaluation can be gotten from the free energy curve and by determining the reactant well solvent relaxation time by

$$\tau_R = \int_0^\infty dt \Delta_R(t), \tag{4.12}$$

(which is identical to that for the product well).

κ_Z is found to give very poor results for the $\beta = 1$ kcal/mol case, for which the barrier is cusped [10, 45]. This can be improved by realizing that the rate of unidirectional crossing the barrier top itself is not infinitely fast, but rather can be estimated from the Marcus TST formula. This gives [10]

$$\kappa'_Z = (1 + \kappa_Z)^{-1} \kappa_Z \tag{4.13}$$

and predicts transmission coefficients much closer to the MD values. But, despite this improvement, there remains a serious problem. For, with either equation, the barrier recrossings responsible for the transmission coefficient are supposed to arise from trajectories as in Figure 8a, i.e., arising from individual single recrossings subsequent to erratic motion in the R and P wells before the trajectory settles down into the bottom of either well. Figure 8b shows that the MD computed trajectories are not at all like this. Instead, the recrossings that occur are completely associated with the barrier top region. Any trajectory that makes it into the P (or R) well proper in fact continues on to the bottom of that well without returning to the transition state and recrossing it. We have to conclude that the solvation dynamics in the wells are not relevant for the rate, despite all the theoretical focus on them.

When the coupling is increased to $\beta = 5$ kcal/mol and the barrier becomes more rounded, the transmission coefficient is smaller ($\kappa_{ET} \approx 0.6$) and there are noticeable departures from the Marcus TST theory, although they are not enormous. The character of the barrier recrossings is illustrated in Figure 8b. Again, these recrossing occur in the immediate neighborhood of the transition state, and not in the wells. It is not fair or even appropriate to apply the Zusman approach or its variants – which assume cusped barriers – to this smooth barrier situation.

For each of the electronic coupling value ET reactions above, the barrier recrossings occur exclusively in the barrier top region. The GH theory is a theory for just such dynamical rate effects and the transmission coefficient is given by Equation (2.5), rewritten here as

$$\kappa_{ET} = \frac{k_{ET}}{k_{ET}^{TST}} = [\lambda + \zeta_{ET}^{\ddagger}(\lambda)]^{-1}\omega_{beq}^{ET}. \tag{4.14}$$

In the present context, GH Theory is based on the assumption that the GLE (cf. Section 2.2)

$$\delta\Delta\ddot{E}(t) = (\omega_{beq}^{ET})^2\delta\Delta E(t) - \int_0^t \mathrm{d}\tau\zeta_{ET}^{\ddagger}(t-\tau)\delta\Delta\dot{E}(\tau) \tag{4.15}$$

holds in the vicinity of the barrier top $\Delta E = \Delta E^{\ddagger} = 0$ [10, 23]. The barrier frequency ω_{beq}^{ET} can be estimated [10] from the formula $\omega_{beq}^{ET} = \omega_R[(2\Delta G^{\ddagger}/\beta) - 1]^{1/2}$, where ω_R is the frequency of the reactant solvent well. The friction $\zeta^{\ddagger}(t)$ appropriate for the transition state is more difficult to obtain than the well friction, but it can be approximated [10] by the time dependent friction for the reference situation of a *neutral* pair. This approximate identification derives from the observation, noted above, that in the ET reaction, the transition state charge distribution is that of a neutral pair. Note that this friction would govern the dynamics in a TDF experiment in which the Franck–Condon transition $A^{-1/2}A^{1/2} \rightarrow A^0A^0$ occurred. This provides, at the most fundamental level, the connection between the TDF solvation dynamics – but for the transition state – and the ET rate. The neutral pair friction $\zeta_{NP}(t)$ can be extracted [10] from the TDF studies, and with the approximation $\zeta_{ET}^{\ddagger}(t) = \zeta_{NP}(t)$, κ_{GH} can be estimated via Equation (4.14) for the ET reaction. The results agree within the error bars with the MD simulation values [10, 45].

All these results suggest that the conventional dynamic approaches to electronically adiabatic electronically adiabatic ET in dipolar aprotic solvents based on well relaxation miss the correct picture, even for fairly cusped barrier reactions. Instead, it is the solvent dynamics occurring near the barrier top, and the associated time dependent friction, that are the critical features. It might however be thought possible that, for cusped barrier adiabatic ET reactions in much more slowly relaxing solvents, the well dynamics could begin to play an important role. However, MD simulations have now been carried out for the same ET solute in a solvent where the solvent molecule internuclear separation is increased [45b]. This lengthening slows down the motion of the solvent molecules and causes it to be decidedly overdamped – the ratio of $\zeta_{\delta\Delta e}/2\omega_{\delta\Delta E}$ is of order 10. Yet once again it is found [45b] that the transmission coefficient is determined by trajectories near the barrier top and that the conventional well relaxation picture for the reaction does not apply.

All of the above considerations were for ET reactions with modest to high barriers where the barrier frequency is reasonably high. A quite different situation can arise for low barrier reactions where there is no obvious intrinsic bias in favor of short time solvent dynamical effects in influencing the reaction rate. Some striking illustrations of this are given in [49] where it is seen that various portions of the solvation dynamics can influence the ET rate for several low barrier exothermic reactions in methanol, depending on the detailed character of the free energy surface. This emphasizes the important lesson that what solvation dynamics are relevant for a reaction is very much a function of the reaction.

4.4 Vibrational Energy Relaxation

There is another aspect of solvation dynamics that, to date, has received little attention – the connection with vibrational relaxation. It was first shown via MD simulation for CH_3Cl in water [50] that the vibrational energy relaxation for a polar molecule in water can be very fast – due to the strong Coulombic solute-solvent interactions. Recent experimental and theoretical work on ionic systems [51] indicates that this basic picture in fact applies more generally. It was also shown that the vibrational relaxation time τ_{vib} is accurately given by a solution phase Landau–Teller formula

$$\tau_{vib}^{-1} \propto \int_{-\infty}^{\infty} dt \exp(i\omega t)\langle \delta F_{vib} \delta F_{vib}(t)\rangle. \tag{4.16}$$

Here ω is the oscillator frequency and δF_{vib} is the fluctuating force, due to the solvent, exerted on the fixed oscillator coordinate.

This largely unexplored probe of solvent dynamics differs in a substantial way both from reactive barrier crossing aspects and from TDF, in that a Fourier component of the vibrational friction at a bound oscillatory motion frequency is required. For the $\sim 700\,\mathrm{cm}^{-1}$ CH_3Cl vibration in water, the relevant solvent dynamics are the water librations, which lie in the appropriate frequency range and which are electrostatically coupled to the vibrational coordinate [50]. On the other hand, for a model hydrogen-bonded complex in a dipolar solvent [52], the first solvent shell motions responsible for rapid vibrational relaxation are more complex and dependent upon which vibrational mode is considered. That is to say, the time dependent vibrational friction on different vibrational modes is different. Of further interest is the question of the role of changing solute electronic structure on the vibrational relaxation [53]. It seems to us that much remains to be discovered and understood in this entire area.

5. Comparisons

In Section 4.3 we discussed the connection between solvation dynamics as probed by TDF and electron transfer reaction rates. Is there a connection between TDF solvation dynamics $\Delta(t)$ and the friction $\zeta(t)$ on the solute nuclear reactive coordinate for heavy particle charge transfer reactions? Inspection of the model results Equations (3.12) suggests that there may be such a connection if it is true that the solvent coordinate s there can be identified with the sort of solvent coordinate $\delta\Delta E$ that figures in the TDF and ET problems. We begin to address this now.

As noted at the beginning of Section 2, the time dependent friction is to be found for the reacting solute fixed at the transition state value $x^{\ddagger}$ of x. By Equation (3.13), its dynamics were related to those of an (unspecified) solvent coordinate s. One strategy to identify the solvent coordinate, its frequency, friction, etc., would be to derive an equation of motion for the relevant fluctuating force δF there. To this end, one can use the same sort of double-membered projection technique employed in Section 2.3, but now in terms of δF and $\delta \dot{F}$. One quickly finds [54]

$$\delta\ddot{F}(t) = -\omega_{\delta F}^2 \delta F(t) - \int_0^t \mathrm{d}\tau \zeta_{\delta F}(t-\tau)\delta\dot{F}(\tau), \tag{5.1}$$

where the associated square frequency is

$$\omega_{\delta F}^2 = \frac{\langle(\delta\dot{F})^2\rangle}{\langle(\delta F)^2\rangle} = \frac{k_{\delta F}}{\mu_{\delta F}}, \tag{5.2}$$

and the time dependent friction for δF, per δF mass,

$$\mu_{\delta F} = \frac{k_B T}{\langle(\delta\dot{F})^2\rangle}, \tag{5.3}$$

involves the projection operator (PO)-modified dynamical tcf of the generalized force

$$\delta f = \mu_{\delta f}[\delta\ddot{F} + \omega_{\delta F}^2 \delta F]. \tag{5.4}$$

Averages fixed at the transition state are to be understood.

Now, by this point, one may be becoming a bit weary of various GLE representations everywhere, so let us hasten to make some immediate connections.

First, if we compare Equations (5.1)–(5.4) with Equations (3.10)–(3.14), a natural identification of the solvent coordinate s in Section 3 is in fact just the fluctuating force δF on x at the transition state. (Note especially that this choice associates the solvent coordinate with a direct measure of the relevant solute-solvent interaction.) The solvent mass, force constant and frequencies in Section 3 would then be given molecular expressions via Equations (5.1)–(5.3), while the solvent friction $\zeta_s(t)$ of Section 3 would be the friction per mass for δF,

$$\zeta_{\delta F}(t) = (\mu_{\delta F} k_B T)^{-1} \langle \delta f \delta f^{\dagger}(t)\rangle, \tag{5.5}$$

in which the dagger denotes PO-modified dynamics. The final joining of the descriptions consists of the (consistent) identification of the square coupling frequency ω_c^2 in Equation (3.6) by

$$\begin{aligned}\omega_c^4 &= \omega_{\delta F}^2 \zeta(t=0) = (\mu k_B T)^{-1} \langle (\delta F)^2 \rangle [\langle (\delta \dot{F})^2 \rangle / \langle (\delta F)^2 \rangle] \\ &= (\mu k_B T)^{-1} \langle (\delta \dot{F})^2 \rangle. \end{aligned} \tag{5.6}$$

Thus, everything maps one to one.

Actually, this sort of identification of the solvent coordinate s with the fluctuating force δF was noticed in [7b], but its generality has not been pursued previously. It is important to remark that this force δF is a highly *nonlinear* function of the coordinates of the solvent molecules, referenced to the solute location [7b]; one should not at all think (as some evidently do) that the intermolecular forces can be untenably linearized.

Even the simplest approximation to this system – totally ignoring the friction $\zeta_{\delta F}$ – gives a Gaussian behavior for the solute friction at short times [1, 5],

$$\zeta(t) = \zeta(t=0) \exp\left[-\frac{\omega_{\delta F}^2 t^2}{2}\right], \tag{5.7}$$

which at least captures the initial behavior evident in the examples of Section 4.1. Since, as noted there, only the short time behavior of the time dependent solute friction is often of importance for the rate, Equation (5.7) could prove useful.

It is especially to be noted that the approximation Equation (5.7) emphasizes the initial inertial contributions to $\zeta(t)$, just as Equation (4.3) emphasizes the inertial contributions to the TDF solvation dynamics. But it cannot account for the subsequent behavior dependent on the intermolecular forces, which for $\zeta(t)$ is evident in Figures 3 and 5, and for $\Delta(t)$ is evident in Figure 6. For that behavior, we need a theory, as noted in Section 4.2. One could pursue a continued fraction approach [55], but in our opinion, some sort of – currently nonexistent – molecular model is required for substantial progress.

We now attempt to push the connections further. If we compare, e.g., the developments of Sections 2 and 3 and this section, it is clear that there would be a *direct* connection between $\zeta(t)$ and the TDF dynamical function $\Delta(t)$ *if* it were true that the fluctuating force on the solute transition state were simply proportional to the fluctuation of the microscopic solvent variable ΔE:

$$\delta F \propto \delta \Delta E = \Delta E - \langle \Delta E \rangle_{\ddagger}. \tag{5.8}$$

The first such attempt [28] along these lines focused on the case of the electrostatic dielectric friction [56] on a nonreactive rotating polar molecule and its connection with the TDF solvation dynamics. For the former, the dielectric friction is determined by the tcf of the generalized force, here the torque, on a rotating dipole μ, equal to $t = \mu \times E$ in the point dipole approximation (with a priviso to be returned to below). On the other hand, $\delta \Delta E$ for a comparable solute would involve the energy $-\mu \cdot E$, in each case E representing an electrostatic reaction field. It it is true that the motion of μ is ignorable on the scale of the lifetimes of the corresponding tcf's, then a connection between the friction and the solvation dynamics should exist. In particular, the various factors involving the solute dipole moment would factor out and the relation

$$\Delta_{rot}(t) \equiv \frac{\zeta_{rot}(t)}{\zeta_{rot}(t=0)} = \Delta(t) \tag{5.9}$$

should hold, where $\Delta(t)$ is the normalized tcf of $\delta \Delta E$, Equation (2.12), defined for the same molecule and $\zeta_{rot}(t)$ denotes the electrostatic rotational friction [56]. The integrated version of this is

$$\frac{\zeta_{rot}}{\zeta_{rot}(t=0)} = \tau_{\delta \Delta E}, \tag{5.10}$$

where the solvent relaxation time would be determined from a TDF study. Experimental tests of Equation (5.10) employing large organic dye molecules [57] have been encouraging, especially when compared to dielectric continuum predictions (which are much less successful [58]).

An MD test [42] of Equations (5.10) and (5.9) for ion pair (IP) and half ion pair (HIP) dipolar solutes in the model dipolar aprotic solvent employed in the reaction studies of Section 4. The diatomic solute here has the same mass and size as the solvent molecules. It is found that Equation (5.10) holds to within a factor of two; in ps: $\tau_{\Delta E} \cong 0.44$ and 0.58, while the friction ratio is $\equiv 0.23$ and 0.34 respectively for the HIP and IP solutes. While this is an encouraging – albeit modest – success for a relation between two superficially different sorts of quantities, it is instructive to examine its deficiencies. Figure 9 shows the MD results [42] for the time dependent relation Equation (5.9). The two major points of note are the fairly good coincidence at short times through the Gaussian regime, but a definite departure at longer times where the solvation dynamics exhibits a tail largely absent in the friction. (There would be no difference between any of these curves in a dielectric continuum description.)

One can advance two reasons for the discrepancy for either solute. First, the solute does not appear as a point dipole to the solvent. From a molecular perspective, $\delta \Delta E$ involves the electrostatic potentials at the solute charge site [cf. Equation (2.10)], while the generalized force δF – e.g., the torque – depends on gradients of those potentials. It is not unreasonable then to expect a faster decay for a tcf of the latter. (Nonetheless this feature seems to have

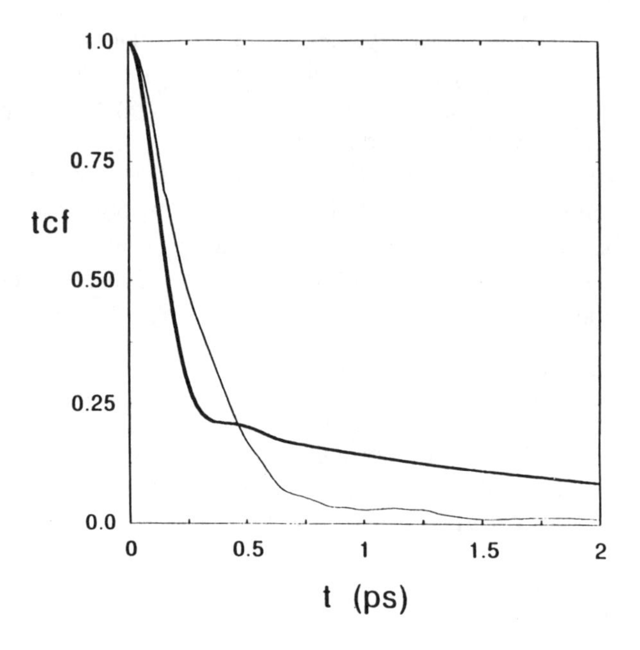

Fig. 9. MD comparison of the relation Equation (5.9) [42] for an IP. (—) $\Delta(t)$; (—) $\zeta_{rot}(t)/\zeta_{rot}(0)$.

little effect on the initial time behavior: the approximation that $\omega^2_{\delta\Delta} \equiv \omega^2_{\delta F}$ has merit.) Second, for solute-solvent mass and size similarity as in the present case, there are certain projection operator-modified dynamics [42] which enter the rotational friction [and not $\Delta(t)$] and obviate a fixed particle perspective. The influence of these would be expected to subside for larger, more massive solutes.

But even if we are concerned with the solute fluctuating force tcf for a fixed solute but for comparably sized and solvent molecules, this long-time discrepancy will remain. This can be illustrated by comparison of Figure 6 and Figure 5 for the friction on the separation coordinate x of an ion pair, in the same solvent. The latter dependent friction $\zeta(t)$ displays an oscillatory aspect not present in $\Delta(t)$. It evidently matters that one is the tcf of $\delta\Delta E$ and the other is the tcf of δF, which is the x gradient of ΔE [cf Equation (2.10)]. It seems that for small reactive solutes, one can only push the reactive solute friction-TDF solvation dynamics connection so far, and we will have to develop the molecular Equation (5.1) via some sort of useful model for the solvent dynamics in intimate interaction with the solute for further progress.

In view of the somewhat sobering discussion above, it is useful for perspective to consider the approximation Equation (5.9) in a reaction context. Figure 10 shows that for a reaction with a modestly sharp barrier, the dominance of

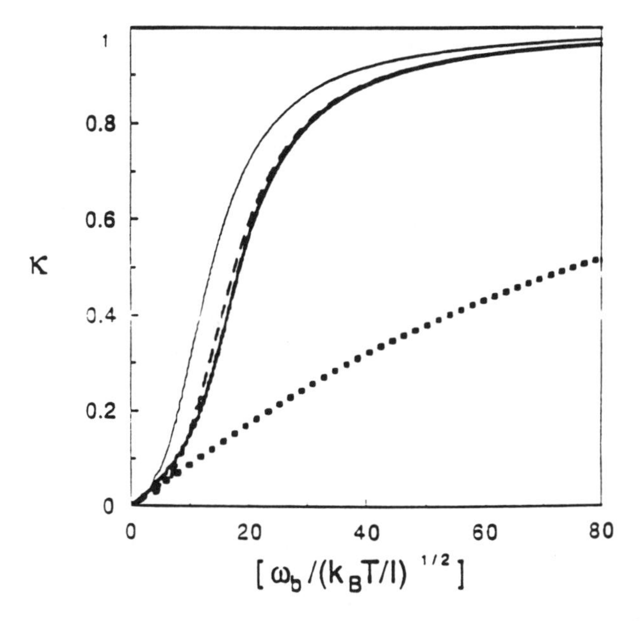

Fig. 10. Reaction transmission coefficient κ for a model isomerization, versus reduced barrier frequency, calculated with a full time dependent IP friction (–) and with the approximation Equation (5.9) (- - -). The Kramers result is displayed for comparison (· · ·). (The lighter solid line refers to another approximation; [42] should be consulted for a more detailed discussion.)

the short time dynamics renders Equation (5.9) quite useful in predicting the rate. This suggests that there is some room for optimism: we can obtain useful measures of the solvent dynamics of relevance for heavy particle charge transfer reactions from the early time dependence of appropriately chosen TDF experimental studies.

6. Concluding Remarks

Here we have discussed some aspects of solvation dynamics and their influence and characterization in connection with chemical reactions. A brief summary would be that in many cases, the short time dynamics of the solvent appears to be necessary and sufficient in a practical sense to account for solvent dynamical effects on the reaction rate. However, it appears that until recently most experimental studies of the solvation dynamics *per se*, via time dependent fluorescence, have instead probed longer time dynamics. There is thus a need to characterize the shorter time dynamics in more detail. Even so, the connection between those probed dynamics and the relevant friction on the reaction coordinates still requires further theoretical elucidation. Beyond this, we still do not have a useful predictive theory to describe time dependent frictions for reactions, and this presents an important opportunity for the future.

It is also worth remarking that there are a number of more complex and interesting reaction situations where the impact of solvation dynamics is only beginning to be explored. Among these are the reactions in electrolyte solutions at finite ionic strength [59], at electrodes [60] and interfaces [61], and in mixtures [62]. Another area is the various issues that arise in connection with the vibrational coherence and relaxation in a reaction context [63]. Much remains to be learned here, but the considerations of this chapter indicate that initial attention should be focussed on the short time, microscopic aspects of these dynamics.

Acknowledgements

We thank our various coworkers, indicated in the references, for their help in our own efforts in the reaction-solvation dynamics field. We also and especially want to acknowledge the inspiration and drive, communicated both in her work and more personally, of the late Teresa Fonseca to create a picture of what the molecules are actually doing during what we term "solvation dynamics"; this challenge remains for others to take up.

This work was supported in part by NSF Grant CHE 88-08852 and by grants of Cray YMP time from the Pittsburgh Supercomputer Center.

References

1. Hynes, J. T., in *The Theory of Chemical Reaction Dynamics*, Vol. IV, M. Bear, e.d (CRC Press, Boca Raton, FL, 1985), p. 171.
2. For some general reviews, see Ref. 1 and Truhlar, D. G., Hase, W. L., and Hynes, J. T., *J. Phys. Chem.* **87**, 2664 (1983); Hynes, J. T., *Annu. Rev. Phys. Chem.* **36**, 573 (1985); Berne, J. B., Borkovec, M., and Straub, J. E., *J. Phys. Chem.* **92**, 3711 (1988); Hänngi, P., Talkner, P., and Borkovec, M., *Rev. Mod. Phys.* **62**, 251 (1990). In addition, the special issue *Chem. Phys.* **152**, nos. 1 and 2 (1991) is devoted to friction in liquid state reactions.
3. Whitnell, R. M. and Wilson, K. R., *Adv. Comp. Chem.* **4**, 67 (1993).
4. For reviews, see, e.g., Simon, J. D., *Acc. Chem. Res.* **21**, 128 (1988); Kosower, E. M. and Huppert, D., *Annu. Rev. Phys. Chem.* **37**, 127 (1986); Barbara, P. F. and Jarzeba, W., *Adv. Photochem.* **15**, (1990); Marconcelli, M., MacInnis, J., and Fleming, G. R., *Science*, **243**, 1674 (1989); Bagchi, B., *Annu. Rev. Phys. Chem.* **37**, 127 (1986); Maroncelli, M., *J. Molec. Liq.*, in press. Some representative experimental studies include Su, S.-G. and Simon, J. D., *J. Phys. Chem.* **91**, 2693 (1987); *Chem. Phys. Lett.* **158**, 423 (1989); Kahlow, M. A., Kang, T. J., and Barbara, P. F., *J. Chem. Phys.* **86**, 3183 (1987); Kahlow, M. A., Jarzeba, W., Kang, T. J., and Barbara, P. F., *ibid.* **90**, 151 (1988); Marconcelli, M. and Fleming, G. R., *ibid.* **86**, 6221 (1987); Castner Jr., E. W., Marconcelli, M., and Fleming, G. R., *ibid.* **86**, 1090 (1987); Simon, J. D. and Su, S.-G., *Chem. Phys.* **152**, 143 (1991).
5. Grote, R. F. and Hynes, J. T., *J. Chem. Phys.* **76**, 2715 (1980). (For some recent discussions of the GH rate equation and especially the associated GLE from a theoretical point of view, see Tarjus, G. and Kivelson, D., *Chem. Phys.* **152**, 153 (1991) and Ref. 7c below.)

6. Bergsma, J. P., Reimers, J. R., Wilson, K. R., and Hynes, J. T., *J. Chem. Phys.* **85**, 5625 (1986).
7. (a) Bergsma, J.P., Gertner, B. J., Wilson, K. R., and Hynes, J. T., *J. Chem. Phys.* **86** 1356 (1987); (b) Gertner, B. J., Bergsma, J. P., Wilson, K. R., Lee, S., and Hynes, J. T., *ibid.* **86**, 1377 (1987); (c) Gertner, B. J., Wilson, K. R., and Hynes, J. T., *ibid.* **90**, 3537 (1989); (d) Gertner, B. J., Whitnell, R. M., Wilson, K. R., and Hynes, J. T., *J. Amer. Chem. Soc.* **113**, 74 (1991).
8. (a) Ciccotti, G., Ferrario, M., Hynes, J. T., and Kapral, R., *J. Chem. Phys.* **93**, 7137 (1990); (b) Rey, R. and Guàrdia, E., *J. Phys. Chem.* **96**, 4712 (1992).
9. Keirstead, W. P., Wilson, K. R., and Hynes, J. T., *J. Chem. Phys.* **95**, 5256 (1991).
10. Zichi, D. A., Ciccotti, G., Hynes, J. T., and Ferrario, M., *J. Phys. Chem.* **93**, 2184 (1989).
11. (a) Zhu, S. B., Lee, J., and Robinson, G. W., *J. Phys. Chem.* **92**, 2401 (1988); (b) Berne, B. J., Borkovec, M., and Straub, J. E., *ibid.* **92**, 3711 (1988); (c) Roux, B. and Karplus, M., *ibid.* **95**, 4845 (1991).
12. Tucker, S. and Truhlar, D., *J. Amer. Chem. Soc.* **112**, 3347 (1990).
13. Bagchi, B. and Oxtoby, D. W., *J. Chem. Phys.* **78**, 2735 (1983).
14. Ashcroft, J., Besnard, M., Aquada, V., and Jonas, J., *Chem. Phys. Lett.* **110**, 430 (1984).
15. Zeglinski, D. M. and Waldeck, D. H., *J. Phys. Chem.* **923**, 692 (1988); Sivakumar, N., Hoburg, E. A., and Waldeck, D. H., *ibid.* **90**, 2305 (1989); Park, N. S. and Waldeck, D. H., *ibid.*
16. McManis, G. E. and Weaver, M. J., *J. Chem. Phys.* **90**, 1720 (1989).
17. Zwan, G. van der and Hynes, J. T., *J. Chem. Phys.* **76**, 2993 (1982).
18. Zwan, G. van der and Hynes, J. T., *J. Chem. Phys.* **76**, 4174 (1983); *Chem. Phys.* **90**, 21, (1984).
19. Kramers, H. A., *Physica* **7**, 284 (1940).
20. See, e.g., Newton, M. D. and Sutin, N., *Annu. Rev. Phys. Chem.* **35**, 437 (1984).
21. For recent deveopments, one may consult: *Dynamics and Mechanics of Photoinduced Electron Transfer and Related Phenomena,* Mataga, N., Okada, T., and Masuhara, H., eds., (Elsevier, Amsterdam, 1992).
22. For an early example of the use of ΔE coordinates, see Warshel, A., *J. Phys. Chem.* **86**, 2218 (1982).
23. Hynes, J. T., *J. Phys. Chem.* **90**, 370 (1986).
24. Marcus, R. A., *J. Chem. Phys.* **24**, 966 979 (1956).
25. Zusman, L. D., *Chem. Phys.* **49**, 295 (1980).
26. Zichi, D. A., Kim, H. J., Carter, E. A., and Hynes, J. T., unpublished.
27. See, e.g., Hynes, J. T. and Deutch, J. M., in *Physical Chemistry, an Advanced Treatise,* Vol. IIB, eds. H. Eyring, W. Jost, and D. Henderson (Academic, New York, 1985), p. 153.
28. Zwan, G. van der and Hynes, J. T., *J. Phys. Chem.* **89**, 4181 (1985).
29. Rips, I. and Jortner, J., *J. Chem. Phys.* **87**, 2090 (1987).
30. Maroncelli, M., Caster Jr., E. W., Bagchi, B., and Fleming, G. R., *Faraday Disc. Chem. Soc.* **85**, 199 (1988).
31. Fonseca, T., *J. Chem. Phys.* **91**, 2869 (1989); *Chem. Phys. Lett.* **162**, 491 (1989).
32. Carter, E. A. and Hynes, J. T., *J. Chem. Phys.* **94**, 5961 (1991); see also, Hynes, J. T., Carter, E. A., Ciccotti, G., Kim, H. J., Zichi, D. A., Ferrario, M., and Kapral, R., in *Perspectives in Photosynthesis,* Jortner, J. and Pullman, B., eds. (Kluwer, Dordrecht, 1990), p. 133.
33. Marconcelli, M. and Fleming, G. R., *J. Chem. Phys.* **86**, 6221 (1987); **89**, 5044 (1987).
34. Fonseca, T. and Landanyi, B. M., *J. Phys. Chem.* **95**, 2116 (1991).
35. Lee, S. and Hynes, J. T., *J. Chem. Phys.* **88**, 6853, 6863 (1988); Zichi, D. A. and Hynes, J. T., *ibid.* **88**, 2513 (1988); Kim, H. J. and Hynes, J. T., *J. Am. Chem. Soc.* **114**, 10508, 10528 (1992).
36. Grote, R. F., Zwan, G. van der, and Hynes, J. T., *J. Phys. Chem.* **88**, 4676 (184).
37. Houston, S. E., Rossky, P. J., and Zichi, D., *J. Am. Chem. Soc.* **111**, 5680 (1989).

38. Hwang, J.-K., King, G., Creighton, S., and Warshel, A., *J. Am. Chem. Soc.* **110**, 5297 (1988).
39. Bader, J. S. and Chandler, D., *Chem. Phys. Lett.* **157**, 501 (1989).
40. (a) Levy, R. M., Kitchen, D. B., Blair, J. T., and Krogh-Jespersen, K., *J. Phys. Chem.* **94**, 4470 (1990), Blair, J. T, Krogh-Jespersen, K., and Levy, R. M., *ibid.* **111**, 6948 (1989); (b) Benjamin, I. and Wilson, K. R., *ibid.* **95**, 3514 (1991).
41. Maroncelli, M., *J. Chem. Phys.* **94**, 2084 (1991).
42. Bruehl, M. and Hynes, J. T., *J. Phys. Chem.* **96**, 4068 (1992).
43. (a) Carter, E. A. and Hynes, J. T., *J. Phys. Chem.* **93**, 2184 (1989); (b) Fonseca, T., Ladanyi, B. M., and Hynes, J. T., *ibid.* **96**, 4085 (1992); (c) Fonseca, T. and Ladanyi, B., *J. Mol. Liq.*, in press; (d) King, G. and Warshel, A., *J. Chem. Phys.* **93**, 8682 (1990).
44. Rosenthal, S. J., Xie, X., Du, M., and Fleming, G. R., *J. Chem. Phys.* **95**, 4715 (1991).
45. (a) Smith, B. B., Kim, H. J., Borgis, D., and Hynes, J. T., in *Dynamics and Mechanisms of Photoinduced Electron Transfer and Related Phenomena,* N. Mataga, T. Okada, and H. Masuhara, eds., (Elsevier, Amsterdam, 1992). p. 39; (b) Smith, B. B., Staib, A., and Hynes, J. T., *Chem. Phys.* **176**, 521 (1993).
46. (a) Alexandrov, I. V. and Gabrielyan, R. G., *Mol. Phys.* **37**, 1963 (1979); Friedman, H. L. and Newton, M. D., *Faraday Discuss. Chem. Soc.* **74**, 73 (1982); Calef, D. F. and Wolynes, P. G., *J. Phys. Chem.* **87**, 3387 (1983); Sumi, H. and Marcus, R. A., *J. Chem. Phys.* **84**, 4272 (1986); Sparpaglioni, M. and Mukamel, S., *J. Chem. Phys.* **88**, 3263 (1988); (b) Rips, I. and Jortner, J., *ibid.* **87**, 2090 (1987); (c) Dakhnovskii, Yu. I. and Ovchinikov, A. A., *Mol. Phys.* **58**, 237 (1986); Zusman, L. D., *Chem. Phys.* **51**, 119 (1988); Murillo, M. and Cukier, R. I., *J. Chem. Phys.* **89**, 6736 (1988); Fonseca, T., *ibid.* **91**, 2869 (1989); Zhou, Y., Friedman, H. L., and Stell, G., *Chem. Phys.* **152**, 185 (1991); McManis, G. E., Gochev, A., and Weaver, M. J., *ibid.* **152**, 107 (1991).
47. Borgis, D., Lee, S., and Hynes, J. T., *Chem. Phys. Lett.* **162**, 19 (1989); Borgis, D. and Hynes, J. T., *J. Chem. Phys.* **94**, 3619 (1991); *Chem. Phys.* **170**, 315 (1993).
48. There are other simulations of aspects of electron transfer reactions, but they do not address the issues considered here. See, e.g., (a) Warshel, A., *J. Chem. Phys.* **86**, 2218 (1982); Hwang, J. and Warshel, A., *J. Am. Chem. Soc.* **109**, 715 (1987); (b) Halley, J. W. and Hautmann, J., *Phys. Rev.* **B38**, 11704 (1988); (c) Kuharski, R. A., Bader, J. S., Chandler, D., Sprik, M., Klein, M. L., and Impey, R. W., *J. Chem. Phys.* **89**, 3248 (1988); (d) Gonzalez-Lafont, A., Lluch, J. M., Oliva, A., and Bertran, J., in *Chemical Reactivity in Liquids,* M. Moreau and P. Turq, eds. (Plenum, New York, 1988), p. 197; (e) Hwang, J. K., Creighton, S., King, G., Whitney, D., and Warshel, A., *J. Chem. Phys.* **89**, 859 (1988).
49. Fonseca, T. and Ladanyi, B. M., to be submitted. The low barrier ET reaction case is generally, though not always, a more faithful adherent to the full spectrum of solvation dynamics: see also Tominaga, K., Walker, G. C., Kang, T. J., Barbara, P. F., and Fonseca, T., *J. Phys. Chem.* **95**, 10485 (1991); Kang, T. J., Jarzeba, W., Barbara, P. F., and Fonseca, T., *Chem. Phys.* **149**, 81 (1990), in *Perspectives in Photosynthesis,* J. Jortner and B. Pullman, eds. (Kluwer, Dordrecht, 1990).
50. Whitnell, R. M., Wilson, K. R., and Hynes, J. T., *J. Phys. Chem.* **94**, 8625 (1990); *J. Chem. Phys.* **96**, 5354 (1992).
51. Johnson, A. E., Levinger, N. E., and Barbara, P. B., **96**, 7841 (1992); Whitnell, R. M. and Benjamin, I., *Chem. Phys. Lett.* **204**, 45 (1993); Owrutsky, J. C., Kim, Y. R., Li, M., Sarisky, M. J., and Hochstrasser, R. M., *ibid.* **184**, 368 (1991).
52. Bruehl, M. and Hynes, J. T., *Chem. Phys.* **175**, 205 (1993).
53. Gertner, B. J., Ando, K., Bianco, R., and Hynes, J. T., *Chem. Phys.*, in press.
54. Hynes, J. T., unpublished.
55. See, e.g., Fonseca, T., Gomes, J. A. N. F., Grigolini, P., and Marchesoni, F., *Adv. Chem. Phys.* **62**, 389 (1985).

56. Nee, T. and Zwanzig, R., *J. Chem. Phys.* **52**, 6353 (1970); Hubbard, J. B. and Wolynes, P. G., *ibid.* **69**, 998 (1978); Wolynes, P. G., *Ann. Rev. Phys. Chem.* **31**, 345 (1980); Madden, P. A. and Kivelson, D., *Adv. Chem. Phys.* **56**, 479 (1984).
57. Simon, J. D. and Thompson, P. A., *J. Chem. Phys.* **92**, 2891 (1990); Alavi, D. S., Hartman, R. S., and Waldeck, D. H., *ibid.* **94**, 4509 (1991).
58. However, the situation improves for the continuum description when finite size effects are accounted for: Alavi, D. S. and Waldeck, D. H., *J. Chem. Phys.* **94**, 6196 (1991); Hartman, R. S., Alavi, D. S., and Waldeck, D. H., *J. Phys. Chem.* **95**, 7872 (1991).
59. Zwan, G. van der and Hynes, J. T., *Chem. Phys.* **152**, 169 (1991); Chapman, C. F. and Maroncelli, M., *J. Phys. Chem.* **95**, 9095 (1991); Ittah, V. and Huppert, D., *Chem. Phys. Lett.* **173**, 496 (1990); Huppert, D., Itah, V., and Kosower, E. M., *ibid.* **159**, 267 (1989).
60. Sebastian, K., *J. Chem. Phys.* **90**, 5056 (1989); Smith, B. B. and Hynes, J. T., *ibid.* in press; Schmickler, W., *J. Electroanal. Chem.* **230**, 43 (1987); Morgan, J. D. and Wolynes, P. G., *J. Phys. Chem.* **91**, 874 (1987).
61. Benjamin, I., *J. Chem. Phys.* **95**, 3698 (1991).
62. Jarzeba, W., Walker, G. C., Johnson, A. E., and Barbara, P. F., *Chem. Phys.* **152**, 57 (1991).
63. Scherer, N. F., Zeigler, L., and Fleming, G. R., *J. Chem. Phys.* **96**, 5544 (1992); Nelson, K. A. and Williams, L. R., *Phys. Rev. Lett.* **58**, 745 (1987).

Index

Understanding Chemical Reactivity

1. Z. Slanina: *Contemporary Theory of Chemical Isomerism.* 1986
ISBN 90-277-1707-9
2. G. Náray-Szabó, P. R. Surján, J. G. Angyán: *Applied Quantum Chemistry.* 1987 ISBN 90-277-1901-2
3. V. I. Minkin, L. P. Olekhnovich and Yu. A. Zhdanov: *Molecular Design of Tautomeric Compounds.* 1988 ISBN 90-277-2478-4
4. E. S. Kryachko and E. V. Ludeña: *Energy Density Functional Theory of Many-Electron Systems.* 1990 ISBN 0-7923-0641-4
5. P. G. Mezey (ed.): *New Developments in Molecular Chirality.* 1991
ISBN 0-7923-1021-7
6. F. Ruette (ed.): *Quantum Chemistry Approaches to Chemisorption and Heterogeneous Catalysis.* 1992 ISBN 0-7923-1543-X
7. J. D. Simon (ed.): *Ultrafast Dynamics of Chemical Systems.* 1994
ISBN 0-7923-2489-7

Kluwer Academic Publishers – Dordrecht / Boston / London